Matter and Methods at Low Temperatures

Frank Pobell

Matter and Methods at Low Temperatures

Third, Revised and Expanded Edition

With 234 Figures, 28 Tables, and 81 Problems

Professor Dr. Frank Pobell
Robert-Diez-Strasse 1
D-01326 Dresden
Germany

Library of Congress Control Number: 2006934863

ISBN-10 3-540-46356-9 Springer Berlin Heidelberg New York
ISBN-13 978-3-540-46356-6 Springer Berlin Heidelberg New York

ISBN 2nd Edition 3-540-58572-9 Springer Berlin Heidelberg New York

Springer is a part of Springer Science+Business Media.

springer.com

Typesetting by author and SPi using a Springer LaTeX macro package.
Cover design: eStudio Calamar S.L., E-17001 Girona, Spain

Printed on acid-free paper SPIN 11801160 57/3100/SPi 5 4 3 2 1 0

Preface to the Third Edition

Success of a product is determined by the market. I am therefore very pleased that the first two editions of this book have been sold out, and that the publisher has asked me to work on a third, revised and expanded edition. Obviously, there is still demand for "Matter and Methods at Low Temperatures", even almost 15 years after publication of the first edition.

Before working on this revision, I had written to more than 20 expert colleagues to ask for their recommendations for revisions. Besides details, the essence of their response was the following (1) Essentially, leave as it is; (2) Add more information on properties of materials at $T > 1\,\mathrm{K}$; (3) Add information on suppliers of low-temperature equipment. Besides following the latter two recommendations, I have, of course, taken into account all relevant new information and new developments that have become available since the second edition was written more than 10 years ago, in 1995. I have found this information in particular in the journals "Journal of Low Temperature Physics", "Review of Scientific Instruments", and "Cryogenics", as well as in the Proceedings of the International Conferences on Low Temperature Physics, which took placc in Prague (1996), Helsinki (1999), Hiroshima (2002), and Orlando, FL (2005), as well as of the International Symposium on Quantum Fluids and Solids, which took place in Ithaca, NY (1995), Paris (1997), Amherst, MA (1998), Minneapolis, MN (2000), Konstanz (2001), Albuquerque, NM (2003), and Trento (2004); the latter proceedings have also been published as issues of the Journal of Low Temperature Physics.

I have included new chapters or sections on:

- Closed-cycle refrigerators
- Methods to measure heat capacity, thermal expansion, and thermal conductivity
- Magnetic properties of some selected materials (replacing the former short appendix) and methods to measure them
- The new temperature scale PLTS – 2000 (replacing the former section on the preliminary scale CTS-2) and the new superconducting fixed-point device SRD 1000

- Oxide compound resistance thermometers
- Coulomb blockade thermometry
- Torsional and translational oscillators
- Low-temperature electronics, in particular SQUIDs (mostly a summary on the relevant literature)
- As well as a list of commercial suppliers of cryogenic equipment and materials needed for low-temperature research.

Furthermore, I have strongly revised or expanded the sections on dilution refrigeration, pressure transducers, cold valves, small superconducting and normal conducting magnets, on thermometry, in particular resistance, paramagnetic and NMR thermometry, as well as electronic microrefrigerators. The new edition also contains more data on properties of materials at low temperatures.

There are still further domains of increasing importance in low-temperature research and technology that I could have included. For example, the exciting developments on magneto-optically trapped and laser as well as evaporatively cooled highly diluted gases and their Bose–Einstein condensation (however, the book still is on condensed-matter physics only); application of low-temperature technology in space research and for particle detection; low-temperature research in very high magnetic fields (partly covered in the sections on calorimetry and on thermometry); or the developments of various scanning tunneling microscopes for applications at low temperatures. These developments show the increasing importance of low temperatures in fields outside of the formerly traditional low-temperature research. However, including them would have surely made this a too voluminous book. It already contains 45 new figures and about 250 additional references compared to the second edition!

I am grateful to A. Bianchi, G. Eska, P. Esquinazi, J. Pekola, R. Rosenbaum, K. Uhlig, A.T.A.M. de Waele, as well as W. Buck and his colleagues at the PTB Berlin for reading and commenting on various parts of my manuscript. Last but not least, it is a pleasure for me to acknowledge my close collaboration with Thomas Herrmannsdörfer for about one and a half decades, which has surely contributed to my understanding and appreciation of several areas covered in this book.

Dresden
August 2006 *F. Pobell*

Preface to the Second Edition

It has been a great pleasure for me to see this book – very often several copies – in almost every low-temperature laboratory I have visited during the past three years.

Low- and ultralow-temperature physics continue to be lively and progressing fields of research. New results have emerged over the four years since publication of the first edition of my monograph. The second edition contains relevant results particularly on thermometry and materials properties, as well as many additional references. Of course, typographical errors I had overlooked are now corrected. I am grateful to J. Friebel for checking and solving the problems I have included in this new edition. And, as for the case of the first edition, I again thank H. Lotsch for the very careful editing.

I hope that this lower-priced paperback edition will continue to be a valuable source for the research and study of many of my colleagues and their students.

Bayreuth
November 1995

F. Pobell

Preface to the First Edition

The aim of this book is to provide information about performing experiments at low temperatures, as well as basic facts concerning the low-temperature properties of liquid and solid matter. To orient the reader, I begin with chapters on these low-temperature properties. The major part of the book is then devoted to refrigeration techniques and to the physics on which they are based. Of equal importance, of course, are the definition and measurement of temperature; hence low-temperature thermometry is extensively discussed in subsequent chapters. Finally, I describe a variety of design and construction techniques which have turned out to be useful over the years.

The content of the book is based on the three-hour-per-week lecture course which I have given several times at the University of Bayreuth between 1983 and 1991. It should be particularly suited for advanced students whose intended masters (diploma) or Ph.D. subject is experimental condensed matter physics at low temperatures. However, I believe that the book will also be of value to experienced scientists, since it describes several very recent advances in experimental low-temperature physics and technology, for example, new developments in nuclear refrigeration and thermometry.

My knowledge and appreciation of low-temperature physics have been strongly influenced and enhanced over the years by many colleagues. In particular, I wish to express my thanks to my colleagues and coworkers at the Institut für Festkörperforschung, Forschungszentrum Jülich and at the Physikalisches Institut, Universität Bayreuth. Their enthusiasm for low-temperature physics has been essential for the work in our groups, and for many achievements described in this book. Among them I wish to mention especially R.M. Mueller, Ch. Buchal and M. Kubota from the time at Jülich (1975–1983), K. Gloos, R. König, B. Schröder-Smeibidl, P. Smeibidl, P. Sekowski, and E. Syskakis in Bayreuth (since 1983).

Above all, I am deeply indebted to my colleague and friend Girgl Eska, who shares with me all the joys and sorrows of our low-temperature work in Bayreuth. He and Bob Mueller read the entire manuscript and made many important comments on it. I am also very grateful to Dierk Rainer for so often sharing with me his deep insight into the physics of many low-temperature phenomena.

It is a pleasure for me to thank Mrs. G. Pinzer for her perfect typing of the manuscript and Ms. D. Hollis for her careful copy-editing. Likewise, I gratefully acknowledge the very thorough and time-consuming checking of the whole manuscript by Dr. H. Lotsch.

Finally, I thank the Institute for Physical Problems, USSR Academy of Sciences, where part of the manuscript was written, for the hospitality extended to me, and the "Volkswagen Stiftung" for a stipend which I held for some time while writing this book.

Bayreuth
September 1991 *F. Pobell*

Contents

1

Introduction

The importance of temperature is very often not fully recognized, the reason probably being that our life is restricted to an extremely narrow range of temperatures. This can be realized if we look at the temperatures existing in nature or accessible in laboratories (Fig. 1.1). These temperatures range from about 10^9 K, the temperature at the center of the hottest stars and necessary to form or destroy atomic nuclei, to 2×10^{-6} K, the lowest temperatures

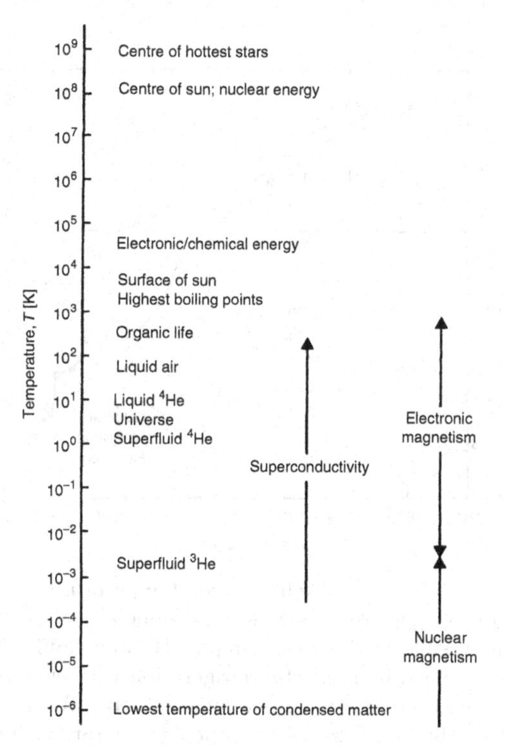

Fig. 1.1. Logarithmic temperature scale with some characteristic phenomena

accessible today in the laboratory in condensed matter physics experiments. This lower limit means that we have been able to refrigerate condensed matter to within two microkelvin of absolute zero ($0\,\mathrm{K} = -273.15°\mathrm{C}$). Indeed, nuclei have been investigated at nuclear-spin temperatures which are another four orders of magnitude lower, to below the nanokelvin temperature range. The nanokelvin temperature range has also been reached for magneto-optically trapped and laser as well as evaporatively cooled highly dilute gases. With these achievements, low-temperature physics has surpassed nature by several orders of magnitude, because the lowest temperature in nature and in the universe is 2.73 K. This background temperature exists everywhere in the universe because of the photon energy which is still being radiated from the "big bang". If we compare low-temperature physics to other branches of physics, we realize that it is actually one of the very few branches of science where mankind has surpassed nature, an achievement which has not yet proved possible, for example, in high-pressure physics, high-energy physics or vacuum physics. The very wide range of temperatures accessible to experiments has made temperature probably the most important among the parameters which we can vary in the laboratory in order to change the properties of matter, to obtain a better understanding of its behavior and to make practical use of it.

The historical development of refrigeration to lower and lower temperatures is illustrated in Fig. 1.2. Air, N_2 and O_2 were liquefied and eventually

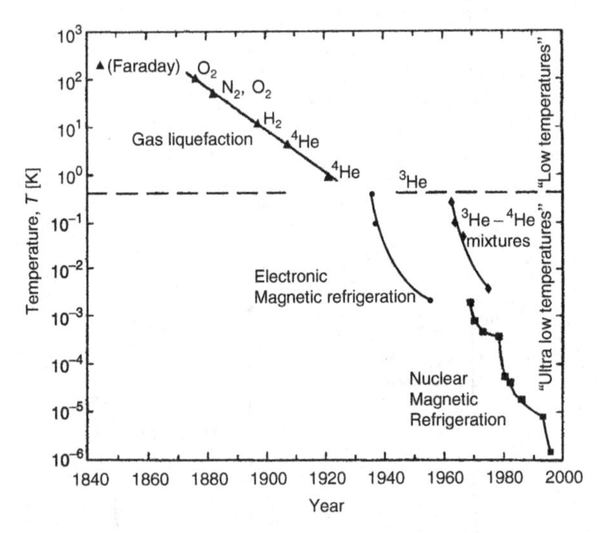

Fig. 1.2. Historical development of refrigeration temperatures of condensed matter, starting about 160 years ago with Faraday's gas liquefaction. The low temperature range was made accessible by liquid air, liquid H_2 and liquid ^{4}He (▲). Ultralow temperatures were attained by magnetic refrigeration with electronic magnetic moments (•) and later with nuclear magnetic moments (■). Refrigeration with liquid ^{3}He and liquid ^{3}He–^{4} He mixtures (♦) developed as the rare helium isotope became available in sufficient quantities

solidified 1883 in Cracow by Z. Wroblewski and K. Olszewski. This was the first time that mankind reached temperatures below 100 K. The scientists and engineers involved in this venture had two aims. A very practical one was to devise a means of refrigerating meat for its journey from other continents to Europe. The scientific aim was to discover whether permanent gases exist, in other words, are there any substances which do not exist in the liquid and/or solid state? This latter aim led to the liquefaction and solidification of hydrogen by James Dewar in 1898; he reached temperatures of 20 K and later 13 K. At about that time the last gaseous element to be discovered, helium, was detected first in spectroscopic investigations of solar protuberances and then as a gas escaping from various minerals on earth. The Dutch scientist Heike Kamerlingh Onnes won the race to liquefy this last element; he liquefied helium-4 at 4.2 K in 1908. Kamerlingh Onnes' work opened up the Kelvin temperature range to science, and scientists at his low-temperature laboratory at the University of Leiden dominated low-temperature physics for at least 20 years, making many fundamental investigations in helium physics, in the physics of metals and in the physics of magnetism, for example. In particular, he discovered superconductivity of metals by refrigerating Hg to 4.2 K in 1911, and in the early 1920s he found some unexpected behavior in the properties of liquid ^{4}He, which later turned out to result from its superfluid transition at 2.2 K. Today Kamerlingh Onnes is considered "the father of low-temperature physics". The gradual introduction of low-temperature physics to hundreds of research laboratories was eventually made possible by the development of commercial ^{4}He liquifiers with counter-flow heat exchangers and Joule-Thompson expansion or/and expansion engines by S.C. Collins after World War II. In the last decade, the style of low-temperature experimentation has again seen a quite substantial change from building most of the equipment at home toward purchasing most of the equipment, like cryostats, closed-cycle refrigerators, ^{3}He-^{4}He dilution refrigerators, magnetometers, SQUID amplifiers, resistance and capacitance bridges, etc. Hence, I have added a list of suppliers of cryogenic equipment.

In 1922, H. Kamerlingh Onnes reached 0.83 K by pumping on the vapour above a bath of boiling ^{4}He; this temperature record was lowered to 0.71 K in 1932 by his successor W.H. Keesom. Because at these temperatures the vapour pressure of ^{4}He becomes extremely small, the development of a fundamentally different technology, magnetic refrigeration, was necessary to attain temperatures appreciably below 1 K. By adiabatic demagnetization of paramagnetic salts, a method proposed in the late 1920s, we can approach absolute zero to within a few millikelvin (Fig. 1.2). An advanced version of this magnetic refrigeration method, adiabatic demagnetization of *nuclear* magnetic moments, is the only method known today by which we can reach temperatures far into the microkelvin temperature range. While these "one-shot" magnetic cooling methods were being perfected, another refrigeration technique, again based on the properties of liquid helium, the dilution of the rare isotope ^{3}He by the common isotope ^{4}He, enabled the development of a continuous refrigeration method to reach the low millikelvin temperature range. This method was proposed in the 1960s and put into practice in the early 1970s. Today it is very well developed and has probably approached its limits; it has largely replaced adiabatic demagnetization of paramagnetic salts. Therefore, I shall discuss dilution refrigeration at length and devote only a short chapter to electronic paramagnetic refrigeration. Nuclear magnetic refrigeration – the refrigeration method for the microkelvin temperature range – will, of course, be discussed in detail. I shall also briefly describe two additional refrigeration methods based on the properties of ^{3}He even though they are of minor importance today: refrigeration by evaporation of liquid ^{3}He and refrigeration by solidification of liquid ^{3}He (Pomeranchuk cooling).

As a result of these developments, three refrigeration methods dominate low-temperature physics today (Table 1.1). Temperatures in the Kelvin range down to about 1 K are obtained by evaporation of liquid ^{4}He, the evaporation of pure ^{3}He for cooling is now only of minor importance. The millikelvin temperature range is completely dominated by the ^{3}He–^{4}He dilution refrigeration method, which in a simple apparatus can reach minimum temperatures of about 20 mK and in a more complicated apparatus 5 mK; the present record is about 2 mK. The other two millikelvin refrigeration methods, Pomeranchuk cooling and adiabatic demagnetization of electronic magnetic moments, are of minor importance today. Finally, in the microkelvin temperature range, we have only one method, nuclear adiabatic demagnetization. This method has opened up the microkelvin temperature range (with a present minimum of 1.5 μK) to condensed-matter physics, and the nanokelvin temperature range to nuclear-spin physics. At these extremely low temperatures it makes sense, as we will see, to distinguish between a temperature of the nuclear magnetic spin system and a temperature of the electrons and of lattice vibrations. When

Table 1.1. Refrigeration techniques. The methods which dominate in the three temperature ranges are in italics

Temperature range	Refrigeration technique	Available since	Typical T_{min}	Record T_{min}
I Kelvin	Universe			2.73 K
	Helium-4 evaporation	1908	1.3 K	0.7 K
	Helium-3 evaporation	1950	0.3 K	0.23 K
II Millikelvin	**^{3}He–^{4}He dilution**	1965	10 mK	2 mK
	Pomeranchuk cooling	1965	3 mK	2 mK
	Electronic magnetic refrigeration	1934	3 mK	1 mK
III Microkelvin	**Nuclear magnetic refrigeration**	1956	100 μK	1.5 μK[a]

[a]The given minimum temperature for the microkelvin temperature range is the *lattice (electronic) equilibrium* temperature. *Nuclear spin* temperatures as low as 0.3 nK have been reached (Table 10.2)

I speak about "temperature" in this textbook, I always mean the latter, unless I explicitly state otherwise.

Low-temperature physics and technology are not possible without knowledge of the relevant properties of liquid and solid matter at low temperatures. And many fundamental properties of matter were only found and/or understood after matter had been cooled to the Kelvin range or to even lower temperatures. Among these properties are the quantization of lattice vibrations (phonons), the electronic excitations leading to the linear temperature dependence of the specific heat of conduction electrons, superconductivity, superfluidity and many aspects of magnetism. Hence, this textbook begins with chapters on the properties of liquid and solid matter at low temperatures relevant for the performance and design of low-temperature experiments. But even though a major part of the book will deal with solid-state physics, it is not intended to replace more general textbooks on this subject. Indeed, readers are assumed to have a basic knowledge of condensed-matter physics as found in [1.1–1.6], for example. I have included extensive information on materials properties at low temperatures and how to measure them.

Experiments at low temperatures make no sense without thermometry. In fact, the measurement of a low or very low temperature and its relation to the thermodynamic temperature scale are as important as the attainment of that temperature itself, and very often just as difficult. This has become increasingly apparent in the last decades as lower and lower temperatures have been reached. Hence, after discussion of the various refrigeration techniques, I shall discuss temperature scales and the various thermometric methods at low temperatures. The book closes with a chapter on various "cryogenic devices and design aids", a discussion of various tools and tricks helpful for low-temperature investigations, as well as some comments on low-temperature electronics.

There are several excellent books on the properties of matter at low temperatures [1.5, 1.6] and on the technology of cryogenics above about 1 K [1.7, 1.8] and below 1 K [1.9–1.12]. The reader should consult them for subjects not included or not discussed in detail in this book; for example, the SQUID and its various applications in low-temperature experiments, for which the relevant literature can be found in Chap. 14. The two most comprehensive monographs [1.9, 1.10] on the subject of this book were published in 1974 and 1976, respectively, but the field of experimental physics at milli- and microkelvin temperatures has been rapidly expanding since then, and therefore a new monograph devoted to this field seemed to be very appropriate. The first two editions of this textbook were written in the years 1991 and 1995, respectively. Its new edition reflects the state of the art of experimental low-temperature physics at the current time, i.e., in 2006. Endeavours in low-temperature physics and technology have been rewarded by a great number of fundamental discoveries that are important for our understanding of matter, in particular of its quantum behavior, and for practical applications. These achievements were only possible by overcoming substantial experimental difficulties, in particular when the range below 1 K was entered, and the (slightly modified) statement from the first page of Lounasmaa's book [1.9] provides a good summary of the essentials:

> An experimentalist wishing to pursue research at low temperatures faces four technical difficulties: how to reach the low temperature, how to measure it, how to reduce the external heat leak so that the low temperature can be maintained for a sufficiently long time, and how to transfer cold from one place to another. Many experimental methods have been developed to provide a satisfactory solution to these problems.

The progress in these areas will be discussed in the present book. I shall restrict the discussion of the properties of matter and experimental techniques mostly to temperatures below about 10 K. Higher temperatures will only be considered if we need them in order to understand what is going on at temperatures of interest in this book. Therefore, some properties of materials are given for temperatures up to 300 K.

References are made preferentially to books or to review articles. Original publications are cited when I consider them necessary for obtaining more details than can be given in a monograph. In many instances I cite not according to priority but to pertinence for the purpose of this book. Since more than a decade has passed since I had worked on the preceding edition of this book, I had to add about 250 new references. I will use cgs and/or SI units, whatever seems appropriate and as is typical for practical work in today's low-temperature laboratories. Equations – as for specific heat or susceptibility – are given for one mole.

2

Properties of Cryoliquids

In this chapter, I shall discuss properties of cryoliquids important for the design and performance of low-temperature experiments. Of course, cryoliquids are very important for low-temperature physics because they are the simplest means of achieving low temperatures. In particular, the properties of liquid helium are essential because all refrigeration methods to $T < 10\,\mathrm{K}$ use liquid helium as a final or intermediate refrigeration stage. I shall not discuss the technology of liquefaction [2.1, 2.2]. For liquifaction the gas has to be isothermally compressed and then expanded to let it perform "external" work (for example, in an expansion engine), or perhaps using the well-known Joule–Thomson effect, which means letting the gas expand and perform "internal" work against the mutual attraction of its atoms or molecules. This latter effect leads to cooling if the starting temperature is below the inversion temperature T_i, which is 6.75 T_c, T_c being the critical temperature for a van der Waals gas. Various properties of cryoliquids are summarized in Table 2.1 and are compared there to the relevant properties of water. Of particular importance for refrigeration are the boiling point T_b (defining the accessible temperature range), the latent heat of evaporation L (defining the cooling power) and – last but by no means least important – the price (defining the availability).

2.1 Liquid Air, Liquid Oxygen, Liquid Nitrogen

Today liquid oxygen is not in common use as a refrigerant because it is extremely reactive. An explosive oxidation reaction can occur if the oxygen comes into contact with organic liquids, like oil used in pumps, or with solid matter having a large surface area, like metal powder. The boiling temperature of oxygen lies above the boiling temperature of nitrogen. Therefore, if one keeps liquid air in a container, the nitrogen evaporates first, resulting in an enrichment of liquid oxygen, and thus eventually leading to this dangerous liquid again. For these reasons, today air is liquefied and separated into oxygen and nitrogen in large liquefaction and separation plants. Liquid

Table 2.1. Properties of some liquids (T_b is the boiling point at $P = 1$ bar, T_m the melting point at $P = 1$ bar, $T_{tr}(P_{tr})$ the triple-point temperature (pressure), $T_c(P_c)$ the critical temperature (pressure), and L the latent heat of evaporation at T_b). The data have mostly been taken from [2.3–2.5, 2.17]

subst.	T_b (K)	T_m (K)	T_{tr} (K)	P_{tr} (bar)	T_c (K)	P_c (bar)	lat. heat, L (kJ l^{-1})	vol% in air
H_2O	373.15	273.15	273.16	0.06*	647.3	220	2,252	–
Xe	165.1	161.3	161.4	0.82	289.8	58.9	303	0.1×10^{-4}
Kr	119.9	115.8	114.9	0.73	209.4	54.9	279	1.1×10^{-4}
O_2	90.1	54.4	54.36	0.015	154.6	50.4	243	20.9
Ar	87.2	83.8	83.81	0.69	150.7	48.6	224	0.93
N_2	77.2	63.3	63.15	0.13	126.2	34.0	161	78.1
Ne	27.1	24.5	24.56	0.43	44.5	26.8	103	18×10^{-4}
n-D_2	23.7	18.7	18.69	0.17	38.3	16.6	50	–
n-H_2	20.3	14.0	13.95	0.07	33.2	13.2	31.8	0.5×10^{-4}
^{4}He	4.21	–	–	–	5.20	2.28	2.56	5.2×10^{-4}
^{3}He	3.19	–	–	–	3.32	1.15	0.48	–

* The exact value is $P_{tr} = 61.1657$ mbar

nitrogen (LN_2) is then sold as a refrigerant at roughly 0.25 € l^{-1} when delivered in large quantities. Very few research organizations still use their own liquefaction plant as was common until the 1960s.

Evaporating nitrogen can cause suffocation if it displaces much of the usual 20% oxygen in an enclosed volume. Therefore, rooms in which evaporating liquid nitrogen is kept have to be fitted with an appropriate ventilation system.

2.2 Liquid Hydrogen

In liquid hydrogen the pair of atoms forming an H_2 molecule are bound by a strong covalent force. The interactions between H_2 molecules, which lead to the liquid and solid states, are the weak van der Waals forces. These weak dipolar forces, as well as the large zero-point motion of the light H_2, result in rather low boiling and melting points (Table 2.1). Because of the large difference in strength between chemical bonds and dipolar forces, liquid and solid H_2 are true molecular fluids and crystals in which the molecules retain many of the properties of free H_2 [2.6, 2.7].

The dangerous feature of hydrogen is its exothermic reaction with oxygen to form water. Therefore, liquid hydrogen should be used in a closed system. However, sometimes the danger associated with hydrogen is over-estimated because the reaction needs a concentration of more than 4% in air in order to occur.

Whereas at one time liquid hydrogen was frequently used as a refrigerant (and its use may become more common in hydrogen energy technologies in the future), it is not often used in laboratories these days because the temperature range between the boiling point of nitrogen (77 K) and the boiling point of helium (4.2 K) is now accessible by means of closed-cycle refrigerators (Sect. 5.3), helium evaporation cryostats (Sect. 5.2.3) or by performing the experiment in helium gas above a liquid-helium bath.

In spite of the minor importance of hydrogen as a refrigerant I shall discuss one of its properties which is important in various low-temperature experiments and which demonstrates some rather important atomic and statistical physics: the *ortho–para* conversion of H_2 [2.6–2.12].

The proton H has a nuclear spin $I = 1/2$. In a H_2 molecule the two nuclear spins can couple to a total spin $I = 0$ or $I = 1$, depending on their relative orientation. Thus, the H_2 molecule can have a symmetric nuclear state ($I = 1$), so-called *ortho*-H_2. This system has a degeneracy of $2I + 1 = 3$, which means that it can exist in three different spin orientational states, namely $m = -1, 0, +1$. For the other total-spin situation ($I = 0$) we have an anti-symmetric state, so-called *para*-H_2, with the degeneracy $2I + 1 = 1$, which means that this system can only exist in one spin state, $m = 0$.

In addition, the non-spherical H_2 molecule can rotate and we have the H_2 rotator with the rotational quantum numbers $J = 0, 1, 2, 3, \ldots$. Because of the weak interaction between H_2 molecules, the behavior of the rotator is free and J remains a good quantum number. The various rotational states are separated by rotational energies:

$$E_{\mathrm{R}} = \frac{\hbar^2}{2\theta} J(J+1)\,, \tag{2.1}$$

where $\theta = 4.59 \times 10^{-48}\,\mathrm{kg\,m^2}$ is the moment of inertia of the H_2 rotator.

Protons are Fermi particles (fermions). As a result, the total wave function of the H_2 molecule has to be anti-symmetric under an exchange of particles. In other words, it has to change its sign if the two nuclei are exchanged. The total wave function is a product of the spin wave function and the spatial wave function (electronic and vibronic excitations are negligible at $T < 1{,}000\,\mathrm{K}$). To get an anti-symmetric total wave function either the spin or the rotator wave function has to be anti-symmetric and the other one has to be symmetric. This results in two possibilities for the hydrogen molecule, which are summarized in Table 2.2.

The energy states of *para*- and *ortho*-hydrogen are depicted in Fig. 2.1. The total degeneracy of each rotational state is given by $d = (2J + 1)(2I + 1)$.

Table 2.2. Properties of *ortho*- and *para*-hydrogen

Molecule	Nuclear spin, I	Symmetry	Deg.	Rotator quantum number, J	Symmetry	Deg.
Ortho-H_2	1	Symmetric	3	$1, 3, \ldots$	Anti-symmetric	$3, 7, \ldots$
Para-H_2	0	Anti-symmetric	1	$0, 2, \ldots$	Symmetric	$1, 5, \ldots$

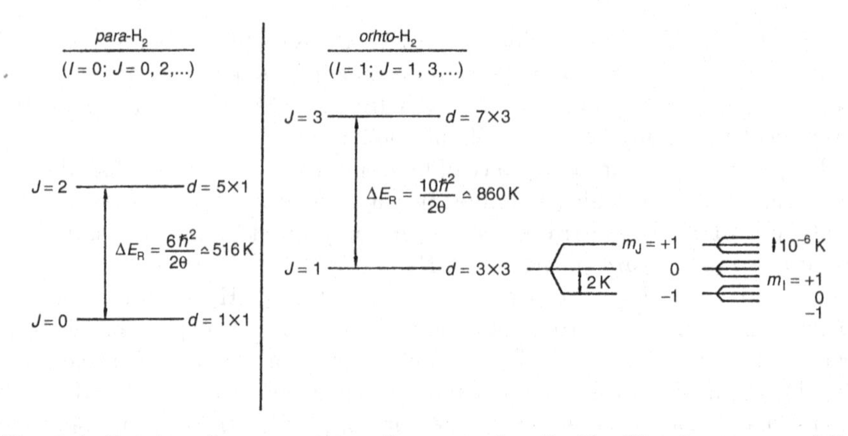

Fig. 2.1. Rotational energy states of *para-* and *ortho*-H_2. The degeneracy of the rotational states of *ortho*-H_2 is lifted by an electric quadrupole interaction between the H_2 molecules which is of order 2 K. A nuclear magnetic dipole interaction of order 10^{-6} K lifts the nuclear magnetic degeneracy of the m_J states

Because of the rather large energies ΔE_R separating the various rotator states, only the lowest rotational energy states are significantly populated at room temperature or lower temperatures. Owing to this fact and because of the different rotational degeneracies as well as of the threefold nuclear degeneracy of *ortho*-H_2, we have 25% *para*-hydrogen and 75% *ortho*-hydrogen in thermal equilibrium at room temperature. Because *para*-H_2 ($J = 0$) has lower energy than *ortho*-H_2 ($J = 1$) (the difference is 172 K if expressed in terms of temperature), a conversion from *ortho-* to *para*-hydrogen occurs if we cool hydrogen to low temperatures. Their ratio is given by

$$\frac{N_{ortho}}{N_{para}} = \frac{3\sum_{J=1,3,\ldots}(2J+1)\mathrm{e}^{-\hbar^2 J(J+1)/(2\theta\cdot k_B T)}}{\sum_{J=0,2,\ldots}(2J+1)\mathrm{e}^{-\hbar^2 J(J+1)/(2\theta\cdot k_B T)}} \,. \tag{2.2}$$

This ratio is illustrated in Fig. 2.2. At 77 K the equilibrium ratio is 1:1, but at 20 K we should have 99.8% *para*-H_2 in thermal equilibrium.

We have to take into account two important aspects of this conversion. First of all, it is an exothermic reaction giving rise to the rather large heat release of $U = 1.06\ (1.42)\,\mathrm{kJ\,mol^{-1}H_2}$ for a starting *ortho* concentration of 75% (100%). Secondly, the conversion between different rotational states is connected with a change of the nuclear spin orientation. For this change of the nuclear quantum state, we need an interaction of the nuclear magnetic moments with each other or with changing magnetic fields in their surroundings. Because of the smallness of the nuclear moments this nuclear dipolar interaction is rather weak. It can only occur in the collision of two H_2 molecules if there is no other magnetic partner available. We then have an autocatalytic reaction and, due to its weakness, the conversion is slow with an *ortho*-to-*para*

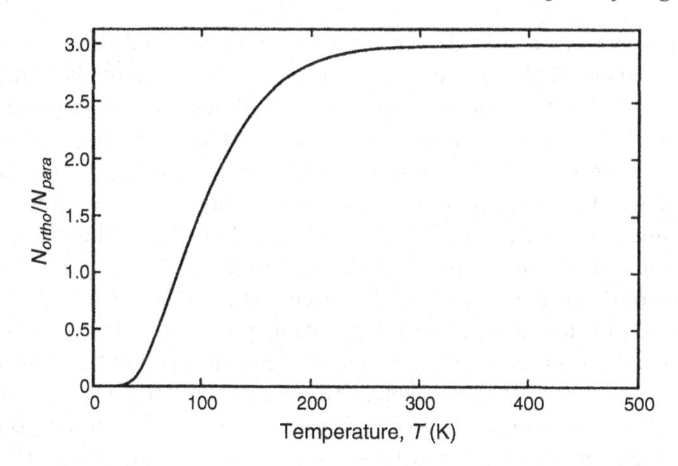

Fig. 2.2. Ratio of *ortho*-to-*para* hydrogen as a function of temperature

rate constant of $k = 1.8\%$ per hour for solid H_2 at melting pressure [2.10,2.13], leading to interesting time effects.

Let us calculate the heat release due to this conversion. The change of the concentration x of the *ortho* molecules with time t for this autocatalytic reaction is given by

$$\frac{\mathrm{d}x}{\mathrm{d}t} = -kx^2 , \tag{2.3}$$

resulting in

$$x(t) = \frac{x_0}{1 + x_0 kt} , \tag{2.4}$$

where x_0 is the starting *ortho* concentration, which is 0.75 for $T = 300\,\mathrm{K}$. This gives rise to the molar heat release

$$\dot{Q} = -U\frac{\mathrm{d}x}{\mathrm{d}t} = U\frac{kx_0^2}{(1 + x_0 kt)^2} . \tag{2.5}$$

There are two situations in which this heat release can give rise to problems in experiments at low or ultralow temperatures. First, we have to keep in mind that the heat release due to *ortho–para* conversion is quite large. Therefore, if one has a liquid consisting mainly of *ortho*-hydrogen it will evaporate due to the *ortho–para* conversion even without any extra external heat being introduced. As a result, one has first to convert the *ortho*-H_2 to *para*-H_2 before the liquid can be used as a refrigerant. The conversion is strongly accelerated if the H_2 is brought into contact with a catalyst containing electronic magnetic moments, e.g., ferric hydroxide, iron or chromic oxide [2.2, 2.11], or oxygen [2.12], because then the small *nuclear* magnetic moment of one of the two interacting O–H_2 is replaced by the much larger *electronic* magnetic moment of the catalyst. The conversion equation is then changed to

$$\mathrm{d}x/\mathrm{d}t = -kx^2 - k_1 x x_1 , \tag{2.6}$$

where $k_1$is the conversion rate constant for the catalyzed reaction and x_1 is the concentration of the paramagnetic impurity. Because of the temperature dependence of the diffusion time of the o–H_2 molecule to the paramagnetic impurity, k_1 is temperature dependent at $T < 10\,\mathrm{K}$. At higher temperatures, k_1=1.15% per second (!) for solid H_2 containing x_1=100 ppm O_2 only, as an example [2.12]. In *pure* liquid and gaseous H_2, the *ortho-* to *para*-conversion rate constant is (3 to 22) 10^{-3} per hour (!) for H_2 densities between 0.02 and $0.1\,\mathrm{g\,cm^{-3}}$ and at temperatures from 16 to 120 K [2.13].

The second problem became apparent with the advent of ultralow-temperature physics, where the refrigeration power of refrigerators can become rather small (in the microwatt range for low millikelvin temperatures and in the nanowatt range for microkelvin temperatures; see Chaps. 7–10). Many metals, such as palladium and niobium, can dissolve hydrogen in atomic form in their lattice. However, many others such as Cu, Ag, Au, Pt and Rh, cannot dissolve hydrogen in a noticeable quantity in their lattice. If these metals contain traces of hydrogen, hydrogen molecules collect in small gas bubbles with a typical diameter of some 0.1 μm [2.14]. In practice, many of these metals contain hydrogen at a typical concentration of 10–100 ppm due to their electrolytic production or from a purification process. The hydrogen pressure in these small bubbles is so high that the hydrogen becomes liquid or solid if the metal is cooled to low temperatures. This then results again in a conversion from *ortho-* to *para*-hydrogen in the small bubbles, and therefore gives rise to

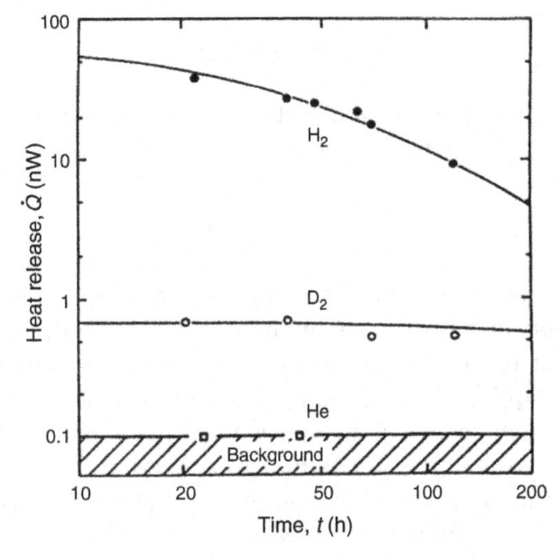

Fig. 2.3. Heat release at $T \leq 0.1\,\mathrm{K}$ from 19 g Cu samples charged at 930°C in 3.05 bar of the indicated gases as a function of time after cooling to $T \leq 4\,\mathrm{K}$. The upper *curve* represents (2.5) for 23 μmol H_2 (76 ppm H_2/Cu). The curve through the D_2 data is the corresponding equation for deuterium [2.9] for 25 μmol D_2 (83 ppm D_2/Cu). The Cu sample heated in an He atmosphere did not give a heat release above the indicate background value of the calorimeter [2.15]

heat release in these metals (Fig. 2.3) [2.15, 2.16]. The heat release is small, typically of the order of $1\,\mathrm{nW\,g^{-1}}$, but it can be very detrimental if such a metal is used in an apparatus designed for ultralow-temperature physics experiments, such as a nuclear refrigerator for the microkelvin-temperature range (Sect. 10.5.3).

In addition to its importance for low- and ultralow-temperature physics and its nice demonstration of atomic and statistical physics, the *ortho–para* conversion also demonstrates in a rather impressive way how very little energy, in this case the nuclear magnetic interaction energy (which is of the order of microkelvins if expressed in temperatures), due to the existence of nuclear spins and in combination with the Pauli principle, can result in rather dramatic effects involving much higher energies, in this case the rotational energy of the order of 100 K.

2.3 Liquid Helium

2.3.1 Some Properties of the Helium Isotopes

In 1868, the two scientists J. Janssen and N. Lockyer detected new spectral lines in the optical spectrum of the sun. First it was suggested that these lines might belong to an *isotope* of hydrogen or to sodium, but very soon it became evident that the spectral lines belonged to a new *element*, later called "helium". In 1895 the British scientist W. Ramsey detected these lines on earth, in a gas escaping from the mineral clevite. Two of the most eminent physicsts of that time, J. Dewar in London and H. Kamerlingh Onnes in Leiden, began a race to liquefy this newly discovered element of the periodic table. H. Kamerlingh Onnes won this race, liquefying helium at a temperature of 4.2 K in 1908. The first commercial helium liquefier built by S.C. Collins in 1947 had a profound impact on the spread of low-temperature experiments because liquid helium is the most important substance for low-temperature physics. All refrigeration methods for temperatures below about 10 K either use liquid helium exclusively or, if they go to very low temperatures, use liquid helium in the pre-cooling stages. At first liquid helium was employed only as a tool, but research in the following decades made it very obvious that liquid helium is a most exotic and interesting liquid exhibiting many unique properties.

Whereas in the first years of helium research the ^{4}He gas was obtained from minerals, today helium is obtained exclusively from helium-rich natural-gas sources. In both cases it is a product of radioactive alpha decay. The gas sources, particularly in the USA, North Africa, Poland, the Netherlands and the former USSR, can contain up to about 10% helium.

The common stable helium isotope is ^{4}He. Its nucleus contains two protons and two neutrons, each with anti-parallel nuclear spin orientation. Therefore, the total nuclear spin of ^{4}He, $I = 0$; it is a Bose particle (boson). The rare helium isotope ^{3}He constitutes a fraction (1–2) $\times 10^{-7}$ of helium gas from natural gas sources and about 1.3×10^{-6} of the helium gas in the atmosphere. Obtaining ^{3}He in a reasonable amount from these two sources by separating it from ^{4}He is very costly. The ^{3}He in use today for low-temperature physics experiments is a byproduct of tritium manufacture in a nuclear reactor:

$$^6_3\mathrm{Li} + ^1_0n \;\Rightarrow\; ^3_1T + ^4_2\mathrm{He}\,,$$
$$^3_1T \overset{12.3y}{\Rightarrow} ^3_2\mathrm{He} + ^{\;\,0}_{-1}\mathrm{e} + \bar{\nu}\,,$$

where $\bar{\nu}$ is the electron anti-neutrino. The helium isotopes are separated from tritium by diffusion processes. Due to this method of production, ^{3}He has only been available in necessary quantities since the late 1950s, and it is expensive (about 200 € l^{-1} of gas at standard temperature and pressure). The ^{3}He nucleus again contains two protons, but only one neutron. Therefore, its total nuclear spin $I = 1/2$, and ^{3}He is a fermion. The different statistics for the boson ^{4}He and for the fermion ^{3}He cause substantial differences in their low-temperature behavior, some of which will be discussed in the following pages. Details can be found in [2.17–2.24].

Besides the stable isotopes ^{3}He and ^{4}He, there exist two unstable helium isotopes with relatively long lifetimes: ^{6}He ($\tau_{1/2} = 0.82$ s), and ^{8}He ($\tau_{1/2} = 0.12$ s); they have not yet been liquified.

In Table 2.3 some important properties of the two stable helium isotopes are summarized and their pressure–temperature phase diagrams are shown in Fig. 2.4. The table and the figure demonstrate some of the remarkable properties of these so-called quantum substances. First, there are their rather low boiling points and critical temperatures. Then, unlike all other liquids these two isotopes do not become solid under their own vapour pressure even when cooled to absolute zero. One has to apply at least about 25.4 or 34.4 bar (for $T \to 0$ K), respectively, to get these two isotopes in their solid state. Consequently, the helium isotopes have no triple point where gas, liquid and solid

Table 2.3. Properties of liquid helium

	^{3}He	^{4}He
Boiling point, T_b (K)	3.19	4.21
Critical temperature, T_c (K)	3.32	5.20
Maximum superfluid transition temperature, T_c (K)	0.0025	2.1768
Density[a], ρ (g cm^{-3})	0.082	0.1451
Classical molar volume[a], V_m (cm^3 mol^{-1})	12	12
Actual molar volume[a], V_m (cm^3 mol^{-1})	36.84	27.58
Melting pressure[b], P_m (bar)	34.39	25.36
Minimum melting pressure, P_m (bar)	29.31	25.32
Gas-to-liquid volume ratio[c]	662	866

[a] At saturated vapour pressure and $T = 0$ K.
[b] At $T = 0$ K.
[c] Liquid at 1 K, NTP gas (300 K, 1 bar).

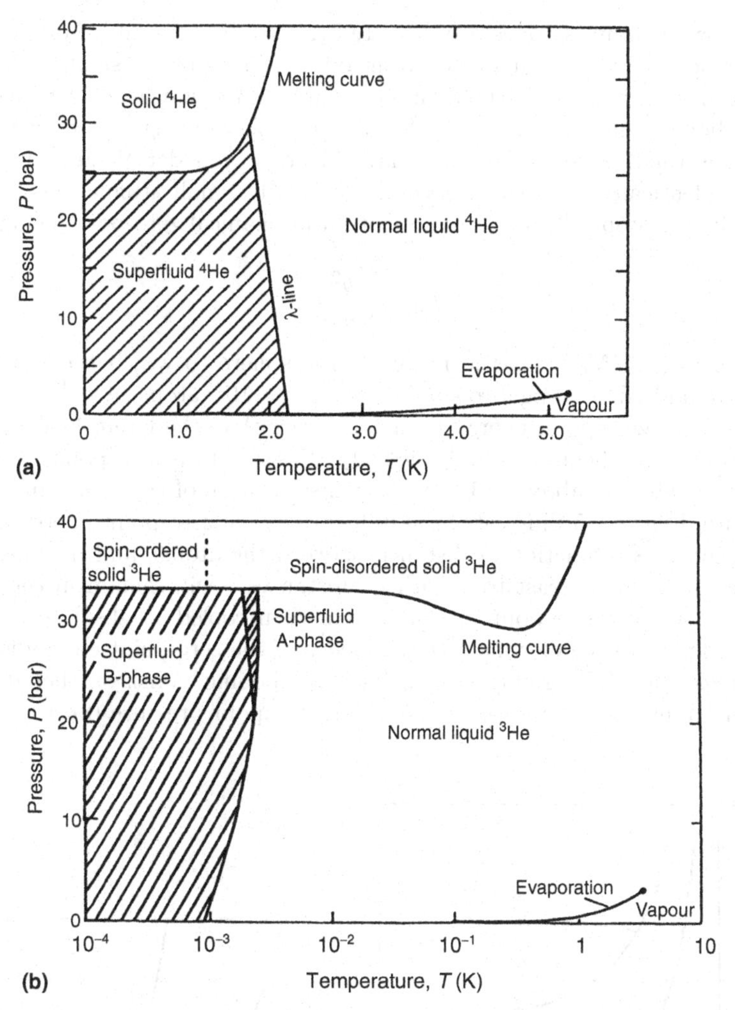

Fig. 2.4. Phase diagrams of (**a**) ^{4}He and (**b**) ^{3}He. Note the different temperature scales

phases coexist. The melting pressure of ^{4}He is constant to within about 10^{-4} below 1 K but for ^{3}He it shows a pronounced minimum at 0.32 K (Chap. 8). Then the two liquids have a rather small density or large molar volume. The molar volume V_m of ^{4}He (^{3}He) is more than a factor of two (three) larger than one would calculate for a corresponding classical liquid.

The origin for all these observations are two essential properties of helium. First, the binding forces between the atoms are very weak. They are van der Waals forces and are weak because of the closed electronic s-shell of helium, giving rise to the absence of static dipole moments and to the smallest known atomic polarizibility $\alpha = 0.1232\,\mathrm{cm^3\,mol^{-1}}$ (the resulting dielectric constants for the two helium isotopes are $\epsilon_4 = 1.0572$ and $\epsilon_3 = 1.0426$). For example, these binding forces are more than an order of magnitude smaller than for hydrogen molecules with their larger polarizibility, which leads to a much higher boiling temperature of H_2. Because the electronic structure of the two helium isotopes is identical, they have identical van der Waals forces and behave identically chemically. Second, due to the small atomic mass m, the two helium isotopes have a large quantum mechanical zero-point energy E_0, given by

$$E_0 = \frac{h^2}{8\,ma^2}\,, \tag{2.7}$$

where $a = (V_m/N_0)^{1/3}$ is the radius of the sphere to which the atoms are confined, and N_0 is Avogadro's number, 6.022×10^{23} atoms mol^{-1}.

The large zero-point energy (which is larger than the latent heat of evaporation of liquid helium, see (Table 2.1)) gives rise to a zero-point vibration amplitude which is about 1/3 of the mean separation of the atoms in the liquid state. Figure 2.5 illustrates the influence of the zero-point energy on the total energy as a function of distance between the atoms, and demonstrates why helium – in contrast to all other substances – will remain in the liquid state under its own vapour pressure even when cooled to absolute zero. Of course, due to its smaller mass the influence of the zero-point energy is more pronounced for ^{3}He, giving rise to its lower boiling point, smaller density, smaller latent heat of evaporation and larger vapour pressure (Sect. 2.3.2).

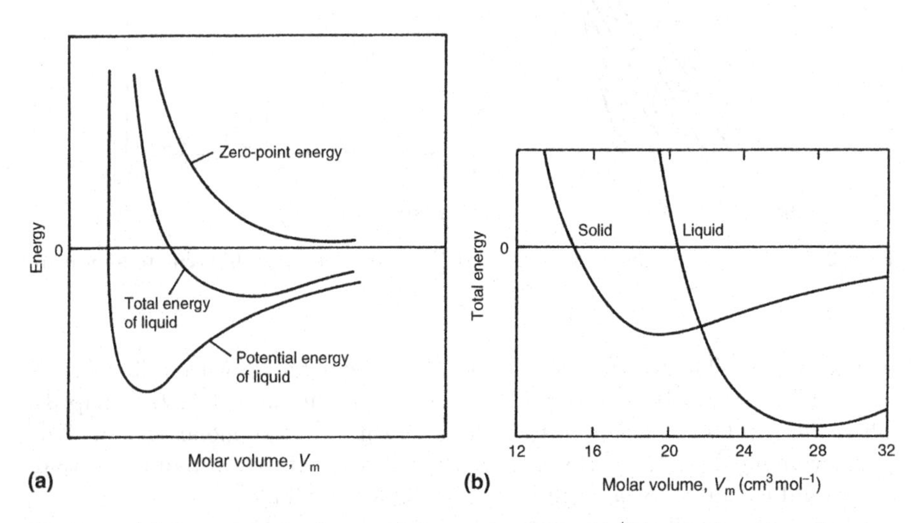

Fig. 2.5. (**a**) Zero-point and potential energies of liquid ^{4}He as a function of molar volume. The total energy is the sum of these two energies. (**b**) Illustration of why the liquid state is the stable one for helium at saturated vapour pressure even at $T = 0\,$K

Because of the strong influence of quantum effects on their properties, the helium liquids are called quantum liquids. In general, this term is used for any liquid whose kinetic (or zero-point) energy is larger than its potential (or binding) energy. To distinguish these liquids from classical liquids one introduces a quantum parameter $\lambda = E_{\mathrm{kin}}/E_{\mathrm{pot}}$. These parameters λ for some cryoliquids are:

liquid:	Xe	Kr	Ar	N_2	Ne	H_2	^{4}He	^{3}He
λ:	0.06	0.10	0.19	0.23	0.59	1.73	2.64	3.05

indicating that hydrogen and the helium isotopes are quantum liquids in this sense.

2.3.2 Latent Heat of Evaporation and Vapour Pressure

The latent heat of evaporation L and the vapour pressure P_{vap} are important properties that determine whether a liquid is suitable for use as a refrigerant. For the helium isotopes both these properties are dramatically different from the values of the corresponding classical liquids due to the large zero-point energy. For example, for ^{4}He the latent heat of evaporation is only about one-third of its value for the corresponding classical liquid. Due to the small heat of evaporation of helium (Fig. 2.6), liquid helium baths have a rather

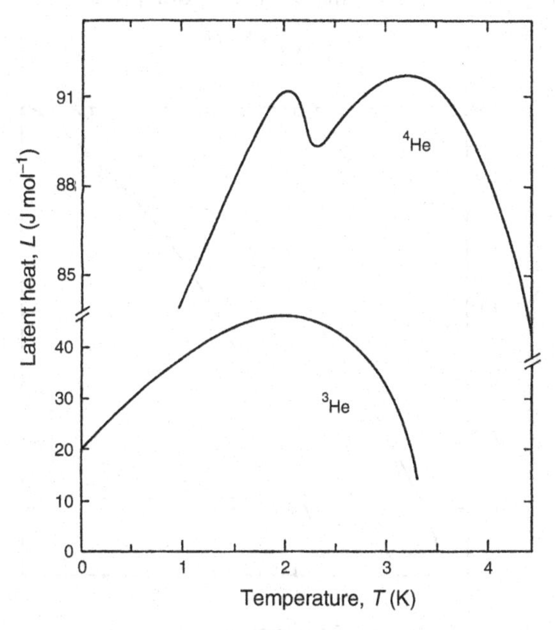

Fig. 2.6. Latent heats of evaporation of ^{3}He and ^{4}He. Note the change of vertical scale

small cooling power (it is very easy to evaporate them). Therefore, all low-temperature experiments require efficient shielding against introduction of heat from the surroundings, e.g., heat due to radiation, heat along supports to the experiments, or heat due to the measurements one wants to perform at low temperatures. This will be discussed later when we discuss the design of low-temperature equipment. The dip in L for ^{4}He at 2.2 K is due to the superfluid transition occurring at this temperature, which will be discussed in Sect. 2.3.3.

The vapour pressure can be calculated, at least to a first approximation, from the Clausius–Clapeyron equation

$$\left[\frac{\mathrm{d}P}{\mathrm{d}T}\right]_{\mathrm{vap}} = \frac{S_{\mathrm{gas}} - S_{\mathrm{liq}}}{V_{\mathrm{m,gas}} - V_{\mathrm{m,liq}}}, \tag{2.8}$$

where S is the entropy and V_{m} the molar volume.

If we take into account that the difference in the entropies of the liquid and gaseous phases is L/T, that the molar volume of the liquid is much smaller than the molar volume of the gas, and that in a rough approximation the molar volume of helium gas is given by the ideal gas equation $V_{\mathrm{gas}} \cong RT/P$, then we obtain

$$\left[\frac{\mathrm{d}P}{\mathrm{d}T}\right]_{\mathrm{vap}} \cong \frac{L(T)P}{RT^2}, \tag{2.9}$$

and eventually arrive at our result for the vapour pressure

$$P_{\mathrm{vap}} \propto \mathrm{e}^{-L/RT}, \tag{2.10}$$

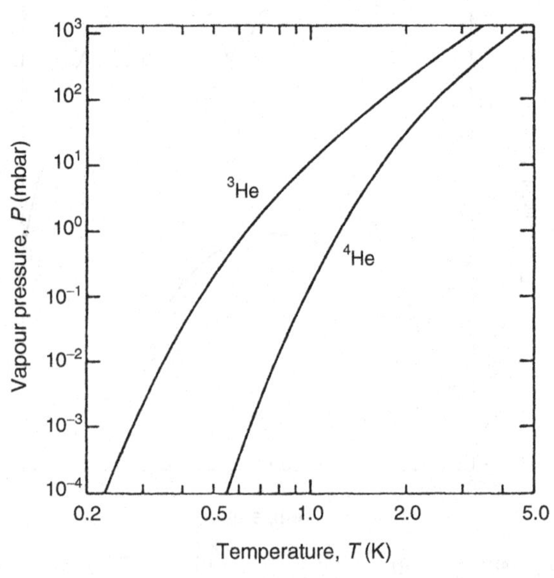

Fig. 2.7. Vapour pressures of liquid ^{3}He and liquid ^{4}He

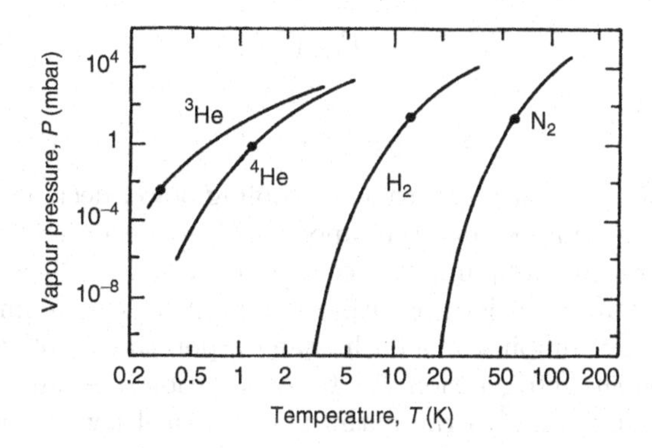

Fig. 2.8. Vapour pressures of various cryoliquids. The dots indicate the practical lower limits for the temperatures which can be obtained by reducing the vapour pressure above these liquids

if we make the further approximation that $L \cong$ const. (Fig. 2.6). Therefore, the vapour pressure decreases roughly exponentially with decreasing temperature, as shown in Fig. 2.7 for the helium isotopes and in Fig. 2.8 for several cryoliquids (for more details on vapour pressures see Sects. 11.2 and 12.2).

One can take advantage of this pronounced temperature dependence of the vapour pressure in several ways. In the Kelvin temperature range, the vapour pressures of all substances except helium are extremely low (Fig. 2.8). Therefore, the surfaces in a low-temperature apparatus cooled to Kelvin temperatures, e.g., by liquid helium, are extremely efficient "pumps". If one has pumped on the vacuum space of a low-temperature apparatus at high temperatures, the valve to the pumping system should be closed when the apparatus has reached the Kelvin temperature range by introducing liquid helium because the cold surfaces can improve the vacuum by several orders of magnitude by condensing the remaining gas molecules. Usually it is much better to do this than to keep the valve to the pumping system open, because after a while the cold surfaces may pump molecules from the pumping system (such as crack products of the pump oil) into the cryostat. This "cryopumping" is utilized commercially in so-called cryopumps available from various suppliers.

Second, one can pump on the vapour above a liquid, for example above a liquid-helium bath, to obtain temperatures below the normal (1 bar) boiling point. If one pumps away atoms from the vapour phase, the most energetic ("hottest") atoms will leave the liquid to replenish the vapour. Therefore the mean energy of the liquid will decrease; it will cool. For a pumped-on liquid bath where $\dot{n}$ particles/time are moved to the vapour phase, the cooling power is given by

$$\dot{Q} = \dot{n}(H_{\text{liq}} - H_{\text{vap}}) = \dot{n}\dot{L}\,. \qquad (2.11)$$

Usually a pump with a constant-volume pumping speed $\dot{V}$ is used and therefore the mass flow $\dot{n}$ across the liquid–vapour boundary is proportional to the vapour pressure

$$\dot{n} \propto P_{\text{vap}}(T)\,, \tag{2.12}$$

giving a cooling power

$$\dot{Q} \propto L P_{\text{vap}} \propto \mathrm{e}^{-1/\mathrm{T}}\,. \tag{2.13}$$

This last equation demonstrates that the cooling power decreases rapidly with decreasing temperature because the vapour pressure decreases rapidly with decreasing temperature and pumping becomes less and less efficient. Eventually there is almost no vapour left, resulting in a limit for the minimum temperature obtainable by pumping on a bath of an evaporating cryoliquid. This limit is reached when the refrigeration due to evaporation of atoms is balanced by the external heat flowing to the bath. The practical low-temperature limits determined by experimental parameters are typically about 1.3 K for ^{4}He and 0.3 K for ^{3}He (Fig. 2.8).

The temperature dependence of the vapour pressure of liquid helium is well known and can be used for vapour-pressure thermometry (Sect. 12.2). By measuring the vapour pressure above a liquid-helium bath (or other liquids at higher temperatures) one can read the temperature from the corresponding vapour pressure table. In fact, the helium-vapour pressure scale represents the low-temperature part of the international temperature scale ITS-90 (Sect. 11.2).

2.3.3 Specific Heat

Many of the basic properties of a material, including liquid helium, are revealed by its specific heat. First of all, the specific heat of liquid helium is very large compared to the specific heat of other materials at low temperatures (Fig. 2.9). For example, at about 1.5 K the specific heat of 1 g of ^{3}He or ^{4}He is of the order of 1 J K^{-1} whereas the specific heat of 1 g of Cu is only about 10^{-5} J K^{-1} at this temperature. This remarkable fact is of great cryotechnical importance for low-temperature physics. It means that the thermal behavior, for example the thermal response time, of a low-temperature apparatus is in most cases determined by the amount and thermal behavior of the liquid helium it contains. In addition, the latent heat of evaporation of liquid helium – even though it is rather small compared to the latent heat of other materials – is large compared to the specific heat of other materials at low temperatures, enabling us to cool other materials, e.g., a metal, by liquid helium. Both properties mean that the temperature of an experiment rapidly follows any temperature change of its refrigerating helium bath.

At low temperatures the properties of materials are strongly influenced by statistical or quantum effects. This is particularly important for the specific heat of the helium isotopes. As Fig. 2.10 demonstrates, the specific heat of ^{4}He has a pronounced maximum at about 2.17 K, indicating a phase transition to a new state of this liquid. The detection of this phase transition came as a great surprise to scientists. It was not expected that anything interesting could happen at low temperatures with this rather simple liquid composed of

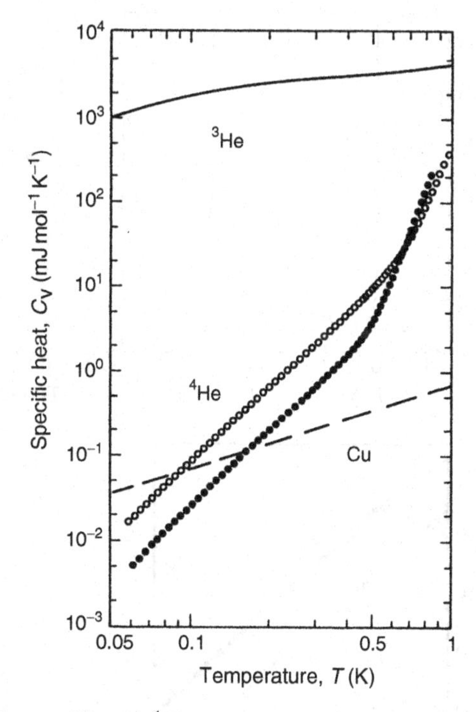

Fig. 2.9. Specific heat of liquid ^{4}He at vapour pressure (27.58 cm^3 mol^{-1}, ○) and at about 22 bar (23.55 cm^3 mol^{-1}, ●) [2.25] compared to the specific heats of liquid ^{3}He at vapour pressure [2.26] and of Cu

inert spherical atoms. Therefore, when L.J. Dana and H. Kamerlingh Onnes measured the specific heat of liquid helium in the 1920s and saw an "anomalous" increase of the specific heat at about 2 K, they did not publish the data points at those temperatures, believing that they resulted from some experimental artefacts. Later W.H. Keesom and K. Clusius (1932) in the same laboratory of the University of Leiden measured the specific heat of liquid helium again and again saw the pronounced peak of the specific heat. They did believe in their data and realized that a phase transition occurred in this liquid at that temperature, the transition to the unique superfluid state of ^{4}He. W.H. Keesom introduced the names He I for the normal liquid above the transition temperature and He II for the superfluid liquid below it. The history of the specific heat of liquid ^{4}He is one of the important examples in physics that one should never disregard any apparently "anomalous" data or data points unless one has good reason not to trust one's own measurements. The specific heat maximum of liquid ^{4}He near 2.2 K has been measured with increasing temperature resolution over the past decades (Fig. 2.10). Today it is known with a sub-nanokelvin temperature resolution, and these very precise measurements are one of the most important testing grounds for modern

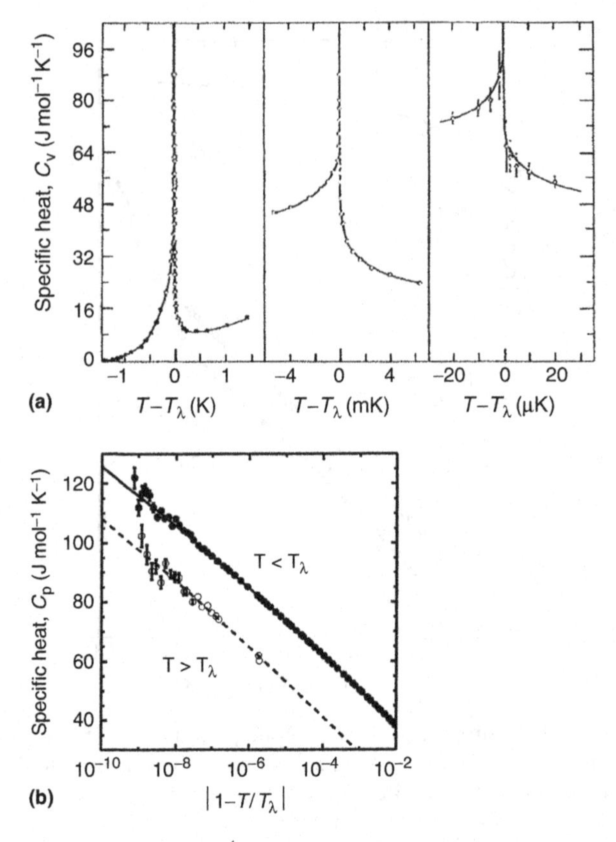

Fig. 2.10. Specific heat of liquid ^{4}He at temperatures close to its superfluid transition. (**a**) With increasing T-resolution on a linear temperature scale [2.27] and (**b**) on a logarithmic temperature scale. Reprinted with permission from [2.30]; copyright (2003) Am. Phys. Soc.. These latter data have been taken in flight on earth orbit to avoid the rounding of the phase transition by gravitationally caused pressure gradients in the liquid sample of finite height. For the applied high-resolution thermometry see Sect. 12.9

theories of phase transitions [2.27–2.30]. It is a demonstration of the model character of the superfluid transition of liquid helium for phase transitions and the very advanced state of low-temperature thermometry (Chaps. 11 and 12, in particular Sect. 12.9). Helium is an extremely good system for the study of properties with a very high temperature resolution near a critical temperature due to its extreme purity (all other materials are "frozen-out"). Therefore the sharpness of the transition. The characteristic shape of the specific heat maximum has led to the term λ-transition for the superfluid transition of ^{4}He occurring at the λ-temperature $T_\lambda = 2.1768$ K at saturated vapour pressure. The λ-temperature decreases with increasing pressure to a value of 1.7673 K at the melting line (Fig. 2.4).

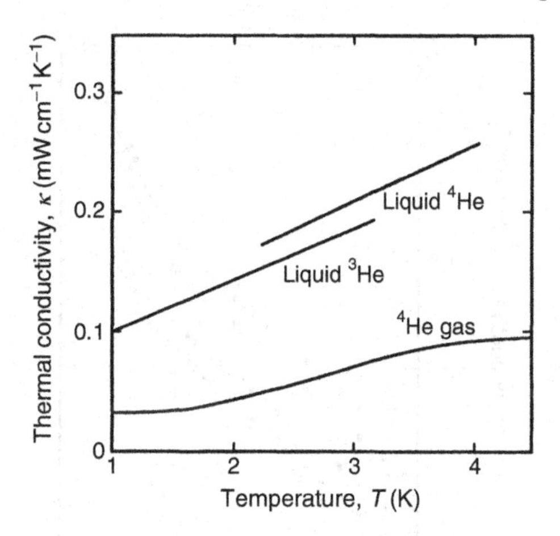

Fig. 2.11. Thermal conductivities of gaseous and liquid helium [2.31]

Above the λ-transition, ^{4}He behaves essentially like a classical fluid or in some respects – because of its low density – almost like a classical gas (Fig. 2.11). But below T_λ due to its (Bose–Einstein like) condensation in momentum space, its entropy and specific heat decrease rapidly with temperature (Fig. 2.9). Between 1 and about 2 K the specific heat of ^{4}He has a strong temperature dependence due to so-called roton excitations. Finally, below 0.6 K, the specific heat decreases with T^3 due to phonon-like excitations, as for an insulating Debye solid (Sect. 3.1.1) [2.17–2.24].

The isotope ^{3}He has a nuclear spin $I = 1/2$. It is therefore a Fermi particle and has to obey Fermi statistics and the Pauli principle. This liquid has many properties in common with the "conducting Fermi liquid" composed of electrons in metals, which are also spin-1/2 particles. For example, the specific heat of ^{3}He obeys $C \propto T$ at low enough temperatures. This result and some other properties of ^{3}He will be discussed in Sect. 2.3.6 and in Chaps. 7 and 8.

2.3.4 Transport Properties of Liquid ^{4}He: Thermal Conductivity and Viscosity

In its normal fluid state above $T = 2.2$ K, liquid ^{4}He, due to its low density, shows transport properties almost like a classical gas (Fig. 2.11). The same applies to liquid ^{3}He at temperatures above 0.1 K. Due to its low thermal conductivity *above* 2.2 K [it is about a factor of 10 (10^4) lower than the thermal conductivity of stainless steel (Cu), see Sect. 3.3] liquid ^{4}He I boils with strong bubbling. When one pumps on liquid ^{4}He I (or liquid ^{3}He), bubbles of vapour form within the liquid and the liquid is agitated when they rise to the surface. The bubbles form because the bulk of the liquid is hotter than its pumped surface. In an experiment employing liquid helium it is very likely that there are large thermal gradients in the liquid at these temperatures.

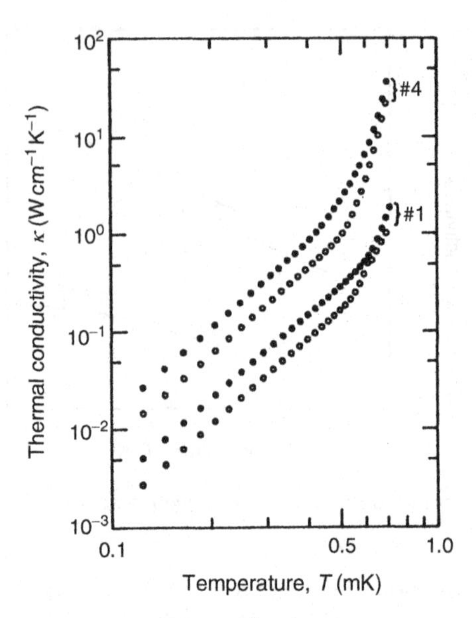

Fig. 2.12. Thermal conductivity of liquid ^{4}He at 2 bar (●) and 20 bar (○) in tubes of 1.38 mm (#1) and 7.97 mm (#4) diameter [2.32]

This is of importance, for example, when the vapour pressure of liquid helium is used for thermometry (Sect. 12.2). Conversely, *below* 2.2 K, in the superfluid state, under ideal experimental conditions (heat flow $\dot{Q} \to 0$) the thermal conductivity of ^{4}He II is infinite; for realistic conditions it is finite but quite large (Fig. 2.12). The very high thermal conductivity of superfluid helium makes this material a very efficient medium for establishing temperature homogeneity or for transporting heat. For example, at these temperatures the liquid does not boil as other liquids do if heated from the bottom or if they are pumped, because for boiling by creation of bubbles and their transport to the surface it is necessary that a temperature gradient is established. This is impossible in liquid helium below T_λ if the heat current is not so large that it "destroys" the superfluid state. In the superfluid state helium atoms evaporate exclusively from the surface and the liquid is "quiet" because no more bubbles are formed. The thermal conductivity of superfluid ^{4}He below 0.4 K, in the phonon regime, and at $P \leq 2$ bar is given by

$$\kappa_4 \propto 20dT^3 (\mathrm{W\,(K\,cm)^{-1}}) , \tag{2.14}$$

where d is the diameter of the ^{4}He column in centimeters (Fig. 2.12) [2.32]. As a result, the thermal conductivity of superfluid ^{4}He is quite high under most experimental conditions, comparable to that of a metal. For the physics of heat transport in superfluid ^{4}He, the reader should consult the literature specializing on this quantum liquid [2.17–2.24].

Because the thermal conductivity of superfluid helium can be very large (or ideally infinitely large), temperature waves or entropy waves can propagate in this unusual liquid. This wave propagation is called second sound to distinguish it from the usual first sound, the density waves. The velocity of second sound in ^{4}He II is about an order of magnitude smaller than the velocity of first sound at $1\,\mathrm{K} \leq T \leq 2\,\mathrm{K}$ [2.17–2.24].

In the superfluid state, liquid helium has a vanishing viscosity, $\eta_s = 0$, for flow through fine capillaries or holes; it is indeed "superfluid" if the flow velocity does not exceed a critical value (corresponding to a critical current in metallic superconductors). The vanishing viscosity allows superfluid helium to flow in a persistent mode as the persistent supercurrents in a metallic superconductor do. Apparatus which seems to be leak-tight for $T \geq T_\lambda$ may show a leak at lower temperatures if it comes into contact with superfluid helium because this liquid can flow through minute cracks or holes which are impermeable to viscous materials. This is a very efficient way of detecting extremely small leaks. On the other hand, it is, of course, a nuisance because apparatus which seems to be leak-tight at higher temperatures may suddenly develop a so-called superleak when in contact with superfluid helium.

The transport properties of liquid ^{3}He will be briefly discussed in Sect. 2.3.6.

2.3.5 Superfluid Film Flow

The walls of a container which is partly filled with liquid helium are coated with a film of helium via the adsorption of atoms from the vapour phase. Due to the rather strong van der Waals interaction which a substrate exerts on helium atoms, the first layer of ^{4}He on a substrate is solid and the liquid film above it is relatively thick, typically 30 nm at SVP (see below). Usually, due to its viscosity, such a film is immobile. However, in the superfluid state with

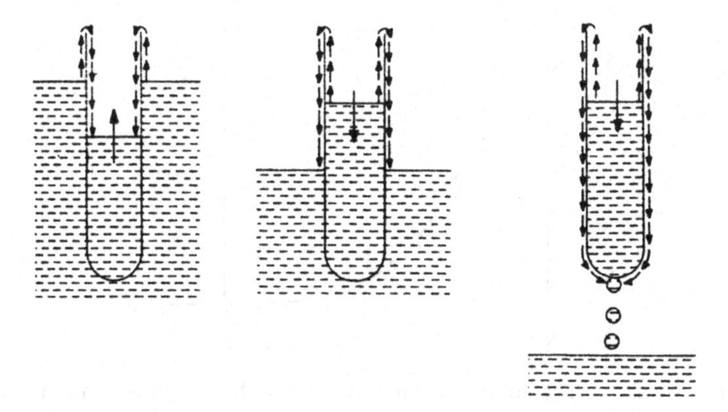

Fig. 2.13. Due to its superfluid properties ^{4}He II can leave a container via superfluid film flow over the walls

$\eta_s = 0$, the film can move. If two containers which are partly filled with liquid He II to different levels are connected, their levels will equalize by means of frictionless flow of the He II film from one container to the other one driven by the difference in gravitational potential (Fig. 2.13); the film acts as though being siphoned. This superfluid film flow [2.17, 2.33] will lead to an enhanced evaporation rate from ^{4}He baths at $T < T_\lambda$ because the superfluid film flows to hotter places and evaporates there.

Here, I will calculate the thickness d of a helium film at a height h above the bulk liquid level (Fig. 2.14). For the chemical potential of the film we have

$$\mu_{\mathrm{film}} = \mu_0 + mgh - \alpha/d^n \, , \tag{2.15}$$

where μ_0 is the chemical potential of the bulk liquid, and mgh is the gravitational term. The last term in the above equation is the van der Waals potential which the substrate supplies to the film (with $n = 3$ for $d \leq 5\,\mathrm{nm}$; $n = 4$ for $d \geq 10\,\mathrm{nm}$). For thin films, the van der Waals constant is $\alpha/k_B = (10\text{–}200\,(\mathrm{K})) \times (\text{no. of helium layers})^3$ for various solid substrates [2.33]. In thermal equilibrium an atom has to have the same chemical potential on the surface of the bulk liquid and on the surface of the film ($\mu_{\mathrm{film}} = \mu_0$); this condition leads to (for $d \leq 5\,\mathrm{nm}$)

$$d = (\alpha/mgh)^{1/3} \, . \tag{2.16}$$

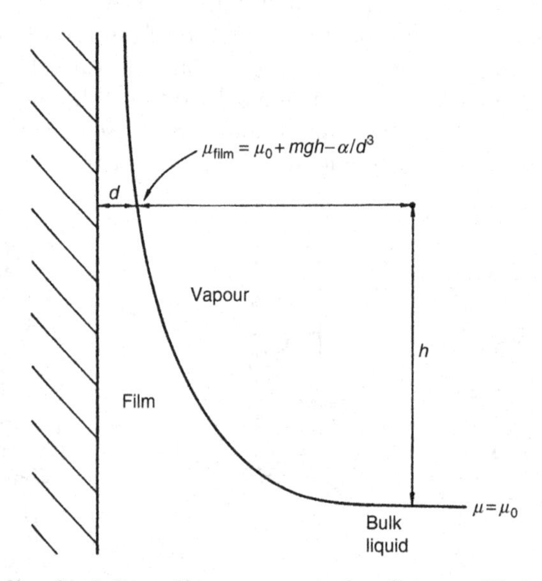

Fig. 2.14. Profile of a helium film on a vertical wall in equilibrium with its bulk liquid and with its saturated vapour. The chemical potential is given for the bulk liquid (μ_0) and for the liquid film (μ_{film}) at a height h above the bulk liquid

A typical value is

$$d \simeq 30\,\mathrm{h}^{-1/3} \tag{2.17}$$

with h in cm and d in nm; but, of course, d strongly depends on the cleanliness and structure of the surface of the substrate.

If there is gas but no bulk liquid present we have an "unsaturated film" whose thickness depends on the gas pressure P. The gas pressure at height h is

$$P(h) = P_{\mathrm{sat}}\,\mathrm{e}^{-mgh/k_{\mathrm{B}}T}. \tag{2.18}$$

Correspondingly, in (2.15), mgh has to be replaced by $k_B T/\ln(P_{\mathrm{sat}}/P)$, and we obtain for the film thickness

$$d = \left[\frac{\alpha}{k_{\mathrm{B}}T\ln(P_{\mathrm{sat}}/P)}\right]^{1/3}, \tag{2.19}$$

showing that the film thickness decreases by lowering the vapor pressure.

For a typical critical superfluid velocity of a ^{4}He film, $v_{\mathrm{s,crit}} \approx 30\,\mathrm{cm\,s^{-1}}$, we determine the volume flow rate out of a beaker of radius $R = 5\,\mathrm{mm}$

$$\dot{V}_{\mathrm{liq}} = 2R\pi d v_{\mathrm{s,crit}} \approx 1\,\mathrm{cm^3\,h^{-1}} \quad \text{or} \quad \dot{V}_{\mathrm{gas}} \approx 1\,\mathrm{l\,h^{-1}}\,. \tag{2.20}$$

We then find for a pump with a volume pumping rate $\dot{V}_{\mathrm{pump}} = 10^4\,\mathrm{l\,h^{-1}}$ at $P = 1\,\mathrm{bar}$ that the minimum pressure to which this pump can pump such a helium bath from which a superfluid film is creeping is

$$P_{\mathrm{min}} = P\dot{V}_{\mathrm{gas}}/\dot{V}_{\mathrm{pump}} \approx 10^{-3}\,\mathrm{bar} \quad (10^{-4}\,\mathrm{bar\ if}\ R = 50\,\mathrm{mm})\,. \tag{2.21}$$

Correspondingly, the minimum helium-bath temperature will be about 1 K, showing that one has to consider the film flow rate quite carefully and possibly has to reduce it, for example by a constriction of about 1 mm diameter in the pumping tube (Chap. 7). In addition, one has to remember that dirt on a surface can offer a very large effective surface area for the film flow.

2.3.6 Liquid ^{3}He and ^{3}He–^{4}He Mixtures at Millikelvin Temperatures

Because ^{3}He atoms are Fermi particles, single ^{3}He atoms cannot undergo the analogue of a Bose momentum condensation into a superfluid state as the bosons ^{4}He do at T_λ. However, there exists a weak attractive interaction between ^{3}He atoms in the liquid which gives rise to pairing of two ^{3}He atoms. Like the paired conduction electrons in a superconducting metal, the paired ^{3}He atoms then behave like bosons and can undergo a transition into a superfluid state. Because the pairing forces are rather weak, this transition occurs only at 2.4 (0.92) mK for $P = P_{\mathrm{melting}}(P_{\mathrm{svp}})$ (see Fig. 2.4 and Table 11.5). The exciting properties of superfluid ^{3}He have been the driving force for the development of refrigeration techniques to $T < 3\,\mathrm{mK}$ and they have been the object of intense investigations since the detection of superfluid ^{3}He in 1972

[2.34,2.35]. They are discussed in several relevant publications [2.19–2.24,2.36–2.43]. As for ^{4}He, where the experiments on the superfluid properties in the late 1920s and 1930s were restricted to the few laboratories which had access to the low Kelvin temperature range, particularly Leiden, the experiments on the superfluid properties of ^{3}He in the 1970s were restricted to the few laboratories which then had access to the low-millikelvin-temperature range, particularly Cornell, San Diego and Helsinki. Because we are interested in this book in the cryotechnological aspects of materials, I will not discuss superfluid ^{3}He but just remind you of those properties of liquid ^{3}He above its superfluid transition that are of importance in cryotechnical applications of this Fermi liquid.

Liquid ^{3}He, due to its smaller mass, has a larger zero-point energy than liquid ^{4}He. It therefore has an even lower density and in the almost "classical" regime, for $T > 0.1\,\mathrm{K}$, behaves even more like a dense classical gas. For example, its specific heat is almost T-independent, as it should be according to the Dulong–Petit law (Sect. 3.1.1), at $T \approx 0.5\,\mathrm{K}$ (Fig. 2.16). Liquid ^{3}He is a spin-1/2 particle; it is a fermion with an anti-symmetric wave function. At lower temperatures, its properties then become in many respects increasingly similar to The well-known Fermi liquid composed of the conduction electrons in metals, obeying Fermi statistics, too. The properties of liquid ^{3}He at low temperatures are therefore fundamentally different from those of ^{4}He. From about 0.1 K down to the superfluid transition temperature they can be accounted for by Landau's Fermi liquid theory [2.17,2.19–2.24,2.36,2.39–2.45]. This theory describes the liquid as a system of free fermions with its properties rescaled by the interatomic interactions. Due to the Pauli principle the ^{3}He atoms have to fill energy states up to the Fermi energy E_F (Fig. 3.3). Because $E_\mathrm{F}/k_\mathrm{B} \approx 1\,\mathrm{K}$ for ^{3}He but $E_\mathrm{F}/k_\mathrm{B} \approx 10^4\,\mathrm{K}$ for electrons, and because the specific heat $C \propto T/T_\mathrm{F}$ for a Fermi liquid (Sect. 3.1.2, Fig. 2.15), the specific heat of ^{3}He at low temperatures is very large compared to the specific heat of metals (Fig. 2.9). This linear T-dependence of C is obeyed by ^{3}He only up to about ten millikelvin (Fig. 2.15, Sect. 7.1). At higher temperatures, the specific heat of ^{3}He shows a plateau and then increases again with increasing temperature (Fig. 2.16). In the low millikelvin temperature range when ^{3}He enters its superfluid state, the specific heat, of course, deviates from a linear temperature dependence.

The transport properties of liquid ^{3}He in the Fermi-liquid temperature range, too, show a distinct T-dependence. For example, the thermal conductivity κ_3 increases as T^{-1} at $T \ll T_\mathrm{F}$ (Figs. 2.17 and 2.18), and its viscosity η_3 increases as T^{-2} in the low millikelvin temperature range (Fig. 2.19), making ^{3}He a very viscous but well-conducting fluid at low temperatures above its superfluid transition. These temperature dependences can easily be understood. The mean free path λ of ^{3}He particles in the liquid is limited by scattering with other ^{3}He particles. A two-body collision for Fermi particles is proportional to $(T/T_\mathrm{F})^2$, because the number of particles and the number of empty states into which they can scatter near to the Fermi energy are both proportional to temperature. Hence, we have $\lambda \propto T^{-2}$, and with $C \propto T$ and

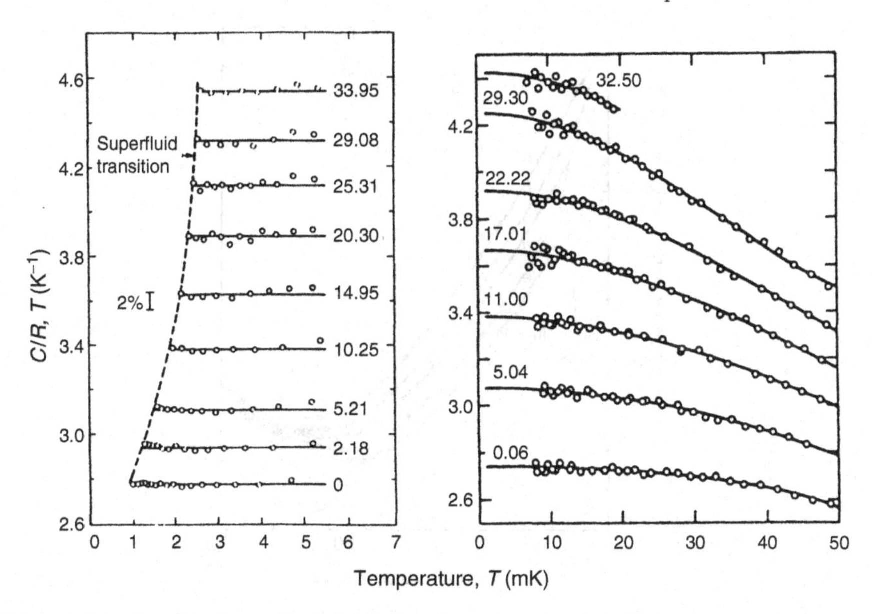

Fig. 2.15. Specific heat C divided by the gas constant R times temperature T of liquid ^{3}He at millikelvin temperatures at the given pressures (in bar). Note the different scales [2.26]

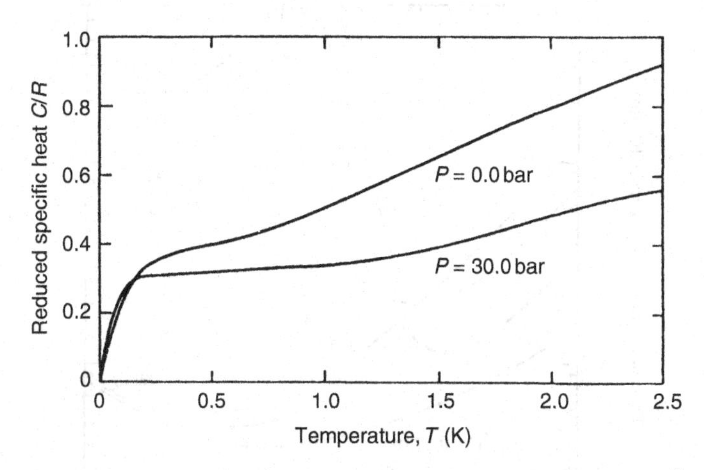

Fig. 2.16. Specific heat C (divided by the gas constant R) of liquid ^{3}He at two given pressures (after data from [2.26])

v_F = const. we find from (3.30) that $\kappa_3 \propto T^{-1}$ (and $\eta_3 \propto T^{-2}$) for $T \ll T_F$. At higher temperatures κ_3 and η_3 show a more complicated behavior.

It is important to remember that the thermal conductivity of this Fermi liquid becomes equal to the thermal conductivity of metallic conductors at

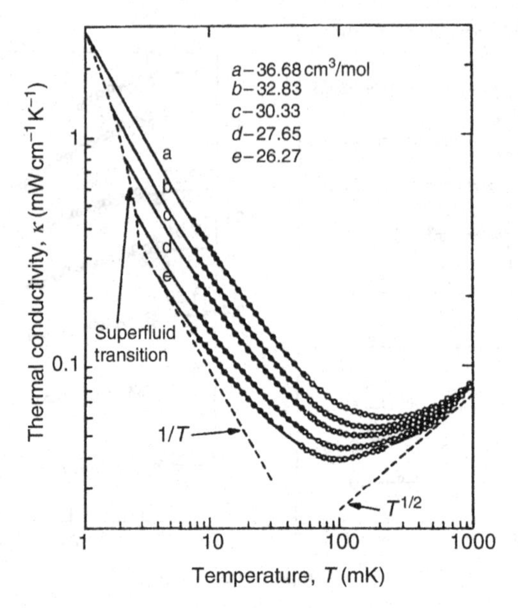

Fig. 2.17. Thermal conductivity of liquid ^{3}He at the given molar volumes [2.48]. At low temperatures, one finds the Fermi-liquid behavior $k \alpha T^{-1}$

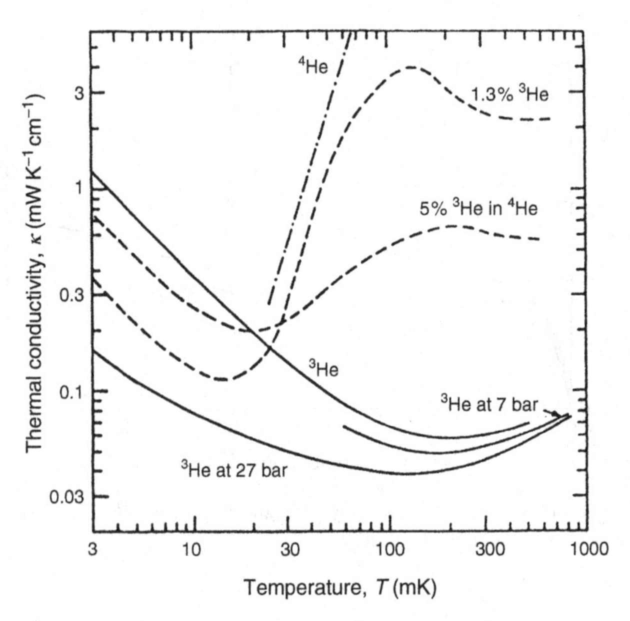

Fig. 2.18. Thermal conductivities of liquid ^{3}He, liquid ^{4}He, and dilute ^{3}He–^{4}He mixtures; unless otherwise indicated, the data are taken at SVP [2.50]. (This reference should also be consulted for references to the original literature). More-recent data for the thermal conductivities of ^{4}He and ^{3}He are given in Figs. 2.12 and 2.17

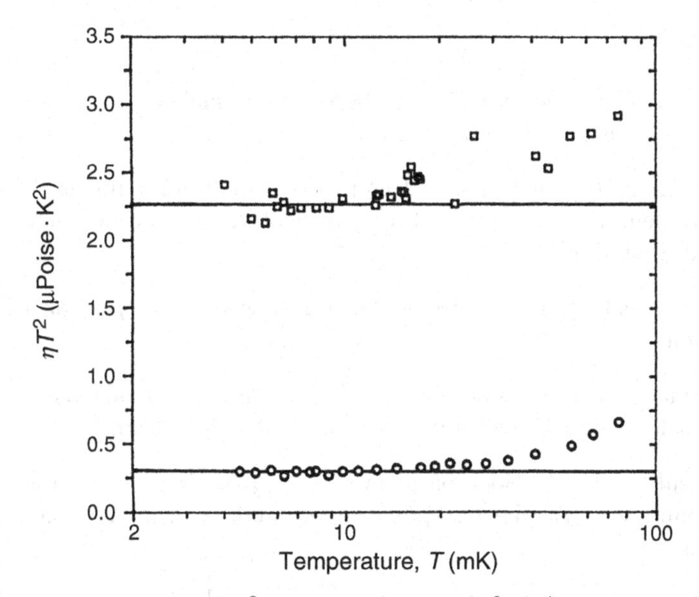

Fig. 2.19. Viscosity η of liquid ^{3}He (□) and of a 6.8% ^{3}He–^{4}He He mixture (○) at 0.4 bar multiplied by T^2 as a function of temperature. The solid lines represent the Fermi liquid behavior $\eta \propto T^2$ [2.52]

about 2 mK! In addition, the viscosity of liquid ^{3}He increases to that of light machine oil just above its superfluid transition temperature!

In liquid ^{3}He–^{4}He mixtures at low temperatures [2.17,2.23,2.24,2.46,2.49, 2.51], say at T <0.1 K, the superfluid ^{4}He component contains very few excitations (phonons and rotons); it acts essentially as an inert background influencing mostly the effective mass of the interacting ^{3}He Fermi particles. Hence, the helium mixtures behave as Fermi liquids, more dilute than pure liquid ^{3}He but with the same Fermi-liquid temperature dependences of their properties but altered absolute values. In particular, C/T (Fig. 7.5), κT (Fig. 2.18), and ηT^2 (Fig. 2.19) are temperature independent at low enough temperatures, $T \ll T_{Fermi}$, as for pure liquid ^{3}He. The advantage of the helium mixtures is the possibility to change their concentration and hence T_{Fermi} and to study the resulting changes of properties for testing theoretical predictions.

This short discussion of the properties of liquid helium relevant for its application in low-temperature experiments, in particular as a refrigerant, should be sufficient for our purpose. For a general discussion of the properties of liquid, in particular, superfluid helium, whose properties can be phenomenologially understood in terms of a "two-fluid model", the reader should consult the special literature on this subject given in the list of references for this chapter. Some of the properties of liquid ^{3}He will be discussed again in connection with Pomeranchuk cooling in Chap. 8, and liquid mixtures of the two helium isotopes will be considered in more detail in Chap. 7 when we discuss the dilution refrigerator.

Problems

2.1. Calculate the energy difference between one mole of normal H_2 at room temperature and one mole of *para*-H_2.

2.2. Determine the nuclear spin I and rotator quantum number J for *ortho*- and *para*-deuterium. What is the *para*-deuterium concentration at room temperature and at 20 K?

2.3. Why does the mixed molecule HD not show an *ortho–para* or *para–ortho* conversion?

2.4. Deduce (2.7) by considering a free particle in a spherical cavity. Compare the result for liquid ^{4}He with its latent heat of evaporation.

2.5. Calculate the de Boer parameter $\lambda = E_{\mathrm{kin}}/E_{\mathrm{pot}}$ (see Table 2.3) for H_2 assuming for E_{kin} the zero-point energy and for E_{pot} the Lennard-Jones potential

$$U(R) = \epsilon[(\sigma/R)^{12} - (\sigma/R)^6] \quad \text{with } \epsilon = 5 \times 10^{-15}\ \text{erg and } \sigma = 2.96\ \text{Å}\,.$$

2.6. Calculate the binding energy of a helium atom at the surface of liquid helium at 1 K (latent heat $L = 2.2\,\mathrm{kJ/l}$).

2.7. Calculate the Debye temperature (3.7) corresponding to the heat capacity of liquid ^{4}He for $T < 60;\ 0.5\,\mathrm{K}$ (Fig. 2.9).

2.8. Calculate the thickness of an unsaturated ^{4}He film on a horizontal glass substrate ($\alpha = k_{\mathrm{B}}\ 30\,\mathrm{K\ layers}^3$) at $P/P_{\mathrm{sat}} = 0.5$.

2.9. To which pressure does one have to pump ^{4}He in a tube of 1 cm diameter and 20 cm length so that the heat transported by the remaining superfluid film is less than 10^{-4} W if the ends of the tube are at 2 and 1 K, see (2.19, 2.20) and Fig. 2.12?

3

Solid Matter at Low Temperatures

The purpose of this chapter is to summarize the basic properties of solid materials at low temperatures [3.1–3.8] that are relevant for the design and construction of low-temperature apparatus and for performing experiments with such apparatus. These properties are, in particular, specific heat, thermal expansion, thermal conductivity and magnetic susceptibility. The latter property will also be discussed in later chapters in connection with magnetic cooling and thermometry.

Quite generally one can say that the properties of materials can be better understood when the temperature is reduced further (except for some exotic cases like solid ^{3}He), because as the temperature is lowered the properties of materials become more and more "ideal" or "simpler"; they approach their theoretical models more closely. At low temperatures the number of excitations decreases and the vibrations of the atoms can be described in the harmonic approximation, which means that the potential V^* as a function of distance $r - r_0$ from the equilibrium position r_0 of the atom can be written as

$$V^*(r - r_0) \propto (r - r_0)^2 \,. \tag{3.1}$$

In this approximation there is no thermal expansion because thermal expansion results from the anharmonic parts of the potential, for which we would have to introduce higher-order terms in the above equation. As a result, the thermal expansion coefficient becomes smaller and smaller, as we approach lower and lower temperatures (Sect. 3.2). A further advantage of low temperatures for the description of the properties of materials is the fact that various "parts" of a material can be treated independently. For example, in many cases one can consider the nuclear spin system (which is of great importance at ultralow temperatures) independently of the electrons and the lattice vibrations (Chap. 10). Of course, this is not true for all the "parts" of a material. For example, the temperature dependence of the electrical resistivity of a metal just results from the interaction of conduction electrons with the lattice. The fact that the specific heat and the thermal conductivity due to lattice vibrations and due to conduction electrons in the metal can be treated independently and can then just be added is a consequence of the large mass difference of the nuclei and electrons. To a very good approximation, their motions are independent, and the Schrödinger equation for the whole crystal can be separated into an electronic part and a lattice part. This is known as the Born–Oppenheimer approximation [3.1–3.5].

3.1 Specific Heat

The specific heat C is one of the most informative properties of a material. It is a measure of how much energy is necessary to increase the temperature of a material. In other words, it is a measure of the excited states of this material in the relevant temperature or energy range.

To calculate the specific heat of a material we therefore have to consider the various excitations that can be excited if we transfer thermal energy to it [3.1–3.8]. Hence, the specific heat carries a lot of information about the material.

3.1.1 Insulators

For nonmagnetic, crystalline insulators the most important, and in most cases only possible, excitations are vibrations of the atoms, the so-called phonons. At high temperatures all possible vibrational states of the atoms are excited. Considering each atom to behave as an independent, classical harmonic oscillator results in the Dulong–Petit law for the high-temperature specific heat of a material of N_0 atoms per mole, each of which has three degrees of freedom for its potential energy and for its kinetic energy. Because each degree of freedom contributes $k_B/2$ to the specific heat (at constant volume of the material),

$$C_v = \frac{6}{2} N_0 k_B = 3R = 24.94 [\mathrm{J\,mol^{-1}\,K^{-1}}] . \tag{3.2}$$

However, when the thermal energy $k_B T$ becomes of the order of the energy necessary for excitation of lattice vibrations one sees deviations from the Dulong–Petit law, because not all of the lattice vibrations will then be excited. The limit for the applicability of the Dulong–Petit law is given by

$$T \approx \hbar\omega_{ph}/k_B , \tag{3.3}$$

where ω_{ph} is the phonon frequency. The limiting temperature is of the order of a few 100 K.

In 1907 A. Einstein showed that a reasonable description of the specific heat of lattice vibrations below this temperature is obtained if the lattice vibrations are considered to be quantized. He performed a calculation of the phonon specific heat by describing the lattice vibrations as quantized phonon "particles", but gave them all the same frequency ω_E. In this Einstein model the material is considered as being composed of independent oscillators with the energies

$$E_n = \hbar\omega_E(n + 1/2)\,, \tag{3.4}$$

where $n = 0, 1, 2, \ldots$ is the excitation number of the modes or of the phonons. Planck then showed that the mean excitation number $\langle n \rangle$ of each oscillator or the mean number of phonons at temperature T is given by the Boson distribution function

$$\langle n \rangle = f_{ph}(\omega) = \frac{1}{\exp(\hbar\omega/k_B T) - 1}\,. \tag{3.5}$$

This model, of course, deviates from reality because the vibrational frequencies of the atoms in a crystal are not independent from each other. Due to the interatomic interactions they are not equal but are distributed over a spectrum, which can have a rather complicated structure as a function of energy (Fig. 3.1a). Again, at low temperatures the situation becomes simpler because the frequency dependence of the phonon density of states goes as ω^2 for low energies. As Debye showed, at low temperatures the phonon density of states g_{ph} can then be described by the following parabolic energy or frequency dependence (Fig. 3.1a):

$$g_{ph}(\omega) = \begin{cases} (3V_m/2\pi^2 v_s^3)\omega^2 = 9N_0\omega^2/\omega_D^3, & \omega < \omega_D\,, \\ 0, & \omega < \omega_D\,, \end{cases} \tag{3.6}$$

V_m being the molar volume, and v_s the average value of the longitudinal and transversal velocities of sound of a crystal: $v_s^{-3} = (v_{long}^{-3} + 2v_{trans}^{-3})/3$.

Of course, this spectrum does not extend to infinity but is cut off at a maximum frequency that is given by the condition that g_{ph} contains all the $3N_0$ phonon frequencies,

$$\int_0^{\omega_D} g_{ph}(\omega)\mathrm{d}\omega = 3N_0. \tag{3.7}$$

This limiting frequency is called the Debye frequency ω_D. The corresponding temperature, the Debye temperature $\theta_D = (\hbar/k_B)\omega_D$ of the material, is a measure of the temperature below which phonons begin to "freeze out". Values for θ_D of several metals are given in Table 10.1. These parameters are material constants with large values for a lattice composed of light atoms which are strongly bound, as in diamond ($\theta_D = 2{,}000$ K), and with small values for a lattice composed of heavy atoms bound by weak forces, as in lead ($\theta_D = 95$ K).

If we apply the Debye model to calculate the internal energy of the lattice vibrations and take its temperature derivative to arrive at the specific heat,[1] we find [3.1–3.8]

[1] In the following I do not distinguish between C_p and C_v, the specific heats at constant pressure and constant volume, respectively; their difference $C_p - C_v = 9\alpha\kappa VT$ (α being the thermal expansion coefficient and κ the compressibility) is negligible for solids at low temperatures, approaching about 1% at $T \approx \theta_D/2$.

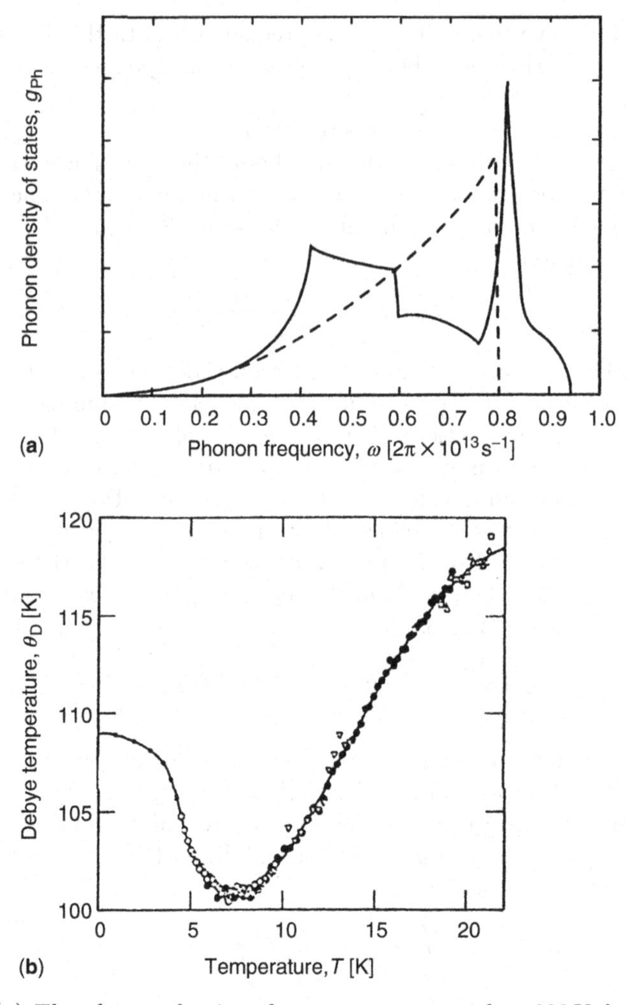

Fig. 3.1. (**a**) The phonon density of states g_{ph} measured at 300 K for aluminum. The dashed curve represents the Debye approximation (3.9) with a value $\theta_{\mathrm{D}} = 382\,\mathrm{K}$ deduced from the specific-heat data [3.9]. (**b**) Variation of the Debye temperature θ_{D} of indium due to the influence of deviations from $g_{\mathrm{ph}} \propto \omega^2$ [3.10]

$$C_{\mathrm{ph}}(T) = 9N_0k_{\mathrm{B}}\left[\frac{T}{\theta_{\mathrm{D}}}\right]^3\int_0^{x_{\mathrm{D}}} x^4\,\mathrm{e}^x(\mathrm{e}^x-1)^{-2}\mathrm{d}x\,, \tag{3.8}$$

with $x = \hbar\omega/k_{\mathrm{B}}T$.

The integral yields

$$C_{\mathrm{ph}}(T) = \frac{12}{5}\pi^4N_0k_{\mathrm{B}}\left[\frac{T}{\theta_{\mathrm{D}}}\right]^3 = 1944\left[\frac{T}{\theta_{\mathrm{D}}}\right]^3 [\mathrm{J\,mol^{-1}\,K^{-1}}] \tag{3.9}$$

for temperatures $T < \theta_D/10$. A deviation of the phonon density of states from the ω^2 dependence can be taken into account by letting $\theta_D = f(T)$, see Fig. 3.1b. The cubic dependence of the phonon specific heat on temperature demonstrates its rather strong decrease with decreasing temperature. Therefore, insulators such as rare-gas crystals very often have a very small specific heat at low temperatures. Examples are illustrated in Fig. 3.2, which shows the good agreement between the Debye model and experimental data at low temperatures.

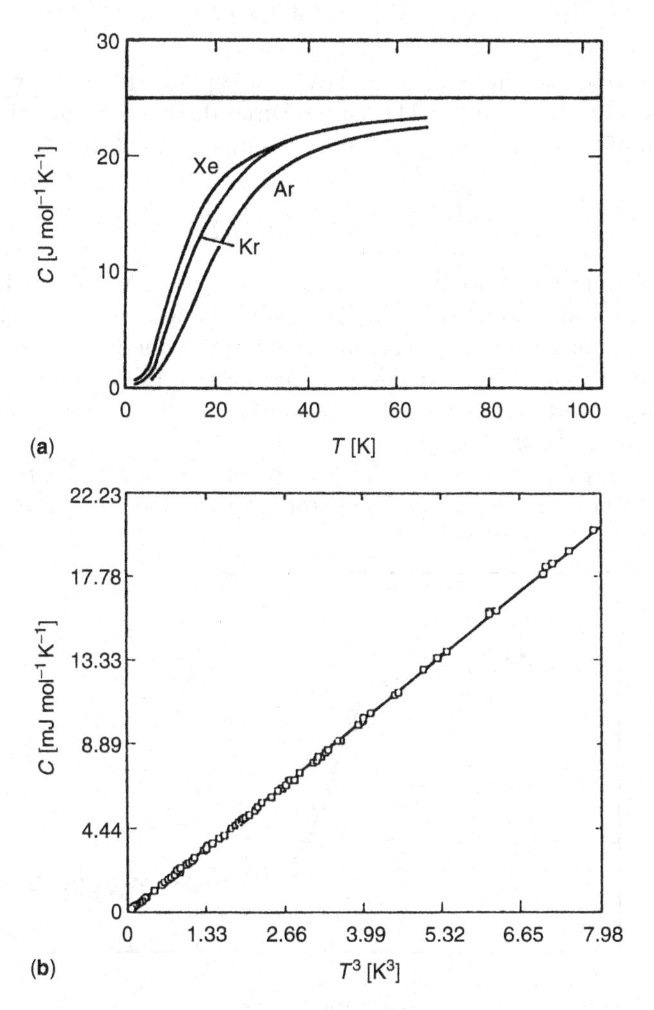

Fig. 3.2. (**a**) Specific heat of solid Ar, Kr and Xe. The horizontal line is the classical Dulong–Petit value [3.11]. (**b**) Specific heat solid of Ar as a function of T^3 at $T < 2\,\text{K}$ [3.1, 3.12]

3.1.2 Metals

Besides the lattice vibrations, in a metal we also have (nearly free moving) conduction electrons, which can be thermally excited. Electrons have the spin 1/2, are fermions and obey the Pauli principle. Therefore, each energy state can be occupied by at most two electrons with antiparallel spin orientation. Putting all our electrons into energy states, we will fill these states up to a limiting energy, the Fermi energy, given by

$$E_{\mathrm{F}} = k_{\mathrm{B}}\, T_{\mathrm{F}} = \frac{\hbar^2}{2m}\left[\frac{3\pi^2 N_0}{V_{\mathrm{m}}}\right]^{2/3} = 3.0 \times 10^5 k_{\mathrm{B}}\, V_{\mathrm{m}}^{-2/3}\,[\mathrm{erg}]\,. \tag{3.10}$$

A typical value for this energy is 1 eV, corresponding to the rather high temperature of about 10^4 K. The Fermi–Dirac distribution function for the occupation of energy states of electrons at temperature T is (Fig. 3.3)

$$f_{\mathrm{e}}\,(E) = \frac{1}{\exp[(E-\mu)/k_{\mathrm{B}}T]+1} \tag{3.11}$$

with the chemical potential $\mu = E_{\mathrm{F}}$ at $T = 0$. Because of the high value of the Fermi temperature $T_{\mathrm{F}} = E_{\mathrm{F}}/k_{\mathrm{B}}$, room temperature is already a "low temperature" for the electron gas, in the sense that here the electron gas is already pretty well in its ground state. And indeed, the properties of metals at low temperatures are determined exclusively by electrons in energy states very close to the Fermi energy.

To fill energy states we first have to determine them. That means, we have to calculate the density $g_{\mathrm{e}}(E)$ of states for conduction electrons. This is

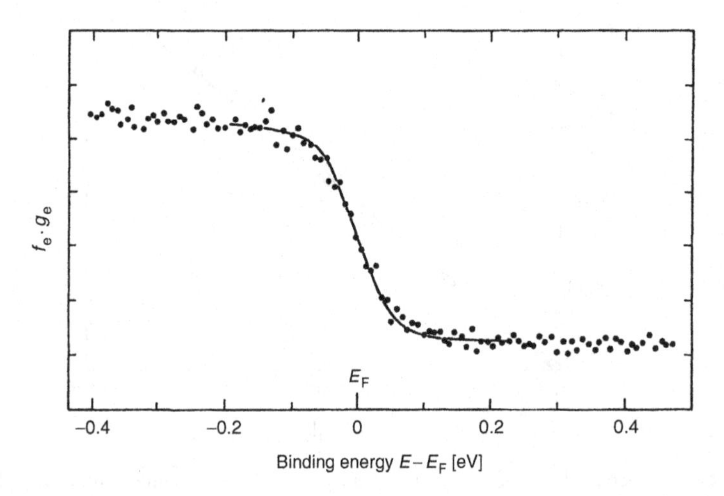

Fig. 3.3. Occupied electronic states of Ag at 300 K near the Fermi energy E_{F} obtained from photoelectron spectroscopy. The full line is the Fermi distribution function (3.11) [3.13]

rather simple within the "free-electron model", which describes many of the electronic properties of metals rather well [3.1–3.7]. The result is

$$g_{\mathrm{e}}(E) = \frac{3N_0}{2E_{\mathrm{F}}^{3/2}} E^{1/2} = \frac{V_{\mathrm{m}}}{2\pi^2} \left[\frac{2m}{\hbar^2}\right]^{3/2} E^{1/2} . \tag{3.12}$$

To find properties of the conduction electrons in a metal we have to multiply the two quantities $g_{\mathrm{e}}(E)$ and $f_{\mathrm{e}}(E)$.

At the temperature T one can thermally excite only electrons near the Fermi energy, within an energy range from about $E_{\mathrm{F}} - k_{\mathrm{B}} T$ to about $E_{\mathrm{F}} + k_{\mathrm{B}} T$; the thermal energy is not enough to excite electrons out of energy states further below the Fermi energy. At T the number of "thermally involved" electrons is then approximately given by $g_{\mathrm{e}}(E_{\mathrm{F}})\ k_{\mathrm{B}}\ T \propto T/T_{\mathrm{F}}$. If we raise the temperature from 0 to T, these electrons experience the energy change $\Delta E \simeq g_{\mathrm{e}}(E_{\mathrm{F}}) k_{\mathrm{B}}^2 T^2 \propto T^2/T_{\mathrm{F}}$ corresponding to an electronic specific heat of $C_{\mathrm{e}} \simeq 2g_{\mathrm{e}}(E_{\mathrm{F}}) k_{\mathrm{B}}^2 T$. If we perform the calculation more rigorously, we have to take into account the Fermi–Dirac distribution (3.11) at finite temperatures, which modifies our result for the electronic specific heat only slightly to

$$C_{\mathrm{e}}(T) = \frac{\pi^2}{2} N_0 k_{\mathrm{B}} \frac{T}{T_{\mathrm{F}}} = \gamma T , \tag{3.13}$$

where T_{F} is given by (3.10). The γ-values (Sommerfeld constants) for some metals are listed in Table 10.1. The result (3.13) is in good agreement with experimental data (Fig. 3.4) and was a great triumph for quantum mechanics, for the free-electron model, and for the Fermi theory of spin-1/2 particles.

Of course, for a real metal we have to give up the free-electron model. We have to consider the mutual interactions of the electrons, their interactions with the ions and the symmetry of the crystal. This can be taken

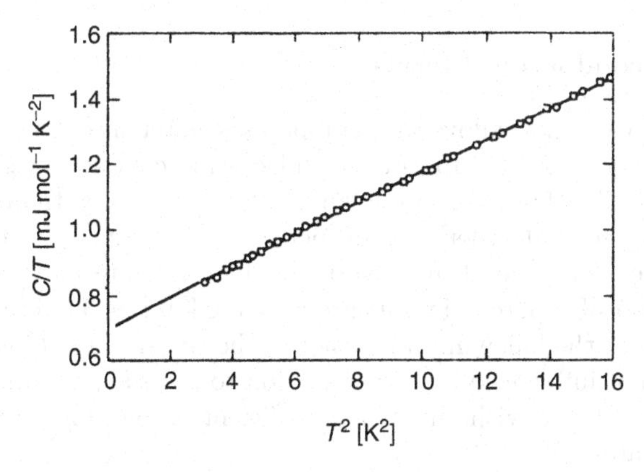

Fig. 3.4. Specific heat C of copper divided by the temperature T as a function of T^2 [3.4]

into account by going from the free-electron model to a so-called quasi-particle model [3.1–3.7]. In this model each electron has an effective mass m^* that deviates from the mass m of bare electrons because the electrons in a metal behave "heavier" or "lighter" than bare electrons owing to their interactions. We still obtain the same equations for the specific heat, in particular $C_e \propto T$; we just have to replace the mass of bare electrons by the effective mass of interacting electrons. For the specific heat, e.g., it means that we have to multiply (3.13) by m^*/m. The ratio of the effective mass m^* to the bare electronic mass m is between 1 and 2 for simple metals, it can be of order ten for transition metals with partly filled electron shells, and it can reach values of several 100 for "complicated" compounds, like the so-called "heavy-fermion" metals [3.5, 3.14, 3.15]. Values for the Sommerfeld constant γ for metallic elements can be found in condensed matter physics books [3.1–3.7].

With these results we arrive at the following equation for the total specific heat of a metal at "low" temperatures:

$$C = \gamma T + \beta T^3 . \tag{3.14}$$

"Low" means "small compared to the Debye temperature" if we consider the phonons and "small compared to the Fermi temperature" if we consider the electrons. If we introduce the appropriate material constants, we find that at room temperature the specific heat of a metal is dominated by the phonon specific heat and usually only at temperatures below 10 K is the electronic specific heat important; it dominates at temperatures below 1 K. The results for the electronic and lattice specific heats of copper are shown in Fig. 3.4, where C/T versus T^2 is plotted for low temperatures. The figure demonstrates how well the data are described by the theory discussed earlier. It also demonstrates that it is very often rather important to choose the right coordinates to get a sensitive indication of whether data follow an expected behavior.

3.1.3 Superconducting Metals

Many metals–elements, alloys and compounds–enter into a new state below a critical temperature T_c. In this so-called superconducting state [3.1, 3.2, 3.5, 3.6, 3.16–3.19] they can carry an electric current without dissipation, and they show several other new properties. As an example, I discuss here the specific heat of a metal in its superconducting state ($T < T_c$). Data for the specific heat of superconductors are presented in Fig. 3.5. Examining these data we arrive at the following conclusions. The specific heat C_{ph} of the lattice vibrations is not influenced by the transition to the superconducting state. It still follows a T^3 law with the same coefficient as one finds in the normal-conducting state (3.9)

$$C_{ph,s} = C_{ph,n} = \beta T^3 . \tag{3.15}$$

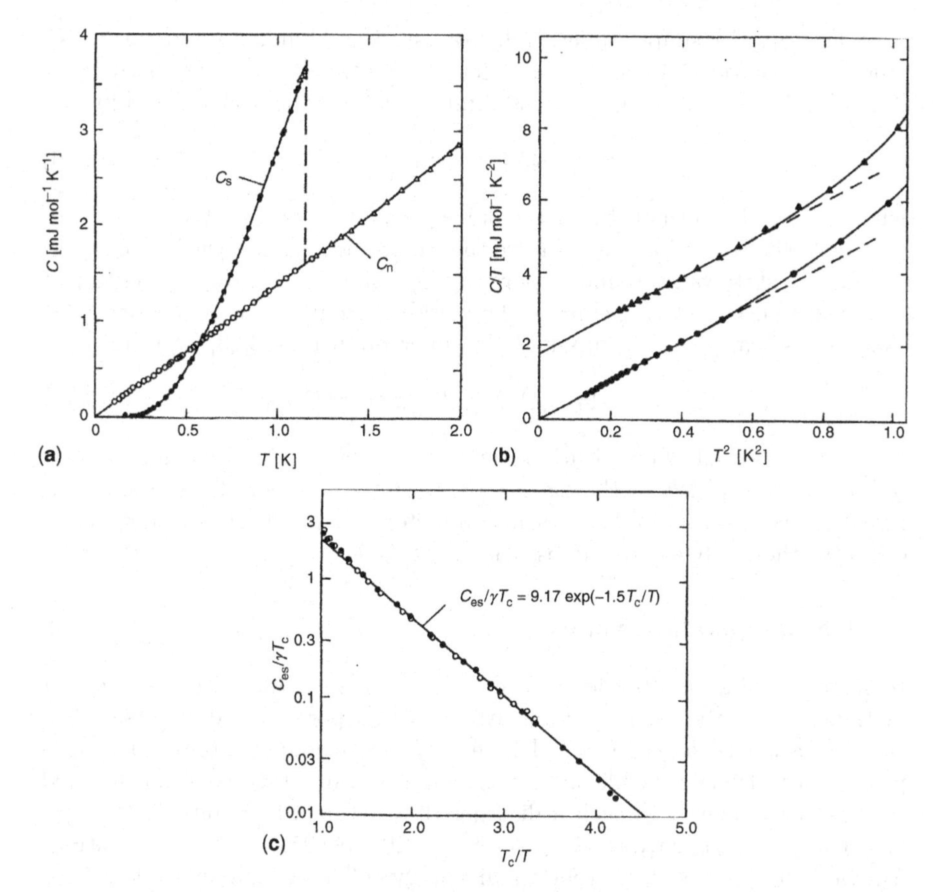

Fig. 3.5. (**a**) Specific heat of Al in the superconducting (C_s) and normal-conducting (C_n) states [3.20]. (**b**) Specific heat C of Hg divided by temperature T as a function of T^2 in the normal (▲) and superconducting (●) states. The straight lines correspond to (3.9,3.13) with $\theta_D = 72\,\mathrm{K}$ and $\gamma = 1.82\,\mathrm{mJ\,mol^{-1}\,K^{-2}}$ [3.21]. For the measurements in the normal conducting states, a magnetic field $B > B_c$ had to be applied to suppress the superconducting state. (**c**) Electronic specific heat C_{es} of superconducting V (●) and Sn (○) divided by γT_c as a function of T_c/T. The full line represents (3.17) [3.22]

The new behavior evident from Fig. 3.5 is entirely due to the altered specific heat of the electrons. First of all, there is a jump of the electronic specific heat at the critical temperature, but no latent heat (the transition is a second-order phase change when the external magnetic field is zero). The jump in C occurs because the metal now has – one might say – a new "degree of freedom" corresponding to the possibility of entering the superconducting state. In the discussion of the Dulong–Petit law for the specific heat at high temperatures we have already seen that each degree of freedom enhances the specific heat. For simple superconductors such as aluminum and tin, which follow the Bardeen, Cooper, Schrieffer (BCS) theory of superconductivity [3.1, 3.2, 3.5, 3.6, 3.16–3.19, 3.23], the jump of the specific heat is given by

$$\Delta C = 1.43\gamma T_{\mathrm{c}}\,, \tag{3.16}$$

where γT_{c} is the normal-state electronic specific heat (3.13) at T_{c}.

Below the transition temperature the electronic specific heat in the superconducting state vanishes much more rapidly than the electronic specific heat in the normal–conducting state of the metal. Its temperature dependence for "simple", so-called weak-coupling BCS superconductors [3.23] is given by

$$C_{\mathrm{e,\,s}} = 1.34\gamma T_{\mathrm{c}}(\Delta E/k_{\mathrm{B}}T_{\mathrm{c}})\exp[-\Delta E/k_{\mathrm{B}}T_{\mathrm{c}}] \tag{3.17}$$

An exponential temperature dependence of the specific heat is indicative of an energy gap ΔE in the density of states, as it occurs for the electrons in a superconductor (and in a semiconductor). This reflects the number of electrons thermally excited across the energy gap.

3.1.4 Non-crystalline Solids

In noncrystalline or disordered solids [3.5, 3.24–3.28] like vitreous silica or metallic glasses the atoms are not arranged in a periodic order; these solids can be visualized as supercooled liquids. As an example, Fig. 3.6 depicts a possible arrangement of silicon and oxygen atoms in vitreous silica compared to crystalline quartz. In such a disordered structure where long-range order is lost many atoms have more than one possible position and these positions can be distinguished by rather small energy differences. Even at low temperatures the atoms can "tunnel" from one position to another. Again, we have a new "degree of freedom" for the material: the possibility of performing

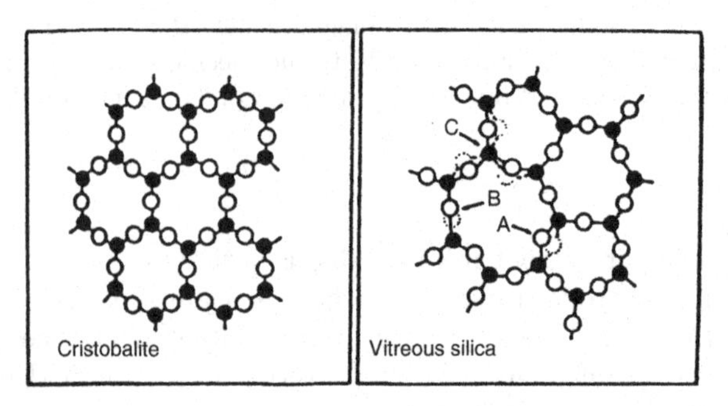

Fig. 3.6. Schematic two-dimensional representation of the structure of cristobalite, a crystalline modification of SiO_2, and of vitreous silica, the amorphous modification of SiO_2. Filled circles represent silicon atoms and open circles oxygen atoms. Three possible types of defects are indicated by arrows [3.5, 3.26]

structural rearrangements. As a result, we observe an additional contribution to the specific heat caused by the tunneling transitions or structural relaxations in a disordered or glassy material. At low temperatures this additional contribution dominates and is given by

$$C_a = aT^n \tag{3.18}$$

with an exponent n which is close to 1 [3.5, 3.24–3.30]. (In addition, one often observes an enhancement of the T^3 specific heat contribution.) This is illustrated for vitreous silica in Fig. 3.7. Disordered insulators, therefore, show an almost linear contribution (from the tunneling transitions between various positions of the atoms) and a cubic contribution (from the vibrations of the atoms) to the specific heat. The same is observed for a metallic glass, but here we have two contributions to the linear part of the specific heat, one from the tunneling transitions and the other one from the conduction electrons. It is very remarkable that the contributions to the specific heat from the non-crystallinity are of very similar size even for quite dissimilar

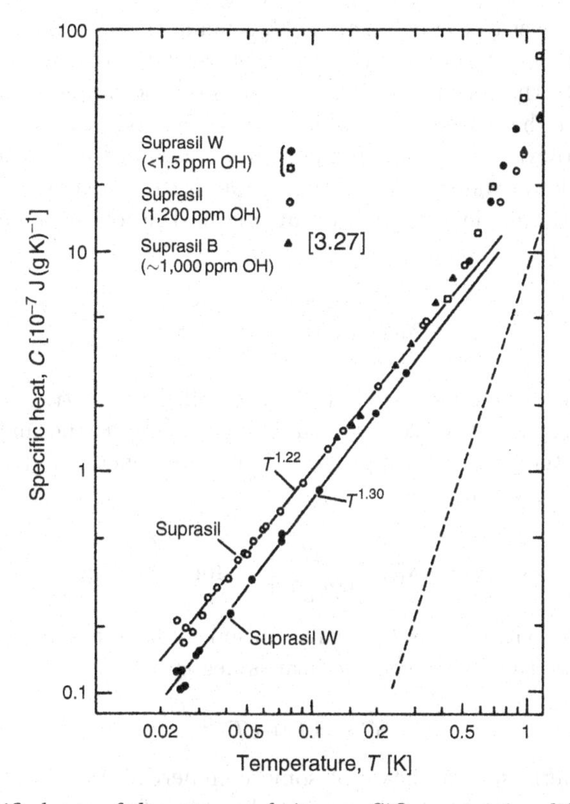

Fig. 3.7. Specific heats of three types of vitreous SiO_2 containing different concentrations of OH^- (as well as metal ions, chlorine and fluorine). The dashed line is the phonon specific heat of crystalline SiO_2 [3.29]

materials [3.5, 3.24–3.30], indicating that the additional excitations in a disordered material are associated with the disorder and do not depend on the type of material very much. Because the specific heat contributed by disorder decreases linearly with decreasing temperature instead of as T^3 as for lattice vibrations, the specific heat of a dielectric in its glassy state is much higher than in its crystalline state at low temperatures (Fig. 3.7). Often, below about 1 K, glasses have a specific heat which is even larger than that of a metal. In fact, at 10 mK the heat capacity of a dielectric glass can be a factor of 10^3 larger than the heat capacity of the corresponding crystal, which can be of great importance for the design of a low-temperature apparatus containing noncrystalline components. The heat capacities of some disordered dielectrics often used in a cryogenic apparatus are given in Sect. 3.1.7.

3.1.5 Magnetic Specific Heat

When a magnetic field is applied to a material whose atoms have magnetic moments, there are $(2I + 1)$ ways the magnetic moments can orient themselves with respect to the magnetic field, I being the spin associated with the magnetic moment. Again, a new "degree of freedom" results in an additional contribution to the specific heat. Let us consider the simplest case, a spin-1/2 system so that there are two possible spin orientations, and they should have equal degeneracy. At very low temperatures most of the magnetic moments will be in the lower energy state. If the temperature increases, transitions from the lower to the upper level will occur, giving the following contribution to the specific heat [3.1–3.6]:

$$C_{\mathrm{m}} = N_0 k_{\mathrm{B}} \left[\frac{\Delta E}{k_{\mathrm{B}} T}\right]^2 \frac{\mathrm{e}^{\Delta E / k_{\mathrm{B}} T}}{(1 + \mathrm{e}^{\Delta E / k_{\mathrm{B}} T})^2} \,. \tag{3.19}$$

This contribution, shown in Fig. 3.8, is called a Schottky anomaly. Very often the energy splitting ΔE is small compared to the thermal energy $k_{\mathrm{B}} T$. In this "high-temperature" approximation the magnetic contribution to the specific heat is given by

$$C_{\mathrm{m}} \Rightarrow N_0 k_{\mathrm{B}} \left[\frac{\Delta E}{2 k_{\mathrm{B}} T}\right]^2 \quad \text{for } \Delta E \ll k_{\mathrm{B}} T \,. \tag{3.20}$$

For a metal with such a T^{-2} contribution we have for the specific heat at $T < 1\,\mathrm{K}$ where the lattice specific heat is negligible

$$C = \gamma T + \delta T^{-2} \,. \tag{3.21}$$

The magnetic specific heats of some commercial alloys containing paramagnetic atoms and being often used as thin wires for low-temperature equipment are exhibited in Fig. 3.9. Again, at $T < 1\,\mathrm{K}$, the specific heat can be strongly enhanced compared to the simple electronic and lattice specific heats.

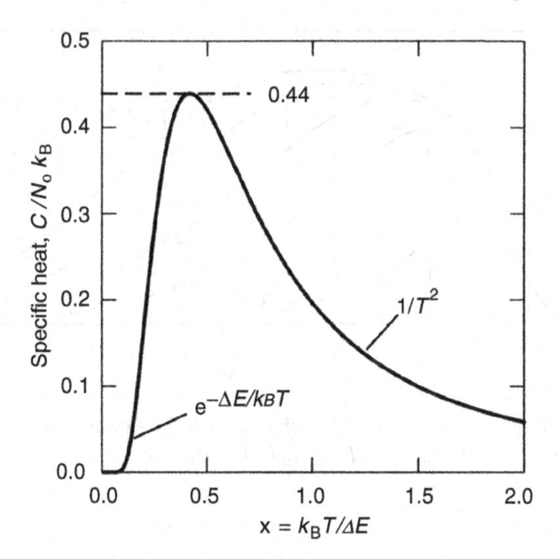

Fig. 3.8. Specific heat C (divided by $N_0 k_B$) of a two-level system with energy separation ΔE

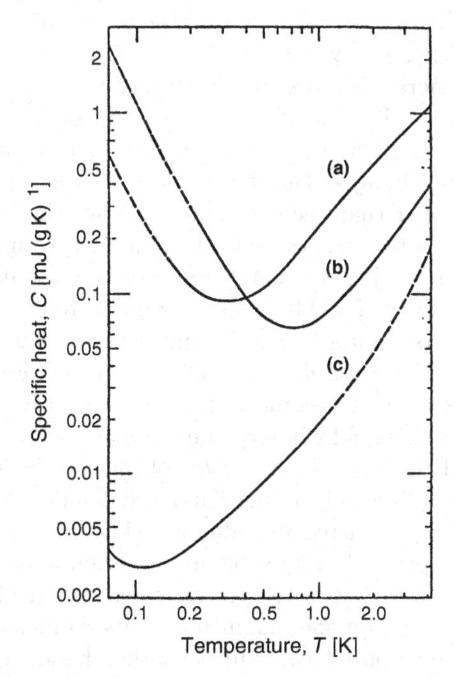

Fig. 3.9. Specific heats of (**a**) Constantan (57% Cu, 43% Ni), (**b**) Manganin (87% Cu, 13% Mn) and (**c**) a 9% W, 91% Pt alloy [3.31]. Similar values, as plotted in (**a**) and (**b**), have been reported for stainless steel in [3.32]; see also Sect. 3.1.7

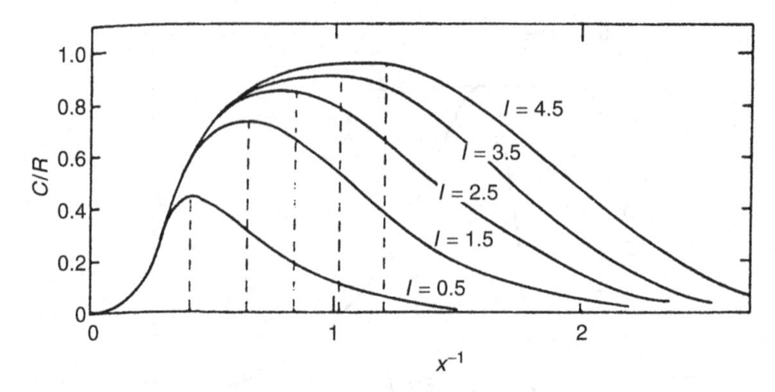

Fig. 3.10. Molar heat capacities C divided by the gas constant R as a function of $1/x_{\mathrm{e}} = k_{\mathrm{B}}T/g_{\mathrm{e}}\mu_{\mathrm{B}}B$ or $1/x_{\mathrm{n}} = k_{\mathrm{B}}T/g_{\mathrm{n}}\mu_{\mathrm{n}}B$ for different values of spin I (see also Table 3.1)

The data in Fig. 3.9 demonstrate that one has to be very cautious in employing wires of manganin (87% Cu, 13% Mn) or Constantan (57% Cu, 43% Ni) at $T < 1\,\mathrm{K}$ and, in fact, it would be better to resort to other commercial materials, like 92% Pt, 8% W.

Of course, in general the spin may be larger than 1/2 and we then have more than two levels. The calculation of the Schottky specific heat using (9.15b) for $I = 0.5$, 1.5, 2.5, 3.5 and 4.5 in Fig. 3.10 reveals that this results only in quantitative changes. But for low-temperature physics it is rather important to remember that the *temperature* at which the maximum of the magnetic contribution to the specific heat occurs is determined by the energy splitting ΔE of the levels. In other words, for nuclear magnetic moments, which are about a factor of 1,000 smaller than electronic magnetic moments, this maximum occurs at much lower temperatures than for the electronic magnetic moments. For example, an electronic magnetic moment of $1\,\mu_{\mathrm{B}}$ in a field of 1 T leads to a maximum in C_{m} at about 1 K, whereas a nuclear magnetic moment in this field will have a maximum in C_{m} at only about 1 mK (Chaps. 9 and 10). But the *maximum value* of the specific heat is *independent* of the energy splitting, it is only a function of the number of degrees of freedom $(2I + 1)$; these values are listed in Table 3.1. This means that an electronic paramagnet with a spin 1/2 in an arbitrary external magnetic field will have $C_{\mathrm{m,max}} = 0.439\,R$ (occurring in the Kelvin range or possibly at even higher temperatures, depending on the magnitude of its moment and the magnetic field this moment is exposed to). On the other hand, a nuclear magnetic moment, again with a spin 1/2, will have the *same maximum value* of the specific heat, but occurring at much lower temperature, in the microkelvin or millikelvin temperature range due to its smaller moment. This fact is of considerable importance for nuclear magnetic refrigeration (Chap. 10).

Table 3.1. Position $x_{\max}$ of $x_e = g_e \mu_B B / k_B T$ or $x_n = g_n \mu_n B / k_B T$ and value ($C_{\max}/R$) of the maximum of the magnetic specific heat divided by the gas constant, as a function of spin (or angular momentum) I (see also Fig. 3.10)

I:	1/2	3/2	5/2	7/2	9/2
$x_{\max}$	2.399	1.566	1.193	0.976	0.831
$C_{\max}/R$	0.439	0.743	0.849	0.899	0.927

A good example for a specific heat composed of a lattice T^3 term and a nuclear–magnetic T^{-2} term is the specific heat of solid ^{3}He depicted in Fig. 8.3.

The above considerations on specific-heat contributions resulting from interactions between a magnetic moment and a magnetic field can be applied analogously to the specific heat resulting from interactions of an electric quadrupole moment with an electric field gradient (see also Sects. 3.1.6 and 10.6).

3.1.6 The Low-Temperature Specific Heat of Copper and Platinum

Copper is a material that is particularly important and often used in a low temperature apparatus. There are several reports in the literature of an enhanced specific heat of Cu at low temperatures [3.33–3.42]. For $0.03\,\mathrm{K} \leq T \leq 2\,\mathrm{K}$ these increases have been traced to hydrogen and/or oxygen impurities, to magnetic impurities, mainly Fe and Mn, and to lattice defects [3.35–3.42]. On the other hand, the increased specific heat of Cu observed in the low millikelvin temperature range may also arise from a nuclear quadrupole Schottky-type contribution, see (10.35), due to Cu nuclei (which have a nuclear quadrupole moment) being located in noncubic neighbourhoods, which occur near lattice defects or in copper oxide [3.41, 3.42] (Fig. 3.11). These anomalies should be considered in the wide-spread low-temperature applications of Cu in calorimetry, thermometry (Chap. 12) and nuclear–magnetic cooling (Chap. 10), and in other applications of this very useful metal. A proper heat treatment of Cu may remove some of these anomalous increases of the specific heat.

Platinum is the "workhorse" of thermometry at low mK and at μK temperatures (Sect. 12.10.3). Hence, it is important to know that its low-temperature properties are strongly influenced by small concentrations of magnetic impurities. As, for example, reported in [3.43], Fe impurities in the ppm range increase the specific heat of Pt at 1 mK and in a field of 11 mT by about an order of magnitude. This effect is caused by the giant moments of 8 μ_B of the Fe impurities in Pt.

3.1.7 Specific Heat of Some Selected Materials

Comprehensive compilations of data and references to specific heat values, including values for technically important materials, can be found in

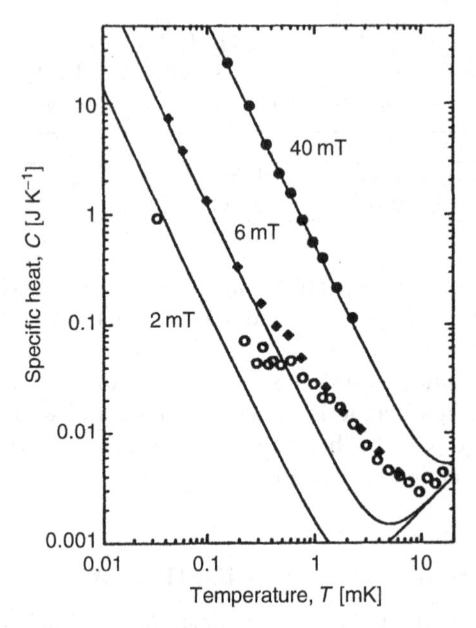

Fig. 3.11. Heat capacity of 275 mol of Cu, of which 104 mol are in the given magnetic fields, as a function of temperature. The full lines are the expected values for the nuclear magnetic heat capacities of 104 mol Cu in the given fields plus – at higher temperatures – the electronic contributions of 275 mol Cu. The deviation of the measured data from the lines is attributed to a heat capacity resulting from a splitting of the $I = 3/2$ nuclear levels of Cu due to a nuclear electric quadrupole interaction, see (10.35), of 11 mmol of Cu which are located in an electric field gradient of about 10^{18} V cm^{-2}. For these Cu nuclei the cubic symmetry must have been destroyed in order to create an electric field gradient [3.42]

[3.7, 3.44–3.48]. Some of these data are depicted in Fig. 3.12. References [3.7, 3.44] contain a compilation of data at 25, 50, 75, 100, 150, 200, 250, and 293 K for several metallic elements as well as for some alloys, non-metals, and polymers often used in the construction of cryogenic apparatus. In the following, I will supplement these compilations with some additional data for often used materials, in particular at lower temperatures. All data are given in units of μJ g^{-1} K^{-1}.

Metals

Be–Cu: $C = 3.58\ 10^{-3}\ T^{-1.64}$ at 0.5–5 mK (!) [3.49]
Constantan: $C = 205\ T + 2.8\ T^{-2}$ at 0.15 – 0.3 K as well as data to 4 K in [3.31]
Cu: see [3.33, 3.34, 3.50]
Manganin: $C = 59.5\ T + 2.94\ T^3 + 11.5\ T^{-2}$ at 0.2–2.5 K as well as data to 4 K in [3.31]
$Pt_{91}-W_9$ heater wire: $C = 17.6\ T + 1.4\ T^3 + 0.12\ T^{-2}$ at 0.07–1.2 K [3.31]

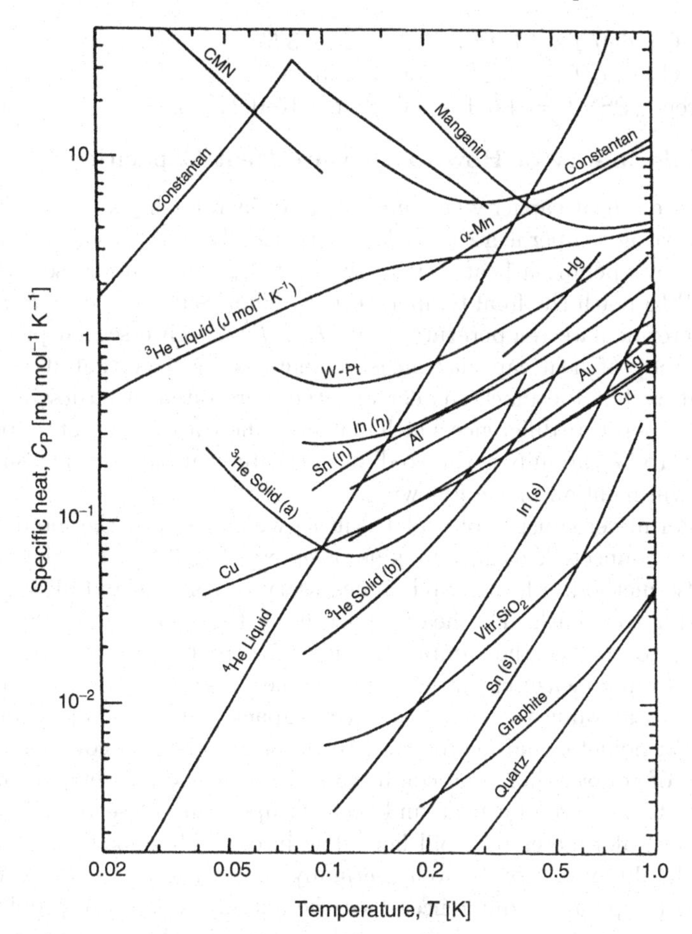

Fig. 3.12. Specific heats of several materials below 1 K [3.45]. (This publication provides references to the original literature)

Stainless steel 304: $C = 465\ T + 0.56\ T^{-2}$ at 0.07–0.6 K [3.32]
$C = 460\ T + 0.38\ T^3$ at 1–10 K [3.51]

Dielectrics

Apiezon N grease: $C = 1.32\ T + 25.8\ T^3 + 0.0044\ T^{-2}$ at 0.1–2.5 K [3.52]; data at 16–319 K can be found in [3.53]
G 10 Composite: $C = 4.74{\times}10^3\ T^{\,0.912}$ at 100–350 K and values to 1.8 K from [3.54]
GE 7031 varnish: $C = 6.5\ T + 19\ T^3$ below 1 K [3.28]
Stycast 1266 epoxy: $C = 2.91\ T + 15.7\ T^3 + 8.98\ T^5$ at 0.1–1 K [3.72]
Stycast 2850FT epoxy: data at 1 to 95 K with $C = 12$ at 1 K [3.55]
Polypropylene: $C = 6.15\ T^{1.33} + 20\ T^3$ at 0.06–1 K [3.56]

PMMA: $C = 3.0\ T + 77\ T^3$ at 0.07–0.2 K [3.57]
$C = 4.6\ T + 29\ T^3$ below 1 K [3.28]
Polystyrene (PS): $C = 4.6\ T + 93\ T^3$ at 0.07–0.2 K [3.57].

3.1.8 Calorimetry or How to Measure Heat Capacities

To measure a heat capacity C, we have to refrigerate the material of mass m to the starting temperature T_i, isolate it thermally from its environment (for example, by opening a heat switch; see Sect. 4.2), and supply some amount of heat P to reach the final temperature T_f. The result is $C/m = P/(T_\mathrm{f} - T_\mathrm{i})$ at the intermediate temperature $T = (T_\mathrm{f} + T_\mathrm{i})/2$. An essential problem for the accuracy of heat-capacity measurements is the fact that it is not the temperature but the much smaller and therefore often much less accurately known temperature difference which matters. There are several other problems that have to be carefully considered for accurate heat-capacity measurements which I will mention in the following.

A calorimeter consists of a platform to which sample, thermometer, and heater are connected, usually by glue or epoxy (Fig. 3.15). For the adiabatic heat-pulse method, a heat switch is necessary to connect and disconnect the calorimeter to a bath. The heat capacities of the addenda should be small compared to that of the sample or they have to be known with sufficient accuracy from a measurement without sample so that they can be subtracted from the total value to obtain the heat capacity of the sample. The leads to the thermometer and heater have to be of low thermal conductivity (the best are, of course, thin superconducting wires at the low-temperature end) and have to be carefully heat-sunk at a temperature close to the temperature of the calorimeter to avoid heat flow into it. Thermometer, heater and sample should, of course, be well thermally coupled to the platform to avoid unknown temperature differences (see the thermal boundary problems discussed in Sect. 4.3). The power needed to read the thermometer should be small to avoid overheating it. Calibration of the thermometer and measurement of the heat capacity should be performed with the same power applied to the thermometer to have in both cases the same unavoidable, hopefully small temperature difference between thermometer and platform. Parasitic heat losses or heat inflow by radiation can be reduced by a thermal shield around the calorimeter with a temperature regulated closely to the temperature of the calorimeter [3.33, 3.58, 3.59]. Heat produced by opening and/or closing a mechanical (Sect. 4.2.1) or a superconducting (Sect. 4.2.2) heat switch has to be small. When exchange gas is used for the cooldown of the calorimeter, it has to be pumped away before performing the measurement, which is usually quite time-consuming. Still, one has to take into account the possibility of adsorption and desorption of residual gas when the temperature is changed. This produces heat of adsorption/desorption, which is on an order of magnitude of the large heat of vaporization. Eventually, time constants for thermal equilibrium within the sample, within the addenda and between sample and addenda have to be considered.

As a result of these problems, heat-capacity data rarely have an accuracy better than 1%, more usual are accuracies of 3–5% (but see below). If a high accuracy is needed or if the parameters of the setup are not well known, the quality of a calorimeter can be checked by measuring the heat capacity of a well-known reference sample like high purity Cu [3.33, 3.34, 3.50].

Substantial advances in calorimetry have been achieved due to the development of high-performance electronics, new thermometers, micro-fabrication techniques, and computer automation. However, one has to keep in mind that the properties of the thermometer (as well as the accuracy of the used temperature scale) is the essential parameter for the accuracy of heat-capacity data. High-resolution heat-capacity data can only be obtained when the relevant precautions are taken and when a calibrated, sensitive, and stable thermometer is used.

There are a myriad of reports on the design, construction, and operation of calorimeters; most of them can be found in the journal Review of Scientific Instruments. Therefore, the selection of calorimeters to be discussed in this section surely is a subjective choice. In the following, I will discuss a few of the more recent elegant or versatile designs for the usual adiabatic heat-pulse method. This will be followed by a discussion of calorimeters for situations where this straightforward adiabatic approach of applying heat and measuring the temperature increase of the thermally isolated calorimeter is inadequate and more sophisticated methods have to be used; for example, for very small samples or in hostile environments, like high magnetic fields or pressures, or for the investigation of sharp features at a phase transition.

Surely, the champion of adiabatic heat-pulse calorimetry with respect to temperature resolution and temperature stability is the measurement of the heat capacity C of liquid ^{4}He near its superfluid transition [3.59]. In these experiments, C has been measured with sub-nanokelvin resolution at temperatures to within about a nanokelvin of the transition temperature $T_\lambda = 2.177\,\mathrm{K}$ (see Fig. 2.10b). Such an extreme temperature resolution is only meaningful for the investigation of a phase transition of liquid helium because only in this substance purity and perfection are high enough so that the phase transition shows the required sharpness. In all other materials, phase transitions are smeared by impurities and by imperfections of the structure. In addition, this measurement had to be performed in flight on earth orbit to reduce the rounding of the transition caused by gravitationally induced pressure gradients and therefore spreading the transition temperature over the liquid sample of finite height. The high-resolution magnetic susceptibility thermometers developed for these experiments are described in Sect. 12.9. In the used experimental setup, four thermal control stages in series with the calorimeter were actively temperature regulated; the last one to a stability of less than $0.1\,\mathrm{nK\,h^{-1}}$. Besides this thermal regulation, the experiment required very careful magnetic shielding, in particular of the electric leads as well as extremely low electric noise levels. Even though this surely is not a typical laboratory heat-capacity experiment, a lot can be learned from the chosen design, the precautions, and detailed considerations of possible error sources [3.59], which could be quite useful for simpler experiments on earth.

Another quite remarkable achievement was the adiabatic measurement of the nuclear magnetic heat capacity of $AuIn_2$ to temperatures of 25 μK in an investigation of its *nuclear* ferromagnetic transition at 35 μK [3.60]. In this temperature range, nuclear magnetic resonance on ^{195}Pt (Sect. 12.10.3) is the only available method of thermometry, making the calorimeter quite demanding; it is shown in Fig. 3.13. This example demonstrates that the addenda can

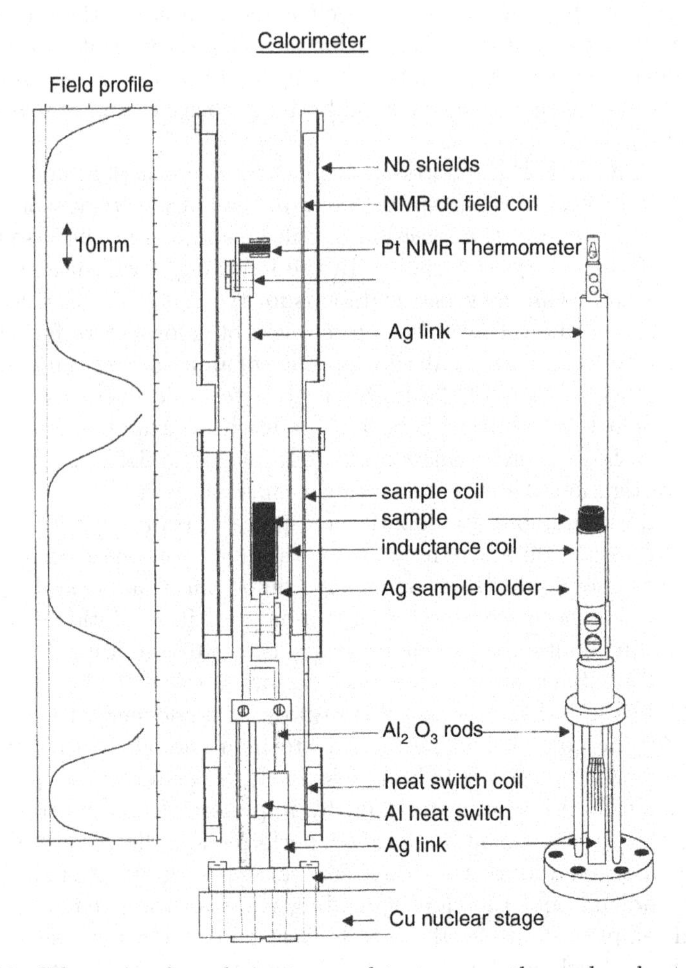

Fig. 3.13. Schematic of a calorimeter used to measure the nuclear heat capacity (Fig. 3.14) as well as the nuclear AC susceptibility of a $AuIn_2$ sample at microkelvin temperatures. The calorimeter is connected via a superconducting Al heat switch (Sect. 4.2.2) to a Cu nuclear refrigeration stage (Sect. 10.8). It contains a Pt NMR thermometer (Sect. 12.10.3). The calorimeter is surrounded by coils for magnetic DC fields for the NMR thermometer, the sample, and the heat switch, as well as by superconducting Nb shields (Sect. 13.5.2), [3.60]

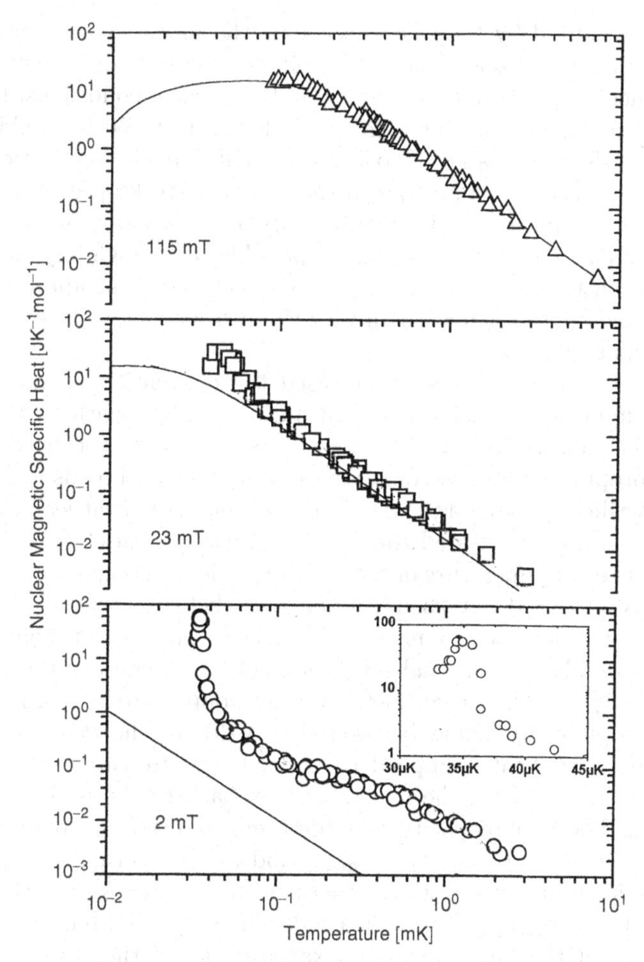

Fig. 3.14. Nuclear magnetic specific heat of In nuclei in $AuIn_2$ per mole of the compound (per 2 moles In) in the three indicated magnetic fields. The full lines are the Schottky curves (see 3.19 and 3.20, as well as Fig. 3.10) for non-interacting In nuclei (I=5/2) in the given external fields. The lower two figures demonstrate the enhancement of the heat capacity by the interaction of the In nuclei which gives rise to a nuclear ferromagnetic ordering of them at 35 μK, shown more clearly in the inset. The heat capacity at the ordering transition is enhanced by the interactions of the small In nuclei by three orders of magnitude at 2 mT; it has about the value as Co has at its Curie temperature of 1390 K (!) [3.60]

be large as long as its heat capacity is small compared to that of the sample, in this case the huge nuclear heat capacity of $AuIn_2$ at microkelvin temperatures. The obtained data shown in Fig. 3.14 also demonstrate how large a magnetic heat capacity can become even at very low temperatures independently of whether it is caused by the electronic or by the much smaller nuclear magnetic moments; it is determined by the spin degeneracy. And eventually, talking about heat-capacity measurements under extreme conditions, heat-pulse calorimetry performed in pulsed magnetic fields up to 60 T should be mentioned [3.61]. Here, it is essential to avoid any material with magnetic moments or containing magnetic impurities in the addenda (to keep its heat capacity small) and to use pure, but electrically badly or non-conducting materials (to reduce eddy current heating) as far as possible, like single crystalline Si or high-purity plastics as platform. Of course, the choice of the appropriate thermometer – mostly a resistor – with very low field dependence (see Sect. 12.5) is of particular importance.

All these mentioned calorimeters used the conventional adiabatic heat-pulse technique. A semi-adiabatic heat-pulse technique without heat switch but with the sample and platform weakly connected to the bath at 30 mK via the appropriately chosen conductance of mounting threads and electrical leads is described in [3.62]. In this calorimeter, parasitic heat leaks or thermal losses have been compensated through an adjustable background heating to assure a constant temperature of the platform. The heat capacity of milligram samples was measured at 0.03–6 K in magnetic fields up to 12 T.

In particular for measurements of the heat capacity of small samples, other methods like the thermal relaxation time technique or the AC calorimetric technique with a calorimeter weakly coupled to a thermal bath at constant temperature should be used (Fig. 3.15). In the *thermal relaxation method* [3.64–3.67] heat is applied for a fixed time to sample and addenda, and then the exponential thermal relaxation back to the bath temperature is recorded. The heat capacity C is determined from the relaxation time $\tau_1 = C/k_1$ with $C = C_{\mathrm{sample}} + C_{\mathrm{addenda}}$ and k_1 the thermal conductance of the weak link to the bath. Both the thermal conductance k_1 and the addenda heat capacity have to be determined independently. The relaxation time τ_1 has to be larger than all internal relaxation times of the system, otherwise the relaxation is not a single exponential curve. This means in particular that the conductance between the sample and the platform has to be much larger than k_1. The sensitivity is determined by the quality of the thermometer and by the (as small as possible) heat capacity of the addenda. This method can have rather high absolute accuracy, however, its relative accuracy is limited.

This is just the opposite for the *AC method*, which can detect very small changes in the heat capacity [3.63, 3.65, 3.68]. Here, heat P is applied sinusoidally and the resulting temperature oscillation at frequency ω is determined. It is $\delta T = (P/\omega C)(1 + \omega^{-2}\tau_1^{-2} + \omega^2\tau_2^2)^{-1/2}$ with τ_2 the relaxation time within the calorimeter assembly and τ_1 the relaxation time of the calorimeter to the bath. In the usual limit $\omega\tau_1 \gg 1 \gg \omega\tau_2$, the heat capacity $C = P/\omega\delta T$. To check whether this limit has been achieved, the temperature response has to be measured at various frequencies – typically between a few and 200 Hz – to determine below which frequency heat escapes through the thermal link to the bath within the measuring period and above which

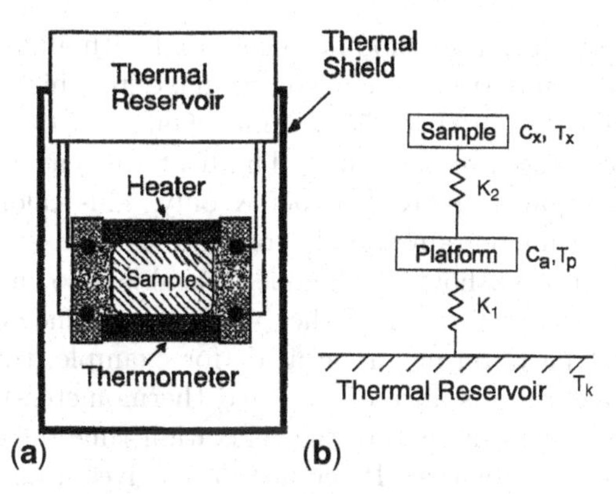

Fig. 3.15. (**a**) Schematic of a setup for heat-capacity measurements by the thermal relaxation or the AC techniques. A platform (for example, a thin sapphire or silicon disc or a membrane) carries sample, heater and thermometer. It is weakly coupled to a thermal reservoir at constant temperature and surrounded by a thermal shield. For high accuracy measurements, the temperature of the shield is regulated close to the temperature of the platform. Leads to heater and thermometer are thermally grounded at the thermal reservoir and then at the platform; they can also serve as weak thermal link between platform and reservoir as well as mechanical support. The set–up can be used for the more conventional heat-pulse technique by adding a heat switch between reservoir and platform. (**b**) Heat flow diagram for the shown setup. The sample of heat capacity C_x and at temperature T_x is connected via a thermal link of conductance k_2 to the platform with heat capacity C_{a} and at temperature T_{p}. The thermal link between platform and thermal reservoir has the conductance k_1. C_a contains the contributions from platform, thermometer, heater, and their mounting (glue, grease, varnish,...); it should be substantially smaller than C_x and has to be measured independently. Similarly, the conductance k_2 should be much larger than k_1

frequency the calorimeter can not follow anymore the heat modulation; i.e., to see in which frequency range a steady state can be reached where $\delta T\omega$ is independent of frequency. A simplified version of the relaxation method is the so-called "dual slope method" [3.69]. Here, the sample plus addenda are continuously heated and cooled, and the specific heat at temperature T is derived from the slopes of these two cycles at T.

In the calorimeters for these techniques, usually the sample holder is a small chip (for example, thin silicon or sapphire) or membrane, to which the heater, thermometer, and sample – with the latter having hopefully a substantially higher heat capacity than all the other parts – are mounted. This device is hanging in vacuum by means of thin supporting wires that provide a weak and controllable thermal link to the bath and can serve also as electrical leads for thermometer and heater (Fig. 3.15). Various micro-thermometers and nonmagnetic, temperature-independent resistive heaters (for example, NiCr or TiCr alloys) have been developed for this purpose (see various references in this section).

In [3.66], the relaxation method has been used with an amorphous silicon-nitride membrane, supported by a silicon frame, onto which thin-film heaters and thermometers (Pt for T > 50 K, amorphous NbSi or boron doped Si for lower temperatures) are patterned. The heat capacity of this addendum is <1 nJ K^{-1} at 2 K and 6 μJ K^{-1} at 300 K only. This calorimeter has been used to investigate microgram samples or thin films in steady fields up to 8 T; according to the authors, it should be usable also in pulsed fields up to 60 T. This is the result of the rather weak dependence of the properties of the calorimeter parts on magnetic field (for example, their conductances and the weak magnetoresistance of the used thermometers), and of the fact that in a very good approximation the relaxation time does not rely on the calibration of the thermometer. Reference [3.66] gives a detailed description of the fabrication and properties of this calorimeter and all its components. A relaxation calorimeter for use in a top-loading ^{3}He-^{4}He dilution refrigerator in high magnetic fields has been described in [3.67]. It has been used for milligram samples from 34 mK to 3 K and in magnetic fields up to 18 T. In order to keep thermal time constants in the magnetic field reasonably short, most of the addenda like the thermal reservoir are made from Ag, which has one of the smallest nuclear heat capacities of suitable materials (see Fig. 10.4 and Table 10.1).

In an elegant version of a calorimeter for microgram samples of [3.65], a modified commercial thin-film ceramic silicon oxynitride chip of $1 \times 0.75 \times 0.1\,\text{mm}^3$ ("Cernox", see Sect. 12.5.3) was used simultaneously as sample holder and temperature sensor, with a $Ni_{80}Cr_{20}$ heater sputtered onto it. This calorimeter of only 0.25 mg weight was used for the relaxation as well as for the AC methods at 1.5–50 K and in magnetic fields up to 11 T (Cernox is rather insensitive to magnetic fields, see Sect. 12.5.3). According to the authors, the sensitivity of this calorimeter over the whole temperature range is a factor of 100 higher than that of a comparable commercial calorimeter (see below), using a silicon-on-sapphire design. The most sensitive microcalorimeters have been built by the group of Chaussy [3.68]. In their latest version, these AC calorimeters for the Kelvin temperature range consist of a 2 to 10-μm-thick monocrystalline silicon membrane substrate produced by etching, taking advantage of the high thermal conductivity and low heat capacity at low temperatures of this material. A 150 nm CuNi heater (with temperature independent resistivity at 1 to 20 K) and a 150 nm NbN thermometer (with slope $\mathrm{d}R/R\,\mathrm{d}T \approx 0.2$ K at 4 K) are deposited onto the substrate. The heat capacity of this addenda in the different versions of these calorimeters varied between less than 0.1 nJ K^{-1} at $T < 1$ K and some nJ K^{-1} at 4 K. The device was used to measure heat capacities of systems of reduced dimensionality like deposited thin films or multilayers or of microgram single crystals and eventually of mesoscopic superconducting loops. The achieved resolution of $\Delta C/C < 5 \times 10^{-5}$ allowed measurements of variations of C as small as 10 fJ K^{-1} [3.68].

Another remarkable achievement reported in [3.70] seems to be the substantial improvement of the conventional differential thermal-analysis (DTA) method by means of using high-precision electronics and careful temperature control. This device was used to measure the heat capacity of milligram samples at Kelvin temperatures and in magnetic fields up to 7 T with a relative accuracy of about 10^{-4}. This makes this method particularly valuable for the investigation of sharp or shallow features at phase transitions. The main ingredient of the method is the continuous comparison of the heat capacity of the sample to the heat capacity of a reference material, which are both weakly connected to a thermal bath and which are simultaneously heated or cooled. The best results with this comparative method are, of course, achieved if thermal links and thermometers are as identical as possible. Temperature differences between bath, reference material, and sample are recorded and inserted in the relevant equations to evaluate the heat capacity of the sample.

An automated heat-capacity measurement system ("Physical Property Measurement System", see "Supplier of Cryogenic Equipment") for samples of about 10–500 mg is available commercially [3.64]. The system also allows performing very sensitive AC-susceptibility, as well as AC- and DC-magnetization and -electric-transport measurements. It employs the thermal relaxation method in the temperature range 1.8–400 K (optional 0.35–350 K with a continuously operating closed-cycle ^{3}He system; see Sect. 6.2). As an option, it can be equipped for measurements in magnetic fields up to 16 T longitudinal or 7 T transverse. Its calorimeter platform consists of a thin alumina square of 3×3 mm^2, backed by a thin-film heater and a bare Cernox thermometer (see Sect. 12.5.3). A heat pulse of duration τ is applied and the platform temperature is recorded for 2τ. With known values for the conductance of the thermal link to the bath, of the heat capacity of the addendum, and of the applied heat, the heat capacity of the sample and the internal time constant of the calorimeter are determined analytically from the T(t) data by numerically integrating the relevant differential equations. The curve fitting is improved by carrying out a number of decay sweeps at each temperature and averaging the results. Two Cernox thermometers are used over the full temperature range and their calibration is based on the ITS-90 temperature scale (Sect. 11.2). The resolution of the system is 10 nJ K^{-1} at 2 K. According to an examination of the system [3.64], the accuracy is 1% at 100–300 K, which diminishes to about 3–5% at $T < 5$ K. The system is quite adequate to investigate reasonably broad second-order phase transitions. However, sharp first-order transitions cannot be investigated properly, mainly because the applied software cannot describe the then non-exponential decay curves. This drawback can be removed by using an alternate analytic approach [3.64]. The system has recently been equipped with a simple, fully automated dilution refrigerator for heat-capacity measurements from 55 mK to 4 K and in fields to 9 T [3.71].

Important new information can be obtained from measurements of the angular dependence of the heat capacity in high magnetic fields. For example, field-angle-dependent specific heat can probe the gap structure in unconventional superconductors such as high-T_c or heavy-fermion superconductors, possibly indicating the nodal structure of anisotropic energy gaps, which is intimately related to the pairing interaction [3.73]. In Sect. 13.7, I will describe various methods how to achieve the necessary rotation of samples at low temperatures in these fields.

Cryogenic calorimeters or more generally devices which measure the temperature increase after a deposition of energy are nowadays not only used to measure heat capacities of liquids and solids but also in a variety of other applications like the detection of weakly interacting massive particles, of x-rays and γ-rays, and in astrophysics; or as bolometers for detection of phonons, of particles or of electromagnetic waves, and in particular for detection of infrared radiation. These applications have emerged from the very high sensitivity of recent microcalorimeters in the range of nJ/K (corresponding to the heat capacity of a monolayer of ^{4}He, for example) or even less. For a recent review on these devices and their applications see [3.74].

3.2 Thermal Expansion

3.2.1 Thermal Expansion of Solids

If the potential which an atom sees in a crystal were parabolic (the harmonic approximation) and therefore given by (3.1), there would be no thermal expansion. However, in reality the potential which an atom experiences is anharmonic due to the electrostatic forces of its neighbours and looks more like that displayed in Fig. 3.16. At low temperatures the amplitudes of the atomic vibrations around their equilibrium position r_0 are rather small and the potential can be approximated by (3.1). As a result the thermal expansion coefficient

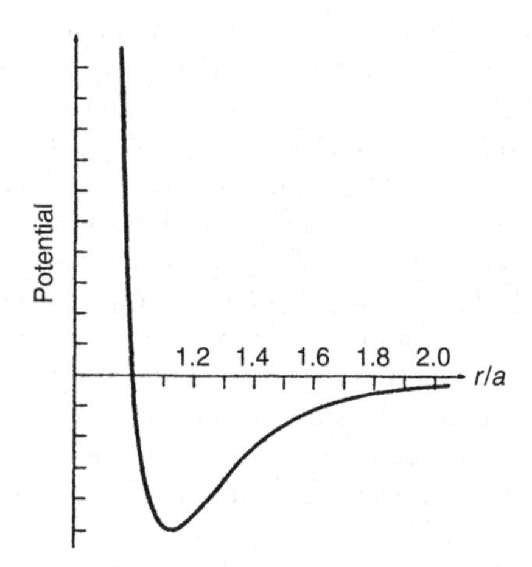

Fig. 3.16. Typical potential which an atom or ion experiences in a lattice as a function of distance r between them (normalized to the lattice parameter a)

$$\alpha = \frac{1}{l}\left[\frac{\partial l}{\partial T}\right]_p \qquad (3.22)$$

does indeed vanish for $T \to 0$ (for Cu: $\alpha = 2.9 \times 10^{-10}T + 2.68 \times 10^{-11}T^3$ at $0.2\,\mathrm{K} < T < 1.9\,\mathrm{K}$ [3.75]).[2] When the temperature is raised the thermal vibrations of the atoms grow and they increasingly experience that the potential is asymmetric: steeper for small distances and flatter for larger distances. The atoms experience the "anharmonic" part of the potential. Due to the shape of the potential the atoms spend more of their time at larger separations, leading to an increased average separation. Therefore the material expands when the temperature is increased. For a potential $V(r - r_0) = V(x)$, the mean deviation from the zero-temperature position $r = r_0$ is given by [3.1–3.7]

$$\langle x \rangle = \frac{\int_{-\infty}^{+\infty} x\mathrm{e}^{-V(x)/k_\mathrm{B}T}\mathrm{d}x}{\int_{-\infty}^{+\infty} \mathrm{e}^{-V(x)/k_\mathrm{B}T}\mathrm{d}x}\,. \qquad (3.23)$$

The thermal contraction of various materials when cooled from room temperature to lower temperatures is displayed in Fig. 3.17. Looking at this figure we can divide the materials quite generally into three groups. First, there is a group of commercial alloys and glasses, which have been specially produced to exhibit an extremely small expansion coefficient. Then we have the groups of metals which contract by about 0.2–0.4% when cooled from room temperature to low temperatures. It is very important to remember that different metals have different expansion coefficients and therefore have to be joined in the proper order (see below). Finally, we have the organic materials with their large expansion coefficients; typically 1–2% length change when cooled from room temperature to the low Kelvin temperature range. These latter materials are rather important for low-temperature purposes as well, because they are used as construction materials on account of their low thermal conductivity, or for bonding, electrical insulation or making leak-tight joints. Here problems can often occur due to their rather large thermal expansion coefficients which can lead to substantial thermal stresses. For some applications the very small expansion coefficients of many glasses of $\alpha \simeq 4 \times 10^{-7}\,\mathrm{K}^{-1}$ at 0°C may be useful. However, unlike the specific heat and other properties, α is not a universal property of glasses at low temperature; it may even have different sign for different glasses at $T \leq 1\,\mathrm{K}$ [3.81].

Joining materials of different thermal expansion in an apparatus whose temperature will repeatedly be changed requires a rather careful selection of the materials and a suitable design, if destruction of the joint by the severe stresses in thermal cycling is to be avoided. This is a very serious consideration because a low-temperature apparatus composed of a variety of materials usually has to be leak-tight. Figure 3.18 illustrates the correct ways of joining tubes of different

[2] Of course, when a material is anisotropic, it may have different expansion coefficients α_i in different directions. The volume expansion coefficient is then $\beta = \sum_{i=1}^{3} \alpha_i/3$.

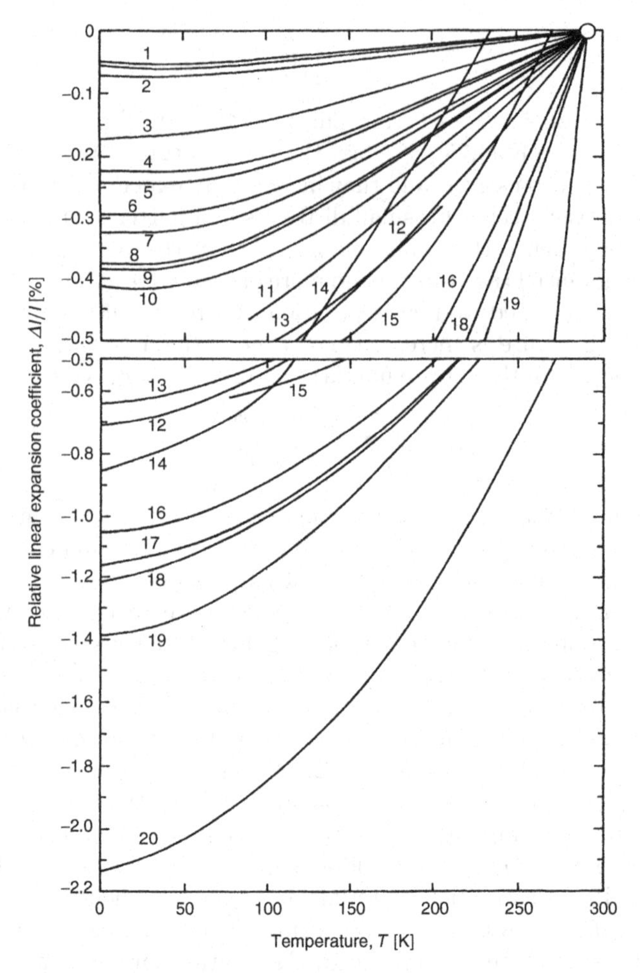

Fig. 3.17. Relative linear thermal expansion coefficient of (1) Invar (upper), Pyrex (lower), (2) W, (3) nonalloyed steel, (4) Ni, (5) $Cu_{0.7}Ni_{0.3}$, (6) stainless steel, (7) Cu, (8) German silver, (9) brass, (10) Al, (11) soft solder, (12) In, (13) Vespel SP22, (14) Hg, (15) ice, (16) Araldite, (17) Stycast 1266, (18) PMMA, (19) Nylon, (20) Teflon [3.76]. Some further data are: Pt similar to (3); Ag between (9) and (10); Stycast 2850 GT slightly larger than (10). The relative change of length between 300 and 4 K is $10^3\Delta l/l = 12, 11.5, 4.4, 6.3$ and 5.7 for Polypropylene, Stycast 1266, Stycast 2850 GT as well as 2850 FT, Vespel SP-22 and solders, respectively [3.44, 3.55, 3.56, 3.76–3.82, 3.114]. Torlon behaves very similar to Stycast 2850FT [3.114]

metals and the design of seals, which appear in almost every piece of low-temperature equipment. (For the proper design of an insulating feedthrough of leads, see Sect. 13.3.) In addition, stresses due to parallel connection by parts of different materials have to be avoided. Very often good thermal contact between different metallic parts of low-temperature apparatus is essential

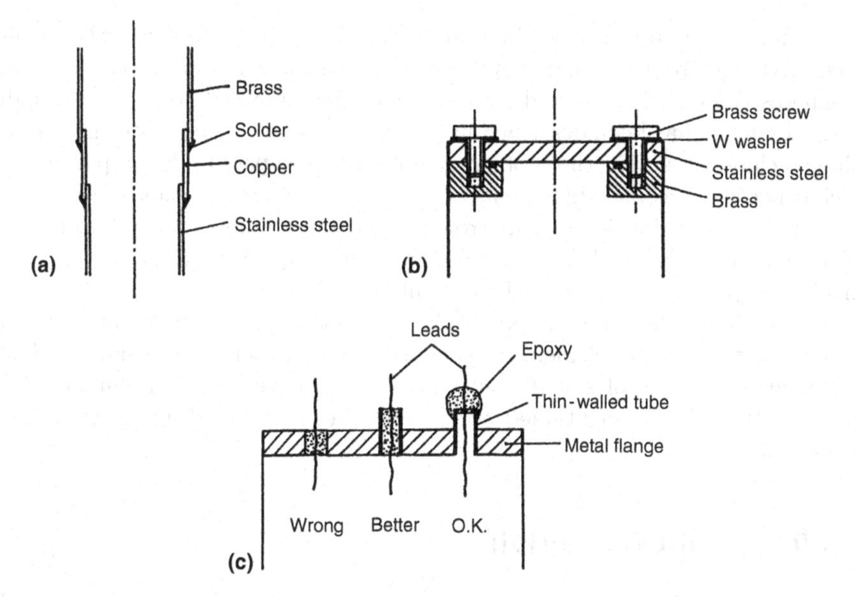

Fig. 3.18. When joining different materials in a cryogenic apparatus one has to take into account the difference in their thermal expansion coefficients. For example: (**a**) The tube with the largest expansion coefficient should be on the outside so that the solder joint is not pulled open during cooldown. (**b**) In an O-ring seal the screw should have a larger expansion coefficient than the flange. The seal will tighten even further during cooldown if a washer with a very small expansion coefficient is used. (**c**) In an epoxy feedthrough for leads the epoxy, with its large expansion coefficient, should contract on a thin-walled metal tube during cooldown rather than pull away. It helps if the tube walls are tapered to a sharp edge. An epoxy with filler should be used to lower its thermal expansion coefficient

(Sect. 4.3.1). Here again the correct selection of the materials for bolts and nuts is important. Frequently, the thermal contact after cooldown can be improved by adding a washer of a material with a low thermal expansion coefficient, such as Mo or W (but they are superconductors at very low temperatures!).

A fairly comprehensive list of references and data on thermal expansion coefficients of solids can be found in [3.7, 3.44, 3.78–3.80].

3.2.2 Dilatometers or How to Measure Thermal Expansions

For measurements of thermal expansion coefficients, capacitive displacement sensors are commonly used because of their sensitivity (up to 0.1 nm) and simplicity (see also Sect. 13.1) [3.7, 3.78, 3.79, 3.82]. In these dilatometers, the change of length of a sample is transferred to the movable part of a capacitor resulting in a capacitance change. The capacitances are measured in a five-terminal capacitance bridge like the one discussed in Sect. 13.1.1 or in a commercial high-resolution capacitance bridge. The resolution of these bridges of up to 10^{-7} or even 10^{-8} translates to resolution changes of up to 0.1 nm.

A miniature capacitance dilatometer of this type with only 22 mm diameter, made from Ag and suitable for measuring thermal expansion and magnetostriction of small and irregularly shaped samples even in very high magnetic fields has been described in [3.83]. The design is based on the tilted-plane technique, so that the problem of holding the capacitor plates parallel to each other is overcome. Its resolution is 0.1 nm or $\Delta l/l = 10^{-6}$ and it has been used in a cryostat inside of the narrow (32 mm) and "noisy" bore of a magnet with fields up to 33 T. For earlier designs using capacitive dilatometers, the mentioned publications should be consulted.

Capacitance dilatometers are only surpassed by dilatometers using a SQUID as the sensing element of elongation or contraction of a sample. The enormous resolution of 2×10^{-5} nm has been achieved and the data for Cu at $0.2 < T < 1.9$ K, mentioned above, have been obtained applying such a device [3.75].

3.3 Thermal Conductivity

Thermal conductivity is a transport property of matter similar to electrical conductivity, viscosity, diffusion, damping of sound, etc [3.1–3.6, 3.84–3.87]. The rate of heat flow per unit area resulting from a temperature gradient in a material of cross-section A is given by

$$\dot{q} = \dot{Q}/A = -\kappa \nabla T \,, \tag{3.24}$$

where κ is the thermal conductivity coefficient. Heat can be carried by conduction electrons or by lattice vibrations; their contributions are additive. These carriers of heat usually do not fly ballistically from the heated end of the material to the other colder end. They are scattered by other electrons or phonons or by defects in the material; therefore they perform a diffusion process. To calculate thermal transport we have to apply transport theory, which in its simplest form is a kinetic gas theory. In this simplified version we consider the electrons or the phonons as a gas diffusing through the material. For the thermal conductivity coefficient this theory gives

$$\kappa = \frac{1}{3} \frac{C}{V_{\mathrm{m}}} v \lambda \,, \tag{3.25}$$

where λ is the mean free path, and v is the velocity of the particles. Intuitively, this is a rather convincing equation identical to those for other transport properties if we choose the corresponding parameters. It tells us that the transport property "thermal conductivity coefficient" is given by the product of "what is transported" [here it is the specific heat C (per unit volume)], "the velocity v of the carriers performing the transport", and "how far the carriers fly before they are scattered again". The factor 1/3 comes from the fact that we are interested in the heat flow in one direction, whereas the motion of the carriers is three-dimensional.

How will the thermal conductivity coefficient look like if the heat is carried by the electron or/and by the phonon gases? In the first part of this chapter we have already calculated the specific heat C of electrons and phonons as a function of temperature. The characteristic velocity of phonons is the velocity of the sound v_s; this is the velocity with which "vibrations" or "phonons" move through the lattice. Typical values for solids, in particular metals, are $v_s = (3–5) \times 10^5\ \mathrm{cm\,s^{-1}}$. The electrons involved in thermal transport can only be electrons with energy near the Fermi energy (Fig. 3.3). Only these can transport heat because they are the only ones which can perform transitions to higher nonoccupied energy states, which is necessary for thermal conductivity. Their velocity is the so-called Fermi velocity v_F determined by the kinetic Fermi energy E_F, see (3.10). Typical values for

$$v_F = (\hbar/m_e)(3\pi^2 N_0/V_m)^{1/3} \tag{3.26}$$

are $10^7–10^8\ \mathrm{cm\,s^{-1}} \gg v_s$. Both the sound velocity and the Fermi velocity are independent of temperature at low temperatures. So we know C and v, and all the problems in calculating transport properties lie in the calculation of the mean free path λ, determined by the scattering processes of the heat carriers.

The main scattering processes limiting the thermal conductivity are phonon–phonon (which is absent in the harmonic approximation), phonon–defect, electron–phonon, electron–impurity, and sometimes electron–electron interactions. The latter process is rather ineffective because it involves four different electron states due to the Pauli exclusion principle. The resistances of the various scattering processes are additive. Because the number of phonons increases with increasing temperature, the electron–phonon and phonon–phonon scattering rates are temperature dependent. The number of defects is temperature independent and correspondingly the mean free path for the phonon–defect and electron–defect scattering do not depend on temperature. As a result, we arrive at the equations for the thermal conductivity given in the following subsections.

3.3.1 Lattice Conductivity: Phonons

The lattice or phonon conductivity which is the dominant and mostly the only conduction mechanism in insulators is given by

$$\kappa_{ph} = \frac{1}{3}\frac{C_{ph}}{V_m} v_s \lambda_{ph} \propto T^3 \lambda_{ph}(T), \quad \text{at } T \leq \theta_D/10\,. \tag{3.27}$$

Intermediate Temperatures: $T \leq \theta_D/10$

In this temperature range the phonon–phonon scattering is dominant and the phonon mean free path increases with decreasing temperature because the number of phonons decreases with decreasing temperature. A quantitative derivation [3.1–3.6, 3.84–3.87] of the thermal conductivity for the phonon–phonon scattering regime is somewhat involved due to the anharmonicity of the potential and of the frequency dependence of the dominant phonons with temperature; it will not be given here. We just state that in this T range the thermal conductivity decreases with increasing temperature (Figs. 3.19 and 3.20).

Low Temperatures: $T \ll \theta_D$

In this temperature range the number of thermally excited phonons is rather small. They are no longer important for scattering, and the phonons which carry the heat are scattered by crystal defects or by crystal boundaries only. Because at low temperatures the dominant phonon wavelength is larger than the size of the lattice imperfections, phonon scattering at crystallite boundaries is the important process. Now the mean free path for phonon transport is, in general, temperature-independent – but see (3.29). The temperature dependence of the thermal conductivity is then given just by the temperature dependence of the specific heat, and decreases strongly with decreasing temperature as

$$\kappa_{ph} \propto C_{ph} \propto T^3 . \tag{3.28}$$

As a result of this consideration we find that the thermal conductivity due to phonon transport goes through a maximum, as illustrated in Figs. 3.19a and 3.20. Owing to differences in the number of defects, the low temperature thermal conductivity of nominally identical samples can vary considerably (see also Fig. 3.19).

I want to mention two particularly important cases of phonon thermal conductivity. First, if we have a rather perfect, large crystal with a very low density of defects and impurities, the mean free path of phonons and therefore the thermal conductivity can become very large, of the order 100 W (cm K)$^{-1}$ (Figs. 3.19a and 3.20). This is comparable to the thermal conductivity of highly conductive metals like copper or aluminum. Second, if we have a strongly disordered insulator, the mean free path determined by the scattering of phonons on defects can become very small, even approaching atomic distances. In particular, if we consider a glass, the tunneling transitions between different structural arrangements of the atoms, which were discussed in the description of the specific heat of noncrystalline materials (Sect. 3.1.4), limit the phonon thermal conductivity by additional scattering of phonons on tunneling states [3.5, 3.24–3.28, 3.94, 3.95]. They reduce the thermal conductivity of glasses considerably, for example by about two (four) orders of magnitude for vitreous silica compared to crystalline quartz at 1 (10 K) [3.27]. This scattering results in an almost universal close to T^2 dependence of the thermal conductivity of dielectric glassy materials below 1 K (Figs. 3.21–3.23) and

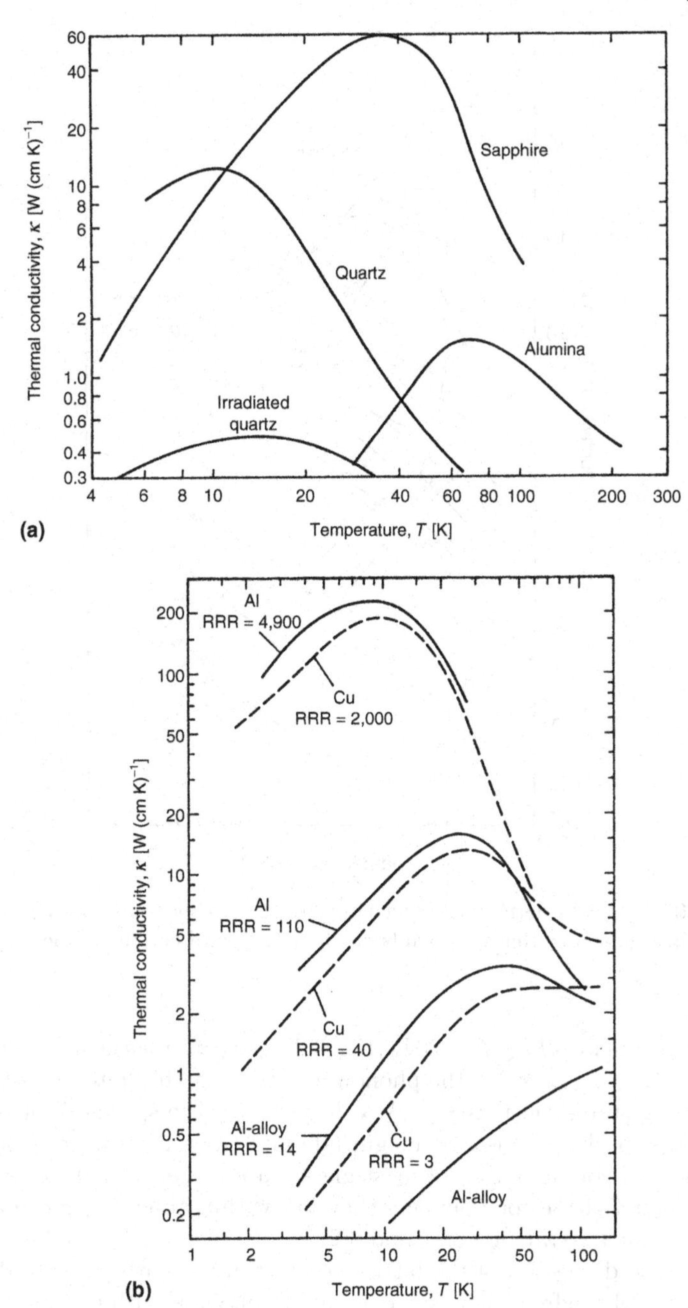

Fig. 3.19. Temperature dependence of the thermal conductivities of (**a**) some dielectric solids and of (**b**) Al and Cu of varying purity [expressed as their residual resistivity ratio (3.43)] [3.88]

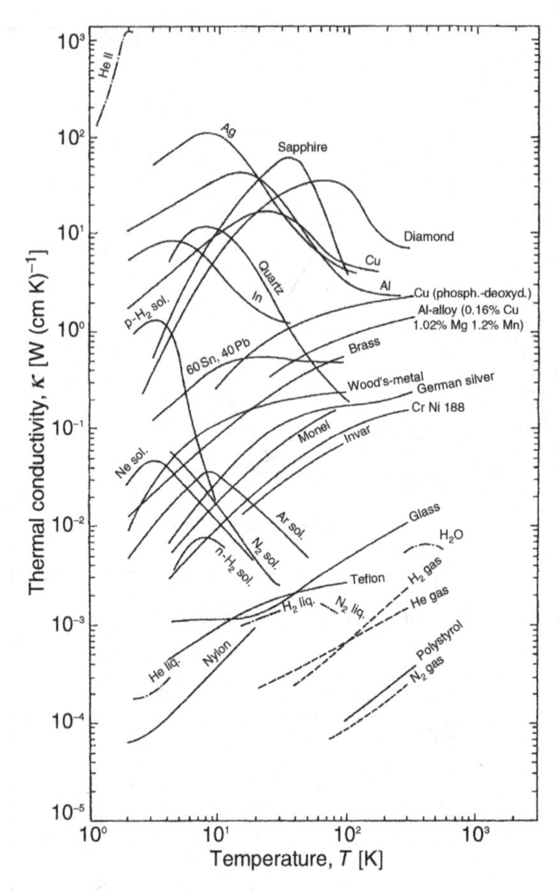

Fig. 3.20. Typical thermal conductivities κ of various materials at $T > 2\,\mathrm{K}$ [3.89–3.93]. Remember that κ depends on the purity and crystalline perfection of a material

a plateau region for $2\,\mathrm{K} \lesssim T \lesssim 20\,\mathrm{K}$ (Fig. 3.22). Because the heat is still carried by phonons with $C_{\mathrm{ph}} \propto T^3$, the phonon mean-free path limited by scattering on tunneling states must vary as $\lambda \propto T^{-1}$. As for the specific heat of disordered solids, both the absolute magnitude and the temperature dependence of the thermal conductivities of most glasses are rather similar; for example, for commercial glasses or polymers they fall within about a factor of two of those given for Pyrex (Figs. 3.21 and 3.22).

A detailed discussion of the lattice conductivity is rather involved due to the variation of the frequency of the dominant phonons with temperature and the various scattering processes. This is particularly true if the conductivity is limited by different lattice imperfections. The main results of such discussions are [3.1–3.6, 3.84, 3.86].

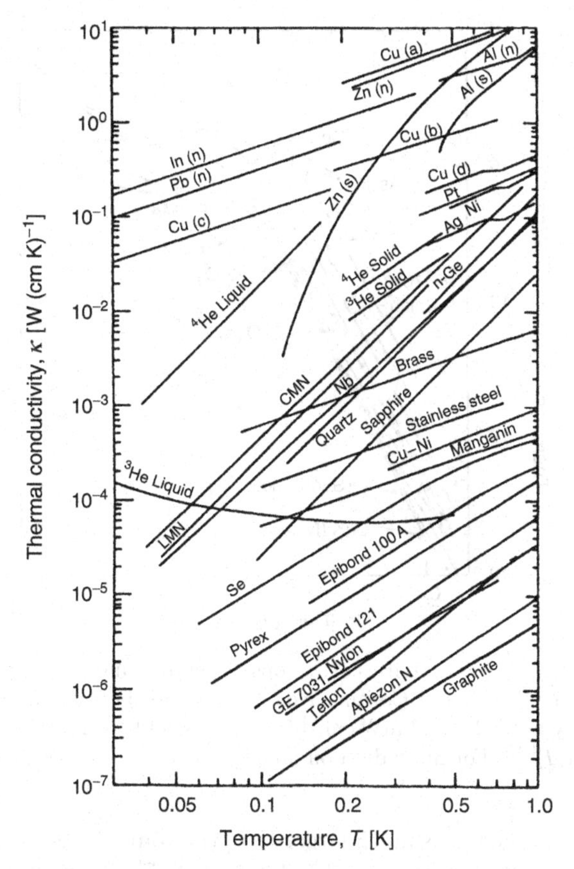

Fig. 3.21. Typical thermal conductivities κ of various materials at $T < 1\,K$ [3.45]. (This publication provides references to the original literature.) Remember that κ depends on the purity and crystalline perfection of a material

$$\begin{aligned}
&\lambda_{\text{ph}} = \text{const.} && \text{for phonon–grain boundary scattering,} \\
&\lambda_{\text{ph}} \propto T^{-1} && \text{for phonon–dislocation scattering, and} \\
&\lambda_{\text{ph}} \propto T^{-4} && \text{for phonon–point defect (Rayleigh) scattering.}
\end{aligned} \tag{3.29}$$

3.3.2 Electronic Thermal Conductivity

For the thermal conductivity due to conduction electrons we have

$$\kappa_{\text{e}} = \frac{1}{3}\frac{C_{\text{e}}}{V_{\text{m}}} v_{\text{F}} \lambda_{\text{e}} \propto T \lambda_{\text{e}}(T)\,. \tag{3.30}$$

Usually, in a metal, this electronic thermal conductivity is considerably larger than the lattice thermal conductivity because the Fermi velocity v_{F} of

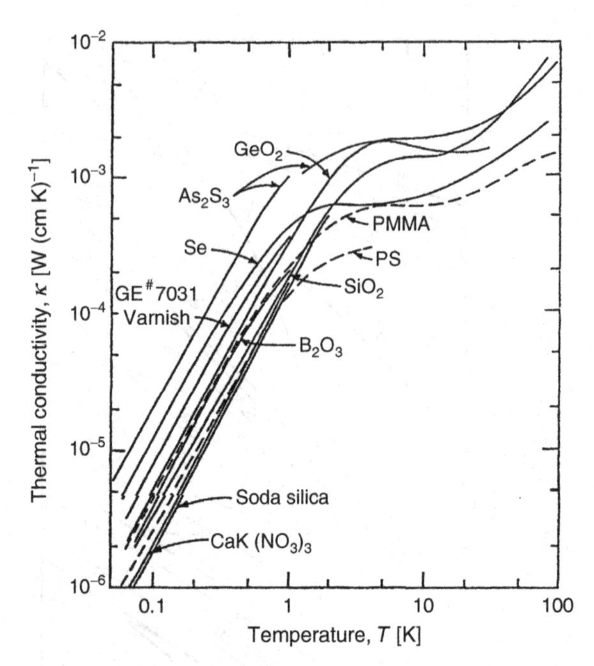

Fig. 3.22. Thermal conductivities of various noncrystalline solids [3.28]. For recent data on the glass BK 7 and Corning No. 2 coverglass between 6 and 700 mK see [3.96,3.97]; they are 0.2 to 0.3 mW cm^{-1} K^{-1} at 1 K and with a temperature dependence of about $T^{1.85}$. For more data on amorphous materials see Table 3.2

the conduction electrons is much larger than the sound velocity v_s of phonons. A detailed theoretical treatment [3.1–3.6, 3.84–3.87] of the electronic thermal conductivity is easier than that of the lattice conductivity because the conduction electrons involved sit within a narrow energy band of width k_BT at the Fermi energy E_F and therefore they all have the same energy.

High Temperatures

At high temperatures the thermally excited phonons are the limiting scatterers for the heat conducting electrons. Because the number of thermally excited phonons increases with temperature we find for the electronic thermal conductivity in the electron–phonon scattering region a thermal conductivity which decreases with increasing temperature (Figs. 3.19b and 3.20).

Low Temperatures

At low temperatures the number of phonons is again small and the scattering of electrons from lattice defects and impurities dominates. We have a

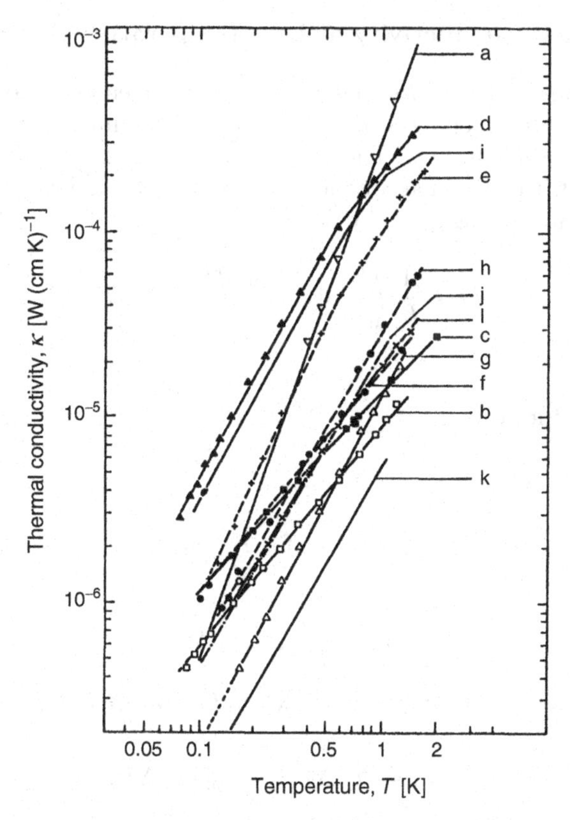

Fig. 3.23. Thermal conductivities of various dielectric materials often used in a cryogenic apparatus (a, sintered Al_2O_3; b, carbon BB5, c, nuclear graphite; d, Araldite CT 200; e, Araldite with talc; f, Vespel SP1; g, Vespel SP22; h, Vespel SP5; i, Epibond 100 A; j, Nylon; k, Graphite AGOT; l, Vespel SP21 with 15% by weight graphite) [3.94]

temperature-independent electronic mean free path resulting in the following equation for the electronic thermal conductivity:

$$\kappa_e \propto C_e \propto T\,. \tag{3.31}$$

We again have two scattering processes dominating in different temperature regions and with opposite temperature dependences. As a result the electronic contribution to the thermal conductivity also goes through a maximum (Figs. 3.19b and 3.20). The value and position of this maximum strongly depend on the perfection of the metal; for pure elements it is located at about 10 K. With increasing impurity concentration, the maximum is diminished and shifted to higher temperature. In a disordered alloy the scattering of electrons by the varying potential can become so strong that electronic and lattice conductivities become comparable.

3.3.3 Thermal Conductivity at Low Temperatures

For our purposes the values and temperature dependences of the thermal conductivity at low temperatures are of particular importance and will be summarized in the following. The heat carried by a material of cross-section A and length L with a thermal conductivity coefficient κ and with temperatures. T_2 and T_1 at its ends is given by

$$\dot{Q} = \frac{A}{L} \int_0^L \dot{q}\,\mathrm{d}x = \frac{A}{L} \int_{T_1}^{T_2} \kappa(T)\mathrm{d}T\,. \tag{3.32}$$

Insulators/Phonons

Here we have

$$\kappa_{\mathrm{ph}} = bT^3 \quad \text{for } T < \theta_{\mathrm{D}}/10 \tag{3.33}$$

and hence

$$\dot{Q} = \frac{Ab}{4L}(T_2^4 - T_1^4)\,. \tag{3.34}$$

For small temperature gradients, $\Delta T = T_2 - T_1 \ll T$, we have

$$\dot{Q} \simeq \frac{Ab}{L}T^3\Delta T = \frac{A}{L}\kappa_{\mathrm{ph}}(T)\Delta T\,. \tag{3.35}$$

Metals/Conduction Electrons

Here we have

$$\kappa_{\mathrm{e}} = \kappa_0 T \quad \text{for } T < 10\,K \tag{3.36}$$

and therefore

$$\dot{Q} = \frac{A\kappa_0}{2L}(T_2^2 - T_1^2)\,. \tag{3.37}$$

Again, for small temperature gradients,

$$\dot{Q} \simeq \frac{A\kappa_0}{L}T\Delta T = \frac{A}{L}\kappa_{\mathrm{e}}(T)\Delta T\,. \tag{3.38}$$

Hence, an accurate determination of thermal conductivity involves also an accurate determination of the relevant sample dimensions.

Low-temperature thermal conductivities of various materials are listed in Table 3.2 and plotted in Figs. 3.19–3.23.

Table 3.2. Thermal conductivity of solids frequently used in low temperature apparatus

material	κ [mW (cm K)$^{-1}$]	T range [K]	ref.
Manganin	$0.94\ T^{1.2}$	1–4	[3.98]
Nb–Ti	$0.075\ T^{1.85}$	4–9	[3.99]
Nb–Ti	$0.15–0.27\ T^{2.0}$	0.1–1	[3.100]
$Cu_{0.70}Ni_{0.30}$	$0.93\ T^{1.23}$	0.3–4	[3.101]
$Cu_{0.70}Ni_{0.30}$	$0.64\ T$	0.05–3.0	[3.100, 3.102]
Pyrex	$0.15\ T^{1.75}$	0.18–0.8	[3.103]
Al_2O_3	$2.7\ T^{2.5}$	2–8	[3.104]
Al_2O_3	$0.29\ T^{2.7}$	0.1–2	[3.94]
Stycast 1266	$0.49\ T^{1.98}$	0.05–0.5	[3.105]
Stycast 1266	$0.39\ T^{1.9}$	0.06–1	[3.100]
Stycast 2850 GT	$78 \times 10^{-3}\ T^{1.8}$	1–4	[3.98]
Stycast 2850 FT	$53 \times 10^{-3}\ T^{1.8}$	2–10	[3.106]
Stycast 2850 FT	$92 \times 10^{-3}\ T^{2.65}$	0.06–1	[3.100]
Vespel SP 1	$18 \times 10^{-3}\ T^{1.2}$	0.1–1	[3.94]
Vespel SP 22	$17 \times 10^{-3}\ T^{2}$	0.1–2	[3.94, 3.100]
Teflon	$30 \times 10^{-3}\ T^{2}$	0.2–1	[3.107]
Teflon	$38 \times 10^{-3}\ T^{2.4}$	0.3–0.7	[3.103]
Nylon	$26 \times 10^{-3}\ T^{1.75}$	0.2–1	[3.94, 3.103]
Macor	$58 \times 10^{-3}\ T^{2.24}$	0.4–1.1	[3.108]
Nuclear graphite	$15 \times 10^{-3}\ T^{1.13}$	0.1–2	[3.94]
AGOT graphite	$5.1 \times 10^{-3}\ T^{1.76}$	0.1–2	[3.94]
	$4.9 \times 10^{-3}\ T^{1.86}$	0.3–3	[3.109]
a-SiO_2	$0.248\ T^{1.91}$	0.06–1	[3.110]
Wood	$9.3 \times 10^{-3}\ T^{2.7}$	0.04–1	[3.111]
Kevlar	$3.9 \times 10^{-5}\ T^{1.17}$	0.1–2.5	[3.112]
Polypropylene	$27.4 \times 10^{-3}\ \mathrm{T}^{1.28}$	0.1–1	[3.56]
PVC	$1.8 \times 10^{-4}\ \mathrm{T}^{2.05}$	0.05–0.12	[3.113]
Torlon	$6.13 \times 10^{-2}\ \mathrm{T}^{2.18}$	0.1–0.8	[3.114]

The given temperature ranges are the ranges where the given equations describe the data; in the cited literature, often data for a much wider temperature range are given.

3.3.4 Superconducting Metals

In superconducting metals some of the electrons are paired to so-called Cooper pairs. They all sit in the same low energy state of zero entropy, which is separated by an energy gap ΔE from the states of the single, unpaired electrons [3.16–3.19, 3.23]. The Cooper pairs cannot leave this ground state to carry heat (they carry no entropy) unless they are broken up into single electrons. Therefore in a superconducting metal only the remaining unpaired electrons can carry heat. Because they are in energy states which are separated from the Cooper ground state by the energy gap $\Delta E(T)$, their number decreases exponentially with T, i.e., as $\exp(-\Delta E/k_B T)$. As a result of this qualitative discussion we find that the electronic thermal conductivity of a metal in the superconducting state is given by the product of the number of remaining unpaired single electrons and their thermal conductivity (which is identical to the electronic thermal conductivity $\kappa_{e,n} = \kappa_0 T$ in the normal–conducting state),

$$\kappa_{e,s} \propto T e^{-\Delta E/k_B T}, \tag{3.39}$$

with $\Delta E = 1.76 k_B T_c$ for most elemental superconductors [3.16–3.19, 3.23].

Hence, the electronic thermal conductivity of a superconductor decreases very rapidly with decreasing temperature [3.115]. Indeed, at low temperatures the electronic thermal conductivity of a superconducting metal can even become smaller than its lattice conductivity, and at sufficiently low temperatures, say at $T < T_c/10$, the total thermal conductivity of a superconductor approaches the thermal conductivity of an insulator, $\kappa \propto T^3$. This is shown in Fig. 4.1 for aluminium.

Because it is rather simple to "switch" a metal from the superconducting to the normal state by applying a large enough magnetic field, one can "switch" its thermal conductivity from one state to the other. We take advantage of this possibility by using a superconducting metal as a thermal switch to disconnect or to connect two parts in a low-temperature apparatus, for example in a magnetic refrigerator (Chap. 10). This application will be discussed in Sect. 4.2.2.

3.3.5 Relation Between Thermal and Electrical Conductivity: The Wiedemann–Franz Law

A correct measurement of the low-temperature thermal conductivity of a metal can be rather cumbersome (see Sect. 3.3.7) and, in general, a measurement of the electrical conductivity is much easier. Fortunately, due to the fact that in a metal usually both conductivities are determined by the flow of electrons and are mostly limited by the same scattering processes, a measurement of the electrical conductivity often gives reasonable information about the thermal conductivity. Let us consider a metal at low temperatures in the defect scattering limit where $\lambda_e = \text{const}$. For the electrical conductivity the electrons conduct charge, which is temperature independent. At low temperatures in the defect scattering limit or in the residual resistivity range (see below) the electrical conductivity σ is therefore temperature independent. For the thermal conductivity κ the electrons carry heat, but the specific heat, and therefore the thermal conductivity, are proportional to temperature in this range. As a result, the ratio of thermal conductivity κ to electrical conductivity σ is proportional to temperature. One arrives at the same result for the temperature range where the conductivity due to electron transport is limited by large-angle elastic electron–phonon scattering ($T \geq \theta_D$). At, say,

4 and 300 K, we then have the Wiedemann–Franz law for the ratio of these two conductivities [3.1–3.6, 3.84–3.87, 3.116],

$$\kappa/\sigma = L_0 T\,, \tag{3.40}$$

where the Lorenz number L_0 is a universal constant, i.e., $L_0 = (\pi\kappa_B/e)^2/3 = 2.4453 \times 10^{-8}\,W\,\Omega K^{-2}$.

Of course, we also arrive at (3.40) if we combine the equation for the electrical conductivity of a metal

$$\sigma = \frac{nN_0 e^2 \lambda_e V_m}{v_F m^*}\,, \tag{3.41}$$

where m^* is the effective mass of the conduction electrons and n is the number of conduction electrons per atom, with (3.30) for the thermal conductivity, assuming that the electronic mean free path is the same for both conductivities.

We can therefore use the measured electrical conductivity together with the Wiedemann–Franz law to calculate the thermal conductivity. In many situations this gives correct results, in particular when the electron scattering is predominantly elastic. It most often holds at low temperatures (impurity scattering; $T < \theta/10$) and at high temperatures (phonon scattering; $T \geq \theta$) but not in between, where energy losses of the order $k_B T$ are associated with electron–phonon collisions. However, cases are known in which the calculated and measured thermal conductivities differ by up to an order of magnitude at low temperatures. Usually the measured thermal conductivity is smaller than the thermal conductivity calculated with the Wiedemann–Franz law from the electrical conductivity. Particularly disturbing is the observation that this can happen at Kelvin and lower temperatures for metals of typical quality commonly used in a low-temperature apparatus (e.g., Al and Ag) while this deviation can be absent for other quite similar metals (e.g., Cu) [3.104], see Fig. 3.24, or has even not been observed for the same metal by other investigators [3.117, 3.118]. The reason for this discrepancy is the fact that our earlier discussion is an oversimplification. In reality scattering processes may contribute with different "effectiveness" to the two conductivities, in particular when inelastic scattering contributes to the thermal conductivity [3.1–3.6, 3.84–3.87].

A formula that is rather useful can be derived from the above equations, namely,

$$v_F \lambda_e = \frac{\sigma}{\gamma}\left[\frac{\pi k_B}{e}\right]^2 \tag{3.42}$$

which allows the electronic mean free path λ_e to be calculated from the measured electrical conductivity σ and specific heat coefficient γ.

In the literature the so-called residual resistivity ratio (RRR) is very often given as a measure of the "purity" of a metal (Fig. 3.19). This is the ratio of the electrical conductivity at low temperatures, e.g., at the boiling point of liquid helium, to the electrical conductivity at room temperature

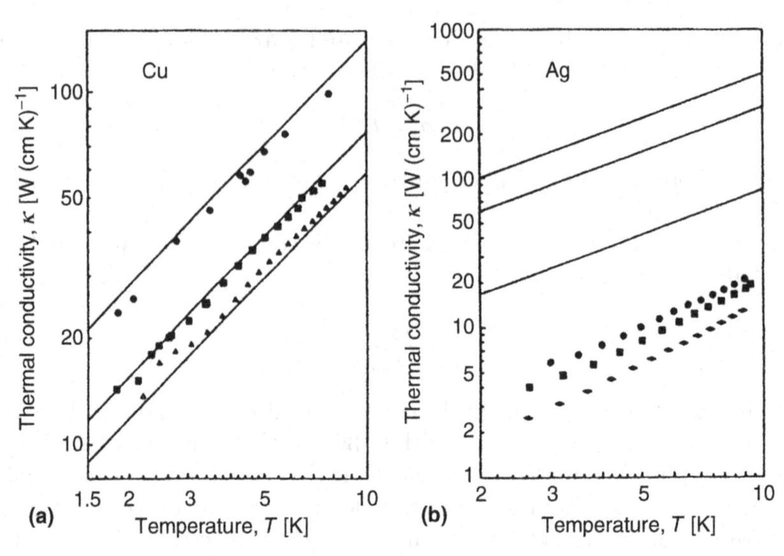

Fig. 3.24. (**a**) Thermal conductivity of a Cu foil 60 μm thick and 20 mm wide after various treatments, resulting in RRRs of 979 (●), 540 (■) and 410 (▲). Lines correspond to thermal conductivities calculated from RRR values (3.43) by applying the Wiedemann–Franz law (3.40). (**b**) Thermal conductivity of a Ag rod 0.86 × 0.86 mm^2 after different treatments, resulting in RRRs of 3,330 (●), 1,988 (■) and 553 (▲). Lines correspond to thermal conductivities calculated from RRR values (3.41) by applying the Wiedemann–Franz law (3.40). The shown results mean that different types of defects may influence the thermal and the electric conductivities differently. Hence, the results may differ from sample to sample [3.104]

$$\mathrm{RRR} = \sigma_{4.2\,\mathrm{K}}/\sigma_{300\,\mathrm{K}} = \rho_{300\,\mathrm{K}}/\rho_{4.2\,\mathrm{K}}\,. \tag{3.43}$$

Because the room-temperature conductivity is determined not by defect scattering but by phonon scattering, whereas the low-temperature conductivity is exclusively determined by the scattering on defects because then there are no phonons left – this ratio is a direct measure of the limiting defect scattering. It indicates how good a material is by stating by what factor its conductivity increases with the vanishing of the phonon scattering at low temperatures.

3.3.6 Influence of Impurities on Conductivity

In Sect. 3.3.5 we have discussed how the thermal conductivity – which is so important for low-temperature experiments – can be calculated from the electrical conductivity and that both conductivities are limited at low temperatures by electron–defect scattering in a metal. In this section I will make some comments on the scattering of conduction electrons on impurity atoms. For this scattering we have to distinguish between nonmagnetic and magnetic impurities.

Electron Scattering by Nonmagnetic Impurity Atoms

The conduction electrons are scattered at the Coulomb potential of the impurity, which may have the valence difference ΔZ compared to the host lattice. The increase of the electrical resistance in this case is often not very large, and for small impurity concentrations is given by the Linde rule

$$\Delta\rho_{\mathrm{nm}} = a + b(\Delta Z)^2 . \tag{3.44}$$

The constants a and b are determined by the host lattice and, in particular, depend on the row of the Periodic Table to which the host atom belongs. Examples for Cu as the host are listed in Table 3.3. These values are typically $1\,\mu\Omega\,\mathrm{cm}\,(\mathrm{at}\ \%)^{-1}$ (impurity concentration); they are comparable to the increase of ρ due to heavy cold working (several 10 $n\Omega\ cm$) or introduction of vacancies ($1\,\mu\Omega\,\mathrm{cm}\,(\mathrm{at}\ \%)^{-1}$) [3.45, 3.87].

Scattering of Electrons by Magnetic Impurity Atoms (in Particular in Copper)

A magnetic or spin-flip scattering of the conduction electrons can occur at localized moments of magnetic impurities [3.5, 3.119–3.121]. For this situation the increase of resistance can be much larger – and much more difficult to understand theoretically – than for the case of nonmagnetic impurities; a theoretical discussion and a calculation of $\Delta\rho_{\mathrm{m}}$ is much more involved. Examples for 3d elements as impurities in Cu are listed in Table 3.4. The strength of the scattering and the resulting resistance increase depend strongly on the properties of the magnetic impurity and it can be very different in different host lattices. For example, iron produces a very large moment of about $12\,\mu_{\mathrm{B}}$ in palladium [3.122], it keeps essentially its bare moment of about $2\,\mu_{\mathrm{B}}$ in noble metals, whereas its moment at low temperatures is very small in rhodium [3.123]

Table 3.3. Relative change $\Delta\rho_{\mathrm{nm}}$ per concentration c of the solute of the electrical resistance of Cu if the given nonmagnetic elements with valence difference ΔZ compared to Cu are introduced as impurities, see (3.44), [3.87]

impurity	As	Si	Ge	Ga	Mg	Zn	Cd	Ag
ΔZ	4	3	3	2	1	1	1	0
$\Delta\rho_{\mathrm{nm}}/c$ [$\mathrm{n\Omega\,cm\,(at.\ ppm)^{-1}}$]	0.65	0.33	0.37	0.14	0.065	0.025	0.02	0.014

Table 3.4. Relative change $\Delta\rho_{\mathrm{m}}$ per concentration c of the solute of the electrical resistance of Cu at 1 K if the given magnetic elements are introduced as impurities [3.87]

impurity	Ti	V	Cr	Mn	Fe	Co	Ni
$\Delta\rho_{\mathrm{m}}/c$ [$n\Omega\,\mathrm{cm\,(at.\ ppm)^{-1}}$]	1.0	0.58	2.0	0.43	1.5	0.58	0.11

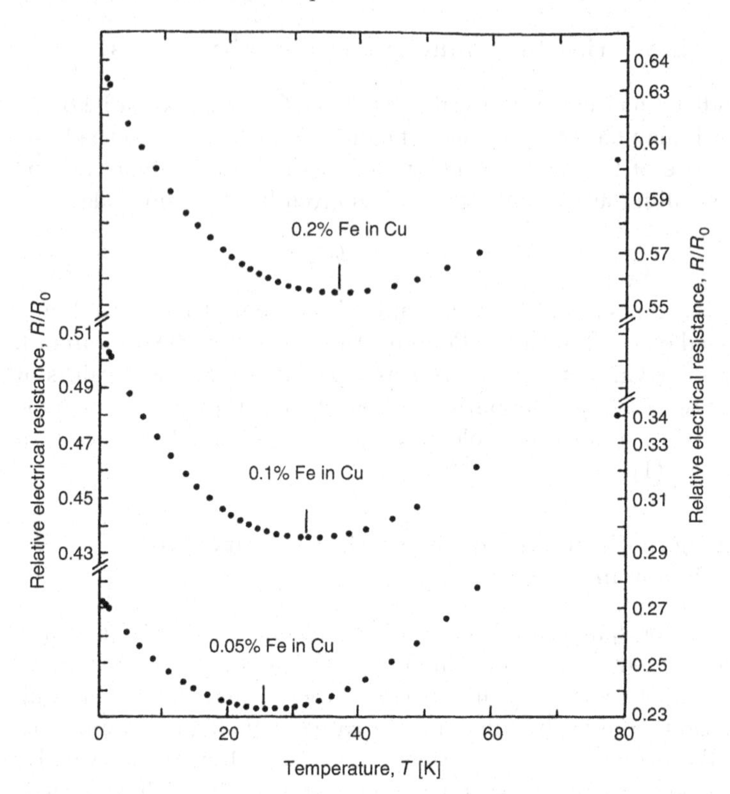

Fig. 3.25. Resistance minima of various dilute alloys of Fe in Cu. R_0 is the resistivity at 0°C. The position of the minimum depends on the concentration of iron [3.125]. For similar, more recent data for Fe in Au see [3.126]

or aluminum [3.124]. Hence, Fe has a strong impact on transport properties in noble and platinum metals, however almost no impact on them in Rh and Al. Furthermore, the strength of the scattering can depend very strongly on temperature due to the so-called Kondo effect, which describes an enhanced inelastic scattering of a cloud of conduction electrons "condensed" around magnetic moments that are localized on impurity atoms [3.5, 3.119–3.121]. The latter experience a temperature-dependent screening by the conduction electrons of the host lattice. As a result the resistivity may not approach a constant value ρ_0 at low temperatures but may rise logarithmically with decreasing temperature T after passing through a minimum whose position depends weakly on the concentration of the impurity (Fig. 3.25). It is then given by

$$\rho = \rho_0 - \rho_{\rm K} \ \ln(T) \tag{3.45}$$

where $\rho_{\rm K}$ denotes the Kondo resistivity.

3.3.7 Thermal Conductivities of Copper, Silver and Aluminum at Low Temperatures

Aluminum and copper (and for some special applications, the more expensive silver; however, it usually contains a substantial amount of O_2) are the most widely used thermal conductors in a low-temperature apparatus. Aluminum has the advantages over Cu that it has a lower density, it is available in higher purity (up to 6 N, whereas Cu seems presently to be commercially available only up to 5 N), and in particular its conductivity is less dependent on deformation (it anneals already at room temperature) and on the amount of impurities (paramagnetic ions have a very small moment in Al opposite to their properties in Cu; see above). However, Cu (and Ag) has the advantage that it is easy to make good thermal contact to it which is much harder for Al due to its tenacious oxidized surface layer (see Sect. 4.2.2). For these metals, the thermal conductivities at low temperatures vary over many orders of magnitude depending on treatment and purity (Fig. 3.19). As a rule of thumb: their RRR is about equal to the thermal conductivity in W/K m at 1 K [3.127]. More exactly

$$\kappa = (RRR/0.76)T[W/K\text{m}] \quad \text{and} \quad \kappa = (RRR/0.55)T[W/K\text{m}] \tag{3.46}$$

for Cu and Ag, respectively (but see the above discussion and Fig. 3.24). Typical values for the thermal conductivity of Cu and Al at different purities are 10^2 to 10^3 (10^2 to 10^4; 10^3 to 4×10^4) W/K m for 4 N (5 N; 6 N) purity, annealed (400°C) Al, and 50 to 200 (200–5,000) for 4 N (5 N) annealed Cu [3.127, 3.128]. Information on the thermal conductivity of a large number of Al alloys for temperatures between 2 K and room temperature can be found in [3.129]. At 1 K, they are typically between 1 and $3\,\text{W}\,\text{K}^{-1}\,\text{m}^{-1}$.

Because of the importance of copper and of other noble and platinum metals for making high-conductivity thermal joints (or for thermometry purposes; see Sects. 12.9, 12.10), I shall discuss a process by which magnetic impurities can be "passivated" in Cu, Ag, Au, Pt, Pd, Rh and possibly some other metals as well, by oxidizing the less noble metal impurities, resulting in a dramatic increase of the low-temperature conductivities. The following discussion applies to the "passivation" of the magnetic scattering of iron in copper [3.130–3.133] (for Ag, see [3.134]).

The typical residual resistivity ratio of a piece of copper which one can buy from a shop is in the range of 50–100. Heating the copper to a temperature of 400–500°C anneals structural lattice defects, and the residual resistivity ratio usually increases to a value of 300–400. A further increase is only possible by passivation of magnetic impurities, in particular by oxidation of iron. The less noble impurities are oxidized by heat treatment of the Cu specimen at a temperature of 900–1,000°C in an atmosphere of oxygen or air at 10^{-5}–10^{-4} mbar (the time required for this treatment depends mainly on the thickness of the Cu sample). This oxidation and passivation is a two-step process. In the first step Fe is oxidized to FeO by oxygen diffusing through the Cu matrix.

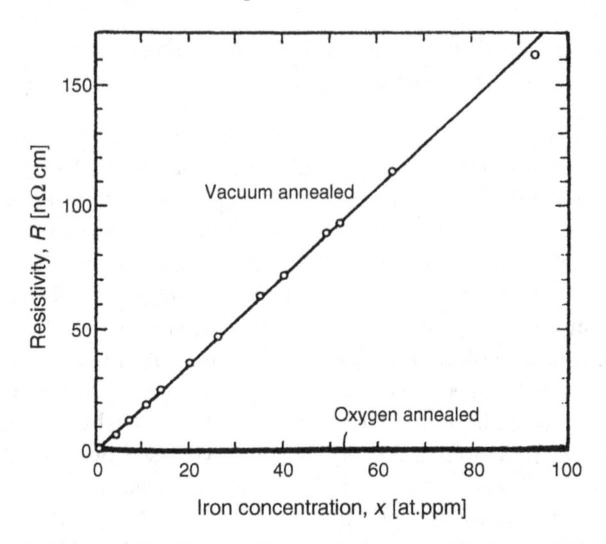

Fig. 3.26. Resistivity of Cu–Fe at 4 K, as a function of Fe impurity concentration, annealed in vacuum and in an O_2 atmosphere (92 h; about 10^{-6} bar air) [3.130]

FeO is stable, whereas copper oxide is unstable at these temperatures. The iron oxides attract more FeO and O and create small (0.1 μm) Fe_3O_4 clusters, reducing the number of magnetic scattering centres. These clusters are magnetically ordered; the conduction electrons do not suffer spin-flip scattering at the fixed iron moments any more. In this sense, the iron moments are magnetically inactive for the scattering of conduction electrons of the host lattice. Thus, although the Cu is not purer after oxygen annealing, the impurities are much less effective as scattering centres (Fig. 3.26).

Another very effective purification of Cu from Fe reported in the literature is an electrolytic process starting from a Cu sulphate solution, followed by a wet-hydrogen treatment to remove nonmagnetic impurities [3.135]. The residual hydrogen has then to be "pumped out" by heating the Cu in vacuum at around 1,000°C. The resulting increased conductivity corresponds to a residual resistivity ratio of at least 1,000 and in extreme cases up to several 10,000, at which point dislocation scattering of the conduction electrons dominates.

Of course, it is an advantage if the starting copper already contains oxygen; oxygen-free high-conductivity (OFHC) Cu is often not well-suited for the production of high-conductivity copper needed for low-temperature experiments. However, it has been shown that oxygen annealing can be detrimental for the conductivity of commercially available ultrapure (7N) Cu [3.136].

Oxygen annealing does not change the conductivity of Al because of the very small magnetic moments of 3d-elements in Al.

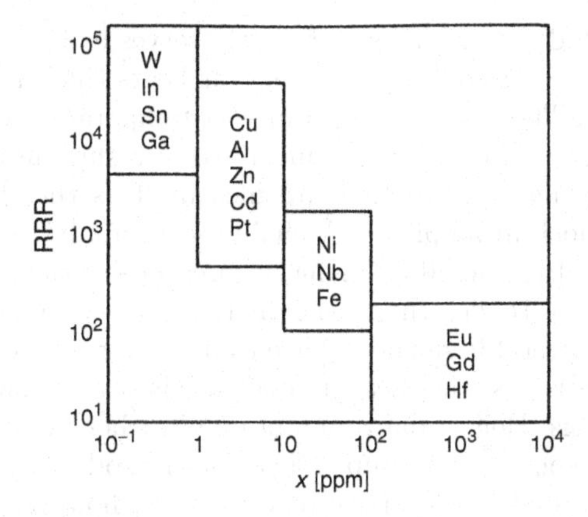

Fig. 3.27. Typical values of the residual resistivity ratio (RRR) of various metals as a function of typical impurity concentrations with which they are available [3.138]

A survey on the transport properties of pure metals in the defect scattering limited range and a compilation of the smallest residual resistivities which have been reported in the literature for all metals except the rare earths can be found in [3.137]. Figure 3.27 gives typical values of the residual resistivity ratio which one can attain in pure metals as a function of the impurity concentration. Comprehensive compilations of data on the thermal conductivity of solids can be found in [3.44, 3.45, 3.90–3.93]. Values for metallic elements at 273 K are compiled in [3.87], p. 28.

3.3.8 How to Measure Thermal Conductivities

Usually, the thermal conductivity of a material is determined by measuring with two thermometers the temperature difference produced by heating one end of it and keeping the other one at the constant temperature of a heat sink (see [3.86, 3.87] and references therein). This simple steady-state method has several pitfalls and problems that are rather similar to the ones in measurements of heat capacities (see Sect. 3.1.8). For example, heat losses by radiation and into eventually still remaining gas molecules in the surrounding, Joule heating in the thermometers, parasitic heat leaks along leads to heater and thermometers, the thermal resistances of them to the sample; etc (see also Sect. 5.1.2). The heat flow along the leads can be strongly reduced by using thin superconducting wires at their low-temperature end and by properly heat sinking them. Radiation losses are particularly problematic at higher temperatures. They can be substantially reduced by using radiation shields around the setup at a temperature close to the samples temperature [3.33, 3.58], or by applying the so-called 3ω-method introduced by [3.139]. This method uses a radial AC heat flow into the sample from a heated narrow metal film deposited onto it. This metal strip serves as heater as well as thermometer. A current of frequency ω through it heats the sample with a power at a frequency 2ω. The resulting resistance or temperature oscillations of the metal strip at 2ω times the driving current at ω results in 3ω-oscillations in voltage across the metal strip that are measured as the third harmonic of the voltage by a lock-in amplifier. A challenge in applying this technique is the measurement of the small 3ω-signal in presence of the much larger background signal at ω. To determine the thermal conductivity of the sample, measurements at several frequencies have to be performed. Besides essentially avoiding radiation losses, the measurement needs only a few periods of temperature oscillations. Hence, data are obtained rather fast and can be taken while warming or cooling the sample. The 3ω-method is also quite suited for measuring the thermal conductivity of thin films, however, it then requires special design and data analysis considerations [3.87, Chap. 2.2].

For choice of appropriate thermometers and heaters the considerations are again rather similar as discussed in the above section for heat-capacity measurements. For the heater, of course, a material with temperature-independent resistance, like phosphor bronce, TiCr, or NiCr alloys, should be chosen.

The problems of parasitic heat dissipation and thermal shunts or losses through the electrical leads to heater and thermometers have been substantially reduced in two recent experiments in which the thermal conductivity of badly conducting glasses have been measured down to 5 mK. In the first one [3.96], the dielectric constant of the glass sample itself was used for thermometry (see Sect. 12.8) and the parasitic power dissipation was reduced to about 0.1 pW. This method avoids any thermal boundary resistance between thermometers and sample. The careful anchoring of the leads (resulting in less than 0.1% of heat loss) and calibration of the dielectric constant thermometers are discussed. In the second experiment [3.97], thermometry was performed by measuring in a second-order axial gradiometer (to avoid magnetic contributions from the sample) inductively the change of DC magnetization of two paramagnetic $Au\mathrm{Er}_{500\mathrm{ppm}}$ strips glued to the sample. The gradiometer leads are directly connected to the input of a SQUID, resulting in extremely low power dissipation (see Sect. 12.9). Because heat is applied optically from a light emitting diode to one end of the sample, the method is contact-free, reducing the parasitic heat leak to below 10^{-15} W (Fig. 3.28). Data obtained by these methods are mentioned in the caption of Fig. 3.22. One may have to resort to these or other more elaborate techniques [3.86, 3.87], particularly if thermal conductivities of small samples or of thin films have to be measured. For these cases, microfabricated heaters and thermometers or optical excitation and possibly also detection methods may have to be used. As for a heat-capacity apparatus, the quality of a thermal conductivity setup can be checked by investigating a well-characterized standard sample. Another appropriate check of the validity of the obtained data is a measurement of the thermal conductivity at various heat flows.

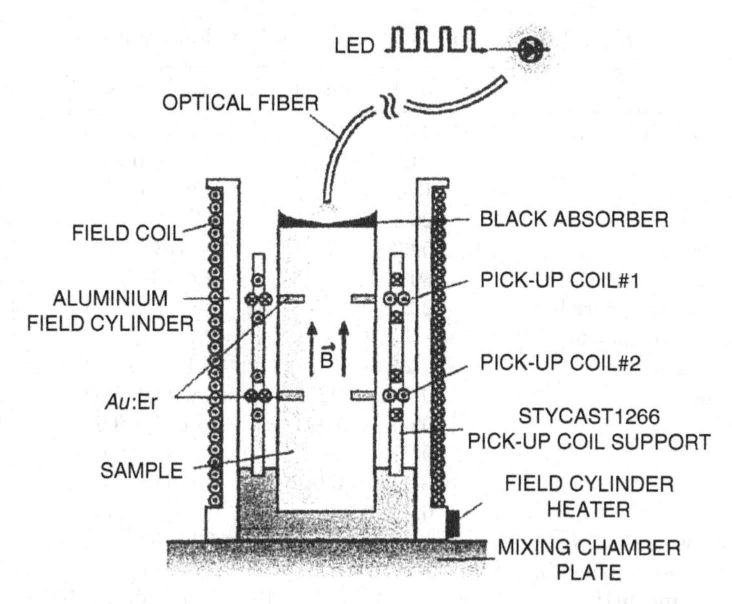

Fig. 3.28. Schematic of a setup to measure contact-free thermal conductivities at millikelvin temperatures. The temperature gradient in the sample is generated by cooling one end with a dilution refrigerator and heating the other one optically with a light emitting diode kept at 1 K. The temperatures are determined by measuring inductively in second order gradiometers the susceptibility of two paramagnetic $Au\mathrm{Er}^+$ pieces (see Sect. 12.9) glued to the sample. The superconducting Al shield can be heated above its critical temperature for a short time to freeze in the measuring field of 0.5 mT. Reprinted with permission from [3.97]; copyright (2004), Am. Inst. Phys.

3.4 Magnetic Susceptibilities

3.4.1 Magnetic Susceptibilities of Some Selected Materials

This section gives representative data for the magnetic susceptibility of various materials, in particular of weakly magnetic materials often used in the construction of cryogenic apparatus where sensitive magnetic measurements are to be performed. In Table 3.5 the parameters x_∞ and C of the equation

$$\chi/\rho = x_\infty + C/T \tag{3.47}$$

for the susceptibility χ per mass density ρ are given. The given data have been obtained in different temperature ranges. One should keep in mind that a variation of χ can be found in the literature for some of the listed materials, particularly weakly magnetic, multicomponent materials. This can be partly explained by slight differences in the composition and/or in the preparation of nominal identical materials. For details, see the original publications [3.140–3.146].

Table 3.5. Susceptibility of some selected materials

material	C [10^{-6} emu K/g]	x_∞ [10^{-6} emu/g]	Refs.
Fused quartz, Suprasil	0.05 ± 0.02	-0.383 ± 0.004	[3.140]
Macor	24.5 ± 0.06	2.87 ± 0.18	[3.140]
Stycast 2850 GT	17 ± 1	2.97 ± 0.15	[3.140]
	26 ± 7	15 ± 4	[3.141]
Stycast 1266	0.63 ± 0.03	-0.18 ± 0.01	[3.140]
Epibond 1210 A/9615-10[1]	80 ± 4	7.0 ± 0.4	[3.140]
GE varnish 7031, cured	2.2 ± 01	-0.54 ± 0.03	[3.140]
Mixed 1:1 with toluene	-7 ± 3	2 ± 1.5	[3.141]
Apiezon N grease	-2 ± 3	0.1 ± 1.2	[3.141]
Mylar tape	0.34 ± 0.02	-0.43 ± 0.02	[3.140]
Teflon tape	0.063 ± 0.003	-0.33 ± 0.02	[3.140]
	0.05 ± 0.01	-0.33 ± 0.01	[3.142]
Nylon	-0.06 ± 0.02	-0.63 ± 0.01	[3.142]
	0.3 ± 3.0	-0.6 ± 1.5	[3.141]
Manganin wire (low Ni)	(318)	92 ± 5	[3.140]
Manganin wire, enamel insulated	-13 ± 3	120 ± 1	[3.141]
$Cu_{0.7}Ni_{0.3}$ (Inconel)	$(-2.3 \pm 0.2) \times 10^5$	$(2.6 \pm 0.2) \times 10^5$	[3.141]
Stainless steels	-120 to ± 10	100 to 300	[3.141]
$Pt_{92}W_8$, alloy 479	0.43 ± 0.02	0.24 ± 0.01	[3.140]
AR glass	36 ± 1	-0.35 ± 0.01	[3.143]
Duran glass	15.8 ± 0.5	-0.40 ± 0.01	[3.143]
Suprasil glass	0.01 ± 0.01	-0.40 ± 0.01	[3.143]
Gelatine	1.0 ± 0.1	-0.41 ± 0.01	[3.143]
AR glass	29 ± 3		[3.143]
Duran glass	5.7 ± 0.6		[3.143]
Suprasil glass	< 0.01		[3.143]

Data are taken at: 2–10 K [3.140]; 4.2 K [3.141, 3.142]; > 5 K for first set, < 0.1 K for second set of data of [3.143]

In addition to the data given in Table 3.5, remanent magnetic moments and susceptibilities for a large number of metals, dielectrics, wires and ribbons as well as for some further materials used in the construction of cryogenic equipment can be found in [3.144] for $T = 2$ K, and in [3.145] for $T = 4.2$ K. The most recent publication on this topic is [3.146], where the results obtained for the magnetization of 13 materials at 2–12 K and at 0.25–4 T have been presented. Some of these data are shown in Fig. 3.29. Data for various austenitic stainless steels (AISI 300 series 304, 310, 316, and AWS 330) at 4.2–414 K are given in [3.147]. Typical parameters for (3.47) are $\chi_\infty = 16(11;0)$ and $C = 3(4;11)$ for AISI 304/316(310S;AWS 330).

It is remarkable that a large number of materials which are usually considered as "nonmagnetic" show quite appreciable values for their susceptibilities and remanent magnetizations at low temperatures, often making them

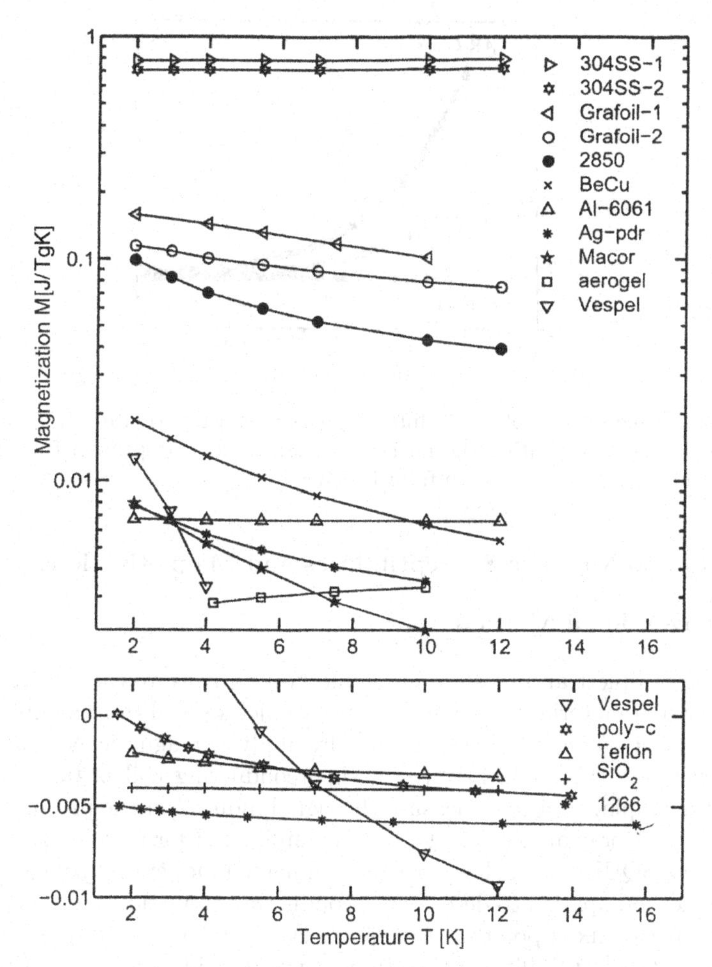

Fig. 3.29. Magnetization per mass (J/TgK) of the indicated materials in a field of 1 Tesla. The upper graph is logarithmic and shows data for paramagnetic samples. The lower graph is linear and shows data for diamagnetic materials [3.146]

unsuitable for use in magnetically sensitive equipment or experiments. From the shown data, one can conclude that the fused glass Suprasil is essentially free of magnetic impurities (substantially less than 1 ppm; quite different from other glasses!) and has a very low susceptibility [3.43, 3.104] (Fig. 3.30), making it a very good candidate for many applications, for example as sample holder in susceptometers or for experiments in high magnetic fields. Other candidates with somewhat higher but still low concentration of magnetic impurities seem to be cotton thread and dental floss, as well as Teflon and Nylon (see Table 3.5).

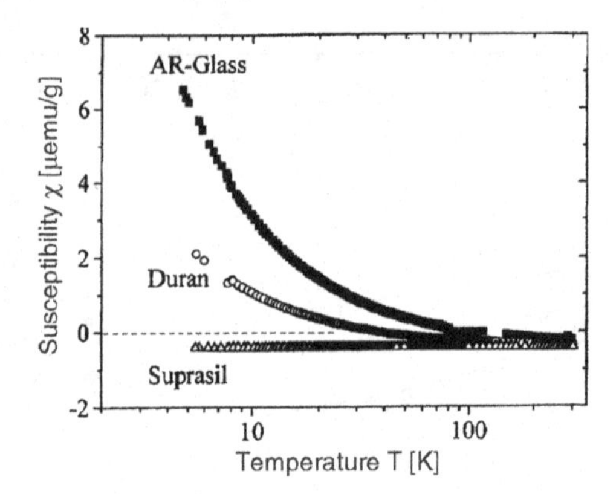

Fig. 3.30. Susceptibility of three different glasses as a function of temperature in a field of 1 T, demonstrating the purity of Suprasil glass. Reprinted from [3.143], copyright (1995), with permission from Elsevier

3.4.2 How to Measure Susceptibilities and Magnetizations

The Conventional Method

Surely, the simplest and most common method to measure the magnetic susceptibility of a material is to wind a sensitive pick-up coil (for example, some 1,000 turns of 25 μm Cu wire) around it, apply a magnetic field at a frequency between 0.1 and 1 kHz (by a superconducting coil of 0.1 mm NbTi wire, for example) to it and measure the signal induced in the pick-up coil by a home-made (for example, using a lock-in amplifier or for more sensitive measurements a SQUID, see below) or by a commercial inductance bridge. This is also the method applied to detect superconducting transitions, for example in superconducting fixed-point-devices (see Sects. 11.4.2 and 11.4.3), or for measurements of susceptibilities of samples down to milli- or even microkelvin temperatures. A setup in which several samples can be investigated in one run is shown in Fig. 3.31. It has been used for measurements to 25 μK [3.122]. Examples of electronic setups for measuring susceptibilities using either a lock-in amplifier in a self-inductance bridge or a SQUID as detector for more sensitive measurements are shown in Figs. 3.32 and 3.33. To avoid heating effects due to relaxation in the sample or from eddy currents, a bridge should be operated at low enough frequency, for example in the range of 20–300 Hz. This will also reduce capacitance leakage effects. Another suitable setup for determination of susceptibilities measures the resonant frequency in the range of 1 MHz of a tank circuit driven by a tunnel diode whose inductance contains the paramagnetic sample [3.148, 3.149]. More sophisticated methods, in particular the application of a SQUID as detector as well as commercial versions of magnetometers will be discussed in more detail below.

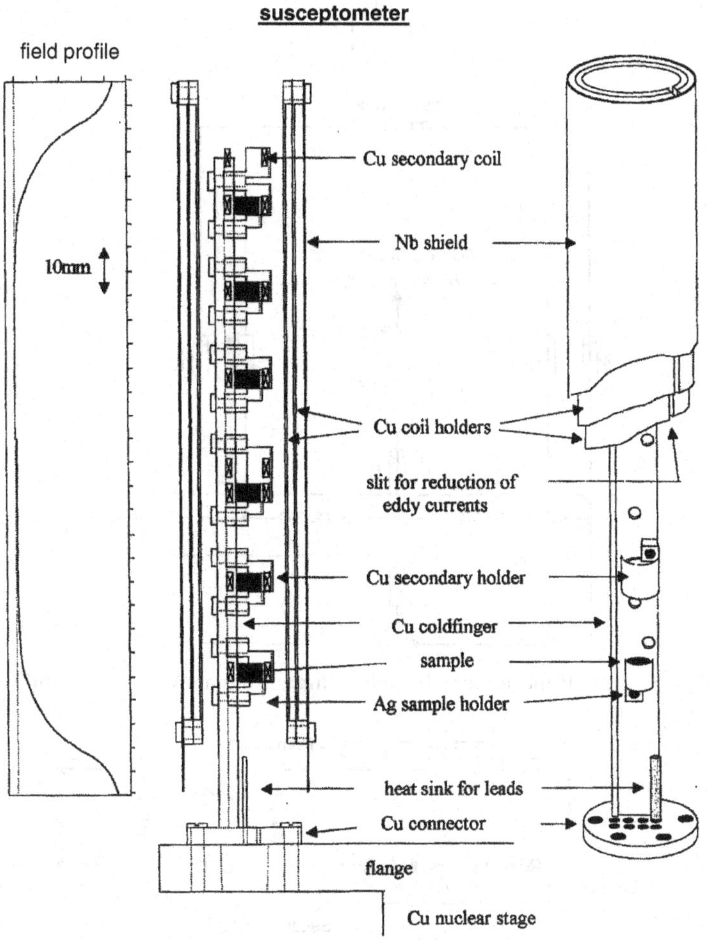

Fig. 3.31. Schematics of a susceptometer cooled by a Cu nuclear refrigeration stage (Sect. 10.8). It is designed for simultaneous investigation of six samples down to microkelvin temperatures. The two empty coils are used as a reference. In addition, the profile of the applied static field is shown; the field coil is surrounded by a superconducting Nb shield (Sect. 13.5.2), [3.122]

If an inductance measurement is performed, then the following relations are relevant. In the primary coil we generate a current $I = I_0 \sin(\omega t)$ which produces a change of magnetic flux

$$\frac{\mathrm{d}\phi_1}{\mathrm{d}t} = \frac{A_1(\mathrm{d}I/\mathrm{d}t)N_1^2}{L_1}, \tag{3.48}$$

where A_1 is the area, N_1 the number of windings and L_1 the length of the primary coil. This flux change induces a voltage in the secondary coil given by

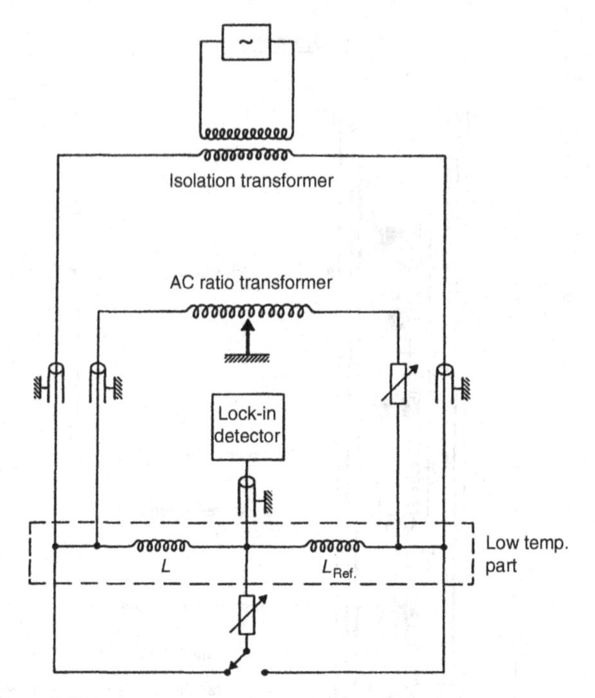

Fig. 3.32. Self-inductance bridge to measure magnetic susceptibilities

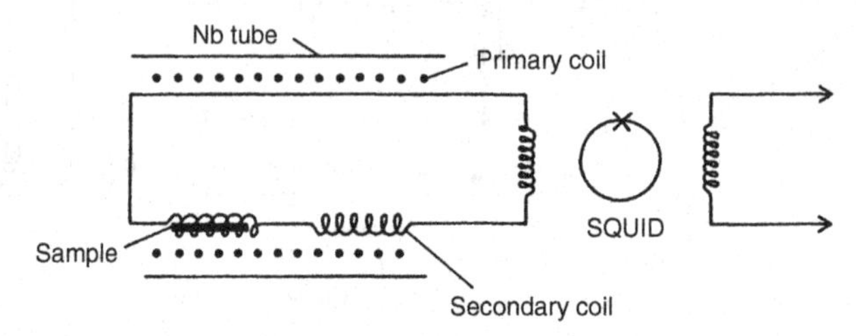

Fig. 3.33. Circuit to measure magnetic susceptibilities with a SQUID. The design uses an astatic pair of pick-up coils with the sample in one of them. Because of the high sensitivity of the SQUID a few milligrams of a paramagnetic material is sufficient. The astatic pair, which can be made with an accuracy of about 1%, reduces the sensitivity of the device to external disturbances and also compensates the influence of the magnetization of the construction materials. The principle of the design is to keep the total flux constant, which makes it necessary to make the low-temperature wiring from a superconducting material [3.45, 3.151]

$$U_2 = \mu_0(1+\chi)\frac{N_1 N_2 A_2 \omega I_0}{L_1}\cos(\omega t)\,, \qquad (3.49)$$

where A_2 is the area, N_2 the number of windings of the secondary coil, and χ is the temperature dependent susceptibility of the sample in the secondary coil.

The mutual inductance of two secondary coils sitting inside of a primary coil and with one of them containing a cylindrical sample of susceptility χ is [3.150]

$$M_\mathrm{s} = \pi \mu_0 n_\mathrm{p} n_\mathrm{s} r_\mathrm{s}^2 f_1 (1 - D + f_2) q \chi \tag{3.50}$$

with $n_{\mathrm{p(s)}}$ being the number of turns per cm of the primary (of one of the secondary) coil, r_s the radius of the sample, D the mean demagnetization factor (typically 1/3), and q the filling factor of the sample in one of the secondary coils (typically 1/2). The geometric factors f_i depend on the lengths of the sample, primary and secondary coils, and were given in [3.150]. A similar equation for a spherical sample in one of the secondary coils can be found in [3.151]. In all these measurements care must be taken to determine the isothermal ($\omega\tau \ll 1$) and not the adiabatic susceptibility.

All the described methods, of course, "see" not only the sample but all other magnetization in the neighbourhood as well hence one should avoid other magnetic materials rather carefully or choose a design so that their contributions are cancelled, for example, as is the case in a carefully wound astatic pair of secondary coils. For very high-sensitivity measurements one may even consider to regulate the temperature of these coils.

For all the discussed methods one should carefully choose a "non-magnetic" sample holder, like Suprasil glass (see Sect. 3.4.1), in a symmetric setup to reduce background contributions to the signal. In addition, in the analysis of the data, one may have to consider demagnetization effects and the contribution from the Weiss field (see Sect. 12.9).

Vibrating – Sample Magnetometers

An alternative method is the use of a vibrating-sample magnetometer. Here, a sample oscillates inside of a coil and the induced voltage is proportional to the magnetic moment of the moving sample. According to my knowledge, the most recent design of an automated vector vibrating-sample magnetometer can be found in [3.152]. It allows rotating the sample against the applied magnetic field so that the angular dependence of longitudinal, the transversal and the total magnetization can be measured. Using lock-in techniques, the sensitivity is 5×10^{-6} emu at 4.2–340 K in fields up to 2 T and at vibration frequencies of 42 Hz. The authors stress the vibration isolation and the detection system consisting of six coils.

Vibrating sample magnetometers are available commercially.

SQUID Magnetometers

With the advancement and the commercial availability of the SQUID, very sensitive SQUID magnetometers have become a quite common instrument to measure susceptibilities and magnetizations. In this device, a superconducting coil (made from 0.1 mm NbTi wire, for example) is wound around the sample to measure its magnetic moment. This coil and the input coil of the SQUID are parts of a closed superconducting loop acting as a flux transformer. The setup is surrounded by a superconducting magnet to provide a small magnetic field for the measurement. Any change in the permeability of the circuit from a change of the magnetic properties of the sample will result in a screening current to keep the total flux constant and will therefore produce a flux change in the input coil to the SQUID. This change can be produced by moving the sample in the pick-up coil – the conventional operation of commercial SQUID magnetometers – or by changing its magnetization due to a temperature change. In general, the pick-up coil is wound as a second-order gradiometer with the two outer loops wound oppositely to the two central loops to reduce spurious influences. Changes as small as 10^{-4} flux quanta can be measured. The part of the superconducting loop connecting the pick-up coil and the input coil of the SQUID should be tightly twisted and shielded by a superconducting capillary of Nb, for example, to reduce flux changes from external disturbances. A further useful precaution is a low-pass filter on the pick-up coil. Usually, the SQUID is used as a nullmeter to increase sensitivity and the dynamic range to several orders of magnitude as well as to avoid currents circulating in the coils. This is achieved by adding a feedback coil, which produces a compensating flux or voltage. This compensating voltage is measured as a function of position or temperature of the sample. A heater at one point of the superconducting loop can raise its temperature above T_c to eliminate unwanted trapped flux when preparing the setup for the next measurement.

A very good recent design of a home-built, fully automated SQUID magnetometer for operation in combination with a ^{3}He-^{4}He dilution refrigerator has been described in [3.153]. The refrigerator contains a specially designed plastic mixing chamber that allows the sample to be thermalized directly by the ^{3}He flow to 10 mK. The measuring equipment is based on a DC-SQUID coupled to a second-order gradiometer coil system. To measure a magnetic moment, the sample is moved through the gradiometer coils by smoothly lifting the whole dilution refrigerator. The noise of the setup in operation translates to a magnetic noise of 10^{-6} emu. The paper includes details on grounding and shielding as well as on minimizing eddy current effects. The relevant equations for converting the flux indicated by the SQUID and its electronics to the magnetic moment of the sample can be found there as well as in [3.45], for example.

Of course, if no special features, like ultra-low temperatures or very high magnetic fields, of the magnetometer are required and if the necessary funds are available, one should buy one of the commercially offered fully automatic, very flexible and powerful SQUID magnetometers or magnetic-property-measurement systems. They are offered for the temperature range of 1.8–400 K (optionally to 0.5 [3.154] and to 800 K) and for fields up to 7 T. These systems allow setting of temperature and field sequences as well as of particular operation modes, for example, for the motion of the sample and/or for the SQUID, and allow fully automated around-the-clock operation. In these systems, the SQUID output, i.e., the pick-up coil signal as a function of the samples position, is fitted to the expected theoretical response of a magnetic dipole. Eventually, this ideal response curve is compared to the systems calibration, and the sign and value of the samples magnetic moment are calculated. The volume or mass susceptibility can then be calculated from this moment. The resolution for magnetic moments reaches $1\times(5\times)10^{-8}$ emu at zero (7 T) field for moments up to 5 emu; optionally, moments up to 300 emu can be measured.

A quite common problem of these instruments has been discussed and investigated in [3.155,3.156]. It arises from the assumption in the analysis that the magnetic moment of the sample does not change during the measurement. However, in its movement through a non-homogeneous field, a superconducting sample, for example, will follow a hysteresis loop, resulting in a position-dependent magnetic moment of it. The moment calculated by the software of the magnetometer will then not represent the actual moment of the sample. An example of such a situation is shown in Fig. 3.34. The shown features arise from field inhomogeneities of the order of 0.1 mT only. This is well within the change of remanent fields of superconducting magnets over the scan length of the magnetometer, and is usually within the range of the field homogeneity

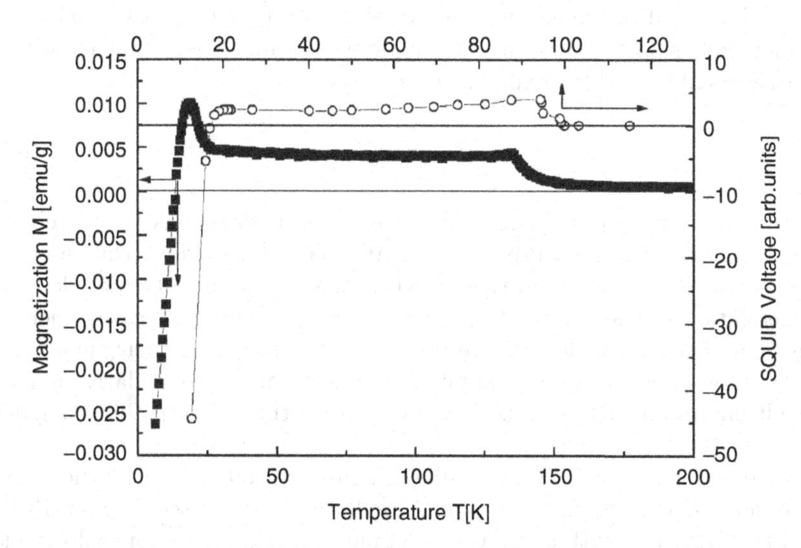

Fig. 3.34. Zero-field cooled DC magnetization measured in a commercial SQUID magnetometer by moving the sample $RuSr_2GdCu_2O_3$ in a field of 0.25 mT through the pick-up coil system of the magnetometer. The magnetization shows an apparent peak at temperatures below $T_c = 30$ K (■). When measuring the output voltage of the SQUID circuit as a function of temperature without moving the sample in the magnetometer no peak like feature is observed (○) [3.155]

claimed by the supplier of the instrument. Of course, the remanent field can change from one cycle to the next one of the magnet. The obtained apparent moment of the sample will then depend on the field profile and its sign as well as on the scan length in the magnetometer and the properties of the sample. A first check whether the mentioned problem occurs is an inspection of the symmetry of the output signal of the SQUID. However, sometimes the deformations of the signal can be rather small but still creating significant errors in the data. The most reliable test is to measure the magnetic moment of the sample without moving it, for example just as a function of temperature, if the magnetometer offers such an option.

Of course, SQUID magnetometers are in general not practical in experiments at very high magnetic fields. This is an area where the magnetometers to be discussed below have become of particular importance.

Micro-Torsional and Cantilever Magnetometers

In Sect. 13.10 I will discuss very sensitive torsional and translational devices that have been developed to mechanically measure a variety of properties of liquid and solid samples. Most of these devices can be used to measure the magnetic properties of a sample as well by just mounting (or evaporating) the sample onto the moving part and measuring the induced voltage inductively by a pick-up coil or capacitively by an electrode (see Fig. 13.19). The measurement relies on the fact that a sample with a magnetic moment m in an external magnetic field B experiences a torque

$$\tau_m = mB. \tag{3.51}$$

Hence, the torque τ in the equations of Sect. 13.10 has to be replaced by $\tau + \tau_m$. For the measurement, an AC magnetic field at the resonance frequency of the oscillator in the low-kHz range is applied at an angle to the magnetization of the sample to drive the oscillator. In most cases micro- or even nano-mechanical devices are used. The advantage of their small size – besides their sensitivity and simplicity – makes them particularly useful in difficult environments, such as very low temperatures or very high magnetic fields.

For example, in [3.157] a two-stage, high-Q silicon torque magnetometer was described that is just a torsional oscillator as discussed in Sect. 13.10.4 (see Fig. 13.19) but with a magnetic sample mounted or evaporated onto the oscillating head. The small torsional spring constant of the oscillator (typically 10^{-2} N m) combined with its high Q-value of typically 10^6 allow detection of magnetic moments as small as 10^{-10} emu (10^{-13} J T^{-1}). These magnetometers can either be used by measuring the deflection, where a sensitivity for displacement of less than 0.1 nm has been achieved [3.157], or by measuring, with an accuracy of 10^{-8}, the shift in resonant frequency due to the

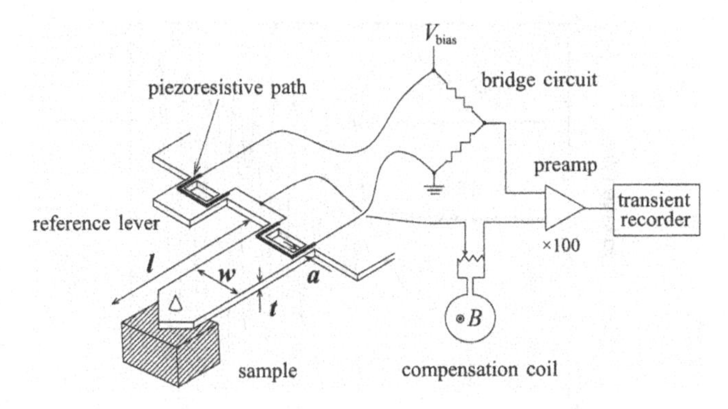

Fig. 3.35. Schematics of a torque magnetometer consisting of a commercial piezoresistive cantilever – to which a sample is connected – with a reference cantilever, a detecting bridge circuit, and a compensation coil. Reprinted with permission from [3.159], see this reference for details; copyright (2002), Am. Inst. Phys.

anisotropy of magnetization of a sample [3.158]. The latter magnetometer was designed for use in magnetic fields up to 25 T with an accuracy of 2×10^{-8} emu (2×10^{-11} J T^{-1}).

In [3.159], a miniature high-frequency torque magnetometer has been described, which has been successfully used to investigate the magnetic properties of sub-µg samples in pulsed (!) magnetic fields up to 38 T with pulse durations of 30 or 60 ms. This magnetometer (Fig. 3.35) consists of a small cantilever with the sample mounted on its free end. Sensitivities – limited by the mechanical noise from the pulsed magnet – of about 10^{-11} N m for the torque and 5×10^{-10} emu for the magnetic moment have been reached; 10^{-11} emu should be possible in a field of 10 T produced by a much more quiet superconducting magnet, which is higher than for a SQUID magnetometer. The authors used commercially available piezoresistive silicon microcantilevers (produced for atomic force microscopy) with eigenfrequencies of 250–300 kHz and spring constants of 30–40 N m^{-1}. These rather high frequencies strongly reduce the disturbances from externally induced vibrations, for example from pulsing the strong magnetic field; a longer cantilever with about seven times lower eigenfrequency did behave considerably worse, for example. The mass of the sample should be small enough to keep the eigenfrequency of the setup higher than the frequency of the signal. A reference cantilever is used to cancel background signals from the temperature and field dependencies of the cantilever (Fig. 3.35). A compensation coil is used to eliminate $\mathrm{d}B/\mathrm{d}t$ signals caused by the rapid field sweep. A result of a measurement using this magnetometer is shown in Fig. 3.36. Of course, the cantilever can also be used as a microbalance to determine the mass of the usually quite small samples by just measuring the shift of the eigenfrequency when the sample is mounted on it (see Sect. 13.10.4).

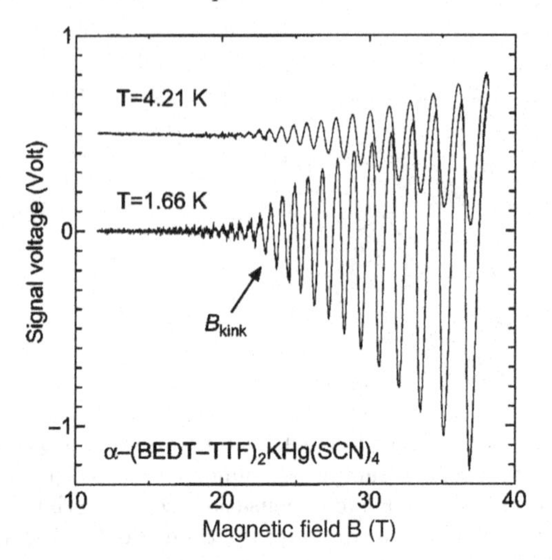

Fig. 3.36. Signal voltage proportional to the magnetic torque of 10 μg of the indicated quasi-two-dimensional organic conductor at two temperatures as a function of magnetic field. The measurements have been performed with the piezoelectric torque magnetometer shown in Fig. 3.35. The oscillations in magnetization – periodic in inverse magnetic field – are due to the de Haas–van Alphen effect resulting from the discrete Landau levels of the electronic energy in an external magnetic field. At low temperatures the compound undergoes a transition ("kink transition") from a charge density-wave to a metallic state at 23 T. Reprinted with permission from [3.159]; copyright (2002), Am. Inst. Phys.

In [3.157–3.159] one can find references to former work using microcantilever magnetometers – usually for much lower frequencies than just discussed – and in particular for applications in steady magnetic fields.

Problems

3.1. Verify that (3.9) is the valid result in the approximation $T < \Theta_D/10$.

3.2. Calculate the Debye temperature of solid Argon (see Fig. 3.2b).

3.3. The atomic weight of Cu is 63.55 and its density is 8.93 g cm^{-3}. Calculate its Fermi temperature.

3.4. Calculate the Fermi temperature of liquid ^{3}He.

3.5. Under which condition does the Fermi–Dirac distribution function (3.11) change to the Maxwell–Boltzmann distribution function?

3.6. At which temperature do the lattice and conduction–electron contributions to the specific heat of copper become equal?

3.7. At which temperature is the magnetic contribution (3.20) in a field of 100 mT equal to the conduction–electron contribution to the specific heat of Ag [for material parameters of Ag, see Table 10.1; use (10.3, 10.4)]?

3.8. Calculate the temperature at which the specific heat of aluminum in its superconducting and its normal state become equal, see Fig. 3.5 a,c, and Table 10.1.

3.9. Calculate the wall thickness of a $Cu_{0.7}Ni_{0.3}$ tube of 6 mm diameter, which has the same thermal conductivity as a 10 mm diameter Teflon rod at 1 K (Table 3.2).

3.10. What should be the inner diameter of a $Cu_{0.7}Ni_{0.3}$ capillary of 1 mm wall thickness filled with solid ^{4}He at 0.3 K so that its thermal conductivity is equal to that of the solid ^{4}He in it (Fig. 3.21)?

3.11. Calculate the temperature at the "hot" end of a cylindrical rod of 20 mm length and 2 mm diameter, if it is heated with 0.1 nW and the cold end is kept at 0 K. Carry out the calculation for a plastic, for brass and for Cu (see Fig. 3.21 for thermal conductivities).

3.12. At which temperature are the thermal conductivities of Cu($\propto T$) and of liquid ^{3}He at SVP ($\propto T^{-1}$) become equal, see Figs. 2.18 and 3.21?

4

Thermal Contact and Thermal Isolation

In any low-temperature apparatus it is necessary to couple some parts thermally very well, whereas other parts have to be well isolated from each other and, in particular, from ambient temperature. The transfer of "heat" (or better "cold") and the thermal isolation are essential considerations when designing a low-temperature apparatus. These problems become progressively more acute at lower temperatures, and they will be discussed in this chapter. A general treatment of the thermal conductivity of materials is given in Sect. 3.3. Besides learning how to take advantage of the very different thermal conductivities of various materials, we have to discuss how to design an apparatus to achieve the desired goals. For example, there are situations, as in low-temperature calorimetry, where two substances have to be in good thermal contact and then have to be very well thermally isolated from each other for the remainder of the experiment. For this purpose we need a thermal switch and I shall place special emphasis on the discussion of superconducting heat switches which dominate the temperature range below 1 K. One of the severest problems in low-temperature technology is the thermal boundary resistance between different materials. This is a particularly severe problem if good thermal contact between liquid helium and a solid is required, as it will be discussed in the final sections of the present chapter.

The nuisance heat transfer by conduction and radiation will be considered at the beginning of Chap. 5. The various heat sources will be treated in Sects. 5.1.2 and 10.5, in connection with refrigeration to extremely low temperatures, where they can be particularly detrimental.

4.1 Selection of the Material with the Appropriate Cryogenic Thermal Conductivity

The low-temperature thermal conductivity of different materials can differ by many orders of magnitude and, fortunately, the thermal conductivity of the same material can even be varied by a great amount just by changing the number of defects or impurities in it (Fig. 3.19). Hence, one has to be careful in selecting the right material for a low-temperature apparatus. The low-temperature thermal conductivities of various materials commonly in use in low-temperature equipment are exhibited in Figs. 3.19–3.24 and are listed in Table 3.2.

For *good thermal conductivity* the right choices are Cu (but: soft; nuclear specific heat at $T < 0.1\,\mathrm{K}$ (Sect. 3.1.6 and Fig. 3.11), Ag (but: soft; expensive) or Al (but: soft, superconducting below 1 K; soldering only possible in an elaborate process, see Sect. 4.2.2). The highest practical conductivities of these metals are $\kappa \approx 10\,T$ [$\mathrm{W\,K^{-1}\,cm^{-1}}$] if they are very pure; more typical is $\kappa \approx T$ [$\mathrm{W\,K^{-1}\,cm^{-1}}$].

For *thermal isolation* the right choices are either plastics (Teflon, Nylon, Vespel, PMMA, etc.), graphite (careful, there exists a wide variety), Al_2O_3, or thin-walled tubing from stainless steel (but: soldering or silver brazing only with aggressive flux, which should be washed off very thoroughly; better is welding) or from $Cu_{0.7}Ni_{0.3}$ (easy to solder). However, the last two can be slightly magnetic at low temperatures (see Sect. 3.4.1) and can interfere with sensitive magnetic experiments. In general, glasses or materials composed of small crystallites (for phonon scattering) and containing a lot of defects and impurities (for electron scattering) are good thermal insulators. For example, the thermal conductivity of quartz glass at 1 K is only about 1% of that of crystalline quartz at the same temperature. The lowest thermal conductivity, $\kappa \cong 5 \times 10^{-6} T^{1.8}$ [$\mathrm{W\,K^{-1}\,cm^{-1}}$] has been observed for AGOT nuclear graphite (Figs. 3.21 and 3.23).

If other properties do not matter too much, aluminium alloys or brass should be used because of their relatively low prices and, above all, because they can be easily machined. If tubes are employed which are filled with liquid ^{3}He, ^{4}He or an isotopic helium mixture, then the conductivity of the helium – which may be rather large (Sects. 2.3.4 and 2.3.6 and Figs. 2.12, 2.17, and 2.18) – usually dominates. The effect can be reduced by using capillaries with a small diameter to reduce the mean free path of the liquid's atoms.

In each low-temperature apparatus one needs *wires* to carry signals from room temperature to the low-temperature part, and back. For low-current leads thin Constantan ($\rho_{300\,\mathrm{K}} = 52.5\,\mu\Omega$ cm, $\rho_{4\,\mathrm{K}} = 44\,\mu\Omega$ cm) or Manganin ($\rho_{300\,\mathrm{K}} = 48\,\mu\Omega$ cm, $\rho_{4\,\mathrm{K}} = 43\,\mu\Omega$ cm) wires should be used because of their low thermal conductivity and the small temperature dependence of their electrical resistivity. However, one has to take into account the increase of their electrical resistance and of their specific heat due to magnetic contributions at $T < 1\,\mathrm{K}$ (Fig. 3.9). This effect is smaller for PtW which has become a favorite heater wire for very-low-temperature applications; here the increase in specific heat due to a minute amount of magnetic impurities starts only below 0.1 K (Fig. 3.9). If large electrical currents – for example, for superconducting magnets – have to be carried to the low-temperature part, the advantage gained by using a good conductor and a large wire diameter to reduce Joule heating, and the disadvantage of the then increased thermal conductivity have to be carefully considered. Often one may end up using Cu wires. Of course, then a proper heat sinking of the wires at various places on their way to low temperatures is of even greater importance. The optimum dimensions of leads carrying large currents into a cryostat were discussed in [4.1–4.6] (see also Sect. 13.4). At $T < 1\,\mathrm{K}$ the use of superconducting wires with their vanishing thermal conductivity for $T \to 0$ (Sects. 3.3.4, 4.2.2) is the right choice. Often it is adequate just to cover a Manganin or Constantan wire with a thin layer of superconducting solder ($T_c \simeq 7\,\mathrm{K}$, see later) to have a lead with low thermal conductivity but zero electrical resistance. At $T < 0.1\,\mathrm{K}$ one can use monofilamentary NbTi without Cu or even without a CuNi matrix if the lowest possible thermal conductivity is required. In extreme cases the fine filaments of multifilament superconducting wires can be used, but soft soldering to these wires is not possible, so spot-welding or squeeze contacts are necessary [4.6,4.7]; these techniques require some practice before they can be applied reliably. If the joint does not have to be superconducting then one can remove the Cu coating with concentrated HNO_3 except at the ends of the wire where the solder joints have to be made. An even better method is to coat the superconducting wire electrolytically with a layer of Cu (use $1\,l$ H_2O with at least 200 g $CuSO_4$, $27\,\mathrm{cm}^3$ H_2SO_4, and a current density of about $40\,\mathrm{mA\,mm}^{-2}$ between NbTi cathode and Cu anode). The same electrolytic process can be applied to cover stainless-steel tubing with a Cu layer to make the subsequent soldering easier. Phillip et al. [4.8] have described two techniques for joining multifilamentary superconducting NbTi wires. These joints have achieved critical currents in vacuum at 4.2 K comparable to the short segment ratio given by the manufacturer for the wires.

Wires for mesurements of small signals have to be twisted pairwise on their way in the cryostat, rigidly fixed and well shielded to keep pick-up signals low (Sect. 12.5). The design of coaxial cryogenic cables and the proper heat-sinking of leads will be discussed in Sect. 13.3.

4.2 Heat Switches

4.2.1 Gaseous and Mechanical Heat Switches

The simplest way to thermally connect and disconnect various parts of a low-temperature apparatus is to use a gas (at such a pressure that it does not condense at the temperatures involved) for thermal coupling and then remove it by pumping. This method is often employed in precooling the inner parts of a cryostat to LN_2 or LHe temperatures (Chap. 5). A gas pressure of 10^{-4} bar is sufficient for an adequate heat transfer. But usually many hours of pumping are then required to reduce the gas pressure for sufficient thermal isolation. The temperature at every place in the cryostat has to be above the condensation temperature of the gas so that efficient pumping is possible. If the exchange gas has not been pumped to a low enough pressure, time-dependent heat leaks due to a continuing desorption and condensation of the remaining gas at the coldest surfaces may result (Sects. 5.1.2 and 10.5.3).

For ^{4}He, the superfluid film contributes to the heat transfer, too. For thermal isolation, ^{4}He has to be pumped very well to make sure that there is not enough of it left to form an unsaturated superfluid film if $T < 2.2\,\mathrm{K}$ (Sect. 2.3.5). The advantage of H_2 as an exchange gas is the fact that it can be totally condensed out ("cryo-pumping") when liquid helium is transferred into the cryostat, so that time-consuming pumping can be avoided. However, one has to remember that the remaining H_2 molecules may undergo ortho–para conversion, giving rise to substantial heating (Sect. 2.2). As a conclusion, ^{3}He with its high vapour pressure, absence of exothermic reaction, and absence of superfluidity in the Kelvin temperature range, is the safest exchange gas for thermal contact.

Heat switches using liquid ^{4}He or ^{3}He have been described in [4.9–4.11]. The sealed ^{3}He gas heat switch of [4.11] uses a small charcoal pump to adsorb or desorb ^{3}He to create the gas pressure for switching between the "on" and "off" states. The thermal conductivity then changes by more than two orders of magnitude (from below 0.1 mW/K to several 10 mW/K) in the temperature range 0.5–3 K for the chosen design.

For many purposes, for example for calorimetry at $T > 1\,\mathrm{K}$, a mechanical heat switch is adequate. Thermal contact is made by metallic, usually gold-plated contacts pressed together mechanically (see below). Here the "open" state really is open, with no residual heat flow. Conductances of $1\,\mathrm{mW\,K^{-1}}$ in the low Kelvin range are typical for the closed state (Sect. 4.3.1) [4.1, 4.12–4.17]. However, a value of $1\,\mathrm{W\,K^{-1}}$ at 15 K has been reported in [4.18]. The main disadvantages of these switches are the large forces (typically 100 N) necessary to make adequate thermal contact and the heat generated when the contact is broken (typically $0.1\text{–}1\,\mu\mathrm{J\,N^{-1}}$). I will not discuss these switches in more detail here because they are being used less and less these days. Readers interested in mechanical thermal switches should consult the literature [4.1, 4.12–4.18].

4.2.2 Superconducting Heat Switches

In Sect. 3.3.4 we concluded that the thermal conductivity κ_s of a metal in the superconducting state can become very small because the number of electrons decreases exponentially with temperature; it can be orders of magnitude smaller than the thermal conductivity κ_n of the same material in the normal state (Fig. 4.1). Because some metals can easily be switched from the superconducting to the normal state by applying a magnetic field, we can build a "superconducting heat switch" as already mentioned in Sect. 3.3.4. Superconducting heat switches are the most common thermal switches at temperatures below about 1 K. Their advantages are that the heat flow in the open state is small, that they are very easy to switch, and that the switching ratio κ_n/κ_s can be very large; but for that we need $T < T_c/10$, which often means $T \ll 1\,\mathrm{K}$ (Fig. 4.2). Very little heat is generated in the switching process if the design

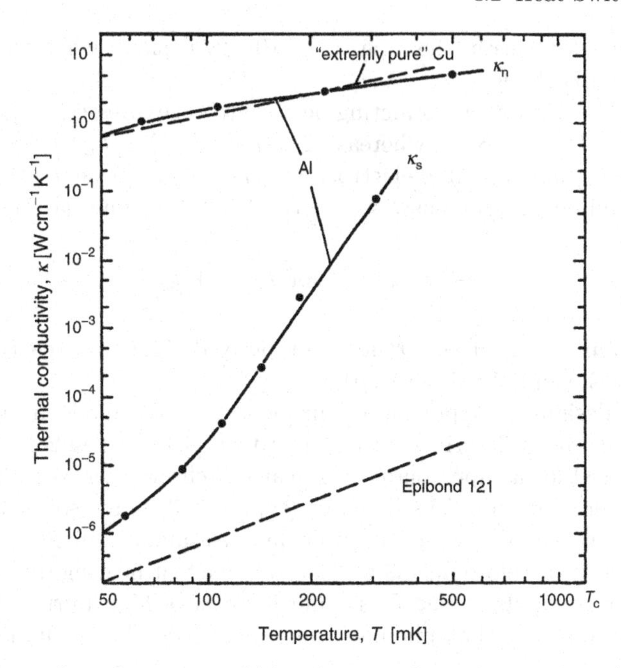

Fig. 4.1. Thermal conductivity κ of Al in the normal–conducting state (compared to κ of Cu) and in the superconducting state (compared to κ of the dielectric Epibond 121) at $T < 50$ mK [4.19]

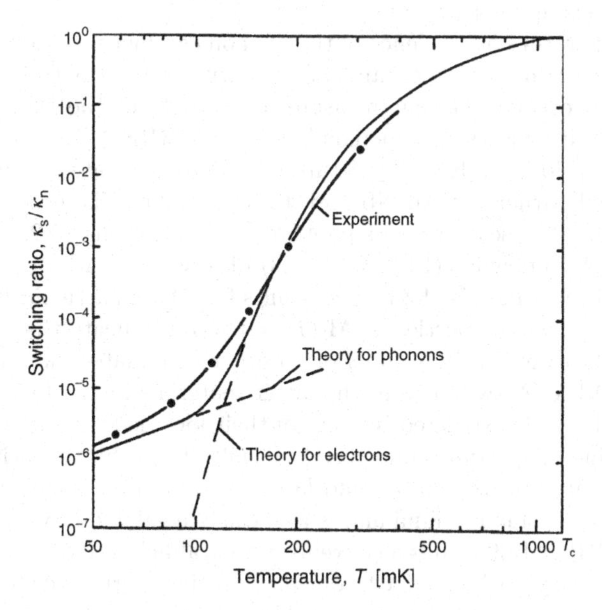

Fig. 4.2. Switching ratio κ_s/κ_n of the thermal conductivity of Al compared to theoretical predictions [4.19]

ensures that eddy-current heating (Sect. 10.5.2) is small when the switching magnetic field is changing.

The quality of a superconducting heat switch is expressed by its switching ratio κ_n/κ_s. Here $\kappa_n \propto T$ whereas $\kappa_s \propto T \exp(-\Delta E/k_B T)$ for $T > T_c/10$ (from the remaining unpaired electrons) and $\kappa_s \propto (T/\theta_D)^3$ at $T < T_c/10$ (from the now dominating phonons), see Sect. 3.3.4. Hence one has the switching ratio

$$\frac{\kappa_n}{\kappa_s} = aT^{-2} \quad \text{for } T < 0.1\ T_c, \tag{4.1}$$

with a constant a of $10^2 - 10^3$ for a properly designed switch (but see the comment at the end of this section).

Various designs of superconducting heat switches made from a variety of metals have been described in the literature [4.12, 4.19–4.28]. High-purity metals are used for a superconducting heat switch in order to make κ_n large. One should employ thin foils or wires (typically 0.1 mm) so that the mean free path of phonons – which is given by the sample dimensions for pure materials – and therefore κ_{ph} as well as eddy current heating during the field change become small. If $T < T_c/10$ and if we have high purity thin foils of a superconductor (so that phonons are only scattered at boundaries), then ideally

$$\frac{\kappa_n}{\kappa_s} \simeq 0.05 \left(\frac{\theta}{T}\right)^2, \tag{4.2}$$

for Al as an example [4.19].

The temperature dependence of the phonon conductivity (3.28) of a metal in the superconducting state and of the switching ratio (4.1, 4.2) should be used with caution. Recent measurements [4.28] of the thermal conductivity of massive pieces of superconducting Al (RRR $\geq 5,000$) have shown that $\kappa_{ph} \propto T^2$ for $10\,\text{mK} \leq T \leq 80\,\text{mK}$ (Fig. 4.3). Deviation from $\kappa_{ph} \propto T^3$ were reported earlier for Al, Nb and Ta [4.29, 4.30]. These deviations may be attributed to a scattering of phonons at dislocations (3.29) [4.31, 4.32], or to the glassy behavior (Figs. 3.21–3.23) observed recently for the acoustic properties of these metals [4.33]. The results for Al are particularly disturbing because the known properties of Al ($T_c = 1.18\,\text{K}$) indicate that for a superconducting heat switch it is superior to other candidates like Sn, In, Zn or Pb at $T \leq 0.1\,\text{K}$. A switch from the latter materials in the form of wires or foils can easily be constructed because of their low melting temperatures and good soldering properties. However, Al usually has a higher switching ratio (Fig. 4.2) because of its high κ_n and large Debye temperature ($\theta_D = 400\,\text{K}$), which makes κ_{ph} small. Aluminum is also easily available in very high purity (5 or 6 N; RRR > 1,000), has a convenient critical field (10.5 mT), good durability and is easy to handle. Of course, there is a serious contact problem due to the tenacious surface oxide on aluminum; various ways of solving this problem have been described in the literature. For a successful but elaborate

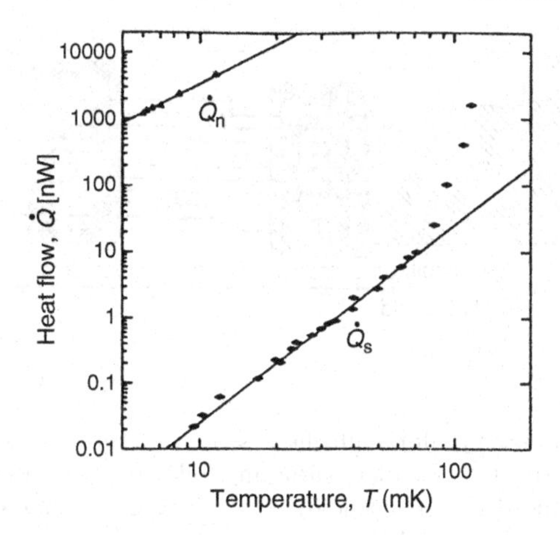

Fig. 4.3. Heat flow $\dot{Q}$ across an Al heat switch in the normal (▲) and superconducting (•) states. The lines correspond to $\dot{Q}_n \propto T^{-2}$ and $\dot{Q}_s \propto T^{-3}$. The thermal conductivity is $\kappa \propto (\dot{Q}T)^{-1}$, resulting in $\kappa_n \propto T$ and $\kappa_s \propto T^2$ at $T < 80$ mK [4.28]

electroplating process see [4.6, 4.19], and procedures for welding Al to Cu or Ag have been described in [4.22–4.25]. In our laboratory recently contact resistances of $\leq 0.1\,\mu\Omega$ were achieved at 4.2 K by screwing well-annealed Ag screws into Al threads. The thread diameter had to be at least 6 mm in order for the necessary forces to be applied to the soft, annealed metals. Because of its larger thermal expansion coefficient the Al will shrink onto the Ag in the cooldown process (Fig. 3.17). Of course, for such a design high-temperature annealing does not help. The metals were chemically cleaned (HNO_3 for Ag; $22\text{g}\,l^{-1}Na_3PO_4 + 22\text{g}l^{-1}Na_2CO_3$ at $T \geq 75°$C, for A; several minutes), and annealed (5×10^{-5} mbar; 5 h; 800°C for Ag, 500°C for Al). The best results for cold welding of Al to Cu are $\rho < 1\,\mu\Omega$ and RRR ≥ 150 for a pressure of $P \geq 100\,\text{N}\,\text{mm}^{-2}$ applied at 500°C.

Often the behavior of a superconducting heat switch is deteriorated by frozen-in magnetic flux from the switching field, which may cause parts of the metal to remain in the normal state when the field is removed. This problem can be avoided by orientation of at least part of the metal perpendicular to the field (Fig. 4.4), so that the normal cores of trapped flux lines will not short-circuit the switch material, and/or by saw-tooth-like cycling of the field during its reduction. If a superconducting heat switch is used in a magnetic refrigerator (Chaps. 9 and 10), then in many cases a superconducting Nb shield should be placed around it to shield the switch from the changing fringe field of the demagnetization solenoid. The shape of the switch should be such that eddy current heating (10.28) produced while the switching field is changing will not cause heating effects, and that closed superconducting rings trapping flux are avoided (put slits perpendicular to the field into bulk Al) [4.28].

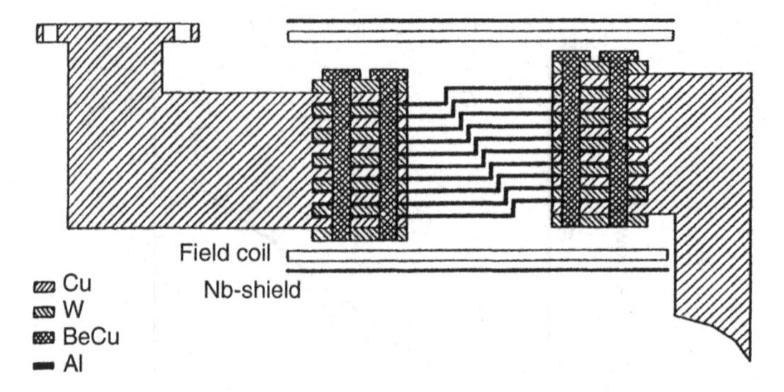

Fig. 4.4. Superconducting aluminum heat switch of the refrigerator in Bayreuth [4.34], see also Fig. 10.14. The superconducting foils are partly perpendicular to the magnetic field produced by the solenoid, thus avoiding that magnetic flux lines are able to penetrate into the switching foil along its entire length

4.3 Thermal Boundary Resistance

4.3.1 Boundary Resistance Between Metals

To achieve thermal equilibrium in a system becomes increasingly more difficult when the temperature is lowered, not only because the thermal conductivity of materials decreases with decreasing temperature but also because the thermal boundary resistance at the interface between two materials becomes increasingly important. If we have two different materials in contact and heat $\dot{Q}$ has to flow from one material to the other, for example in a cooling process, there will be a temperature step at the boundary between them. This temperature step is given by

$$\Delta T = R_{\mathrm{K}}\dot{Q}, \tag{4.3}$$

where R_{K} is the thermal boundary resistance, or Kapitza resistance, named after the Russian physicist P. Kapitza who discovered this thermal boundary resistance in 1941 for the case of liquid helium in contact with solids. This is still a problem which is not fully understood, at least for very low temperatures (Sect. 4.3.2). The boundary resistances between several materials are shown in Fig. 4.5.

Between metals the actual contact area often is only about 10^{-6} of the nominal contact area due to the microscopic irregularities of the opposing surfaces; the conductance therefore does not scale with the nominal contact area. The actual contact area can be considerably increased by the application of pressure close to the yielding stress of the materials. The thermal conductance across the boundary between the two metals is often proportional to the applied force used to press them together. The disadvantage of this procedure is a deformation of the lattice with a reduction in bulk conductivity. This problem is reduced by joining surfaces via diffusion welding because it uses high temperatures ($0.6\,T_{\mathrm{melting}}$, for example) annealing lattice defects.

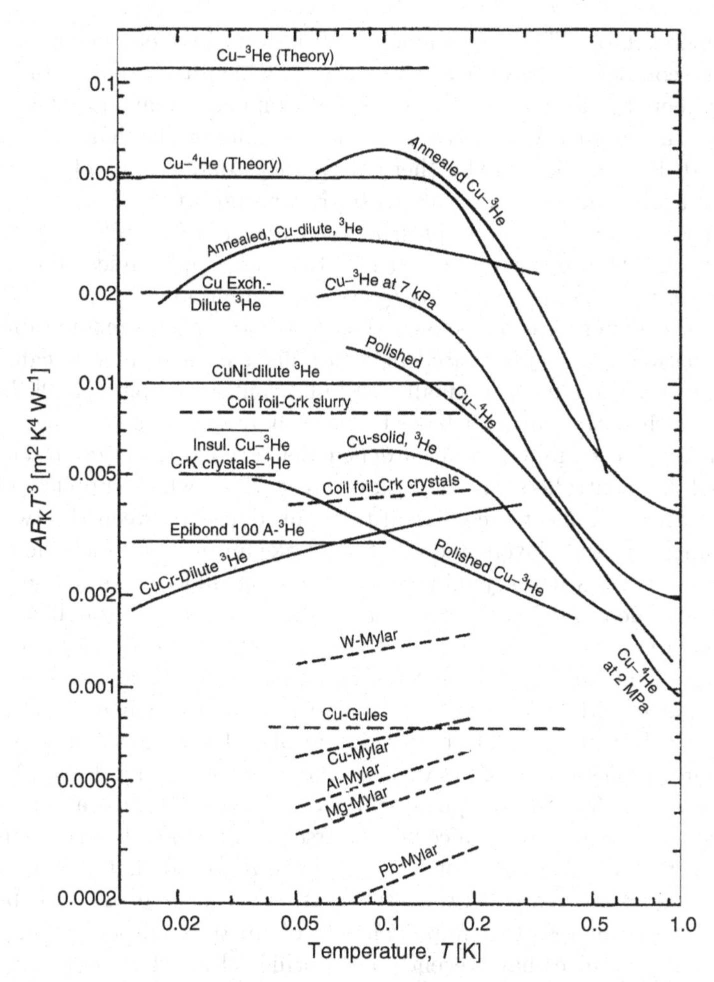

Fig. 4.5. Thermal boundary resistance R_K multiplied by AT^3 between liquid helium and various solids, and also between various metals and various dielectrics [4.12]. (This book provides references to the original literature); for more recent helium data see Fig. 4.9

The boundary resistance can be kept reasonably small, if the surfaces are clean, possibly gold-plated, and pressed together with a high force. We should then have an overlap of the electronic wave functions of the two metals, giving a good electric and thermal flow between them.

The heat transfer across the contact between two metals – similar or dissimilar – is a common problem in cryogenics. No unique solution can be found in the literature, even though it is full of recipes [4.1, 4.12, 4.35, 4.37]. In Sect. 3.2, I mentioned how important it is to correctly join two dissimilar materials, if good thermal contact is the goal. Predicting values for the conductance across a real joint is in general not possible; one has to resort to experimental data. The largest amount of data on low-temperature contacts relates to contacts between two Cu surfaces, which are usually gold plated. Gold prevents the formation of an oxide layer on Cu; it may also play a role in enlarging the contact area because it deforms more easily than Cu. In [4.35], a detailed investigation of the thermal contact conductance of several gold plated Cu joints measured at about 0.4 K was reported. They were bolted together or clamped with torques between about 50 to 150 N cm. Most of the data fell into the range of 0.1–0.2 W/K at 1 K, and a linear temperature dependence was found, confirming electronic conductance.

Our own experience has shown that a well-designed demountable press contact between two gold plated or well-polished, clean metals can have a thermal resistance almost as small as a bulk, continuous part [4.19]. In order to achieve this, the surfaces have to be well prepared and strong enough bolts, made, for example, from hardened BeCu, have to be used. These are tightened in a controlled way until they almost yield, which supplies sufficient force for what is almost a cold weld to be produced between the two parts. This can rip up oxide layers and can then make an intimate metallic contact. Sometimes it helps if the joining surfaces are sprinkled with a fine soft Ag powder. As mentioned in Sect. 3.2, a washer with a very small expansion coefficient (e.g., W or Mo) improves the contact after cool down by taking advantage of the different thermal contractions. Rather good results have also been obtained with a joint of two tapered metals. Extremely small contact resistances of 10 nΩ at 4.2 K between gold-plated Cu discs bolted together with 4 mm stainless-steel screws with a tightening torque of at least 4 Nm have been reported [4.36] (more typical, still good values are 0.1 μΩ at 4 K, [4.37]). The measured contact resistance was inversely proportional to the tightening torque on the screw. Salerno et al. [4.38] have reported that the addition of In foil or apiezon grease between the contact surfaces of Cu, Al, brass or stainless steel can result in improvements of up to an order of magnitude. Very small resistances have recently been achieved for electron-beam welded Cu–Ag joints with values of 0.2 μΩ mm^2 at 4.2 K [4.39].

The mechanical and electrical contact between two metals often is made by soldering them together. Unfortunately, most solders, in particular soft solders, become superconducting at low temperature (Table 4.1) and eventually behave like a dielectric with regard to thermal conductivity [4.1, 4.40–4.45] (Fig. 4.1); then one could just as well use a dielectric glue! Of course, this problem can be avoided if the solder joint can be exposed to a magnetic field high enough to suppress the superconducting state. Here, Ga with a melting temperature of 30°C and a critical field of only 2 mT ($T_c = 1.1$ K) may be quite appropriate [4.46]. In some cases the low superconducting thermal conductivity may not reappear even after removal of the field because magnetic flux may be trapped in the solder, keeping enough of it in the normal state to provide good thermal contact. The transition temperatures to the superconducting state and melting temperatures of various solder alloys are given in Table 4.1. In some cases solder joints with non-superconducting Bi may be

Table 4.1. Melting temperatures T_m and superconducting transition temperatures T_c of some solders [4.1, 4.40–4.46]

solder	T_m [°C]	T_c [K]
12–14% Sn, 25–27% Pb, 50% Bi, 10–13% Cd (Wood's metal)	70	8–9
50–52% In, 50–48% Sn	120	7.1–7.5
30–60% Sn, 70–40% Pb	257–183	7.1–7.8
97% Sn, 3% Ag	240	3.7
95.5% Sn, 3.5% Ag, 1% Cd	220	3.05
26% Sn, 54% Bi, 20% Cd	103	3.7
43% Sn, 57% Bi	140	2.25
82.5% Cd, 17.5% Zn	265	1–1.6
70% Au, 30% Sn (eutectic)	280	1.17
Ga	30	1.1
60% Bi, 40% Cd		<0.8
40% Ag, 19% Cu, 20% Cd, 21% Zn	610	<0.064
56% Ag, 22% Cu, 17% Zn, 5% Sn	650	<0.064
60% Ag, 30% Cu, 10% Sn	(700)	<0.057
50% Ag, 15.5% Cu, 16.5% Sn, 18% Cd	630	<0.057

appropriate [4.44]. These problems can be avoided by welding or using non-superconducting hard solders, but sometimes the alloy that is produced leads to a rather high thermal resistance at the interface between the two connected metals.

4.3.2 Boundary Resistance Between Liquid Helium and Solids

Acoustic Mismatch

Between dielectrics, for example a nonmagnetic dielectric in contact with liquid or solid helium, the transfer of energy can only occur via phonon transmission. We then have to match the acoustic properties of the two materials to optimize the transmission of phonons from one material to the other. The temperature step ΔT at the interface arises from the acoustic mismatch [4.12, 4.47–4.50] of the two materials, which I will treat in analogy to optics. In the following I will consider the case of transfering heat from liquid helium to another body with which it is in contact, because this is the most important case in low-temperature physics. For helium/solid interfaces the situation is particularly grave because acoustic impedances are $\rho_s v_s \approx 10^6\,\mathrm{g\,(cm^2\,s)^{-1}}$ for solids but $\rho_h v_h \approx 10^3\,\mathrm{g\,(cm^2\,s)^{-1}}$ for liquid helium. The importance of this heat transfer for low-temperature physics arises first of all from the fact that, except for magnetic refrigeration (Chaps. 9 and 10), all low-temperature refrigeration methods use helium as the working substance (Chaps. 5–8). Therefore, the cold produced by changing the thermodynamic state of helium has to be transferred to solid bodies to be useful.

Second, helium itself is a material of high scientific interest and in order to refrigerate it to the lowest possible temperatures by magnetic refrigeration, cold has to be transferred to it from a solid body, and its temperature has to be measured by a thermometer in intimate thermal contact with it. As a result, the thermal boundary resistance or energy transfer between liquid helium and solids is a matter of concern in a majority of low-temperature experiments, and it is a very interesting piece of physics in its own right.

If the velocity of phonons in helium is v_{h} and that in the solid is v_{s}, we have *Snell's law*

$$\frac{\sin \alpha_{\mathrm{h}}}{\sin \alpha_{\mathrm{s}}} = \frac{v_{\mathrm{h}}}{v_{\mathrm{s}}} \tag{4.4}$$

for the angles α at which the phonons cross the boundary. Because $v_{\mathrm{h}} \approx 238\,\mathrm{m\,s^{-1}}$ for ^{4}He at $T \lesssim 1\,\mathrm{K}$ whereas $v_{\mathrm{s}} \approx 3{,}000$ to $5{,}000\,\mathrm{m\,s^{-1}}$ for metals, the critical angle of incidence at which phonons from helium may enter the solid is very small,

$$\alpha_{\mathrm{crit}} = \arcsin\left(\frac{v_{\mathrm{h}}}{v_{\mathrm{s}}}\right) \approx 3^\circ. \tag{4.5}$$

The fraction of phonons hitting the interface that fall into the critical cone is

$$f = \frac{\pi \sin^2(\alpha_{\mathrm{crit}})}{2\pi} = \frac{1}{2}\left(\frac{v_{\mathrm{h}}}{v_{\mathrm{s}}}\right)^2 \approx 2 \times 10^{-3}. \tag{4.6}$$

However, because of the difference in acoustic impedance $Z = \rho v$, not even all of these phonons are transmitted. The energy transmission coefficient under the assumption of perpendicular incidence of the phonons on the interface (which is well fulfilled because $\alpha_{\mathrm{int}} \approx 3^\circ$) is given by (with $Z_{\mathrm{s}} \gg Z_{\mathrm{h}}$)

$$t = \frac{4 Z_{\mathrm{h}} Z_{\mathrm{s}}}{(Z_{\mathrm{h}} + Z_{\mathrm{s}})^2} \simeq \frac{4 \rho_{\mathrm{h}} v_{\mathrm{h}}}{\rho_{\mathrm{s}} v_{\mathrm{s}}} \simeq 3 \times 10^3. \tag{4.7}$$

Therefore only a fraction

$$ft = 2\frac{\rho_{\mathrm{h}} v_{\mathrm{h}}^3}{\rho_{\mathrm{s}} v_{\mathrm{s}}^3} < 10^{-5} \tag{4.8}$$

of the phonons will enter the solid; hence the two bodies are rather well isolated from each other. The combination of acoustic mismatch and a small critical angle severely limits the energy exchange between helium and other materials.

The transmitted energy flux $\dot{Q}$ of phonons impinging on the contact area A per unit time is given by

$$\frac{\dot{Q}}{A} = \frac{\pi^2 k_{\mathrm{B}}^4 T^4 \rho_{\mathrm{h}} v_{\mathrm{h}}}{30 \hbar^3 \rho_{\mathrm{s}} v_{\mathrm{s}}^3}. \tag{4.9}$$

The boundary resistance is then (for $\Delta T \ll T$)

$$R_{\mathrm{K}} = \frac{\Delta T}{\dot{\mathrm{Q}}} = \frac{\mathrm{d}T}{\mathrm{d}\dot{\mathrm{Q}}} = \frac{15 \hbar^3 \rho_{\mathrm{s}} v_{\mathrm{s}}^3}{2\pi^2 k_{\mathrm{B}}^4 T^3 A \rho_{\mathrm{h}} v_{\mathrm{h}}}. \tag{4.10}$$

In all the above equations, v_{s} is the transverse sound velocity.

A more rigorous consideration of the problem introduces corrections of order 2 depending on the materials properties. But the essential result is $R_K \propto (AT^3)^{-1}$; the boundary resistance increases strongly with decreasing temperature. This "acoustic mismatch prediction" is in reasonable agreement with most experimental data for $0.02\,\mathrm{K} < T < 0.2\,\mathrm{K}$ with typical values of $AR_K T^3 \simeq 10^{-2}\,\mathrm{m^2\,K^4\,W^{-1}}$ for liquid and solid helium in contact with metals, but deviates considerably both in the Kelvin temperature range (Figs. 4.5 and 4.6) and at $T < 10\,\mathrm{mK}$ (see "Acoustic Coupling Between Liquid Helium and Metal Sinters"). Another unexplained result is the observation that R_K seems to be about the same for ^{3}He and ^{4}He in the liquid as well as in the solid state at $T \approx 1\,\mathrm{K}$ (Fig. 4.6). Without question, the physics of the anomalously good thermal coupling at $T \geq 1\,\mathrm{K}$ is still not well understood, even though very detailed frequency-, angle-, and surface-condition-dependent studies have been performed [4.50–4.54] employing even modern high-frequency spectroscopic techniques [4.55]. The above results apply to annealed, bulk and clean metal surfaces. Of course, the experimental results depend strongly on the surface condition of the body in contact with helium, in particular surface roughness or mechanical damage of the surface. It can easily be

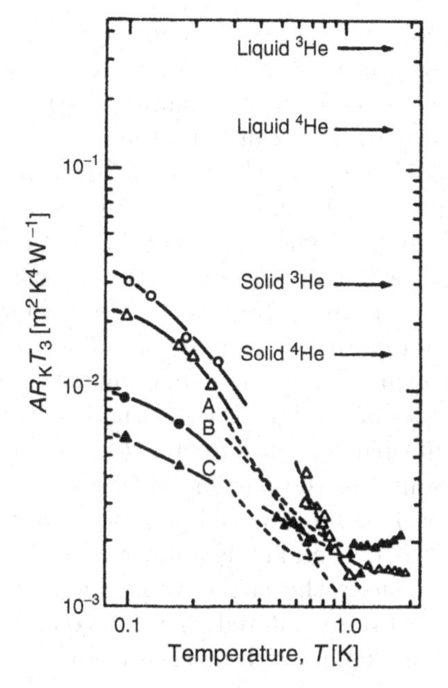

Fig. 4.6. Thermal boundary resistance R_K multiplied by AT^3 between ^{3}He or ^{4}He and copper as a function of temperature. (*open circle*): liquid ^{3}He; (*filled circle*): solid ^{3}He; (*open triangle*): liquid ^{4}He; (*filled triangle*): solid ^{4}He. The dashed curves A, B and C are for liquid ^{3}He, liquid ^{4}He and solid ^{3}He, respectively. The arrows at the right indicate the prediction from the acoustic mismatch theory (4.10) for an ideal Cu surface [4.50], where references to the original work are given

changed by an order of magnitude, for example, for ^{3}He–Cu interfaces from $AR_KT^3 = 4 \times 10^{-3}$ to $4 \times 10^{-2}\,\mathrm{m^2K^4W^{-1}}$ for $10\,\mathrm{mK} \leq T \leq 100\,\mathrm{mK}$ for different surface treatments (sand blasting, machining) [4.50–4.54] (Figs. 4.5 and 4.6). A rigorous treatment of the boundary resistance due to acoustic mismatch has to take into account the structure and properties of the solid and of helium near to the interface, and also the excitations there [4.47]. Surface excitations as well as a deviation from crystalline structure and the compression of helium resulting from the van der Waals attraction of the solid may influence the transmission coefficient for phonons.

The thermal boundary resistance $R_K \propto 1/(AT^3)$ is the most severe obstacle for establishing thermal contact between helium and other substances at $T < 1\,\mathrm{K}$. The common approach to improving this contact is by increasing the contact area A. This is mostly done by using heat exchangers made from sintered-metal powders; they will be discussed in Sects. 7.3.3 and 13.6. In addition, experiments reveal that the boundary conductance is considerably improved at $T < 20\,\mathrm{mK}$, as compared to the predictions of the acoustic mismatch prediction; this will be discussed in the next two Sections.

Acoustic Coupling Between Liquid Helium and Metal Sinters

In the preceding section on Acoustic Mismatch we saw that the mechanism of energy transfer across a metal-to-liquid interface is not completely understood; it is even less so if the metal is a sinter. Neither the electrons nor the single-particle helium excitations can cross the interface. The transfer has to be mediated by phonons and – in the case of ^{3}He – possibly by a magnetic coupling (see next section). As shown in the preceding section, the phonon coupling varies as T^3 and becomes very weak at low temperatures. One can compensate for this weakening by increasing the contact area A, i.e., by using metal sinters. Those used have surface areas of up to a few $100\,\mathrm{m^2}$ (Sects. 7.3.3 and 13.6). But this complicates the understanding even more because the vibrational modes of a sinter made of submicrometer particles will differ from the corresponding modes of the bulk metal. The lowest vibrational frequency ν for bulk phonons in a particle of diameter d is of order $(0.1\text{–}1)v_s/d$, where v_s is the velocity of sound in the particle [4.50, 4.56–4.59]. This corresponds to several gigahertz or $T \approx h\nu/3k_B \approx 10\,\mathrm{mK}$ for a particle of $d = 1\,\mu\mathrm{m}$. At higher frequencies the particle and its Kapitza resistance will behave bulk-like, whereas at lower frequencies the latter would increase exponentially. However, when metal particles are sintered they are connected by narrow elastic bridges (Sect. 13.6). This sponge with its elastic, open structure can have low-frequency continuum modes with a density of states which may be two orders of magnitude larger than the corresponding bulk density of phonon modes at $T \approx 10\,\mathrm{mK}$ [4.56–4.59]. These soft modes are assumed to couple well to

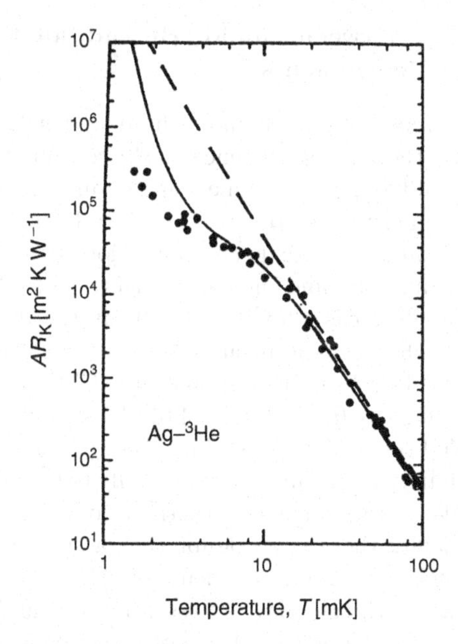

Fig. 4.7. Thermal boundary resistance R_K multiplied by the surface area A between liquid ^{3}He and a sinter of Ag particles. Experimental data from [4.60]. The dashed line represents the prediction of the acoustic mismatch theory (4.10) for bulk Ag. The full curve shows the prediction for a coupling of (zero) sound modes of liquid ^{3}He to soft modes with a characteristic energy of 15 mK k_B of the Ag sinter [4.50]

the helium phonon modes with their small velocity of sound, as shown for a sinter of 1 μm Ag powder and liquid ^{3}He in Fig. 4.7. As a further result, the boundary resistance will show a temperature dependence which is weaker than T^3 below a temperature where the dominant phonon wavelength becomes comparable to the size of the sinter particles; in [4.56] this dependence was found to be $R_K \propto T^{-1}$. Nakayama [4.50], in particular, has reviewed the heat transfer due to (zero) sound in liquid ^{3}He from ^{3}He to a metal sinter, the most important contact medium in the problem of energy transfer to liquid helium. But the situation is now more complicated than one would expect just from the change of the vibrational modes. The mean free paths of the electrons (whose influence on R_K is not understood) and phonons are limited by the sinter grain size and the mean free paths of the ^{3}He particles, and ^{3}He phonons are limited by the open dimensions of the sinter. The thermal resistance between excitations *in* the solid and/or *within* liquid helium can then become comparable to or even larger than the thermal boundary resistance at $T < 10$ mK, and it certainly has to be taken into account when the thermal resistances inside of the liquid helium and/or inside of the metal sinter are considered. These effects have not yet been fully investigated.

Magnetic Coupling Between Liquid ^{3}He and Solids Containing Magnetic Moments

If the thermal coupling between liquid helium and solids were limited to phonon transfer or if the T^{-3} dependence of R_K continued to low millikelvin temperatures, it would require extremely large surface areas to refrigerate liquid ^{3}He into its superfluid states at $T < 2.5$ mK. However, in the middle of the 1960s a completely unexpected behavior of the thermal boundary resistance between liquid ^{3}He and the paramangetic salt CMN (Sect. 9.3) was reported [4.61–4.65]. Whereas above about 20 mK the data were in reasonable agreement with the acoustic mismatch theory, for $2\,\text{mK} \leq T \leq 20\,\text{mK}$ the thermal resistance between ^{3}He and powdered CMN was strongly reduced and *decreased* according to $R_K \propto T$ (Fig. 4.8). This result has inspired much research and provided the means of coupling ^{3}He well; and therefore cooling it into the low millikelvin temperature range, actually to 0.1 mK (see Table 10.2). Even though details of this enhanced thermal coupling are still not quite understood theoretically there is no doubt that a surface magnetic interaction between the nuclear magnetic moments of ^{3}He and electronic moments in the solid in contact with ^{3}He plays an essential part and short-circuits the acoustic mismatch [4.50, 4.66, 4.67]. The most convincing support for this interpretation comes from the observation that the enhanced coupling dramatically decreases if the solid surface is "plated" by a layer of (nonmagnetic) ^{4}He [4.62, 4.64, 4.65] (Fig. 4.8). The ^{4}He atoms coat walls preferentially

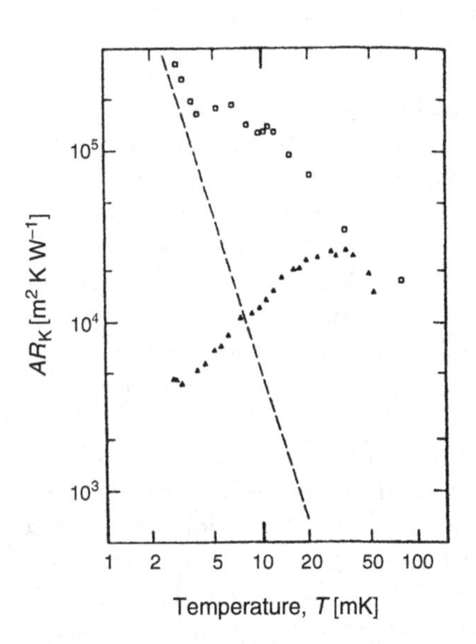

Fig. 4.8. Thermal boundary resistance R_K multiplied by the surface area A between the paramagnetic salt CMN (Sect. 9.3) and ^{3}He (*filled triangle*) and a 6% ^{3}He–^{4} He mixture (*open square*). The dashed line indicates the prediction of the acoustic mismatch theory, (4.10) [4.65]

because – due to their smaller zero-point energy compared to ^{3}He – they sit deeper in the van der Waals potential exerted by the wall. This interpretation is also supported by theoretical treatment of the magnetic dipole–dipole coupling between ^{3}He nuclear spins in the liquid and electronic spins in CMN, leading to $R_K \propto T$ [4.50, 4.66, 4.67], but the theory contains a number of significant assumptions [4.49, 4.50].

Later, qualitatively similar results –reasonable agreement with the acoustic mismatch theory in magnitude and temperature dependence for 20 mK < $T <$ 100 mK, and an enhanced thermal coupling at lower temperatures – was found for liquid ^{3}He in contact with various metals, particularly ("dirty") sinters. However, now the temperature dependence is $R_K \propto T^{-1}$, with typical values of $AR_K T \approx$ several 10^2 to 10^3 m^2 K^2 W^{-1} for sintered Cu and Ag powders or metal foils containing magnetic impurities [4.60, 4.68–4.73] (Figs 4.6, 4.7, and 4.9). Strong support for the magnetic coupling explanation is the

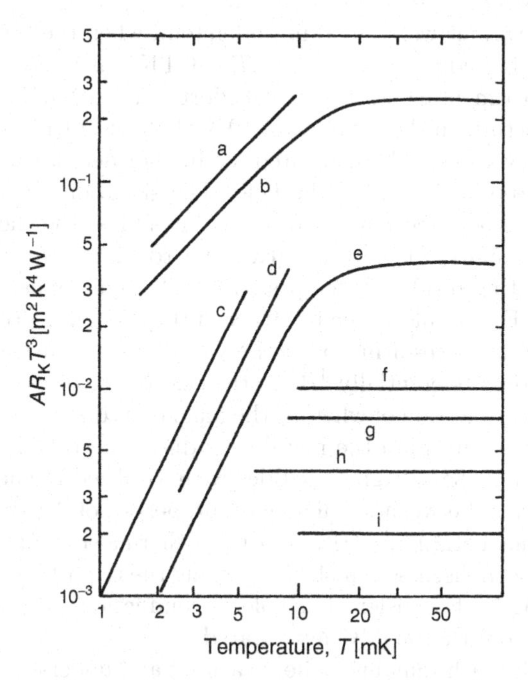

Fig. 4.9. Thermal boundary resistance R_K (multiplied by AT^3) between liquid ^{3}He or a liquid ^{3}He $-^4$He mixtures and various solids as a function of temperature. The data are for (**a**) mixture to 400 Å Ag sinter ($25\,T^{-2}$); (**b**) mixture to 700 Å Ag sinter ($16\,T^{-2}$ at <10 mK); (**c**) ^{3}He to 700 Å Ag sinter ($1{,}000\,\mathrm{T}^{-1}$); (**d**) ^{3}He to 400 Å Ag sinter ($470\,\mathrm{T}^{-1}$); (**e**) ^{3}He to 1 μm Ag sinter ($200\,\mathrm{T}^{-1}$ at <10 mK); (**f**) mixture to CuNi ($10^{-2}\,\mathrm{T}^{-3}$); (**g**) mixture to brass ($7 \times 10^{-3}\,\mathrm{T}^{-3}$); (**h**) mixture to 7 μm thick Kapton foils ($4 \times 10^{-3}\,\mathrm{T}^{-3}$); (**i**) mixture to Teflon tubing with 0.1 mm wall thickness ($2 \times 10^{-3}\,\mathrm{T}^{-3}$); (the data in brackets are AR_K) [4.73, this paper gives listings of the original works]

apparent absence of a contribution from the electron–phonon resistance in the metal and from the phonon–^{3}He-quasiparticle resistance in the liquid [4.50]; this indicates a direct ^{3}He spin-metal spin/electron coupling bypassing the phonons.

The possibility of changing the Fermi temperature T_F of liquid ^{3}He–^{4}He mixtures by changing their concentration (7.23a) offers an opportunity to gain more insight into the coupling between ^{3}He–^{4}He mixtures and solids. Indeed, for this combination, too, an enhanced coupling was found [4.72–4.77], but now with a dependence on T_F, with typical values $AR_K \approx 6T^{-2}T_F^{-1}\,\mathrm{m^2\,K\,W^{-1}}$ at $T \leq 20\,\mathrm{mK}$ [4.50, 4.76, 4.77]. The T^{-2} dependence is obtained by assuming a magnetic dipole coupling between the nuclear magnetic moments of ^{3}He and the electronic magnetic impurities moment S localized in the solid surface [4.50]. The dependence on the Fermi temperature T_F of the mixtures indicates the quantum and magnetic character of the coupling. For saturated ^{3}He–^{4}He mixtures (Chap. 7), values of $AR_K = (12\text{–}35)T^{-2}\,\mathrm{m^2\,K\,W^{-1}}$ have been reported [4.72–4.77].

These observations were far from understood when the first review article on the thermal boundary resistance at $T < 0.1\,\mathrm{K}$ was written in 1979 [4.49]. Substantial experimental and theoretical effort enabled Nakayama to present a more concise picture on the problem in 1989 [4.50] (Fig 4.10). These two comprehensive reviews should be consulted for further details and, in particular, for references to other works in the field. The situation still cannot be considered as understood and many more experiments on the thermal boundary resistance, particularly at $T \leq 20\,\mathrm{mK}$, seem to be necessary. For example, there are conflicting results on the pressure and magnetic field dependence of R_K [4.78–4.80]. Furthermore, the influence of the changing structure of helium near to the substrate resulting from the van der Waals attraction (localization, compression, preferentially ^{4}He in the case of mixtures) is an interesting topic. Finally, the question of whether the magnetic coupling results from electronic magnetic moments in the metal or is due to absorbed impurity layers, particularly various paramagnetic oxides with localized moments [4.50], is of interest. A relation between the magnetic properties of Ag sinters containing a few ppm of magnetic impurities and their thermal boundary resistance to liquid ^{3}He has been discussed in [4.81] stressing the importance of a "magnetic channel" for the heat transfer. The role of conduction electrons for R_K also remains to be more thoroughly investigated.

Even though much remains to be measured and understood, the Kapitza-resistance problem can clearly be divided into three distinct temperature regimes:

(a) Above 1 K: R_K is essentially the same for liquid and solid ^{3}He and ^{4}He; it is at least an order of magnitude smaller than predicted by the acoustic mismatch theory, and it is not understood.

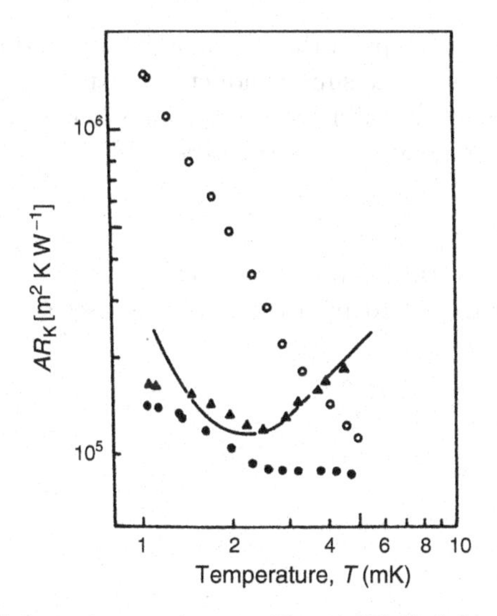

Fig. 4.10. Thermal boundary resistance R_K multiplied by the surface area A between liquid ^{3}He and a sinter of Ag particles. The experimental data for zero magnetic field (*filled circle*) and for a measurement in $0.385\,T$ (*open circle*) are from [4.78]. The field is assumed to suppress the magnetic contribution to R_K, leaving only the acoustic channel for heat transfer. Therefore the difference between the two sets of experimental data should given the magnetic contribution (*filled triangle*) to R_K. The solid line represents the theoretical prediction for this magnetic contribution [4.50]

(b) For $20\,\text{mK} \leq T \leq 100\,\text{mK}$: $R_K \propto T^{-3}$ and behaves as predicted by the acoustic mismatch theory if well characterized, clean, bulk metallic surfaces are used.

(c) At $T \leq 10\,\text{mK}$: $R_K \propto T^{-2}$ or T^{-1} between liquid ^{3}He or helium mixtures and metals, and is again much smaller than predicted by the acoustic mismatch theory; here a magnetic dipole coupling between the ^{3}He nuclear moments and electronic moments in (or on) the solid together with a coupling of helium phonon modes to soft vibrational modes, if a sintered metal is used, seem to determine the energy flux. These effects have turned out to be extremely important for refrigeration in the low millikelvin and sub-millikelvin temperature ranges.

Problems

4.1. Calculate the temperature at which the contributions from phonons and from electrons to the switching ratio of a superconducting heat switch of Al become equal (Fig. 4.2; take material parameters from Table 10.1).

4.2. The difference in entropy of the superconducting and normal–conducting state at the transition of a superconducting metal is $S_s - S_n = (V/4\pi) B_c(dB_c/dT)$. Calculate the latent heat occuring when a superconducting heat switch is changed from one state to the other.

4.3. Deduce (4.9).

4.4. To which temperature can one refrigerate a sample of liquid ^{3}He in a sintered heat exchanger of $10\,m^2$ if the liquid releases $\dot{Q} = 0.1\,nW$?

5

Helium-4 Cryostats and Closed-Cycle Refrigerators

Here, and in the five chapters to follow, I will discuss the equipment needed for reaching and maintaining low temperatures, and start with cryostats for the Kelvin temperature range in which the low temperatures are obtained by evaporating liquid helium [5.1–5.3]. An important aspect of our consideration will be keeping the evaporation rate of the cryoliquid as low as possible by optimizing the cool-down process and by minimizing the heat transferred from the warm surroundings to the cold parts of the equipment.

The main isolation of cold parts from warm parts is done by vacuum isolation. Therefore, cryotechnology always involves vacuum technology. A low-temperature physicist necessarily needs experience in vacuum physics: in the design of vacuum-tight equipment, in techniques of avoiding leaks, in the technologies of soldering, welding and gluing; or, if leaks occur, in ways of finding and fixing them. These technologies are very important because low-temperature equipment can experience extreme mechanical stresses due to the large temperature gradients and different thermal expansion coefficients of the various materials used in it.

Low-temperature experimentalists have to design and construct their equipment very carefully with a very high reliability because it is built and tested at room temperature, then closed, evacuated and cooled down. After having been cooled, defects may develop due to vacuum leaks resulting from thermal stresses, or because electrical leads may not have been properly connected and may open up or may short to ground after cool-down, etc. These defects cannot be removed at low temperature. Therefore the equipment has to be warmed up for repair. Even worse, there are defects which occur only at low temperatures and disappear again when the equipment has warmed up to room temperature. Often these defects cannot be located easily and time-consuming cross checks may be the only way to find out what went wrong and where. Therefore, careful design and careful construction is of utmost importance for low-temperature equipment.

In addition, we have to remember that cold surfaces adsorb gases. If the apparatus has a small leak to atmosphere, for example, air may leak in and will condense at helium-cooled surfaces. If the apparatus has been at low temperature for a long time, the cold surfaces may be covered by a large amount of air. During warm-up the condensed gases will desorb and a substantial pressure may build up in the "vacuum" space. Pressure will also build up when an accidentally closed valve or an air plug in a tube stops the escape of the evaporating cryoliquid. Let me remind you that all containers which at some time will be filled with a cryogenic liquid either have to be able to withstand a rather high pressure or, better, must contain safety features so that following an accident, the pressure cannot rise to dangerous values. One should therefore have surveillance pressure gauges as well as safety valves or bursting discs on the vacuum vessel as well as on the containers for the cryogenic liquids. For all these vessels certain safety regulations have to be followed, which in some countries are enforced by law.

5.1 Use of Liquid ^{4}He in Low-Temperature Equipment

Every refrigeration process to a temperature of 10 K or lower uses liquid helium in either the final or the pre-cooled stage. In the following I will stress some points that are important so that the consumption of this expensive cryoliquid can be minimized. For such an optimization we have to distinguish between the cool-down and the running period of the experiment after cooling.

5.1.1 Cool-Down Period

The heat of evaporation for helium transforming from the liquid to the gaseous phase is about 2.6 kJ l^{-1} at its normal boiling temperature of 4.2 K (Fig. 2.6). This is a rather small number compared to the enthalpy of helium gas between 4.2 K and 300 (77) K, which is about 200 (64) kJ l^{-1}. When 1 W is applied, 1.4 lHe h^{-1} will evaporate if the heat of evaporation is used, but only 0.017 lHe h^{-1} if the enthalpy of the cold gas between 4.2 and 300 K is utilized as well. As a result, it is very important to make use of the enthalpy of the cold helium gas after the liquid has undergone the transition to the gaseous state when cooling the equipment, and the gas should leave the cryostat with a temperature as close as possible to room temperature. In addition, pre-cooling with liquid nitrogen from 300 to 77 K will save a large amount of liquid helium. Liquid N_2 has about 60 times the latent heat of evaporation of liquid helium, it is almost an order of magnitude cheaper, and at 77 K most of the heat capacity of the materials to be refrigerated has been removed already. Table 5.1 lists how many liters of a cryoliquid we need to refrigerate 1 kg of aluminium, stainless steel or copper with liquid N_2 from room temperature to 77 K, or with liquid He from room temperature to 4.2 K, or only from 77 to 4.2 K. In the table, I distinguish between the two situations in which (a) only the heat of evaporation and (b) in addition the much larger enthalpies of the cold gases are used. The table demonstrates that it is clearly desirable to:

Table 5.1. Amount of cryoliquid [*l*] necessary to refrigerate 1 kg of aluminum, stainless steel (SS) or copper if only the latent heat (latent heat plus enthalpy of the gas between its boiling point and room temperature) is used

cryoliquid	temperature change [K]	Al	SS	Cu
N_2	$300 \to 77$	1.0 (0.63)	0.53 (0.33)	0.46 (0.28)
^{4}He	$77 \to 4.2$	3.2 (0.20)	1.4 (0.10)	2.2 (0.16)
^{4}He	$300 \to 4.2$	66 (1.6)	34 (0.8)	32 (0.8)

- Use liquid N_2 to first pre-cool the equipment from room temperature to 77 K, and only then utilize liquid helium for the lower temperatures
- Make use of the enthalpy of the cold helium gas as much as possible to refrigerate the equipment

Otherwise the experimentalist may waste more than two orders of magnitude of expensive liquid helium.

5.1.2 Running Phase of the Experiment

Now let us consider the situation where the experiment has reached a stationary state, which means that the equipment has reached the required low temperature and we can perform an experiment at this constant temperature. In this situation the heat transferred from external sources has to be compensated by the cooling power provided by the cryoliquid, which now mainly comes from the heat of evaporation. The cold gas can only be used for cooling, for example, the support structure of the experiment or the walls of the cryostat. Let us consider the main sources of heat for experiments in the Kelvin temperature range.

Heat Conduction

Heat conduction may be along leads used for performing measurements on the experiment or along the walls, pumping tubes and support structure of the cryostat, for example. The transfer of heat by conduction is determined by the equations and the material properties considered in the discussion of thermal conductivity (Sects. 3.3 and 4.1). To minimize heat conduction we have to choose the right material and we have to give it the correct dimensions. In practice that means we should use strongly disordered or organic materials because of their low thermal conductivity, or, if we have to employ metals, they should be low-conductivity stainless steel or Cu–Ni alloy tubes with thin walls.

Heat Radiation

The radiation of heat is determined by the Stefan–Boltzmann equation, which, in its simplest form, reads

$$\dot{Q}[\mathrm{W}] = 5.67 \times 10^{-12} A[\mathrm{cm}^2](T_1^4 - T_2^4) \tag{5.1}$$

Table 5.2. Typical emissivities of various solids [5.1][a]

solid	emissivity, ϵ
Au; Ag	0.01–0.03
Cu; brass	0.02–0.6
Al	0.04–0.3
stainless steel	0.1–0.2
glass/organic	0.1–0.9

[a] A detailed report on the total hemispherical absorptivity and emissivity of chemically polished surfaces of Cu and Al as well as of their alloys, and of stainless steel, held at 5 K and with the temperature of the absorbed radiation varied between 30 and 140 K can be found in [5.4].

for the case that the radiating and the absorbing areas, A, are equal, and that the emissivities and absorptivities of these two areas are equal to 1 ("blackbody radiation"), which means we have dirty surfaces with very low reflectivity. In practice, of course, the situation can be improved by using polished metal surfaces [5.4]. For this situation the emissivities ϵ can be reduced to the values given in Table 5.2. For $\epsilon \neq 1$, (5.1) has to be multipied by $\epsilon_1\epsilon_2/(\epsilon_1 + \epsilon_2 - \epsilon_1\epsilon_2)$ [5.1]. If the emissivities of the radiating and absorbing surfaces are equal but much smaller than 1 this factor is $\epsilon/2$. If the heat due to radiation has to be reduced, surfaces should be gold-plated to avoid oxidation and to keep them clean and shiny. This is particularly important for apparatus used at $T < 100$ mK (Chaps. 7–10). We have to keep in mind that the dominant wavelength radiated by a body at $T \leq 300$ K is in the infrared (10 μm at 300 K), therefore it is the emissivities at these wavelengths that matter rather than those in the visible. Highly polished metals are highly reflective in the infrared as well.

Equation 5.1 tells us that parts cooled to LHe temperatures should not be exposed to surfaces which are at room temperature, i.e., they should not be able to "see" surfaces at room temperature. If we calculate the heat radiated from a 10 cm^2 surface at 300 K we find 0.5 W, which means that this would evaporate 0.7 l He h^{-1}. If we reduce the temperature of this 10 cm^2 area to the temperature of boiling nitrogen (77 K), the radiated heat would be reduced to 2 mW and only 3 cm^3 of liquid helium would be evaporated per hour. The inner part of a cryostat cooled by liquid helium should be surrounded by radiation shields or baffles at intermediate temperatures, to shield it from radiation from room temperature. These baffles should be either gas-cooled or, even better, directly thermally anchored to an intermediate temperature, for example to the LN_2 reservoir (see later).

Conduction by Gas Particles Remaining in the Vacuum Space

Because low-temperature surfaces act as cryopumps (Sect. 2.3.2) the vacuum in a cryostat is usually good enough to make the conduction by remaining gases sufficiently small. For the possibly remaining helium gas the heat transferred from the warmer to the colder surfaces by this process is [5.1,5.2]:

$$\dot{Q}[\mathrm{W}] \approx 0.02aA[\mathrm{cm}^2]P[\mathrm{mbar}](T_2 - T_1)[\mathrm{K}] \tag{5.2}$$

if the mean free path of the gas particles is large compared to the dimensions of the container so that they travel from wall to wall without particle collisions on their way, which is the usual low-pressure case.[1] The heat transported by them is then determined by their number and their temperature (velocity). Here a is the accommodation coefficient for the gas particles on the walls; it is 1 as an extreme case but can become as small as 0.02 for a clean metallic surface exposed to helium gas. If we consider an apparatus with an area of 100 cm^2, take $a = 1$, and calculate how much heat is transported from room temperature to low temperatures if the remaining pressure is 10^{-6} mbar, then we arrive at about 1 mW, which is negligibly small for the Kelvin temperature range but may be detrimental at lower temperatures (Sect. 10.5.3). To improve the cryopumping action, often a gettering material is attached to the cold wall of a cryostat, which can very effectively adsorb remaining gas (Fig. 6.5).

Thermoacoustic Oscillations

The geometry of the neck tubes of dewars or of other gas filled tubes connecting cold parts and warm parts is often suitable for the excitement of thermoacoustic (or *Taconis*) oscillations in the vapour above the helium bath or in the tubes [5.5]. These oscillations may cause excessive heat flux into the dewar and may trigger errors in measurements. They may also generate disturbing vibrations for sensitive apparatus like nuclear refrigerators (Sect. 10.5.1). Often, such oscillations can be removed by changing the geometry by simply adding an additional volume, and/or putting a damping element (valve) in the vapour exhaust line [5.5]. A microphone in the vapour phase can usually detect the efficiency of the changes.

Further Heat Sources

A real cryostat, of course, is connected to pumps as well as to the building and it contains experimental setups with electrical leads to them. All this will cause further heating, which can be quite troublesome, in particular for experiments at very low temperatures where the refrigerator has a rather small cooling power or for sensitive experiments, like calorimetry (Sect. 3.1.8) or with a torsional or translational oscillator (Sect. 13.10).

Vibrational heating from the movements of the building, from bubbling of the cryoliquid (Sect. 5.2.4), or, in particular from the pumps via the pumping tubes can be a serious problem if the dominant frequency of one of these

[1] The full equation can be found in [5.1, 5.2]

sources (typically in the range of 1 to several 10 Hz) is close to a resonance frequency of the cryostat or the experimental setup in it. This can give rise to heat leaks in the range of 0.1–1 μW. To reduce this energy input, a rigid construction inside the cryostat and a strong mounting of it to a heavy foundation may be necessary (Sect. 10.5.1). In addition, flexible pumping tubes connected to some heavy mass or to walls of the building somewhere between the pumps and the cryostat may also be in order (Sect. 10.5.1).

Joule heating can easily be calculated. Its removal requires a proper thermal coupling of the sample to be investigated and, in particular, of the electric leads to it. This is of special importance for thermometers, in calorimetry (Sect. 3.1.8), or when measuring thermal conductivities (Sect. 3.3.8). One has to decide for the leads between a good conductor to give small Joule heating but good thermal conduction or a bad conductor with the opposite properties. Of course, the use of superconducting wires with their vanishing thermal conductivity (Sect. 3.3.4) and zero electrical resistivity is a substantial advantage. In any case, leads from room temperature to an experiment should be properly heat sunk on as many temperature stages as are accessible on their way to the experiment by winding them around posts, for example, and gluing them there. Pick-up of RF radiation can be reduced by shielding the cryostat if necessary (Sect. 10.5.1). In particular for leads to thermometers, it may be necessary to add a filter to reduce RF pick-up (Sect. 12.5.4). Special considerations are, of course, required for the leads to superconducting magnets which usually carry a large current (Sect. 13.4). Here, one has to decide on the optimum diameter-to-length ratio for a particular material and current [5.1, 5.6].

Further sources of heat may be gas adsorption or desorption when the temperature is changed, superfluid film flow (Sect. 2.3.5), or eddy current heating from changing electromagnetic fields (Sect. 10.5.2). The latter can be reduced by using materials with low conductivity or by using small dimensions like foil or "coil foils" made from wires.

Each cryostat and each low-temperature experiment is of a particular design, so that it is not appropriate to give a more detailed discussion of these obvious sources of heat. They have to be considered and can usually be calculated in a straightforward way for the particular setup. For experiments in the low millikelvin or even microkelvin range, these as well as some additional heat sources become particularly important; they will be discussed in Sect. 10.5.

5.2 Helium-4 Cryostats

In London, at the end of the last century, the Scottish scientist J. Dewar presented public demonstration lectures on his low-temperature experiments. For these demonstrations as well as for the experiments in his laboratory, he had to improve the storage vessels for cryogenic liquids. After various trials he eventually arrived at a double-walled vacuum isolation vessel, now commonly called a "dewar". The dewar in its simplest form is nothing but the double-walled flasks which are used to keep coffee warm on a camping trip.

5.2.1 Double-Walled Glass Dewars

The typical setup for experiments using liquid ^{4}He as the cryogenic liquid is a nested, double-walled glass dewar system as shown in Fig. 5.1. The advantages of such a system are its reasonably low price, the low thermal conductivity of glass and the ease with which the level of the cryogenic liquid can be seen. A disadvantage, of course, is the ease with which glass can be broken. A particularly dangerous situation may arise if there is a small leak to the vacuum space between the two glass walls. Then air, for example, may enter this space and will condense onto the cold surfaces. A small leak may stay undetected. On warming the system, the condensed air will evaporate. If the air cannot

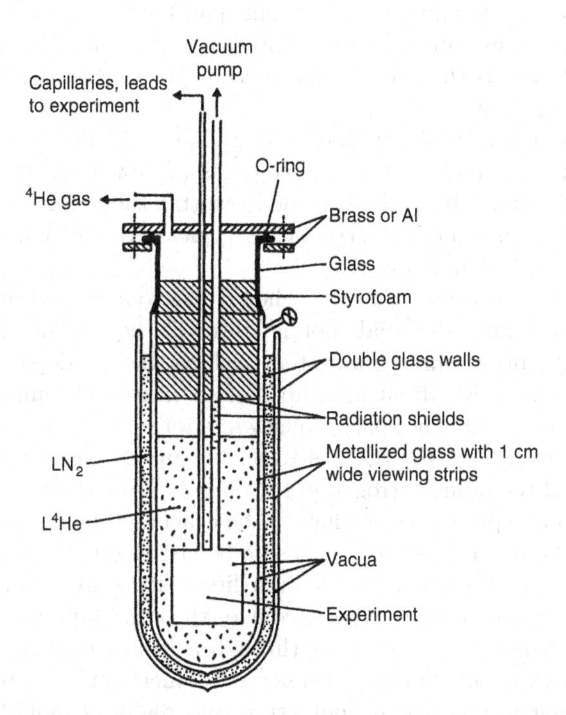

Fig. 5.1. A double-walled glass dewar for LN_2 and L^4He. The inner surfaces of the glass are covered with aluminium or silver films to block thermal radiation (except for a vertical 1 cm wide viewing strip). Four or five radiation baffles reduce the thermal radiation; these baffles force the exiting cold gas to flow along the dewar walls, thus cooling it as well as the pumping tubes. The effectiveness of the baffles is increased by filling the spaces between them with Styrofoam

escape fast enough through the tiny leak, pressure will build up in the vacuum space and the glass dewar may explode. To avoid serious consequences of such an explosion one has to follow two rules:

1. Never take a cryogenic system apart before all of it has been warmed up to room temperature.
2. A glass dewar system has to be surrounded by a protecting container, so that pieces of glass cannot fly through the laboratory.

One also has to keep in mind that at room temperature, helium diffuses through the glass walls (typically 10^{-12}–10^{-10} $cm^3\,s^{-1}$ through 1 cm^2 of glass of 1 mm thickness if $\Delta P = 1$ bar, depending on the glass, with the largest of the given numbers referring to Pyrex; the diffusion constant of ^{4}He gas through plastics is typically $D \simeq 10^{-6}\,cm^2\,s^{-1}$ at 300 K [5.7]). The same is true for fibreglass often used in metal dewars (see later). To keep the vacuum one should pump the helium gas from the inner volume immediately after warming to room temperature. In any case, the vacuum space of a glass dewar should be evacuated from time to time, at least every few weeks when operating with liquid helium, because some parts of the dewar may have been in contact with room-temperature helium gas even during the running period. Another possibility is the use of some special type of glass with a very small gas diffusion constant.

Such a dewar has to be equipped with radiation baffles to keep the radiation heat leak small and to force the cold gas to flow along the dewar walls, thereby cooling them. In a well-designed cryostat the helium gas is at 300 K when it reaches the top of the cryostat. A typical evaporation rate of a glass dewar should be 0.1 l/h at most.

In Sect. 5.1.1 we learnt that one should pre-cool everything with liquid nitrogen before taking the final cooling step with liquid ^{4}He. There are two methods of LN_2 pre-cooling in a system such as the one depicted in Fig. 5.1. One possibility is to let about a millibar of air into the vacuum space of the LHe dewar and also into the space which will later be filled with liquid helium. This pressure is enough for the gas to act as an exchange gas between the experiment and the liquid nitrogen in the outer container for pre-cooling. The air in the vacuum space does not have to be evacuated before liquid helium is allowed into the inner glass dewar; it simply freezes out. Another possibility is to fill the space for the liquid helium first with some liquid nitrogen to pre-cool the experimental setup. Of course, then the liquid nitrogen has to be blown out from this inner space through a tube which runs all the way down to the lowest point in the LHe dewar by increasing the pressure above the liquid nitrogen. One has to make sure that really *all* liquid nitrogen has been removed before liquid helium is transferred, because liquid nitrogen has a rather large specific heat and it would take a sizeable quantity of the liquid helium to cool nitrogen to the low Kelvin temperature range. A regulated and protected heater system to evaporate the remaining part of the LN_2 or to warm up a dewar system has been described in [5.8].

5.2.2 Metal Dewars

More common nowadays are commercial metal dewars for which examples are shown in Fig. 5.2. A metal dewar has the advantage that such a setup is much more rugged and can withstand higher pressures or stresses than a glass dewar. In addition, it has a greater flexibility, so much more complex designs are possible and it does not have the helium diffusion problem. A disadvantage compared to the glass dewar system is the higher price of a metal cryostat, which is usually made from stainless steel or, more often now, from a combination of aluminium and fibreglass.

In such a system one does not usually have two vacuum spaces as in a double-walled glass dewar system. The vacuum space cannot be filled with air in order to provide thermal contact of the inner parts with the liquid nitrogen vessel since one would then also cool the outer room-temperature wall. Liquid nitrogen has to be put into the helium vessel for pre-cooling and then removed before transferring liquid helium into the cryostat.

Nowadays many of these metal ^{4}He cryostats do not use LN_2 vessels for pre-cooling and radiation shielding; this is particularly important when vibrations produced by the constantly boiling LN_2 may interfere with the experiment. In such a situation many layers of so-called *superinsulation* are wrapped around the LHe vessel and possibly around radiation shields. This superinsulation is a thin plastic foil onto which a reflective layer of aluminium has been evaporated to give an emmissivity coefficient of about 0.06. The sheets of superinsulation have decreasing temperature from the outermost to the inner ones, so they act as radiation shields at continuously decreasing temperatures. For further improvement there can be a metallic radiation

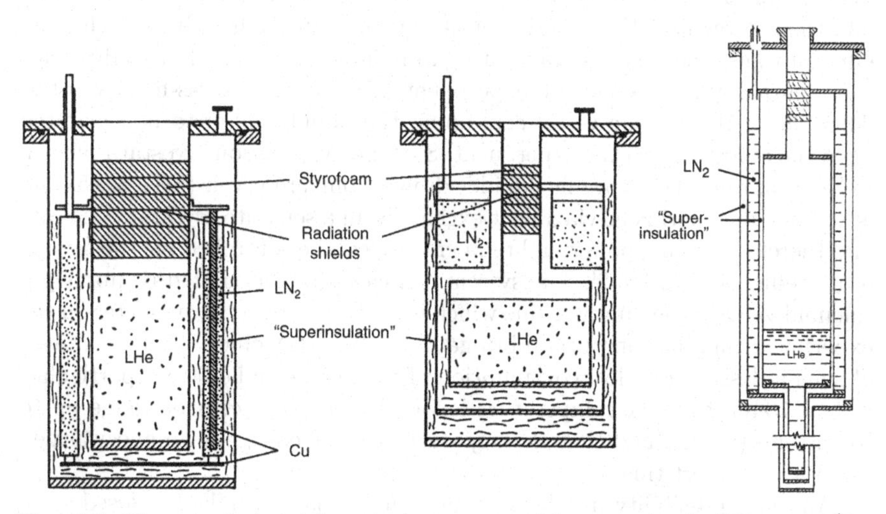

Fig. 5.2. Three typical stainless steel cryostats with reservoirs for LN_2 and L^4He (see text and caption on Fig. 5.1)

shield between the room-temperature vessel and the LHe vessel, and to these radiation shields tubing may be soldered, through which the evaporating cold ^{4}He gas is vented before leaving the cryostat. This makes further use of the enthalpy of the cold gas by pre-cooling the radiation shield [5.9]. A good, simple helium dewar should have an evaporation rate of not more than $0.1\,\mathrm{l\,h^{-1}}$; but for large dewars, such as are necessary for a powerful nuclear magnetic refrigerator (Chap. 10), this figure may go up to about $1\,\mathrm{l\,h^{-1}}$.

The design of ^{4}He dewars for various applications has been discussed in [5.1–5.3]. When possible, soft solder joints should be avoided on cryogenic equipment. They are less reliable, develop "cold leaks", fatigue sooner and are weaker than silver solder joints. Of course, a welded design is superior to even silver soldering. Where flux has to be used in soldering, it has to be washed off very thoroughly to avoid later corrosion.

In addition to the discussed cryostats made from glass or metal, many of them on the market are nowadays made from plastics. They usually do not use LN_2 for precooling and radiation shielding but use in an efficient way the enthalpy of the evaporating ^{4}He plus radiation reflecting superinsulation. The widespread use of ^{4}He as a cooling agent in applications of superconducting magnets, for example for nuclear magnetic resonance in otherwise non-cryogenic environments like hospitals, has led to the development of cryostats with extremely low evaporation rates, requiring replenishing of the cryoliquid once a week or even less often.

5.2.3 Cryostats for $T > 5\,\mathrm{K}$

For temperatures above the normal boiling point of liquid ^{4}He it is rather uneconomical to use the main ^{4}He bath at $4.2\,\mathrm{K}$ as a temperature reservoir and then to regulate the experiment at higher temperatures. It is much more efficient to use the cold gas evaporating from liquid helium and take advantage of its enthalpy for cooling the experiment in a continuous gas-flow cryostat. This allows the storage vessel containing the liquid helium to be separated from the cryostat with the experiment. Such an evaporation cryostat is shown in Fig. 5.3. Here the cryogenic liquid is drawn from a reservoir (Sect. 5.2.5) and is cooling an experiment via a heat exchanger in a separate cryostat. The rate and therefore cooling power and temperature can be controlled via the setting of a needle valve [5.1, 5.2]. The system can easily be automated by appling a solenoid valve in the pumping line which is controlled by a thermometer on the experiment and an appropriate bridge controller electronics. The advantages of such a design are a low consumption of the cryogenic liquid (in particular, if $T > 10\,\mathrm{K}$), that the temperature is variable in a rather wide range up to room temperature, and that the apparatus can be cooled down and warmed up in a very short time.

Another possibility is the use of commercially available closed-cycle Gifford-McMahon or pulse-tube refrigerators which will be discussed in Sect. 5.3.

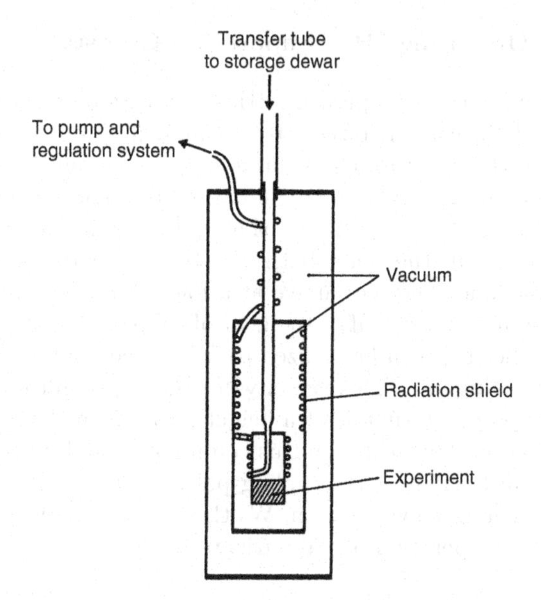

Fig. 5.3. Evaporation cryostat for the temperature range $T > 5\,\mathrm{K}$. A mixture of liquid and gaseous ^{4}He is transferred via a transfer tube (Fig. 5.8) from the storage dewar (Fig. 5.7) to the cryostat. The mixture is pumped through a spiral tube which is first connected to a chamber on which the experiment is mounted and then soldered to a radiation shield before it leaves the cryostat. The temperature of the experiment can be regulated by a heater and/or a valve which regulates the helium flow

5.2.4 Cryostats with Variable Temperature for $1.3\,\mathrm{K} \leq T \leq 4.2\,\mathrm{K}$

The temperature range $1.3\,\mathrm{K} \leq T \leq 4.2\,\mathrm{K}$ is determined by the normal boiling point of ^{4}He and the temperature at which its vapour pressure has become very small (Fig. 2.8). There are two ways to access this temperature range using pumped liquid ^{4}He.

Pumping on the Main ^{4}He Bath

Of course, one can just pump the vapour above the liquid ^{4}He bath away to decrease its temperature. This is very uneconomical because about 40% of the liquid ^{4}He has to be evaporated to cool it from 4.2 to 1.3 K, due to the large change of its specific heat in this temperature range. On the other hand, the specific heat of solids is rather small in this temperature range (see the figures in Sect. 3.1); to cool them from 4.2 to 1.3 K we have to evaporate only a small fraction of liquid ^{4}He. It is therefore much more efficient to leave the main part of the liquid at its normal boiling point of 4.2 K and just pump on a small fraction of it in a separate container to reach the lower temperature for the experiment. This idea is realized in the design described in the following section.

Continuously Operating ^{4}He Evaporation Cryostat

The design of a continuously operating ^{4}He evaporation cryostat [5.10] is presented in Fig. 5.4. In such a refrigerator a small fraction of the liquid from the main 4.2 K bath flows through a suitable flow impedance (see figure) into a small vessel of several cm^3 located in a vacuum cane inside the cryostat. Through the central tube we pump on the liquid arriving in this evaporation vessel. The liquid from the main bath at 1 bar is isenthalpically expanded through the impedance and will arrive at a lower temperature in the evaporation vessel. Again, almost half of the heat of evaporation is used for cooling the liquid; the other half can be utilized to fill up the inner vessel with liquid and to cool something else. This vessel will continue to fill until the level of the liquid in the pumping tube is at a height h at which the heat transferred from the main helium bath through this column of liquid, plus the heat from the experiment, just balances the cooling power of the refrigerator available from the latent heat L of evaporation. We then obtain the following equation for the steady-state operation of the refrigerator:

$$\dot{Q}_{\text{He}}(\simeq \frac{1}{2}\dot{n}L) = \dot{Q}_{\text{tube}}(h) + \dot{Q}_{\text{ext}} \,. \tag{5.3}$$

The refrigerator is self-regulating; if we increase the external load, the level of the liquid in the pumping tube will drop, so that its contribution to the heat transferred to the inner vessel is reduced. The temperature of the continuously evaporating ^{4}He refrigerator remains fairly constant at about 1.3 K when the heat load supplied to the vessel is varied (Fig. 5.5). Of course, the externally supplied heat may be so large that all the liquid is evaporated

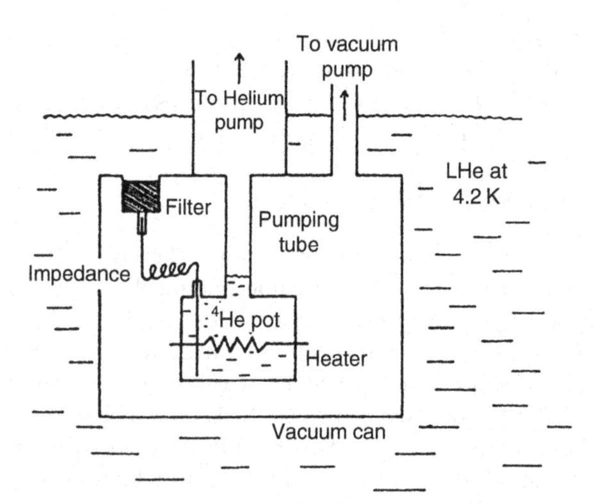

Fig. 5.4. A continuously operating ^{4}He refrigerator for the temperature range between 1.3 and 4.2 K (see text)

from the vessel, resulting in a rapid temperature increase. The continuous ^{4}He evaporator today is a standard condensation or pre-cooling stage for ^{3}He cryostats and, above all, for ^{3}He–^{4}He dilution refrigerators, which will be discussed in the two chapters to follow. In this case one has to make sure that its refrigeration capability is sufficient to remove the heat of condensation of the circulating ^{3}He.

The equation for the required impedance is

$$Z = \Delta P/\dot{V}\eta\,, \tag{5.4}$$

where ΔP is the pressure drop (1 bar to about 1 mbar) required to cause a volume flow rate $\dot{V}$ of a medium with viscosity η, see also (7.47). The typical value required (several $10^{11}\,\mathrm{cm}^{-3}$) can be obtained from a one to several meters long capillary with an inner diameter of about 0.05 or 0.1 mm. Another possibility is to use a shorter piece (10–20 cm) of a capillary and insert a tightly fitting wire into it [5.11]. With too large Z the ^{4}He chamber will run dry and with too small Z the required temperature will not be reached. During cooldown the refrigerator should be connected to a volume with pressurized very pure ^{4}He gas in order to prevent N_2 or air from entering and blocking the fill capillary. Sometimes problems arise because impurities in the main liquid helium bath (e.g., frozen air) block the fine capillary used for the impedance. One therefore has to put a filter (of Cu powder, for example) in front of the capillary and keep the main ^{4}He bath clean. Using the heat of evaporation of liquid ^{4}He (Fig. 2.6) and a typical flow rate of $\dot{V} = 10^{-4}\,\mathrm{mol\,s^{-1}}$ (obtainable with a mechanical pump of moderate size) one arrives at a cooling power of about 5 mW for such a ^{4}He evaporation cryostat (Fig. 5.5).

Substantially lower temperatures can be reached if the temperature of the main bath is lowered into the superfluid state of liquid ^{4}He. The best result reported in [5.12] is 0.95 K at the continuously filling pot with the main bath pumped to about 1.6 K, and with 1.89 K at a point on the fill line at one-third of the distance from the main bath; both, higher and lower temperatures at the fill line increase the temperature of the pot. The cooling power is 0.1 mW at 1.2 K. While this is much lower than with the bath held at 4.2 K, it is adequate for many experiments.

The continuously operating ^{4}He refrigerator substantially reduces the consumption of liquid ^{4}He. It also requires a much smaller pumping capacity as compared to pumping on the main ^{4}He bath. And, as another advantage, the main ^{4}He bath is kept at 4.2 K and 1 bar; it can be refilled without interrupting the operation of the evaporation refrigerator and the experiment as long as the level of the main bath does not drop below the inlet to the evaporator.

Formerly, these evaporators were operated in a discontinuous mode by replacing the impedance by a valve. In this mode the vessel can be filled once with liquid from the main bath through the valve. The valve is then closed and pumping on the inner vessel can begin. Now, of course, the experiment has to be interrupted when the liquid in the vessel is used up, in order to refill it.

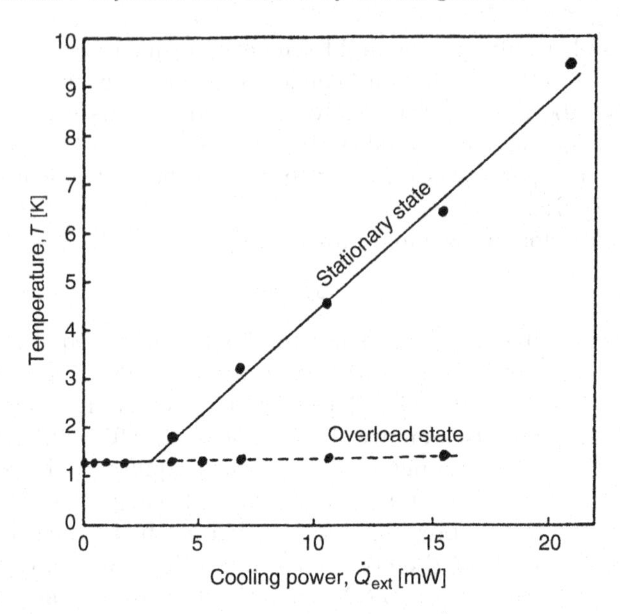

Fig. 5.5. Temperature of a continuously operating ^{4}He evaporation refrigerator. The refrigerator for which the data are shown can operate at a constant temperature up to a heat input of about 3 mW (this value can be changed by changing the impedance – see text and Fig. 5.4). At higher heat inputs the refrigerator will work in the overload state for a short time until the pumped ^{4}He pot (Fig. 5.4) is empty. In the stationary state at $T > 1.3$ K, the refrigerator operates with an empty ^{4}He pot at higher temperatures

Recently, various continuously operating ^{4}He cryostats designed for insertion into storage dewars (32 or 50 mm neck diameter, for example) have been described [5.6, 5.13–5.16, 5.18, 5.19]. A cryostat of this type employed in our laboratory is shown in Fig. 5.6. It can work from room temperature to $T \simeq$ 1.3 K, and has extremely low helium consumption, fast cool-down and warm-up times, low price, easy construction and excellent temperature stability. The small diameter of the cryostat is made possible by using a tapered (7°) grease or "glycerine-and-soap" [5.17] seal on the vacuum can (use non-aqueous silicon grease). However, an In ◯-ring seal is more reliable. The cryostat can even be equipped with a superconducting magnet of 2.5 (2) cm i.d., 4 cm o.d. for a 2 (4) T field, for example. These very useful cryostats are available commercially but can easily be constructed in the laboratory workshop according to the detailed description in [5.6, 5.13].

The vibrational noise observed on continuously filling pots can be detrimental for sensitive experiments, like with high-Q audio frequency torsional oscillators (Sect. 13.10), high-precision thermometry (Sect. 12.9), or for sensitive cryogenic detectors. These problems can be reduced by regulating the helium flow from the main bath and adjusting the helium level in

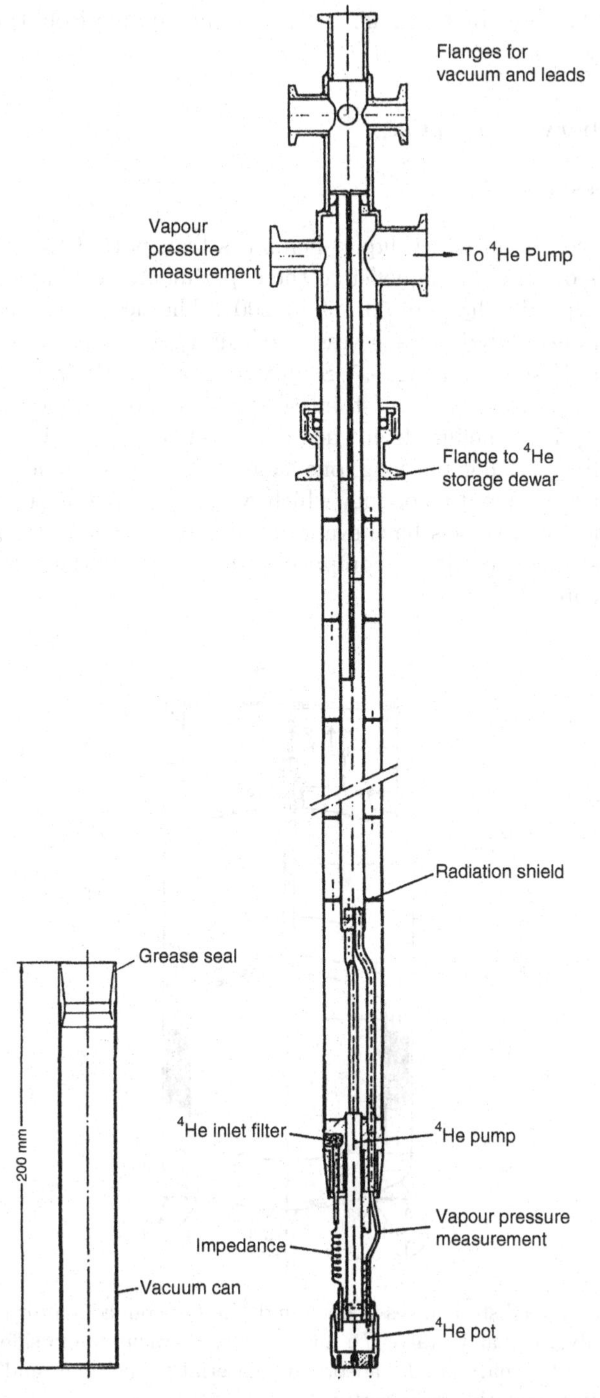

Fig. 5.6. Design of the low-temperature part of a ^{4}He dipstick cryostat for the temperature range $T > 1.3$ K (see text) (Courtesy of P. Sekowski, Universität Bayreuth)

the 1-K pot [5.18], or by thermalizing the helium coming from the main bath appropriately [5.19].

5.2.5 Auxiliary Equipment

Storage Vessel

The storage vessels in which liquid helium is transported from the supplier or from one's own helium liquefier to the experiment are commercially available with a typical volume of 50, 100 or 200 *l*. The design of a modern commercial vacuum-isolated storage vessel containing superinsulation instead of LN_2 shielding is shown in Fig. 5.7. Such vessels are made from aluminum to keep their weight low, or from stainless steel for more rugged applications. After drawing liquid helium from such a storage vessel, provided the periods between refilling are not too long, one should always leave a few liters of the cryoliquid in it to avoid recooling, which would consume a large amount of liquid helium. Such vessels have evaporation rates of about 1% per day. All commercial storage vessels are equipped with the appropriate safety features to avoid overpressure.

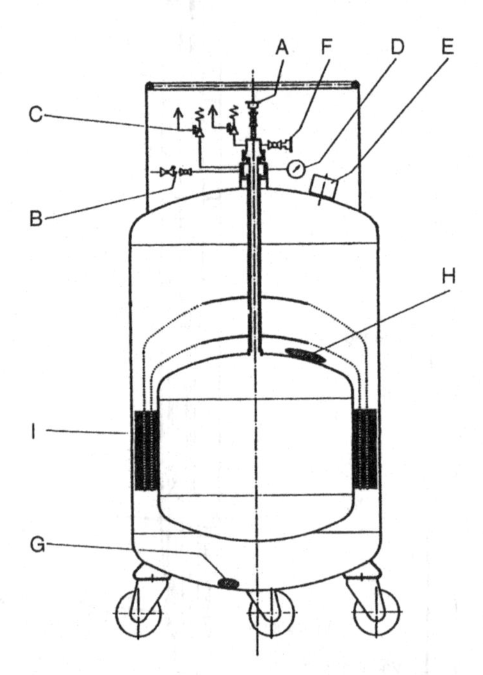

Fig. 5.7. Commercial storage vessel for liquid ^{4}He (*A*: connection for transfer tube, *B*: overflow valve, *C*: safety valve, *D*: manometer, *E*: vacuum and safety valves, *F*: gas valve, *G*: getter material, *H*: adsorbent material to maintain and improve the vacuum, *I*: superinsulation only partly shown)

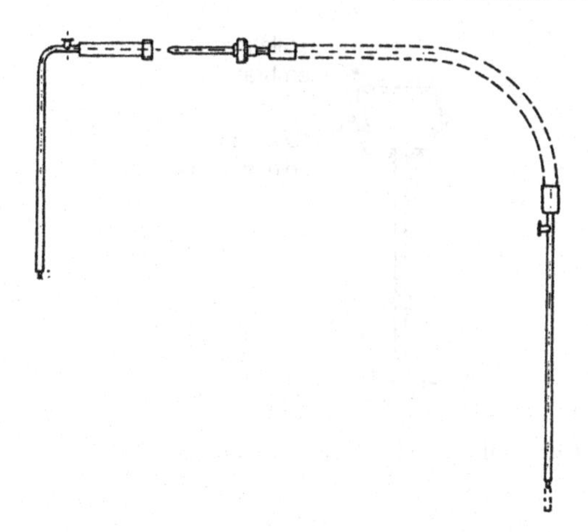

Fig. 5.8. Double-walled vacuum-isolated transfer tube for liquid ^{4}He. The *dashed part* is flexible. There should be a filter on the transfer tube to keep frozen air out of the cryostat (Courtesy of P. Sekowski, Universität Bayreuth)

Transfer Tube

For transferring liquid helium from the storage vessel to the experimental cryostat we need a double-walled vacuum transfer tube [5.1, 5.2], which can be either bought or made in the laboratory workshop, if the necessary experience in welding and machining stainless steel is available. A typical design is presented in Fig. 5.8. If bending of the two concentric tubes separated by spacers is necessary, then this should be done with ice as a filler between them to avoid collapse. Bending is usually necessary in order to take care of the differential contraction of the two tubes, which will be at different temperatures.

Level Detectors and Automatic Refilling

In containers for cryogenic liquids we need some means of determining the level of the liquid [5.1, 5.2]. This is, of course, rather easy in a glass system where one can detect the level optically. However, a problem arises for metal containers, where we have to apply other means of determining the liquid level.

Acoustic Level Detection

A simple design for an acoustic level detector is depicted in Fig. 5.9. When the thin stainless steel tube of this device is lowered into the cold gas or into the cryogenic liquid, some of the liquid evaporates, the pressure in the vapour space suddenly increases, and the liquid level in the container oscillates. These oscillations have a different amplitude and frequence according

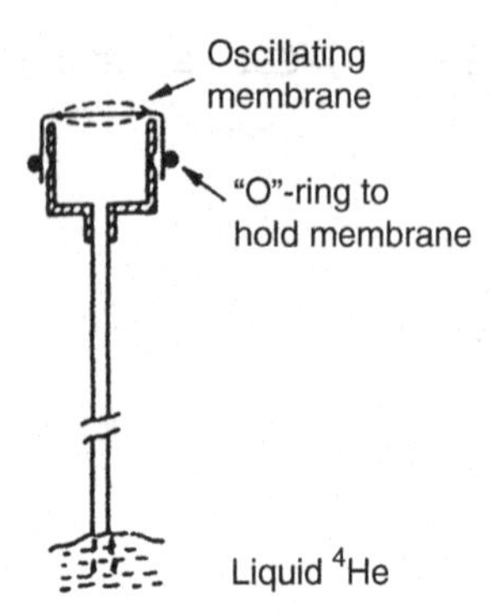

Fig. 5.9. Acoustic level detector for liquid ^{4}He

to whether the tube has been lowered into the cold gas or into the liquid. The oscillations are transferred to a diaphragm (thin rubber or plastic sheet) on top of the tube and the high amplitude/high frequency oscillations when the tube is lowered just into the gas can easily be distinguished from the low amplitude/low frequency oscillations when the tube has been lowered all the way into the liquid.

Resistive Level Detection

Many conductors have a very distinct temperature dependence of their resistance; examples are discussed in Sect. 12.5. If one sends a current of about 0.1 mA through a semiconductor or a carbon resistor with a low-temperature resistance of some kΩ, this resistor stays at the temperature of its surroundings if it is in a cryogenic liquid. But the current can overheat the resistor if it is raised into the gas phase, because now the thermal contact with its surroundings is rather weak. The resistance and therefore the temperature of the element will change quite a bit as the resistor is lowered from the gas phase into the liquid, and this change can easily be detected with a Wheatstone bridge.

Another possible detecting element for the liquid level is a commercial superconducting wire of NbTi or a normal-conducting wire with a superconducting coating (PbSn solder can be used) and a transition temperature between 5 and 10 K (Table 4.1) [5.20, 5.21]. If this wire is held vertically in the gas and liquid phases then the part which is in the liquid at 4.2 K will be superconducting and show zero resistance. The part of the wire in the gas phase will very rapidly attain a temperature above its transition temperature if an appropriate current is sent through it. It will become resistive and the total resistance of the wire is a direct measure of how much of its length is outside the liquid. Unlike the two-level detection means discussed earlier, this detector is stationary and does not have to be moved up and down to detect the liquid level; it can give a continuous recording of the level height. Actually it is better to turn off these detectors except while taking a reading because the heat generated in the normal section of the wire can raise the helium consumption considerably.

Capacitive-Level Detection

In this case the liquid level is determined by measuring the capacitance of two concentric tubes of length L immersed partly in the liquid. The height h of the liquid in the capacitor is related to the capacity C according to

$$\frac{C - C_0}{C} = (\epsilon_4 - 1)\frac{h}{L}, \tag{5.5}$$

with the dielectric constant of liquid ^{4}He being $\epsilon_4 = 1.0572$, the capacity of the partly filled capacitor

$$C = \frac{2\pi\epsilon_0\epsilon_4 L}{\ln(d_\mathrm{i}/d_\mathrm{o})}, \tag{5.6}$$

and C_0 the capacity measured at 4.2 K in vacuum.

A discussion of capacitive level meters for cryogenic liquids with the appropriate electronics for continuous readout can be found in [5.22, 5.23].

Of course, several of these level indicators can also be used to actuate the transfer of the cryoliquid by pressurizing the reservoir in the storage vessel or opening a valve in the transfer tube. In [5.24], an automated liquid helium transfer system has been described that directs helium gas elsewhere, letting only liquid helium entering the cryostat.

5.3 Closed-Cycle Refrigerators

In recent years, so-called "closed-cycle refrigerators" have been developed to a very high degree of simplicity, reliability, and efficiency. They avoid the annoying and expensive use of a cryoliquid in a low-temperature apparatus. Different concepts have been used: Joule-Thompson expansion, Stirling cycles, Gifford-McMahon cycles, and in particular pulse-tube coolers. They have in common that cooling is achieved by letting the operating gas – usually helium at 10–20 bar – perform work against internal or external forces. These refrigerators are available commercially. It would be beyond the scope of this book to describe their thermo- and hydrodynamics, as well as their technologies, and the variety of versions on the market. Short descriptions of their working principles can be found in [5.1,5.25], while more detailed overviews are [5.26–5.29], for example.

Just in short: A Gifford–McMahon cooler consists essentially of a regenerator containing a porous, usually magnetic material of high heat capacity like Er–Ni compounds and a gas displacer. A varying pressure is obtained by connecting the system via a set of (rotary) valves to the high- and the low-pressure sides of a compressor. The heat of compression is removed by cooling water. The expansion leads to cooling. One attractive feature of the Gifford–McMahon coolers is the separation of the compressor unit and expansion unit (the cold head), which are operated at different frequencies. With a single-stage Gifford–McMahon cooler, a temperature of 10–15 K can be reached. Commercial two-stage Gifford–McMahon coolers reach 4.2 K with a cooling power of 1.5 W [5.25].

In a pulse-tube cooler [5.25–5.29] the cooling effect relies on smooth (no pulses!) periodic, close to adiabatic pressure variations and displacement of the working gas in the "pulse" tube (Fig. 5.10). This is achieved by a pressure wave generator which can be either a high-frequency pressure oscillator (as in a Stirling cooler) or a low-frequency compressor with a rotating valve distributor (as in a Gifford–McMahon cooler). The second part consists of a traditional regenerator – again containing a porous magnetic material of high heat capacity – as heat reservoir, which is connected at its cold end to the pulse tube, which is a simple hollow tube. At its ambient temperature end, this tube is connected through a flow impedance or "orifice" to a reservoir; the buffer volume of the latter is large enough that negligible pressure oscillations occur in it. In this so-called "orifice-pulse-tube-cooler", the oscillating gas flow through the impedance or orifice separates the heating and cooling effects. The heat of compression (Q_o) at the generator is removed by a heat exchanger to

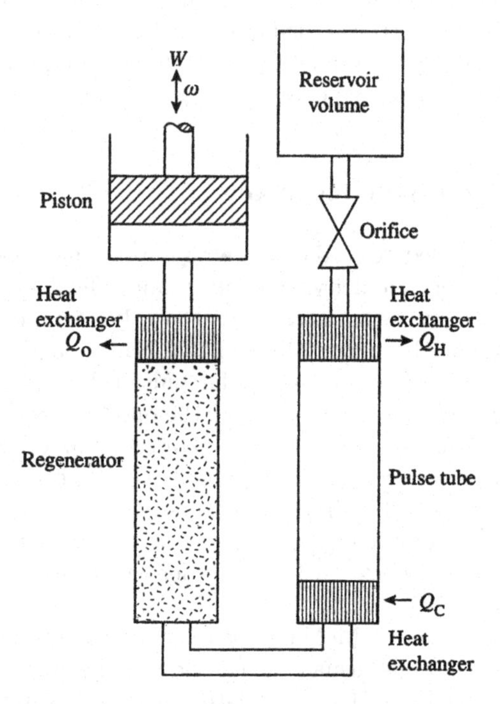

Fig. 5.10. Schematic diagram of a Stirling-type, single-orifice pulse – tube refrigerator (for details and for its operation see text)

the surrounding, usually by cooling water. During the compression phase, the direction of the gas flow in the system is basically from the piston to the reservoir. The regenerator takes up heat from the gas that flows from the compressor into the pulse tube. At the hot end of the pulse tube, part of the gas in the pulse tube (about one-third) flows through the orifice into the reservoir and gives off heat (Q_h) to the surrounding. During the expansion phase, the flow is reversed. The gas that flows from the pulse tube toward the compressor takes up heat from the regenerator. At the cold end of the pulse tube the gas that flows into the regenerator provides the desired cooling power (Q_c). For maximum cooling efficiency, the phase between the pressure wave generated by the compressor and the gas velocity in the tube has to be adjusted. This is achieved by properly adjusting the impedance to the buffer volume and the size of this reservoir. The function of the pulse tube is mostly to insulate the processes at its two ends. As a result, it must be large enough that gas flowing from its end at room temperature traverses only part way through the pulse tube before flow is reversed. Likewise, flow from the cold end never reaches the warm end. The gas in the middle portion of the pulse tube never leaves the tube, and forms a temperature gradient that insulates the two ends. Even though, this all sounds rather trivial and occurs mostly in a simple tube, the involved thermodynamics and the understanding of the various loss mechanisms are rather complex. The attractive feature of this refrigerator – besides simplicity, reliability and efficiency – is the absence of any moving parts at its cold end, unlike for the Stirling or Gifford–McMahon refrigerators. This strongly reduces the otherwise annoying influence of vibrations to the connected experiment and enhances its reliability. The cooler and its various modifications (of the Stirling type for $T > 50$ K with a frequency of typically 50 Hz, or of the Gifford-McMahon type with frequencies of typically 50 Hz at the compressor and 1–2 Hz at the tube for $T < 20$ K) have gone through a variety of improvements over the last decade (for example, a second orifice between compressor and warm end of the pulse tube has been introduced). Its principles, various designs, and the present state are described in [5.25–5.29]. Present pulse-tube coolers reach around 15–20 K as single-stage units and about 2 K as double-stage units. Their cooling power goes to zero when the thermal expansion coefficient of the used material goes to zero. This occurs, for example, at around 2 K for ^{4}He and at around 1 K for ^{3}He at a pressure of 15 bar, giving the lower limits in using these two isotopes. The minimum achieved temperature using ^{3}He as the working agent is 1.27 K [5.30]. Even though these temperatures are below the low-pressure condensing temperatures of the helium isotopes, there is no liquid-gas interface in the tube because the operating pressures are above the critical pressures of 2.28 and 1.15 bar, respectively, of the isotopes. The commercial units deliver 0.5–1 W at 4.2 K. Some suppliers offer multistage versions, for example in combination with a Joule-Thompson circuit, which work to 1.5 K.

Closed-cycle refrigerators are particularly important for cooling superconducting magnets ("cryogen-free superconducting magnets") in a variety of applications, in particular for users without other cryogenic facilities and experience, for example in hospitals or industrial applications. This technology has been mostly advanced for research purposes at Tohoku University in Sendai where cryogen-free superconducting magnets for fields up to 20 T have been built and where a cryogen-free hybrid magnet (normal conducting inner coil, superconducting outer coil) for fields up to 30 T is on its way [5.31]. A small commercial cryogen-free superconducting magnet providing a 6 ppm homogeneous field of 2.5 T in a bore of 51 mm has been presented at LT 24 [5.32]. Other important applications are cooling cryopumps or infrared sensors, for example in space missions and military applications, or for cooling superconducting electronic devices [5.29]. Closed-cycle refrigerators, in particular pulse-tube coolers, are nowadays also used as precooling stages for ^{3}He-^{4}He dilution refrigerators and nuclear demagnetization cryostats (see Sect. 7.4). Recently, a commercial system consisting of a pulse-tube cooler in combination with a paramagnetic adiabatic refrigeration stage reaching 100 mK did appear on the market (see Sect. 9.4).

5.4 Temperature Control

In many experiments, the experimenter wants to hold the temperature of his sample or experimental setup constant within some accuracy and for some length of time. For a ^{4}He bath cryostat, this can simply be achieved to about 1 mK by regulating the vapor pressure above the liquid, for example by a diaphragm device [5.33]. In a flow cryostat, the flow of the cryogenic liquid or of its gas can be regulated. In a charcoal pumped setup, one can regulate the temperature of the charcoal. However, in most cases, controlled heat is applied electrically by a heater. In most cases, it is much more efficient not to regulate the refrigerator's temperature but to let it float and regulate the temperature of the sample or of the experimental setup weakly coupled to the bath. To achieve the regulation, the signal from the thermometer at the place to be regulated, usually taken from a bridge, is compared to some reference signal corresponding to the desired temperature. The difference is amplified and fed back to the heater on the place to be regulated. The time constant for a stable and accurate regulation is $\tau = R\,C$ with the thermal resistance R between regulated place and bath, and C the heat capacity of the regulated setup. For designing an electronic regulation circuit, one has to consider [5.1] the change of C and R with temperature, choice of time constant, loop gain and its dynamic range, time constant of electronics, and the appropriate thermometer (mostly resistance thermometers) with some temperature dependent often non-linear characteristic, which lead to the necessity of using a proportional, integral, derivative (PID) control (Fig. 5.11). I do not discuss these matters here in detail because there are rather good electronic temperature regulators of varying degrees of quality and capability on the market, if a lock-in amplifier in a feedback loop alone is not sufficient. Another important

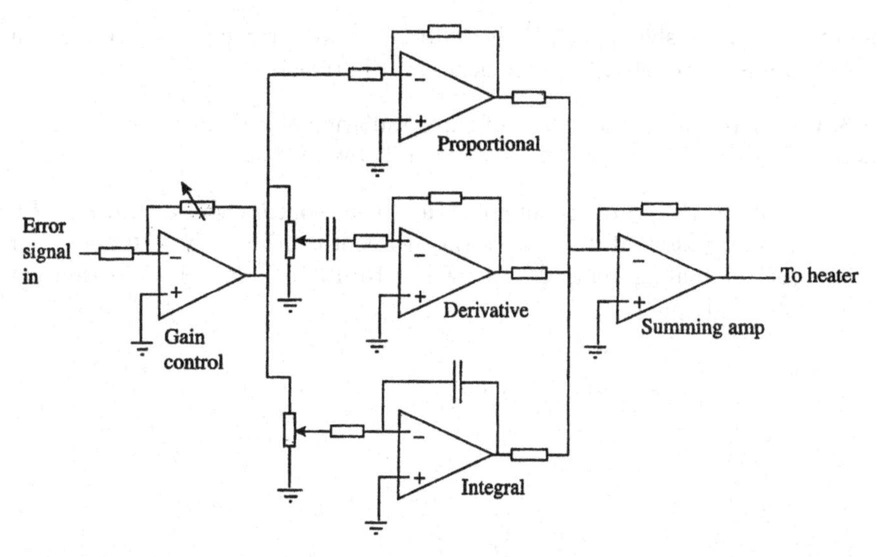

Fig. 5.11. A circuit to provide proportional, integral and derivative (PID) signals for feedback temperature control [5.1]. The error signal, usually the difference between the signal from a thermometer and a reference signal, is amplified and given to a heater for temperature control. Copyright (2002), Oxford University Press

consideration – of which industry cannot take care off – is, of course, the correct location of heater and thermometer; for example, to avoid heat flow from the heater directly to the thermometer. If very high accuracy and stability in temperature regulation is required like in measurements of heat capacity or thermal conductivity, it may be necessary to put a thermal shield around the experiment whose temperature may be regulated in addition.

Problems

5.1. How do the numbers in the second line of Table 5.1 change if the three metals are cooled by H_2 from 77 to 20 K, the normal boiling temperature of liquid hydrogen (for material parameters see Tables 2.1 and 10.1)?

5.2. Calculate the relative contributions of heat transferred to a container filled with liquid ^{4}He at 2 K from heat radition, from heat conduction along the supporting tube as well as from conduction by gas particles in the vacuum space. For the calculation assume that the surface area A of the vessel is 100 cm^2 and that it is surrounded by a radiation shield at 77 K. The supporting tube is made from $Cu_{0.7}Ni_{0.3}$ of 40 mm diameter and 0.2 mm wall thickness, and its length between the helium vessel and a thermal sink at 77 K is 150 mm (assuming that the thermal conductivity given in Table 3.2 extends to 77 K). The remaining gas pressure in the vacuum space between the helium vessel and the radiation shield at 77 K should be 10^{-5} mbar (take $a = 1$). How much liquid ^{4}He is evaporated by these heat contributions?

5.3. Calculate the temperatures of three radiation shields in the glass dewar depicted in Fig. 5.1 (neglect the styrofoam between them).

5.4. Calculate the throughput of ^{4}He in a continously operating ^{4}He evaporation cryostat at 1.5 K necessary to cool continuously 50 μmol s^{-1} of ^{3}He from its boiling temperature of 3.2 to 1.3 K in a ^{3}He–^{4}He dilution refrigerator (Chap. 7).

6

Helium-3 Cryostats

The temperature range accessible with a liquid ^{4}He bath is typically 4.2–1.3 K, but this temperature range can be extended to about 0.3 K if the rare isotope ^{3}He is used instead of the common isotope ^{4}He. The main reason is that ^{3}He has a substantially larger vapour pressure than ^{4}He at the same temperature; the ratio P_3/P_4 is 74 at 1 K but about 10^4 at 0.5 K (Figs. 2.7 and 2.8). A further advantage of using liquid ^{3}He instead of liquid ^{4}He at temperatures below their normal boiling point is due to the fact that the specific heat of liquid ^{3}He varies much less between, for example, 2 and 0.5 K than the specific heat of liquid ^{4}He does (Fig. 2.9). One therefore has to evaporate only about 20% of ^{3}He to cool this liquid from 1.5 to 0.3 K by using its own heat of evaporation. Furthermore, the specific heat of liquid ^{3}He is larger than the specific heat of liquid ^{4}He below 1.5 K, resulting in a larger heat reservoir in this temperature range. Finally, liquid ^{3}He is not superfluid in the temperature range of concern in this chapter. One therefore does not have the heat transfer problems sometimes arising from the superfluid film flow of liquid ^{4}He (Sect. 2.3.5).

There are two rather serious disadvantages of liquid ^{3}He. Firstly, its latent heat of evaporation is substantially smaller than that of liquid ^{4}He, see Fig. 2.6. Secondly, and more importantly, ^{3}He is much more expensive than ^{4}He. A typical price is about 200 € per liter gas or about 150 € per cm^3 liquid (a liter of liquid ^{4}He, if bought commercially, can be obtained for about 6 €). Due to this high price, which results from the expensive production method of ^{3}He (Sect. 2.3.1), one can only use ^{3}He in a closed gas handling and cryogenic system, making sure that no gas is lost. As a second consequence, one never has enough ^{3}He gas to liquefy it on a technological scale in a liquefaction plant. The ^{3}He is transformed from the gaseous to the liquid state in the cryostat in which it will be used for doing experiments just by bringing it in contact with a ^{4}He bath at a temperature of, say, 1.3 K. This temperature can be obtained by the continuously evaporating ^{4}He refrigerator discussed in Sect. 5.2.4. As the critical temperature of ^{3}He is 3.3 K, the ^{3}He gas will condense on surfaces which are below this temperature if the gas pressure is high enough.

Because the use of liquid ^{3}He as a refrigerant occurs in a ^{4}He cryostat anyway, all the parts refrigerated by the ^{3}He can be surrounded by shields kept at ^{4}He temperature. In addition, all the tubing and wiring going to the experiment can be thermally heat sunk at the ^{4}He bath. Therefore, the heat transferred from the outside world to parts below 1 K can be absorbed by the ^{4}He bath, which has a substantially larger volume and also a larger heat of evaporation than the ^{3}He bath. The ^{3}He bath is then only used for cooling from the temperature of the ^{4}He bath to the temperature obtained by the ^{3}He refrigerator. In any case, the use of an evaporating ^{3}He bath is the simplest way of reaching temperatures between 0.3 and 1 K, and several suppliers offer ^{3}He cryostats commercially. Reports on ^{3}He cryostats can be found in [6.1–6.16].

6.1 Helium-3 Cryostats with External Pumps

Figure 6.1 schematically presents typical designs of ^{3}He cryostats in the order of increasing sophistication. In each design a few cm^3 of ^{3}He are liquefied by bringing the ^{3}He in thermal contact with a ^{4}He bath, which is pumped to $T \leq 1.5$ K. Figure 6.1a shows a setup where the pre-cooling stage is the main ^{4}He bath pumped to a temperature of about 1.3 K and which absorbs the latent heat of condensation of the incoming ^{3}He. The ^{3}He gas condenses on the cold surfaces of the thin-walled pumping tube leading to and supporting the ^{3}He pot (surrounded by vacuum), which will slowly cool and eventually collect liquid ^{3}He. When all the ^{3}He has condensed we pump on this liquid to reduce its temperature from about 1.3 K to the desired temperature; the minimum is typically 0.3 K.

In the second setup the main ^{4}He bath is not pumped but is left at normal pressure and 4.2 K, and to condense the ^{3}He we use a continuously evaporating ^{4}He refrigerator, as discussed in Sect. 5.2.4. Both designs utilize the single-cylce discontinuous refrigeration method for the ^{3}He part because eventually all the ^{3}He is evaporated (and is hopefully recovered!) and we have to recondense it to start again.

Finally, in the third design the ^{3}He refrigerator, too, is run in a continuous mode by introducing a recondensing tube. The ^{3}He vapour that we pump away from the liquid ^{3}He bath at its vapour pressure will leave the room-temperature pump at a pressure of several 0.1 bar, it is pre-cooled by the ^{4}He pot and eventually recondensed into the ^{3}He pot. For this purpose the ^{3}He gas will run through a heat exchanger in the 4.2 K bath as well as a second heat exchanger in the 1.3 K ^{4}He bath where it will, of course, condense. The now liquid ^{3}He will then be isenthalpically (not isothermally) expanded through an impedance of order 10^{12}–10^{13} cm^{-3}, which maintains a pressure sufficient for condensation before the ^{3}He arrives as a low-pressure, low-temperature liquid at the pumped ^{3}He pot. In such a design we have introduced several features (like the heat exchanger and impedances) which

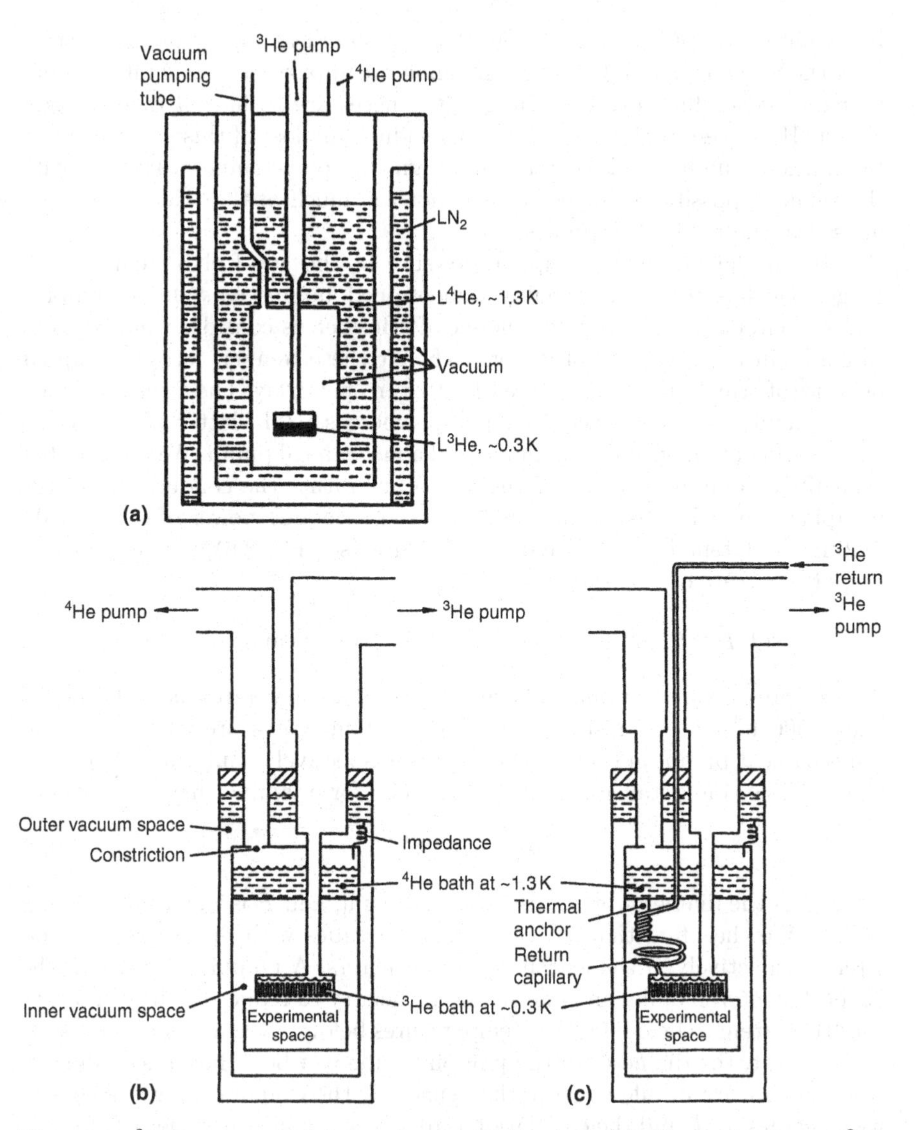

Fig. 6.1. ^{3}He cryostats of increasing sophistication. (**a**) Non-recirculating ^{3}He refrigerator with a pumped main ^{4}He bath. (**b**) Non-recirculating ^{3}He refrigerator with a continuously operating ^{4}He evaporator. (**c**) Recirculating ^{3}He refrigerator with a continuously operating ^{4}He evaporator [6.4]

will be discussed in more detail in Chap. 7 on ^{3}He–^{4}He dilution refrigerators. The narrow capillaries used as impedances can easily be blocked if there are impurities such as frozen air in the ^{4}He or ^{3}He entering them. One therefore has to be careful to avoid these impurities and, in particular, one may use a LN_2 cooled trap (Fig. 7.25) after the ^{3}He room-temperature pump to freeze out oil vapour or oil crack products possibly leaving the pump together with the ^{3}He gas, as well as any remaining air impurities. It is quite obvious that due to the high price of ^{3}He gas it is necessary to have a vacuum-tight closed ^{3}He system with a sealed pump avoiding any loss of this expensive gas. In addition, one has to be careful to design the room-temperature part with the smallest possible volumes, so that not too much of the ^{3}He gas remains unused in these "dead" volumes.

Due to the low helium-vapour pressure in the sub-Kelvin temperature range, one has to use reasonably dimensioned pumps as well as pumping tubes to circulate the required amount of ^{3}He. Let us consider what we need to maintain temperatures of 0.5 or 0.3 K. At these temperatures the vapour pressure of ^{3}He is about 0.2 mbar and 2 μbar, respectively. The former pressure can be maintained by a mechanical pump, whereas for the latter one we need a combination of an oil-diffusion pump and a mechanical pump. We will see that very often the limitation is not the pump but rather the conductance of the pumping tubes. Let us assume that we need a cooling power of $\dot{Q} = 1\,\mathrm{mW}$. Taking the latent heat of evaporation of ^{3}He (see Fig. 2.6), this requires an evaporation rate of

$$\dot{V} = \dot{Q}/L \simeq 5\,\mathrm{cm^3\ liq\ ^3He\,h^{-1}} \simeq 3\,\ell\,\mathrm{gas\ ^3He\,h^{-1}} \text{ at } P = 1\,\mathrm{bar}\,. \tag{6.1}$$

At the pressures mentioned above, the volume flow rates have to be 15 and $1,500\,\mathrm{m^3\,gas\,h^{-1}}$, respectively. These volume rates are no problem for a mechanical pump and a diffusion pump, respectively. But what about the tubing? The conductance $L[\mathrm{m^3\,h^{-1}}]$ of a tube for a laminar flow is given by

$$L = 486\bar{P}d^4/l\,, \tag{6.2}$$

where d is the diameter [cm], l is the length [cm], and $\bar{P}$ is the mean pressure [mbar]. We then find that we need pumping tubes with diameters of 3 and 10 cm, respectively, if the length is several meters. A pumping tube with the latter diameter is bulky and one should rather try to reduce the heat input to the ^{3}He system to below 1 mW if temperatures below 0.5 K are required. Inside the cryostat, the diameter of the pumping tube can be reduced according to the temperature profile because the density of the evaporating gas increases with decreasing T and the circulation rate $\dot{n}$ is the same everywhere [6.4,6.14].

6.2 Helium-3 Cryostats with Internal Adsorption Pumps

One can avoid the room-temperature pump as well as the often-bulky pumping tubes by inserting a cold adsorption pump inside the cryostat. Gases adsorb at cold surfaces if their temperature is low enough. If we keep a large surface at a low enough temperature above our ^{3}He bath, this surface will pump the helium vapour and keep the liquid ^{3}He at a low temperature [6.6–6.16]. There are various suitable materials (e.g., charcoal, zeolites or fine metal powder) with surface areas of at least several $m^2\,g^{-1}$ [6.10,6.11,6.17] (see also Sect. 13.6). If we fill a volume of several cm^3 with such an adsorbent with large surface area at low temperature, it will very effectively pump the liquid ^{3}He bath. When all the ^{3}He has been pumped away, so that the ^{3}He pot is empty, we just have to lift the charcoal pumping system into a space at higher temperature in the cryostat to desorb the helium, which will then enter the gas phase, condense at the cold surfaces of the cryostat and eventually drip back down into the ^{3}He pot. Such a cryostat with a hermetically sealed ^{3}He system reaching 0.25 K is shown in Fig. 6.2. An alternative way of switching between the pumping and the releasing state of the adsorbent (instead of lifting and lowering this part) is to switch the thermal contact of the adsorbent with the surrounding ^{4}He bath on and off. This can be done by using helium exchange gas in the

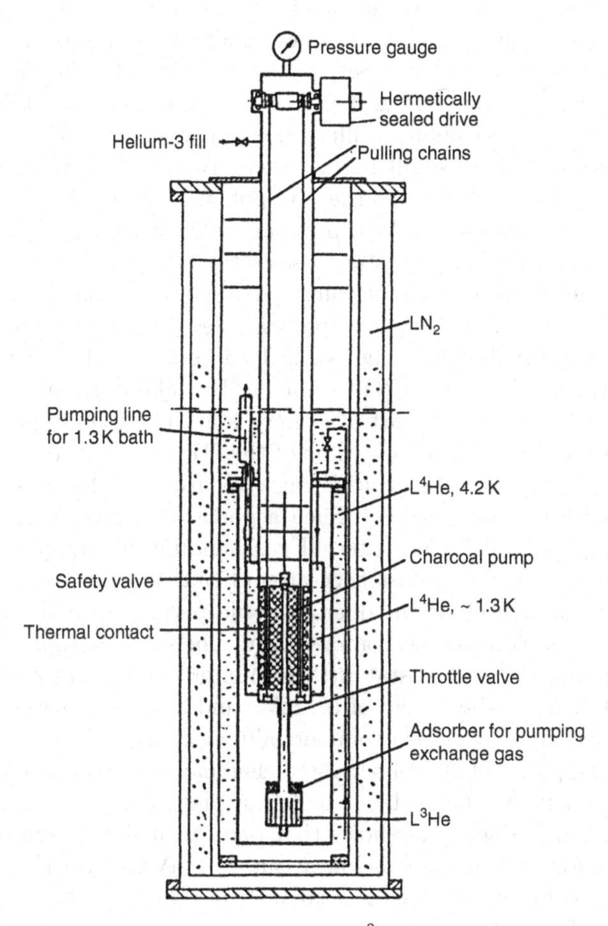

Fig. 6.2. Hermetically sealed, charcoal pumped ^{3}He refrigerator. The charcoal pump can be raised or lowered by means of a chain drive operated via a hermetically closed rotating seal [6.11]

space around the vessel containing the adsorbent and pumping this exchange gas away when the adsorbent has to be thermally isolated and heating it by a heater to a higher temperature to release the ^{3}He gas. We have to bear in mind that the heat of adsorption is rather large – of the order of the latent heat of evaporation – so the adsorbent has to be in good thermal contact with the ^{4}He bath at about 1.4 K to remove the heat of adsorption and avoid an unwanted high temperature of the adsorbent. The pumping speed of charcoal pumps is not only a function of its temperature but also, in practice, a complicated function of geometry, thermal coupling and pressure [6.10, 6.11, 6.17, 6.18].

Cold charcoal pumping systems are also more efficient than room-temperature mechanical pumps because they are connected via a short, cold pumping tube to the ^{3}He pot, taking advantage of the very high pumping speed of the adsorbent (see below).

A very versatile ^{3}He cryostat which can be inserted into a storage ^{4}He dewar (50 mm neck diameter) has been designed by Swartz [6.14, 6.15], as well as in our laboratory at the University Bayreuth (Fig. 6.3). It is an extension of the "^{4}He dipper cryostat" mentioned in Sect. 5.2.4 (Fig. 5.6). The first stage of the cryostat is a continuously filling ^{4}He pot at about 1.3 K, on which the ^{3}He condenses out of its small room-temperature storage volume on top of the cryostat and then drips into the ^{3}He pot. Cooling of the ^{3}He to 0.3 K is achieved (within a few minutes!) by pumping with an activated charcoal pump which is located inside the cryostat close to the ^{3}He pot. The charcoal pump is equipped with its own (charcoal pumped) vacuum/exchange gas space and heater so that its temperature can be regulated between about 5 and 25 K to adsorb or desorb the ^{3}He. The small dead volumes allow the quantity of ^{3}He to be restricted to just 1 STP liter of ^{3}He (about 1.5 cm^3 of liquid) in the permanently sealed ^{3}He part of the cryostat. The pump and heaters can be computer controlled. The hold-time of the ^{3}He charge in the pot depends, of course, on the heat load; typical times are from 3 h for $\dot{Q} \approx 0.1$ mW to about 20 h with no external load. Such a portable ^{3}He cryostat requires no transfer of liquid ^{4}He and no external gas handling or pumping system for the ^{3}He part, it has a very fast turn-around time and low cost. The design and construction, as well as the operation and performance, are described in detail in [6.15]; it is available commercially. The novel design of a compact adsorption pumped ^{3}He cryostat with a minimum temperature of 0.24 K, a hold time of 30 h at a heat load of 0.1 mW, and a cooling power of 1 mW at 0.3 K described in [6.16] is also commercially available.

The cooling power of 20 g of activated charcoal as a function of the heating power is shown in Fig. 6.4, while Fig. 6.5 displays the "pumping power" of activated charcoal. Figure 6.4 shows that one can maintain temperatures as low as 0.25 K for heating rates below about 0.01 mW, and 0.4 K at 1 mW.

Of course, cold materials with a large surface area are not only useful as a pump for reducing the temperature above an evaporating cryogenic liquid. They are also quite useful in the vacuum space of cryogenic vessels, where they can substantially improve the vacuum by adsorbing remaining gas particles

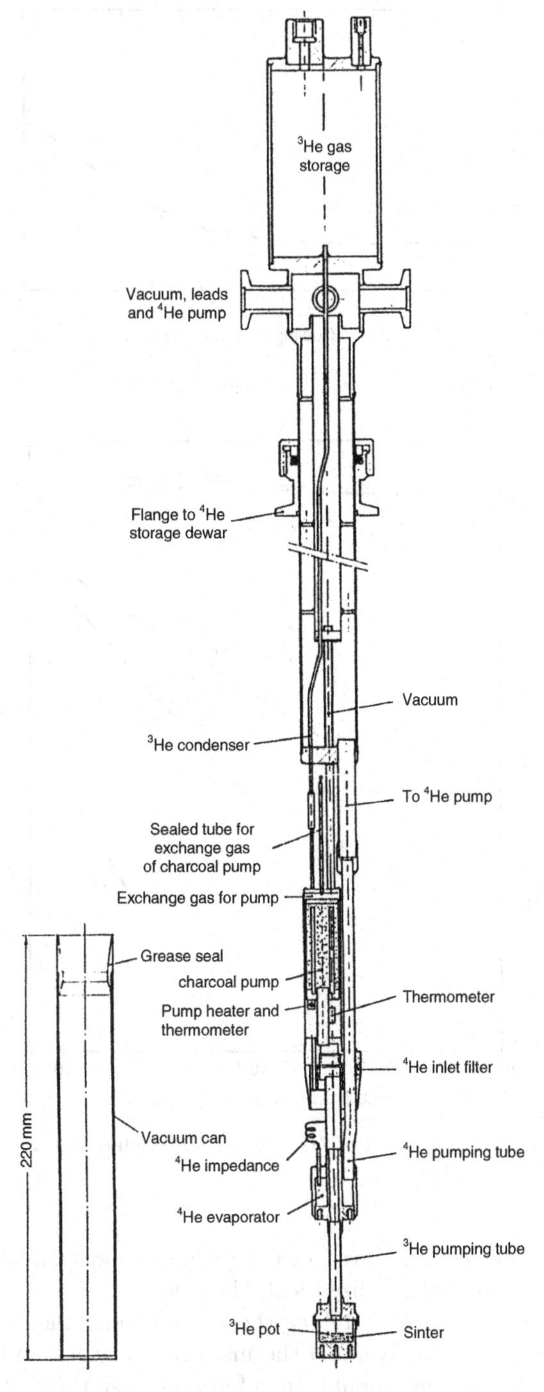

Fig. 6.3. ^{3}He dipstick refrigerator with ^{3}He gas storage, charcoal pump, continuously operating ^{4}He refrigerator and ^{3}He refrigerator; for details see text and [6.15] (courtesy of P. Sekowski, Universität Bayreuth)

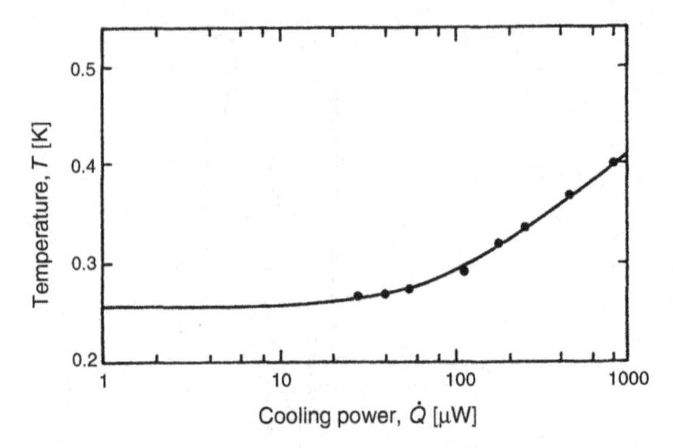

Fig. 6.4. Cooling power of a ^{3}He refrigerator pumped by 20 g of charcoal [6.11]

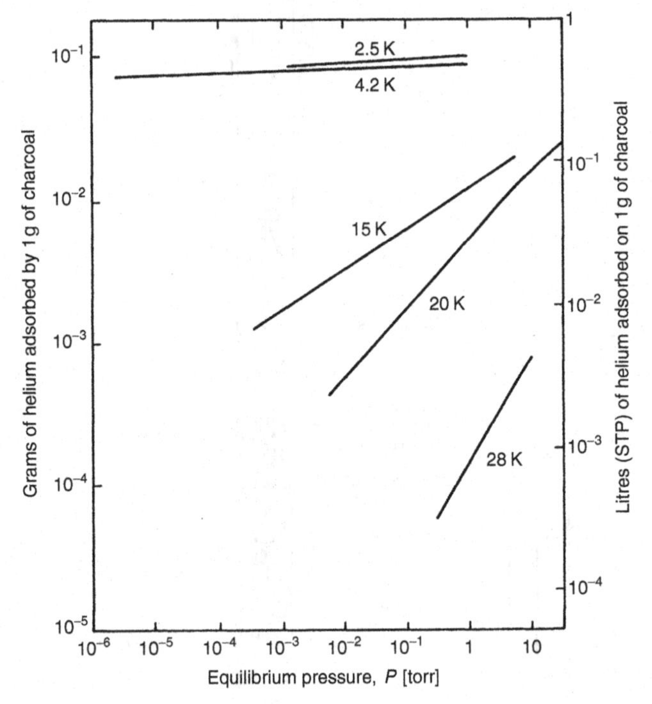

Fig. 6.5. Adsorption isotherms of ^{3}He on activated charcoal as a function of the helium-gas pressure [6.11]

after cooling to low temperatures, or even make the gas entering the vacuum space through a tiny, not localized leak "harmless".

With a ^{3}He cryostat we are taking the first step into the range of $T < 1$ K. At these temperatures the Kapitza thermal boundary resistance (Sect. 4.3.2) can be of importance. One should, therefore, increase the surface area of the ^{3}He pot to improve the thermal coupling between the cryoliquid and the wall of its container. This can be done by making grooves in the inside of the ^{3}He pot or, better, by sintering fine metal powder (Sect. 13.6) with a total surface area of about 1 m^2 to the bottom of the container.

Helium-3 cryostats were quite popular until about the end of the 1960s, when the ^{3}He–^{4}He dilution refrigerator (see Chap. 7) was invented. This type of refrigerator reaches substantially lower temperatures. Today, ^{3}He refrigerators are mainly used if only a simple setup is possible or necessary, or when one needs a very high cooling power at temperatures between about 0.4 and 1 K.

Problems

6.1. Calculate how much of a liquid ^{3}He bath has to be evaporated to cool it from 1 to 0.3 K.

6.2. How much liquid ^{3}He has to be evaporated to refrigerate 1 kg of Cu from 1 to 0.3 K?

6.3. Calculate the cooling power of a ^{3}He refrigerator at $T = 0.3$ K, if a $100\,l\,s^{-1}$ pump is used (see Fig. 2.6 for the latent heat of evaporation and Fig. 2.7 for the vapour pressure).

6.4. Calculate the surface area of a sintered Ag heat exchanger necessary for a ^{3}He cryostat operating at 0.3 K and $\dot{Q} = 0.1$ mW for the data displayed in Fig. 4.6 and extrapolated from Fig. 4.7.

6.5. How many grams of charcoal does one need to keep a ^{3}He refrigerator running at a heat input of 0.2 mW at 0.5 K?

7

The ^{3}He–^{4}He Dilution Refrigerator

Until about the 1950s the only method of achieving temperatures below 1 K was magnetic refrigeration using demagnetization of a paramagnetic salt (Chap. 9). This method was then replaced for the temperature range down to 0.3 K by the ^{3}He refrigerator, as discussed in Chap. 6. But in 1962 a new proposal for continuous refrigeration with liquid helium to even lower temperatures was published by H. London, G.R. Clarke, and E. Mendoza, based on an idea proposed by H. London about ten years earlier [7.1]. In contrast to the helium refrigerators discussed earlier, where the latent heat of evaporation is used for cooling, it was suggested to use the heat of mixing of the two helium isotopes to obtain low temperatures. A group at Leiden University built the first refrigerator based on this principle in 1965 and reached a temperature of 0.22 K [7.1], the lowest temperature obtained by the helium liquids up to that time. Only one year later, B.S. Neganov and co-workers in Dubna, and H.E. Hall and co-workers in Manchester published their results with an improved design of a "^{3}He–^{4}He refrigerator" by which they could reduce this temperature by about a factor of three; the final version of the Dubna refrigerator soon reached 25 mK [7.1]. The ^{3}He–^{4}He dilution refrigeration method to be discussed in detail in this chapter is the only continuous refrigeration method for temperatures below 0.3 K. In addition, magnetic fields, often needed in low-temperature experiments, have negligible effects on its performance. Today it is the most important refrigeration technology for the temperature range between about 5 mK and 1 K, and it is the base from which lower temperatures can be reached.

The complexity of a ^{3}He–^{4}He refrigerator for the temperature range above 30 mK is comparable to the complexity of a ^{3}He cryostat. Actually, existing ^{3}He cryostats can easily be modified to ^{3}He–^{4}He refrigerators. If the temperature range has to be extended to about 15 mK, the system, of course, will become more involved, but it can still be built in the laboratory workshop. If the temperature has to be reduced to values even lower than 10 mK, substantial experience, time and manpower are necessary to achieve this goal, if you want to build the refrigerator yourself. Fortunately, these days commercial dilution refrigerators are available for temperatures down to about 4 mK, (more typical is T_{min} = 8–10 mK) but they are expensive. The minimum temperature obtained by the method to be discussed is slightly below 2 mK, achieved by Frossati and co-workers, formerly at Grenoble and then at the University of Leiden [7.2,7.3], as well as by Pickett and co-workers at Lancaster University [7.4].

Today, ^{3}He–^{4}He dilution refrigerators are part of virtually all apparatus reaching $T \leq 0.3$ K. Detailed discussions of the principle and methods of dilution refrigeration have been published by Wheatley et al. [7.5–7.7], Radebaugh and Siegwarth [7.8], and Frossati [7.2, 7.3] and their status in the seventieth of last century were reviewed in [7.9–7.11]. I will start with a discussion of some relevant properties of ^{3}He–^{4}He mixtures.

7.1 Properties of Liquid ^{3}He–^{4}He Mixtures

In this section I will discuss the properties of isotopic liquid helium mixtures – the working fluid of the refrigerator – that are relevant for the design and operation of a ^{3}He–^{4}He dilution refrigerator. For more details the reader is referred to the specialized literature [7.2–7.18].

In the following, the respective concentrations of the two helium isotopes are expressed as

$$x = x_3 = \frac{n_3}{n_3 + n_4} \quad \text{and} \quad x_4 = \frac{n_4}{n_3 + n_4}, \tag{7.1}$$

where $n_3(n_4)$ is the number of ^{3}He(^{4}He) atoms or moles.

7.1.1 Phase Diagram and Solubility

The x–T phase diagram of liquid ^{3}He–^{4}He mixtures at saturated vapour pressure is depicted in Fig. 7.1. This figure shows several of the remarkable features of these isotopic liquid mixtures. First we consider the pure liquids. We are reminded that liquid ^{4}He becomes superfluid at a temperature of 2.177 K. On the other hand, the Fermi liquid ^{3}He does not show any phase transition in the temperature range considered in this chapter (actually this liquid also becomes superfluid in the low millikelvin temperature range; see Sect. 2.3.6). The temperature of the superfluid phase transition of liquid ^{4}He is depressed if we dilute the Bose liquid ^{4}He with the Fermi liquid ^{3}He. Eventually the ^{4}He superfluidity ceases to exist for ^{3}He concentrations above 67.5%. At this concentration and at a temperature of 0.867 K the λ-line meets the phase separation line; below this temperature the two isotopes are only miscible for certain limiting concentrations which depend on the temperature. The shaded phase separation region in the figure is a non-accessible range of temperatures and concentrations for helium mixtures. If we cool a helium mixture (with

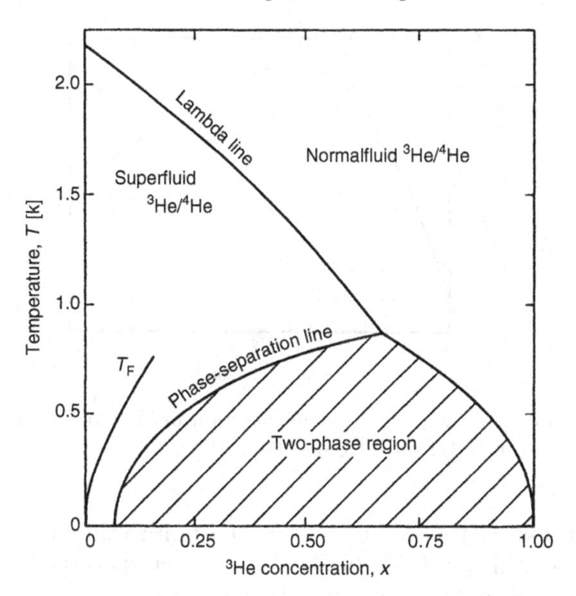

Fig. 7.1. Phase diagram of liquid ^{3}He–^{4}He mixtures at saturated vapour pressure. The diagram shows the lambda line for the superfluid transition of ^{4}He, the phase separation line of the mixtures below which they separate into a ^{4}He-rich and a ^{3}He-rich phase, and the line of the Fermi temperatures T_F of the ^{3}He component (From [7.11, 7.17] which give references to the original work from which data were taken to construct this phase diagram)

$x > 6.6\%$) to temperatures below 0.87 K, the liquid will eventually separate into two phases, one rich in ^{4}He and the other rich in ^{3}He. Because of its lower density, the ^{3}He-rich liquid floats on top of the ^{4}He-rich liquid. If the temperature is decreased to close to absolute zero, we see that the ^{3}He-rich liquid becomes pure ^{3}He. But the great surprise occurs at the ^{4}He-rich side. Here the concentration of the dilute isotope, ^{3}He, does not approach zero for T approaching zero, but rather reaches a constant concentration of 6.6% ^{3}He in ^{4}He at saturated vapour pressure even for $T = 0$ K. This finite solubility is of utmost importance for ^{3}He–^{4}He dilution refrigeration technology. The limiting concentrations of the diluted isotopes on the phase separation line at low temperatures and saturated vapour pressure are given by

$$x_4 = 0.85T^{3/2}\,\mathrm{e}^{-0.56/T} \quad [7.12, 7.19]\,, \tag{7.2}$$

$$x = x_3 = 0.066(1 + 8.3T^2) \quad (\text{for } T < 0.1\,\mathrm{K}) \quad [7.12, 7.20, 7.21]\,. \tag{7.3}$$

The solubility of ^{3}He in ^{4}He can be increased to almost 9.5% by raising the pressure to 10 bar (Fig. 7.2) [7.20, 7.21].

As we will see below, cooling in a ^{3}He–^{4}He dilution refrigerator is achieved by transferring ^{3}He atoms from the pure ^{3}He phase to the diluted, mostly ^{4}He containing phase. The cooling capacity in this cooling process is the heat of

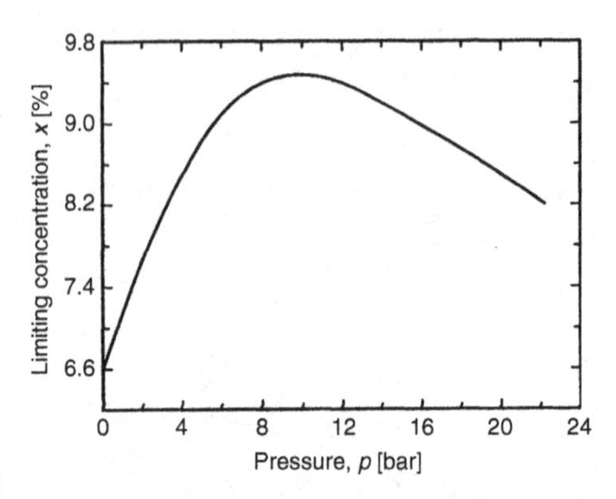

Fig. 7.2. Limiting low-temperature concentration of ^{3}He in ^{4}He at $T = 50\,\mathrm{mK}$ as a function of pressure (after data from [7.20, 7.21])

mixing of the two isotopes. But why is the finite solubility of ^{3}He in ^{4}He so important for this method? In Sect. 2.3.2 we studied the equation for the cooling power of an evaporating cryogenic liquid

$$\dot{Q} = \dot{n}\Delta H = \dot{n}L\,. \tag{7.4}$$

If we make use of the latent heat L of evaporation as we did in the previous chapters on refrigerators, for example by pumping with a pump of constant volume rate $\dot{V}$ on a ^{3}He or ^{4}He bath with vapour pressure P, we obtain

$$\dot{Q} = \dot{V}P(T)L(T)\,. \tag{7.5}$$

Because the latent heat of evaporation changes only weakly with temperature (Fig. 2.6), the temperature dependence of the cooling power (2.13) is essentially given by

$$\dot{Q} \propto P(T) \propto \mathrm{e}^{-1/T}\,. \tag{7.6}$$

For ^{3}He–^{4}He dilution refrigeration the corresponding quantities are the enthalpy ΔH of mixing, which is given by the integral of the differences of the specific heats of the two phases,

$$\Delta H \propto \int \Delta C\,\mathrm{d}T\,, \tag{7.7}$$

and the concentration x of the dilute ^{3}He phase, which is almost constant at $T \leq 0.1\,\mathrm{K}$, 6.6% ^{3}He, see (7.3), in contrast to the vapour density in the refrigerator discussed earlier, where the number of atoms decreases exponentially with temperature. The high ^{3}He particle density in the dilute phase is essential for ^{3}He–^{4}He dilution refrigeration because it permits a high ^{3}He molar flow rate. Because the specific heats of concentrated and diluted ^{3}He are proportional to T at low enough temperatures (Figs. 2.15 and 7.5), we end

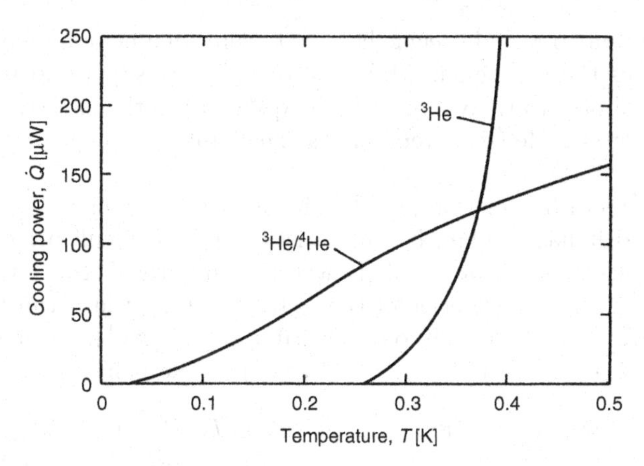

Fig. 7.3. Cooling power of a ^{3}He evaporation cryostat and of a ^{3}He–^{4}He dilution refrigerator, assuming that the same pump with a helium gas circulation rate of $5\,l\,\mathrm{s}^{-1}$ is used [7.9]

up with the following temperature dependence of the cooling power of this dilution process:

$$\dot{Q} \propto x\Delta H \propto T^2\,; \tag{7.8}$$

this result is shown in Fig. 7.3.

This substantial advantage of the ^{3}He–^{4}He dilution process, that the temperature dependence of the cooling power is weaker than that of the evaporation process, was only realized after researchers detected the finite solubilityof ^{3}He in ^{4}He even for T approaching absolute zero. Before this experimental discovery it was believed that the liquid helium isotopes – like other two-component liquids – have to fully separate into the two pure liquids when the temperature is low enough to fulfil the third law of thermodynamics, so that the entropy of mixing is zero for $T = 0$. Of course, the helium liquids have to fulfil this law as well, and they can do so even if they do not separate completely for $T \to 0$ because they are quantum liquids. For $T = 0$ the ^{3}He–^{4}He mixtures are in their fully degenerate Fermi momentum ground state for the ^{3}He part ("one ^{3}He particle per state") and with ^{4}He being superfluid, both with $S = 0$, whereas for a classical system, a finite solubility means $S > 0$.

7.1.2 ^{3}He–^{4}He Mixtures as Fermi Liquids

The isotope ^{4}He has a nuclear spin $I = 0$ and therefore in its liquid state it obeys Bose statistics. Such a Bose liquid will undergo a so-called *Bose condensation* in momentum space at low-enough temperature, and for liquid ^{4}He this corresponds to its transition to the superfluid state at 2.177 K. At $T < 0.5$ K liquid ^{4}He is almost totally condensed into this quantum mechanical ground state; there are essentially no excitations (phonons, rotons) left.

Its viscosity, entropy and specific heat go to zero. In a helium mixture at these temperatures the component ^{4}He acts as an "inert superfluid background", which contributes to the volume of the liquid and to the effective mass of the dissolved isotope ^{3}He (see later) but has negligible heat capacity, for example (Fig. 2.9).

The rare and lighter isotope ^{3}He with its nuclear spin $I = 1/2$ is a Fermi particle, and it has to obey Fermi statistics and the Pauli principle like the conduction electrons in a metal. However, unlike the Fermi temperature of conduction electrons, which is of the order of 10^4 K, the Fermi temperature of liquid ^{3}He is of order 1 K only (because $T_F \propto m^{-1}$). In analogy to the conduction electrons, the specific heat of liquid ^{3}He behaves as

$$\text{Fermi-degenerate}: C_3 = (\pi^2/2)(T/T_F)R \quad \text{at} \quad T \ll T_F \tag{7.9}$$

or

$$\text{classical}: C_3 = (5/2)R \quad \text{at} \quad T > T_F \quad \text{and} \quad P = \text{const.} \tag{7.10}$$

This means the behavior is classic-gas-like at $T > 1$ K but Fermi-gas-like at $T \ll 0.1$ K.

Of course, if ^{3}He is diluted by ^{4}He in a liquid–helium mixture, it still has to obey Fermi statistics, but now it is diluted. Therefore, its Fermi temperature and its effective mass (which is a measure of its interactions with the surroundings) are altered. Many experiments confirm that the properties of liquid ^{3}He–^{4}He mixtures can be described by the laws for an interacting Fermi gas [7.5, 7.6, 7.12–7.18, 7.22–7.24]; because of the dilution by ^{4}He, this description is much better for dilute ^{3}He–^{4}He mixtures than for concentrated ^{3}He. That the liquid helium mixtures are not a *dilute non-interacting* gas is taken into account by replacing the bare ^{3}He mass m_3 in the equations by the effective mass m^*. The effective mass is slightly dependent on ^{3}He concentration, with $m^*/m_3 = 2.34$ for $x \to 0$ and $m^*/m_3 = 2.45$ for $x = 6.6\%$ at $P = 0$ [7.21, 7.25–7.28] ($m^*/m_3 = 2.78$ for pure ^{3}He at SVP [7.29], see Fig. 2.15), and more strongly dependent on pressure; these values are calculated with (7.22) from the measured specific-heat data. In this description ^{3}He is treated as an interacting quasi-particle Fermi gas with a pressure equal to the osmotic pressure of ^{3}He in ^{4}He (Sect. 7.1.5).

7.1.3 Finite Solubility of ^{3}He in ^{4}He

Let us consider whether a ^{3}He atom would prefer to be in a vessel filled with liquid ^{3}He or in a vessel filled with liquid ^{4}He. Each atom that hits the phase separation line of a phase-separated ^{3}He–^{4}He mixture is asked this question and has to decide whether to stay in the upper ^{3}He phase or go into the lower ^{4}He rich phase. The ^{3}He atom will go into the phase where it has the larger binding energy.

In the following I shall discuss the situation for $T = 0$. An extension to finite temperatures does not change the essentials of the results. A discussion for finite temperatures can be found in the books of Lounasmaa [7.9] and Betts [7.10], for example.

^{3}He in Pure ^{3}He ($x = 1$)

The chemical potential[1] of pure liquid ^{3}He is given by the latent heat of evaporation,

$$\mu_{3,\mathrm{c}} = -L_3\,, \tag{7.11}$$

corresponding to the binding energy of ^{3}He in liquid ^{3}He. So L_3/N_0 is the energy one has to supply to remove one ^{3}He atom from liquid ^{3}He into vacuum.

One ^{3}He Atom in Liquid ^{4}He ($x \simeq 0$)

For the dilute phase, the binding energy of a ^{3}He atom in liquid ^{4}He ($x \to 0$) is given by

$$\frac{\mu_{3,\mathrm{d}}(0)}{N_0} = -\epsilon_{3,\mathrm{d}}(0)\,. \tag{7.12}$$

We have already discussed in Sect. 2.3 that due to the identical electronic structure of the helium isotopes the van der Waals forces between them are identical. But due to its smaller mass the ^{3}He atom has a larger zero-point motion than the ^{4}He atom. Therefore, in the liquid phase ^{4}He atoms occupy a smaller volume than ^{3}He atoms. The ^{3}He atom will be closer to the ^{4}He than it would be to ^{3}He atoms or, in other words, its binding – due to the smaller distance or larger density – is stronger if it is in ^{4}He than it would be in ^{3}He. Because the ^{3}He atom will be more strongly bound in ^{4}He it will prefer to stay in liquid ^{4}He. We have the inequalities

$$\begin{aligned}
&\mu_{3,\mathrm{d}}(0) < \mu_{3,\mathrm{c}}(0),\\
&|\epsilon_{3,\mathrm{d}}(0)| > |L_3|\,/N_0,\\
&-\epsilon_{3,\mathrm{d}}(0) < -L_3/N_0\,.
\end{aligned} \tag{7.13}$$

Of course, the finite solubility of ^{3}He in ^{4}He at $T = 0$ is itself an indication that a ^{3}He atom is more strongly bound in ^{4}He than in ^{3}He. This stronger binding is shown in Fig. 7.4.

Many ^{3}He Atoms in Liquid ^{4}He ($x > 0$)

When we put more and more ^{3}He atoms into liquid ^{4}He the situation will change due to two effects. Firstly, there is an attractive interaction between the ^{3}He atoms in liquid ^{4}He. This attraction arises from a magnetic interaction

---

[1] From now on I shall use the subscripts "c" for the upper, concentrated ^{3}He phase and "d" for the lower, dilute ^{3}He phase

due to the nuclear magnetic moments of ^{3}He as in pure ^{3}He, and, in addition, in the mixtures only from a density effect. Remember that the ^{3}He, due to its larger zero-point motion, needs more space than a ^{4}He atom. Therefore, the liquid near to a ^{3}He atom is more dilute than the liquid near to a ^{4}He atom. This low-density region around a ^{3}He atom is felt by another ^{3}He atom in the liquid; it would like to be combined with the first ^{3}He atom, because then it does not have to push so hard against the ^{4}He atoms to make enough space for itself. Due to this attractive interaction between ^{3}He atoms the binding energy of a ^{3}He atom in ^{4}He should increase with increasing ^{3}He concentration x (Fig. 7.4),

$$|\epsilon_{3,\mathrm{d}}(x)| > |\epsilon_{3,\mathrm{d}}(0)|\,,$$
$$-\epsilon_{3,\mathrm{d}}(x) < -\epsilon_{3,\mathrm{d}}(0)\,. \tag{7.14}$$

But now we have to remember that the ^{3}He atoms have to obey the Pauli principle: If we put additional ^{3}He atoms in the liquid they have to go into successively higher energy states so that eventually all the energy states up to the Fermi energy $E_\mathrm{F} = k_\mathrm{B}T_\mathrm{F}$ are filled with two ^{3}He atoms of opposite nuclear spin. Therefore, the binding energy of the ^{3}He atoms has to decrease, due to their Fermi character, if their number is increased. Eventually we arrive at the following equation for the chemical potential of a ^{3}He atom in dilute ^{3}He–^{4}He liquid mixtures (at $T = 0$):

$$\frac{\mu_{3,\mathrm{d}}(x)}{N_0} = -\epsilon_{3,\mathrm{d}}(x) + k_\mathrm{B}\, T_\mathrm{F}(x)\,, \tag{7.15}$$

where $k_\mathrm{B}\, T_\mathrm{F} = (\hbar^2/2m^*)(3\pi^2 x N_0/V_\mathrm{m})^{2/3}$ is the Fermi energy of a Fermi system with effective mass m^* of the particles with the concentration x. We have $T_\mathrm{F} \propto x^{2/3}$ because V_m and m^* depend only weakly on the ^{3}He concentration x. The result (7.15) is illustrated in Fig. 7.4. If we continue to increase the ^{3}He concentration, the binding energy of a ^{3}He atom in a liquid isotopic mixture will eventually reach the binding energy of a ^{3}He atom in pure liquid ^{3}He. The chemical potentials of the two liquids become equal and we arrive at the limiting concentration of ^{3}He in liquid ^{3}He–^{4}He mixtures. The limiting concentration is 6.6% for the liquids under their saturated vapour pressure at $T = 0$, but depends on pressure (Fig. 7.2). Thus, we have for the equilibrium concentration

$$-\epsilon_{3,\mathrm{d}}(6.6\%) + k_\mathrm{B}T_\mathrm{F}(6.6\%) = -\frac{L_3}{N_0} = -2.473[\mathrm{K}] \cdot k_\mathrm{B}\,. \tag{7.16}$$

We could ask the ^{4}He atoms, too, whether they would rather stay in a ^{4}He environment or whether they prefer to be in a ^{3}He environment. Of course, they prefer to stay in ^{4}He for the same reasons as the dilute ^{3}He atoms: they feel a stronger binding when surrounded by ^{4}He, and because they do not have to obey the Pauli principle there is no reason to decrease their binding

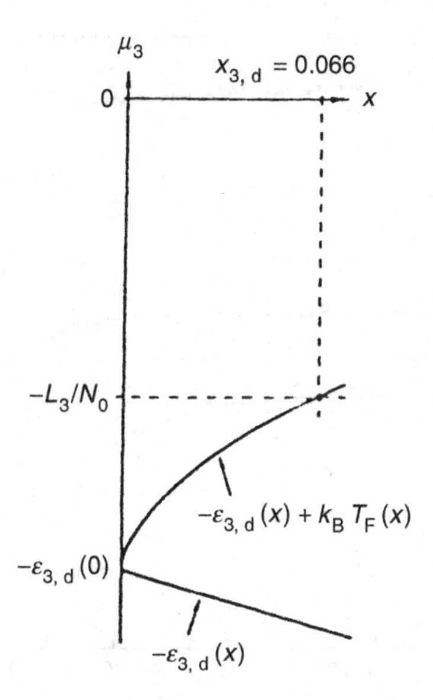

Fig. 7.4. Chemical potential of a ^{3}He atom in pure ^{3}He ($L_3/N_0 = 2.473[\mathrm{K}] \cdot k_B$) and in ^{3}He–^{4}He mixtures as a function of the ^{3}He concentration x. In the latter case the chemical potential is enhanced by $k_B T_F(x)$ due to the Pauli principle (see text). At the limiting concentration $x_{3,d} = 6.6\%$ the chemical potential of ^{3}He in ^{4}He reaches the chemical potential of pure ^{3}He

energy if we increase their concentration. As a result, the concentration of ^{4}He will approach zero rapidly in the upper ^{3}He-rich phase if the temperature is lowered to zero, (7.2). The reason for the finite (zero) solubility of ^{3}He (^{4}He) in liquid ^{4}He (^{3}He) even at absolute zero is that a single ^{3}He (^{4}He) atom is more strongly bound to liquid ^{4}He than to liquid ^{3}He.

7.1.4 Cooling Power of the Dilution Process

From measurements of specific heats (Fig. 7.5), we know that the enthalpy of ^{3}He in the dilute phase is larger than the enthalpy of ^{3}He in the concentrated phase; we have the *heat of mixing*,

$$\dot{Q} = \dot{n}_3 [H_d(T) - H_c(T)] \,. \tag{7.17}$$

If we transfer ^{3}He atoms at the molar flow rate $\dot{n}_3$ from the concentrated phase into the dilute phase of a phase-separated mixture, cooling will result according to the enthalpy difference of the two phases. Because the enthalpy is given by

$$H(T) - H(0) = \int_0^T C(T) \mathrm{d}T \tag{7.18}$$

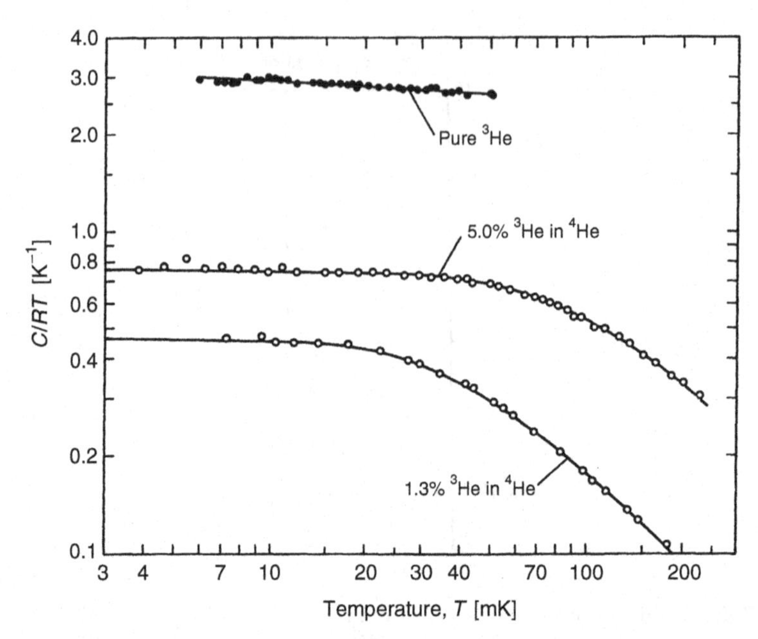

Fig. 7.5. Specific heat C/RT of ^{3}He and of two dilute solutions of ^{3}He in ^{4}He at saturated vapour pressure [7.5, 7.25]. In this figure C is the specific heat per *total* number of helium moles (not: per ^{3}He moles!). (For more recent data for pure ^{3}He see Fig. 2.15; they are somewhat smaller than the shown data, giving $C/RT = 2.78\,\mathrm{K}^{-1}$ for $T < 10\,\mathrm{mK}$ [7.29])

(neglecting PV terms which do not matter here or in the following), we must have

$$C_{3,\mathrm{d}}(T) > C_{3,\mathrm{c}}(T) \tag{7.19}$$

for the molar heat capacities of ^{3}He in the two phases to obtain cooling by the dilution process.

In both phases, of course, the specific heat varies linearly with temperature at low enough temperatures because both are Fermi liquids (Figs. 2.15 and 7.5). But the prefactor is different for the two phases. Liquid ^{3}He is a strongly interacting Fermi liquid and present theories do not allow calculation of the specific heat reliably enough for our purpose. We therefore have to take the specific heat from the accurate experimental data of Greywall [7.29], who, for temperatures below about 40 mK and at saturated vapour pressure, gives

$$C_3 = 2.7RT = 22T\ [\mathrm{J(mol\ K)^{-1}}]\,. \tag{7.20}$$

C_3/T is temperature independent at $T < 40\,\mathrm{mK}$ to within about 4%.[2] Remember that this is a very large specific heat due to the small Fermi

[2] From [7.29], $C_3/RT = 2.78$ at $T < 10\,\mathrm{mK}$, which is reduced by about 4% at $T = 40\,\mathrm{mK}$ and by about 8% at $T = 60\,\mathrm{mK}$; therefore, I will use 2.7 as a mean value.

temperature of ^{3}He. For example, the molar heat capacity of copper is about a factor of $3 \cdot 10^4$ smaller, due to its correspondingly larger Fermi temperature (Fig. 2.9; Sect. 3.1.2). We then have for the enthalpy of liquid ^{3}He

$$H_3(T) = H_3(0) + 11T^2 [\mathrm{J\,mol^{-1}}] . \tag{7.21}$$

If we need the specific heat and enthalpy at temperatures above about 50 mK we cannot use such simple equations because $C_3/T \neq$ constant[2]; we have to integrate the experimental data for C_3 to obtain $\mathrm{H}_3(T)$ [7.29].

To calculate the enthalpy of the mixtures we can, in principle, follow the same procedure. But in our dilution refrigerator we have mixtures at various concentrations (Sect. 7.2), and the specific heat has only been measured at a few low concentrations of ^{3}He [7.25–7.28]. Calculated thermodynamic data for liquid ^{3}He–^{4}He mixtures at $x \leq 8\%$ and $T < 250$ mK have been published in [7.30]. However, very little is known about mixtures at high ^{3}He concentrations which are circulating in a dilution refrigerator (see later). Of course, most important is the mixture in the chamber where the cooling occurs (see later) and there we have 6.6% ^{3}He. Fortunately, in the mixtures where the ^{3}He is diluted by ^{4}He we have a weakly interacting Fermi liquid, for which it is a good approximation to take the equations for a Fermi gas and replace the bare ^{3}He mass m_3 by the effective mass m^*. This substitution reflects the influence of the neighbouring ^{3}He and ^{4}He atoms with which the ^{3}He interacts. We then have at $T < T_\mathrm{F}/10$ the following relation for the specific heat of the mixture per mole of ^{3}He:

$$C_{3,\mathrm{d}} = N_0 k_\mathrm{B} \frac{\pi^2}{2} \frac{T}{T_\mathrm{F}} = 0.745 \frac{m^*}{m_3} \left(\frac{V_\mathrm{m}}{x} \right)^{2/3} T , \tag{7.22}$$

with the equation for the Fermi temperature

$$T_\mathrm{F} = \frac{\hbar^2}{2m^* k_\mathrm{B}} \left(\frac{3\pi^2 N_0 x}{V_\mathrm{m}} \right)^{2/3} = 55.2 \frac{m_3}{m^*} \left(\frac{x}{V_\mathrm{m}} \right)^{2/3} [\mathrm{K}] , \tag{7.23a}$$

and for the molar volume of the mixtures

$$V_\mathrm{m} = V_{\mathrm{m},4} (1 + 0.284x) , \tag{7.23b}$$

where $V_{\mathrm{m},4} = 27.589\,\mathrm{cm^3\,mol^{-1}}$ is the molar volume of pure ^{4}He [7.20].

This means that the specific heat *per* ^{3}He *atom* in the mixture increases with decreasing concentration! For one mole of *mixture* of ^{3}He concentration x we have

$$C'_{3,\mathrm{d}} = x C_{3,\mathrm{d}} \propto x^{1/3} T \quad \text{at } (T < T_\mathrm{F}/10) , \tag{7.24}$$

where the weak concentration dependences of m^* and V_m have been neglected. Equation (7.24) indicates that the specific heat $C'_{3,\mathrm{d}}$ of ^{3}He–^{4}He mixtures decreases not proportional to the ^{3}He concentration x but proportional to $x^{1/3}$ only. The effective mass m^* is weakly dependent on the ^{3}He concentration, and for a mixture of 6.6% at saturated vapour pressure $m^* \simeq 2.5 m_3$. This then gives for the Fermi temperature $T_F(6.6\%) = 0.38\,\mathrm{K}$, and we find for the specific heat of a 6.6% mixture below 40 mK [7.25–7.28]

$$C_{3,\mathrm{d}}(6.6\%) \simeq 106T \;[\mathrm{J(mol^3He\;K)^{-1}}]\,. \tag{7.25}$$

When the two phases are in thermodynamic equilibrium we must have for the chemical potentials

$$\mu_{3,\mathrm{c}}(x_\mathrm{c}, T) = \mu_{3,\mathrm{d}}(x_\mathrm{d}, T)\,. \tag{7.26}$$

With

$$\mu = H - TS\,, \tag{7.27}$$

we find

$$H_3 - TS_3 = H_{3,\mathrm{d}} - TS_{3,\mathrm{d}}\,. \tag{7.28}$$

This then gives with (7.20, 7.21, 7.25)

$$\begin{aligned} H_{3,\mathrm{d}}(T) &= H_3(0) + 11T^2 + T\int_0^T \left(\frac{C_{3,\mathrm{d}}}{T'} - \frac{C_3}{T'}\right)\mathrm{d}T' \\ &= H_3(0) + 95T^2 \;[\mathrm{J\;(mol^3He)^{-1}}]\,. \end{aligned} \tag{7.29}$$

Combining this with (7.21) for the enthalpy of pure ^{3}He we find for the cooling power occurring at the phase separation line when n_3 moles of ^{3}He per unit time are transferred from the concentrated to the dilute phase

$$\dot{Q}\,(T) = \dot{n}_3[H_{3,\mathrm{d}}(T) - H_3(T)] = 84\dot{n}_3T^2 \quad [\mathrm{W}]\,. \tag{7.30}$$

With a value of $\dot{n}_3 = 100\,\mathrm{\mu mol\;s^{-1}}$ and $T = 10(30)\,\mathrm{mK}$, we find

$$\dot{Q} \simeq 1(10)\;\mathrm{\mu W}\,. \tag{7.31}$$

For the above calculation I have used the accurate specific heat data of Greywall [7.29] for the properties of pure ^{3}He. The older data on the specific heat of mixtures [7.25–7.28] are much less reliable. Therefore, the figures given earlier may have to be revised slightly when more accurate specific heat data for the mixtures become available.

7.1.5 Osmotic Pressure

Before describing the design of a ^{3}He–^{4}He dilution refrigerator we have to discuss one more property of liquid–helium mixtures: their osmotic pressure [7.12, 7.31, 7.32]. As we shall see later, we have isotopic helium mixtures at varying concentrations and temperatures in our refrigerator. In such a situation an osmotic pressure develops in a two-fluid mixture.

We can approximately calculate the osmotic pressure π in ^{3}He–^{4}He mixtures by considering them as ideal solutions (which is valid for liquid–helium mixtures in the classical regime of $T > T_F$, e.g., for $T \gtrsim 0.15\,\mathrm{K}$, $x \lesssim 0.03$). In this situation we have van't Hoff's law

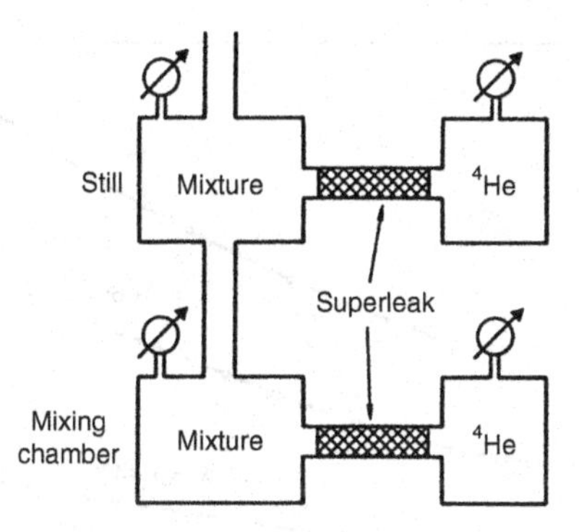

Fig. 7.6. Schematic set-up for a *Gedanken experiment* to measure the osmotic pressure of ^{3}He–^{4}He solutions in the still and in the mixing chamber of a ^{3}He–^{4}He dilution refrigerator

$$\pi V_{m,4} \simeq xRT\,, \tag{7.32}$$

where $V_{m,4}$ is the molar volume of ^{4}He. In a dilution refrigerator a tube connects the mixture in the mixing chamber (where we have the phase separation line) with the mixture in the still (where ^{3}He is evaporated), see Figs. 7.8 and 7.9. Let us assume that both the mixture in the mixing chamber and the mixture in the still are connected via a superleak to separate second vessels, which in each case contain pure ^{4}He (Fig. 7.6). A semipermeable membrane or "superleak" for the helium mixtures can easily be made from tightly compressed powders. Such a superleak with very small pores ($\leq 1{,}000$ Å) is permeable for superfluid ^{4}He but not for the Fermi liquid ^{3}He with its rather large viscosity (Fig. 2.19). For that reason, an osmotic pressure develops across the superleaks, so that the pressures of the liquids are higher on the mixture side than on the pure ^{4}He side. The osmotic pressure results from the "desire" of the liquids to exist at equal concentrations on either side of the semipermeable wall. The difference of the osmotic pressures due to the different concentrations and temperatures in the mixing chamber and the still is given by

$$\pi_{mc} - \pi_{st} \simeq \frac{(x_{mc}T_{mc} - x_{st}T_{st})R}{V_{m,4}}\,, \tag{7.33}$$

where st stands for still and mc for mixing chamber. If no ^{3}He is pumped from the still, then there is no difference in osmotic pressure between the still and the mixing chamber. Assuming the mixing chamber with 6.6% mixture to be at 10 mK and the still to be at 0.7 K, we then have

$$x_{st} = x_{mc}\frac{T_{mc}}{T_{st}} \approx 0.1\%\,. \tag{7.34}$$

If we pump ^{3}He from the still, the concentration of ^{3}He will decrease there, and an osmotic pressure difference $\pi_{mc} - \pi_{st}$ will develop that will drive ^{3}He

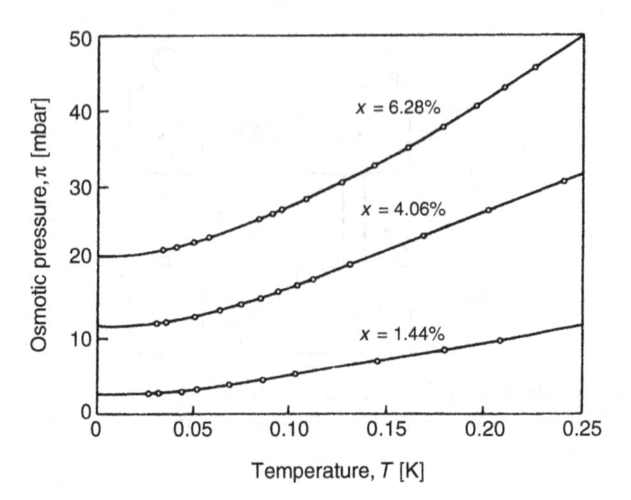

Fig. 7.7. Osmotic pressures of some dilute ^{3}He–^{4}He mixtures at a pressure of 0.26 bar (from [7.11] who used the data of [7.31])

from the mixing chamber into the still and therefore "sucks" ^{3}He from the concentrated into the dilute phase in the mixing chamber. The maximum osmotic pressure will be obtained when the concentration of ^{3}He in the still tends to zero. We then have (Fig. 7.7)

$$\Delta\pi_{\text{max}} = \frac{x_{\text{mc}} R T_{\text{mc}}}{V_{\text{m},4}} \simeq 20\,\text{mbar}\,, \tag{7.35}$$

for $T \lesssim 0.1\,\text{K}$, below which π becomes T-independent for $x = 6.6\%$ (see later and Fig. 7.7). This pressure corresponds to the hydrostatic pressure of about 1 m of liquid helium. In other words, the osmotic pressure will be large enough to drive the ^{3}He from the mixing chamber into the still even if they are separated by a vertical distance of about 1 m.

A more correct treatment takes into account that, at the temperature of the mixing chamber, the ^{3}He in both phases is in the Fermi degenerate state and for $T < T_{\text{F}}/3$ we have to use

$$\pi V_{\text{m}.4} = 0.4 x R T_{\text{F}} \tag{7.36}$$

instead of (7.32). This results in a temperature-independent osmotic pressure $\pi \propto x^{5/3}$ at $T \leq 10\,\text{mK}$; its value is about 20 mbar for a 6.6% mixture (Fig. 7.7).

In the earlier discussion fountain pressure effects ($\Delta P = \rho S \Delta T$) which result from the superfluidity of ^{4}He [7.15–7.18], have been neglected because they are small in comparison to the osmotic pressure.

7.2 Realization of a ^{3}He–^{4}He Dilution Refrigerator

The realization of a dilution refrigerator can be understood if we compare the cooling process with cooling which occurs when liquid is evaporated; this comparison is shown in Fig. 7.8. But we have to keep in mind that the *physics* of these cooling processes is quite different. In evaporation we rely on the classical heat of evaporation for cooling. In dilution refrigeration we rely on the enthalpy of mixing of two quantum liquids. Here the cooling results from quantum mechanical effects: the different zero-point motions of the two helium isotopes and the different statistics which we have to apply to understand their properties at low temperatures.

In the closed ^{3}He circulation cycle of a dilution refrigerator shown in Fig. 7.8, the cooling occurs when ^{3}He atoms are transferred from the ^{3}He-rich to the ^{3}He-poor side. The ^{3}He is then driven up along the liquid mixture column by a pressure caused by the osmotic pressure difference. It eventually reaches the "^{3}He pump", which in this scheme is our still. From the still we can pump and evaporate the ^{3}He, if it is operated at an appropriately high temperature, e.g., 0.7 K, where the vapour pressure of ^{3}He is much larger than the vapour pressure of ^{4}He (Fig. 2.7). For example, at $T = 0.7$ (0.6) K for an $x = 1.0\%$ (1.2%) mixture, we have $P_3 + P_4 = 88$ (46) µbar and $P_3/(P_3 + P_4) = 97\%$ (99%), which are rather convenient conditions. The vapour pressure of ^{3}He–^{4}He mixtures at various T and x can be found in [7.10, 7.33]. The still operates like a destillation chamber, evaporating almost pure ^{3}He. As a result, ^{3}He will flow from the dilute phase in the mixing chamber to the still driven by the osmotic pressure difference. This flow of cold ^{3}He from the mixing chamber to the evaporation chamber is utilized to precool the incoming ^{3}He in a heat exchanging process. The ^{3}He concentration in the dilute phase of the mixing chamber will stay constant because ^{3}He

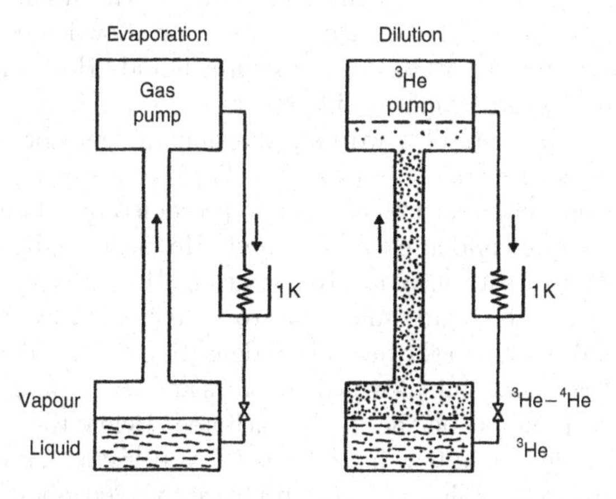

Fig. 7.8. *Gedanken experiment* for comparison of a helium evaporation refrigerator and a ^{3}He–^{4}He dilution refrigerator

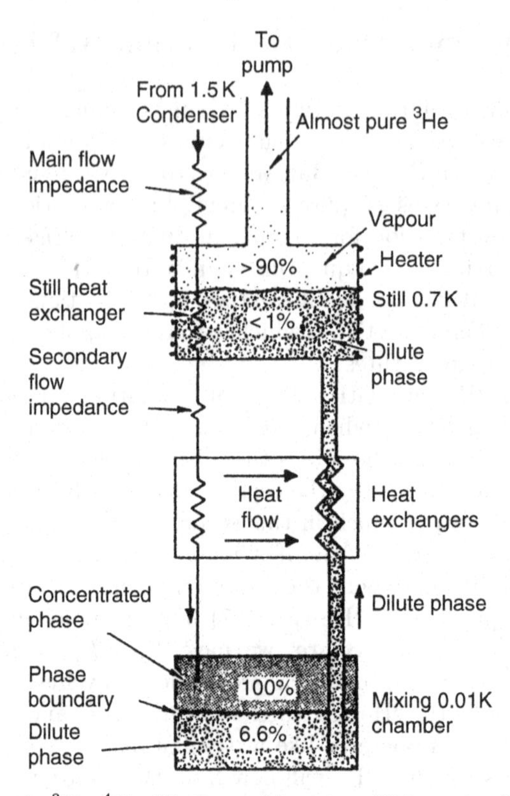

Fig. 7.9. Schematic ^{3}He–^{4}He dilution refrigerator. This part will sit in a vacuum chamber that is immersed in a ^{4}He bath at 4.2 K. The incoming ^{3}He gas is condensed on a continuously operating ^{4}He pot at 1.5 K (Sect. 5.2.4)

atoms are continuously crossing the phase separation line from the concentrated to the dilute phase, producing cooling due to the latent heat of mixing. Of course, the circuit cannot operate in the way shown, because we would have the heavier mixture on top of the lighter liquid ^{3}He. In reality we have to design the refrigerator slightly differently.

The main components of a working dilution refrigerator and a flow diagram for its liquids are depicted in Fig. 7.9. The ^{3}He gas coming from the exit of a pump at room temperature will first be precooled by a liquid ^{4}He bath at 4.2 K. It will then be condensed in a second ^{4}He bath at about 1.5 K, which we can obtain by using a continuously operating ^{4}He refrigerator (Sect. 5.2.4). The heat transfer surface area may have to be increased by using a sintered metal (Sect. 13.6) to absorb the heat of condensation of ^{3}He. This ^{4}He evaporator is also used as a heat sink at which all tubes and leads going to colder parts of the refrigerator should be thermally anchored. Below the ^{4}He refrigerator we need the so-called main flow impedance ($Z \approx 10^{12}\,\mathrm{cm}^{-3}$, see Sect. 5.2.4 for its design [7.34]), to establish sufficient pressure (30–200 mbar) for the incoming ^{3}He so that it will indeed condense at 1.5 K. The now liquid ^{3}He will flow through a heat exchanger which is in thermal contact with (or even inside) the still at a temperature of about 0.7 K. Below the still we have a secondary flow impedance ($Z \approx 10^{11}\,\mathrm{cm}^{-3}$) to prevent reevaporation of ^{3}He. After leaving this secondary flow impedance the liquid ^{3}He will flow through one or several heat exchangers (Sect. 7.3.3) to precool it to a low enough temperature before it enters the upper, concentrated phase in the mixing chamber.

A wider tube for the dilute phase in the refrigerator leaves the lower, dilute mixture phase of the mixing chamber, and then goes through the heat exchanger to precool the incoming ^{3}He. It enters the dilute liquid phase in the still, where we have a liquid ^{3}He concentration of less than 1%. The vapour above the dilute liquid phase in the still has a concentration of typically 90% ^{3}He due to the high vapour pressure of ^{3}He at the temperature of the still. If we then pump on the still and resupply the condensation line continuously with ^{3}He gas we have a closed ^{3}He circuit in which ^{3}He is forced down the condensation line. Then, again after liquefaction and precooling, it enters the concentrated phase in the mixing chamber. It will cross the phase boundary, giving rise to cooling, and will eventually leave the mixing chamber and be pushed up into the still, where it will evaporate. Circulation of the ^{3}He is maintained by a pumping system at room temperature.

Each component of a dilution refrigerator has to be carefully designed to achieve the required properties, but the quality of its heat exchangers is particularly critical, as will become obvious in Sect. 7.3.

7.3 Properties of the Main Components of a ^{3}He–^{4}He Dilution Refrigerator

7.3.1 Mixing Chamber

We will now perform an enthalpy balance for the mixing chamber where cooling is produced by the transfer of $\dot{n}_3$ ^{3}He moles from the concentrated (pure ^{3}He) to the dilute phase of $x = 6.6\%$. The cooling power will be used to precool the warmer liquid ^{3}He coming from the heat exchangers as well as to cool an experiment or to balance other external heat inflow. This enthalpy balance can be easily established by considering Fig. 7.10. We then have

$$\dot{n}_3[H_{3,\mathrm{d}}(T_{\mathrm{mc}}) - H_3(T_{\mathrm{mc}})] = \dot{n}_3[H_3(T_{\mathrm{ex}}) - H_3(T_{\mathrm{mc}})] + \dot{Q}\,, \tag{7.37}$$

resulting in a cooling power, using (7.21 and 7.29), of

$$\dot{Q} = \dot{n}_3(95T_{\mathrm{mc}}^2 - 11T_{\mathrm{ex}}^2) \quad [\mathrm{W}]\,. \tag{7.38}$$

If the liquid leaving the last heat exchanger already has a temperature equal to the temperature of the mixing chamber or if we operate the refrigerator in

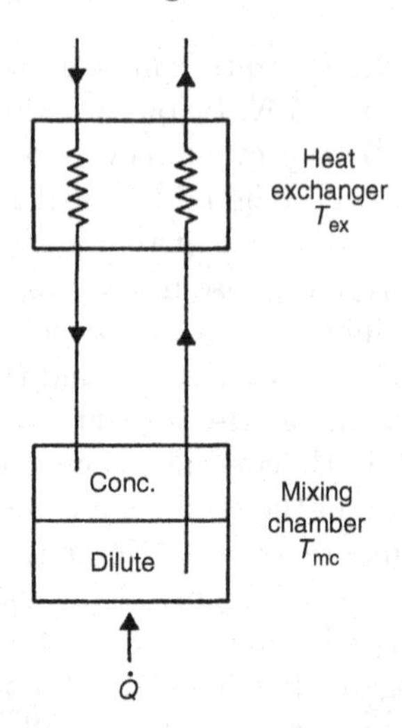

Fig. 7.10. Schematic for calculating the enthalpy balance in the mixing chamber of a ^{3}He–^{4}He dilution refrigerator (see text)

a discontinuous mode, which means we do not supply any more incoming ^{3}He which has to be precooled, then we obtain the maximum cooling power

$$\dot{Q}_{max} = 84\dot{n}_3 T_{mc}^2 , \tag{7.39}$$

or the minimum achievable temperature, but see (7.48, 7.51, 7.52, 7.54),

$$T_{mc,min} = \left(\frac{\dot{Q}}{84\dot{n}_3} \right)^{1/2} . \tag{7.40}$$

On the other hand, we can calculate the maximum temperature allowed for the liquid leaving the last heat exchanger and entering the mixing chamber, which we obtain for the case that there is no cooling power from the dilution refrigerator available, $\dot{Q} = 0$, resulting in

$$T_{ex} \leq 3\, T_{mc} . \tag{7.41}$$

This gives a rather severe requirement for the efficiency of the heat exchangers, telling us that the liquid has to leave the last heat exchanger at a temperature which is *at the most* a factor of three higher than the temperature which we want to obtain in the mixing chamber. This is a very important result and immediately demonstrates the importance of the efficiency of the heat exchangers in a dilution refrigerator.

Assume that we are circulating 100 µmol ^{3}He s^{-1} and we would like to have a cooling power of 1 µW at the mixing chamber. We then obtain

$$T_{mc} = 12\ (15)\,\text{mK}$$

for

$$T_{\text{ex}} = 18\ (30)\,\text{mK}\,,$$

demonstrating again the importance of adequately precooling the incoming ^{3}He in the heat exchangers, particularly the last one (Sect. 7.3.3).

To provide adequate thermal contact from the liquid in the mixing chamber to the walls of it and to an attached experiment, the mixing chamber has to contain a heat exchanger made from sintered metal powder (see Sect. 13.6) with surface areas between several m^2 and several 100 m^2 (see below).

Let me emphasize the important result $\dot{Q} \propto \dot{n}_3$, with $\dot{n}_3$ independent of T below about 0.1 K. This demonstrates the advantage of dilution refrigeration in comparison to evaporation refrigeration, where we have $\dot{n} \propto P_{\text{vap}} \propto \exp(-1/k_{\text{B}}T)$ (Fig. 7.3).

7.3.2 Still

One of the main requirements on the design of the still is that the ratio of ^{3}He to ^{4}He in the vapour phase of the still should be as large as possible. In Sect. 7.2 we argued that a value of 0.6–0.7 K is reasonable for the temperature of the still. In this range the ^{3}He vapour pressure is still large enough to keep the circulation rate reasonably large with a typical pump, and the concentration of ^{4}He in the vapour phase is only some percentage.

We now perform an enthalpy balance for the still, where ^{3}He will arrive with the temperature of the condenser (Fig. 7.9) and will leave with the temperature of the still. Cooling is produced by the heat of evaporation of ^{3}He at the temperature of the still,

$$\dot{n}_3\ L_3(T_{\text{st}}, x_{\text{d,st}}) = \dot{n}_3[H_3(T_{\text{cond}}) - H_3(T_{\text{st}})] + \dot{Q}_{\text{st}}\,. \tag{7.42}$$

If we again assume a circulation rate of 100 μmol s^{-1}, a condenser temperature of 1.3 K and a still temperature of 0.7 K, we find that the required rate of heat supply $\dot{Q}_{\text{st}}$ to the still is several milliwatts; more generally, a good rule is $\dot{Q}_{\text{st}}[\text{W}] \approx 40\dot{n}_3$ [mol]. This heat will be partly supplied by a conventional heater. In addition, we can thermally anchor to the still a radiation shield surrounding all the lower, colder parts of the dilution refrigerator to reduce the heat of radiation; this is done in most modern dilution refrigerators. Of course other parts, like electrical leads and capillaries, could and should be heat sunk at the still as well.

Unfortunately, in many dilution refrigerators the circulating gas contains 10–20% ^{4}He if no special precautions are taken; it is then not pure ^{3}He which circulates but a mixture. This results in a deterioration of the ^{3}He circulation rate $\dot{n}_3$ because it loads the pumping system, which has a constant *total volume* flow rate. It puts also a heat load on the heat exchangers because the specific heat of ^{3}He in mixtures is substantially higher than for pure ^{3}He (Fig. 7.5). In addition, separation of ^{3}He from ^{4}He will result in heating effects, so that the temperature of the heat exchangers will increase. An increased

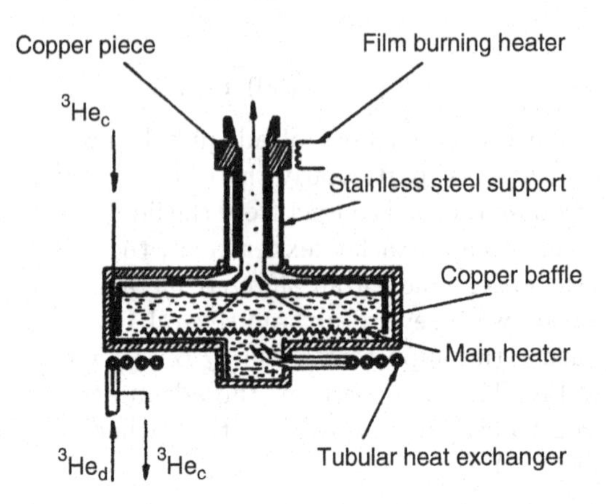

Fig. 7.11. A film-flow inhibiting still of a ^{3}He–^{4}He refrigerator [7.2]

^{4}He circulation rate has a negligible effect on the behavior of the liquids in the mixing chamber because at these low temperatures $\mathrm{d}x/\mathrm{d}T$ is essentially independent of T (Fig. 7.1). Of course, the mixing-chamber temperature and the cooling power of our refrigerator will deteriorate. To keep the amount of ^{4}He pumped from the still reasonably low, one has to suppress, in particular, the superfluid ^{4}He film flow (Sect. 2.3.5) up the still pumping tube to a point warm enough to evaporate there. The film flow can result from a combination of siphoning (typically 0.25 μmol s^{-1} per mm of inner circumference of the pumping tube) and of evaporation at the upper, warmer parts of the tube. The introduction of a small orifice ($\leq$ 1 mm diameter) with sharp edges in the pumping tube or, for more powerful refrigerators, a so-called ^{4}He film burner [7.2–7.11, 7.35, 7.36] in the tube can keep the film flow and hence the amount of ^{4}He in the circulating gas low enough (Fig. 7.11).

The still should be rather large so that the liquid in it can act as a buffer volume to arrange the liquid level over a wide range. It should contain a level gauge, for example a capacitor of two concentric tubes or two parallel plates, to measure and adjust the liquid level adequately [7.3], and, of course, heater and thermometer. In a good refrigerator, the incoming ^{3}He is efficiently cooled by a heat exchanger (see below) in the still.

7.3.3 Heat Exchangers

The purpose of the heat exchangers [7.2–7.11, 7.36–7.40] is to bring the temperature of the incoming ^{3}He as close to the temperature of the mixing chamber as possible by using the cold mixture leaving the mixing chamber to precool the incoming warmer ^{3}He. For this purpose, in general, one needs several heat exchangers between the still and the mixing chamber to achieve the best performance of the refrigerator. An enthalpy balance for the heat exchangers similar to the one described earlier for the still and for the mixing chamber is much more involved. Firstly, because we need to know the enthalpy of the concentrated phase (assuming that it is pure ^{3}He) as a function of temperature and of the dilute phase as a function of temperature and concentration. Secondly, a heat exchanger is a very complicated system with temperature gradients in the viscous fluid streams as well as in the heat exchanger body, both in the direction of the flow and perpendicular to it. We can take the available data on the specific heat of ^{3}He (Fig. 2.15) and on helium mixtures (Fig. 7.5) to construct an enthalpy diagram [7.8, 7.9]. But, of course, the values for the mixtures are only an approximation, because the concentration changes along the heat exchangers from a value of 6.6% in the mixing chamber to a value of about 1% in the still.

For the heat exchangers we assume ideal behavior, in the sense that the exit temperatures of both fluid streams are the same, and neglect viscous heating and axial conduction effects. The enthalpy balance for the Nth heat exchanger is then given by

$$\dot{Q}_{\mathrm{N}} + \dot{n}_3[H_3(T_{N-1}) - H_3(T_N)] = \dot{n}_3[H_{3,\mathrm{d}}(T_N) - H_{3,\mathrm{d}}(T_{N+1})] \,. \tag{7.43}$$

Fortunately, $H_{3,\mathrm{d}} > H_3$, so that the incoming ^{3}He can, in principle, always be precooled effectively by the outgoing mixture if the exchangers are designed properly. To compute the temperatures of the various heat exchangers with the mentioned assumptions, one first starts by setting T_{N+1} equal to T_{mc}, the temperature of the mixing chamber. Then the temperature of the Nth heat exchanger is calculated and so on, until one arrives eventually at a temperature equal to the temperature of the still. This procedure determines the number of heat exchangers for a certain desired temperature of the mixing chamber. The result of such a procedure with the additional assumption that the external heat flow to mixing chamber and to each heat exchanger is negligible can be found in [7.8, 7.9]. It shows that three ideal heat exchangers are needed to arrive at $T_{\mathrm{mc}} = 10\,\mathrm{mK}$. In reality, due to the non-ideality of the heat exchangers, the finite conductivities and viscosities of the flowing liquids, and the heat load from some circulating ^{4}He, one or two additional heat exchangers are needed.

The requirements on the heat exchangers can be quite severe for a powerful dilution refrigerator [7.2–7.4]. Besides coming close to the ideal behavior mentioned they should have small volumes, so that the liquids attain temperature equilibrium quickly, small impedances, so that the viscous heating due to flow is small, and there should be a small thermal resistance between the streams to obtain good temperature equilibrium between them. This last requirement is of particular importance and leads us back to the thermal boundary resistance between different materials already discussed in Sect. 4.3.2.

The thermal boundary resistance between helium and other materials is given by

$$R_{\mathrm{K}} = \frac{a}{A} T^{-3} \quad [\mathrm{K\,W^{-1}}] \,, \tag{7.44}$$

for $T > 20\,\mathrm{mK}$, where $A\,[\mathrm{m}^2]$ is the contact area. Typical values for the constant a obtained from experiments are $a_\mathrm{c} \cong 0.05$ (^{3}He) and $a_\mathrm{d} \cong 0.02$ (helium mixtures), see Sect. 4.3.2; but, of course, they depend on the properties of the body in contact with liquid helium. In a heat exchanger of a dilution refrigerator the two resistances are in series, and because $a_\mathrm{c} > a_\mathrm{d}$, we should have $A_\mathrm{c} > A_\mathrm{d}$.

As an example, for $\dot{Q} = 0.1\,\mathrm{mW}$, $A = 0.1\,\mathrm{m}^2$ and $T = 100\,\mathrm{mK}$, we would have $\Delta T \approx 50\,\mathrm{mK}$ for ^{3}He. Or, at lower T, with $\dot{Q} = 0.01\,\mathrm{mW}$, $A = 10\,\mathrm{m}^2$ and $T = 20\,\mathrm{mK}$, we would have $\Delta T \approx 6\,\mathrm{mK}$. Both temperature steps are unacceptably large, so that we either have to decrease the heat flow $\dot{Q}$ or increase the surface area A. Usually one requires many square meters of surface area to overcome the Kapitza boundary resistance in the heat exchangers for $T < 0.1\,\mathrm{K}$. Because the thermal boundary resistance increases with decreasing temperature, it is particularly serious in the mixing chamber, where one needs surface areas between several $10\,\mathrm{m}^2$ and several $100\,\mathrm{m}^2$, depending on $\dot{n}_3$ and T_mc [7.2–7.11, 7.36–7.40].

Heat exchangers are the most important and most critical part of a dilution refrigerator; they determine its minimum temperature, for example. For the design of heat exchangers we have to take into account the problems and considerations discussed earlier. In particular, we have to take into account that the thermal boundary resistance as well as the liquids' viscosities and conductivities increase with decreasing temperature. When designing dilution refrigerators, former practical experience with a successful refrigerator [7.2–7.11, 7.36–7.41] is sometimes more important than a recalculation of various parameters.

For a rather simple dilution refrigerator with a minimum temperature of about 30 mK, one continuous counterflow tube-in-tube heat exchanger consisting of two concentric capillaries (with diameters of order 0.5 and 1 mm, and with 0.1 or 0.2 mm wall thickness) several meters long is sufficient [7.2–7.11, 7.36–7.41]. These concentric capillaries should be of low thermal conductivity (CuNi, stainless steel or brass) and are usually coiled into a spiral so they do not take up too much space. In such a heat exchanger the temperatures change continuously along the exchanger. Heat is transferred across the body of the wall, and conduction along the capillary and along the liquid streams should be negligible. The dilute phase moves in the space between the tubes and the concentrated liquid moves in the inner tube. In a more elaborate design, the inner capillary of up to 3 mm diameter is not straight but spiralled, to increase the surface area, and then put into the outer, somewhat larger capillary (Fig. 7.12) [7.2–7.4, 7.36, 7.38, 7.41]. In spite of their simple design and small inner volumes, continuous heat exchangers can be quite effective. If one uses the appropriate dimensions, a refrigerator with a well-designed continuous heat exchanger can reach minimum temperatures of 15–20 mK. The warmer end of a continuous heat exchanger can also be designed as a secondary flow impedance by reducing its diameter or by inserting a wire into it to prevent re-evaporation of the ^{3}He.

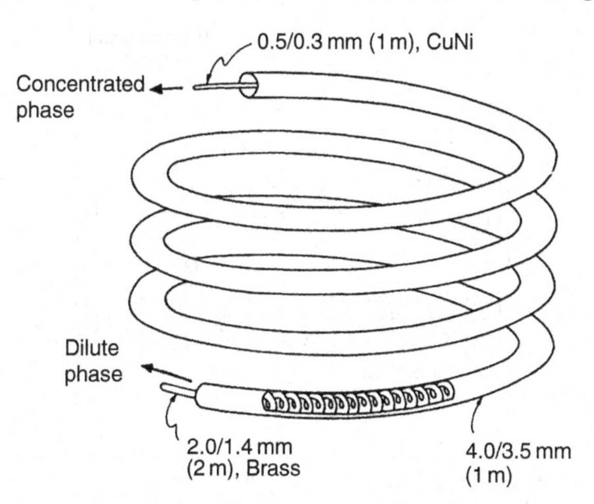

Fig. 7.12. Schematic of a concentric heat exchanger. The design used in [7.36] had for the dilute phase: $Z \approx 10^7\,\mathrm{cm}^{-3}$, $A_{\mathrm{brass}} = 126\,\mathrm{cm}^2$, $A_{\mathrm{CuNi}} = 16\,\mathrm{cm}^2$, $V_{\mathrm{liq}} = 6\,\mathrm{cm}^3$; and for the concentrated phase: $Z_{\mathrm{brass}} \approx 5 \times 10^8\,\mathrm{cm}^{-3}$, $Z_{\mathrm{CuNi}} \approx 8 \times 10^9\,\mathrm{cm}^{-3}$, $A_{\mathrm{brass}} = 88\,\mathrm{cm}^2$, $A_{\mathrm{CuNi}} = 9\,\mathrm{cm}^2$, $V_{\mathrm{liq,brass}} = 2\,\mathrm{cm}^3$, $V_{\mathrm{liq,CuNi}} \approx 0.07\,\mathrm{cm}^3$. For concentric heat exchangers of different designs and dimensions see [7.2–7.4, 7.38, 7.41]

For a more powerful dilution refrigerator with a larger cooling power and lower minimum temperature we need more than one heat exchanger. These are, first a continuous heat exchanger below the still, as described earlier, and then several step heat exchangers in series [7.2–7.11, 7.36–7.41]. At lower temperatures the earlier-discussed continuous heat exchanger can no longer be used because it does not provide enough surface area to defeat the increasing thermal boundary resistance. This can only be provided by step exchangers, which in their simplest design can be machined from Cu blocks into which two holes have been drilled (Fig. 7.13). To reduce the thermal boundary resistance, each channel has to be filled with pressed and/or sintered metal powder to provide the required surface area for adequate heat exchange (Sect. 13.6). Other design criteria are a low flow impedance (to reduce $\dot{Q}_{\mathrm{visc}}$, see later) and a small size (to economize on the amount of ^{3}He needed and to give a short thermal time constant). For example, the flow channel diameter in the heat exchangers has to be increased with decreasing T. To produce a large surface area but small flow impedance, the metal powder can first be sintered, then broken up, and the sinter blocks are then sintered together to leave large open flow areas. Another possibility is to drill an open flow channel through the sintered powder (Fig. 7.13). This large flow area is important not only in order to keep viscous heating low but also to take advantage of the high thermal conductivity of ^{3}He (Sect. 2.3.6, which is then always in good thermal contact with the almost stationary liquid in the sinter. Table 7.1 lists the dimensions used for the six-step heat exchangers of a rather successful dilution refrigerator with a minimum temperature of about 3 mK inside the mixing

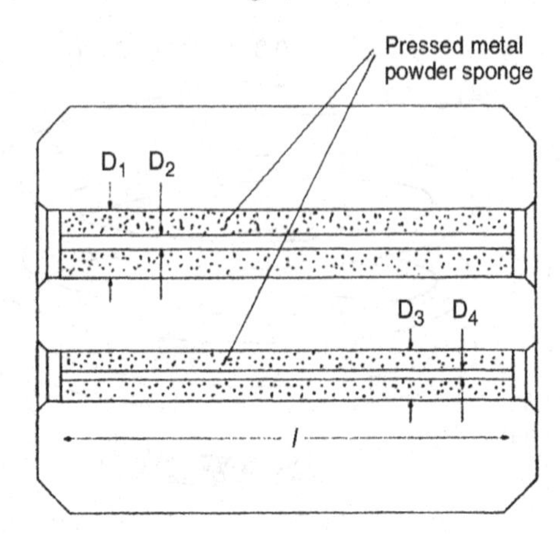

Fig. 7.13. Schematic of a step heat exchanger made from Cu with compressed and sintered metal powder in the two channels into which flow channels for the counter-flowing liquids have been drilled. The dilution refrigerator of [7.36] with $T_{\min} = 3\,\mathrm{mK}$ and a maximum value of $\dot{n}_3 = 700\,\mu\mathrm{mol\,s^{-1}}$ used six such step exchangers with the parameters given in Table 7.1

Table 7.1. Data for the Cu-sinter heat exchangers used in the ^{3}He–^{4}He dilution refrigerator of [7.36], which has reached $T_{\min} = 3\,\mathrm{mK}$. For the symbols, see Fig. 7.13. (The filling factor f is for sinter made from powder of grain size d). For data on Ag-sinter heat exchangers see [7.2–7.4, 7.38, 7.41]

T_N	[mK]	83	48	32	19	13	10
l	[mm]	62	94	94	94	94	88
D_1	[mm]	8	13	13	18	18	22
D_2	[mm]	2.7	2.7	3.2	4.8	6.5	8.0
D_3	[mm]	6	10	10	14	14	17
D_4	[mm]	2.0	2.0	2.4	3.2	4.8	6.5
Sintered		Cu	Cu	Cu	0.2 Cu	0.4 Cu	0.4 Cu
Metal powder					0.8 Ag	0.6 Ag	0.6 Ag
d	[μm]	100	50	30	44 Cu	44 Cu	44 Cu
					2.5 Ag	0.07 Ag	0.07 Ag
f		0.55	0.65	0.65	0.75	0.70	0.70
A_c	[m^2]	0.05	0.27	0.45	6.5	100	135
A_d	[m^2]	0.09	0.46	0.76	10.6	164	230
V_c	[cm^3]	1.6	7.1	7.0	13.7	12.8	17.1
V_d	[cm^3]	2.8	11.9	11.7	22.2	20.8	29.0

chamber at $\dot{n}_3 = 300\,\mu\mathrm{mol\,s^{-1}}$. The simple step exchangers were filled with compressed Cu and Ag powders [7.36]. The art of designing heat exchangers [7.8] has been greatly advanced by Frossati [7.2, 7.3], see also [7.4, 7.38]. His welded, rectangular heat exchangers filled with submicrometer sintered silver

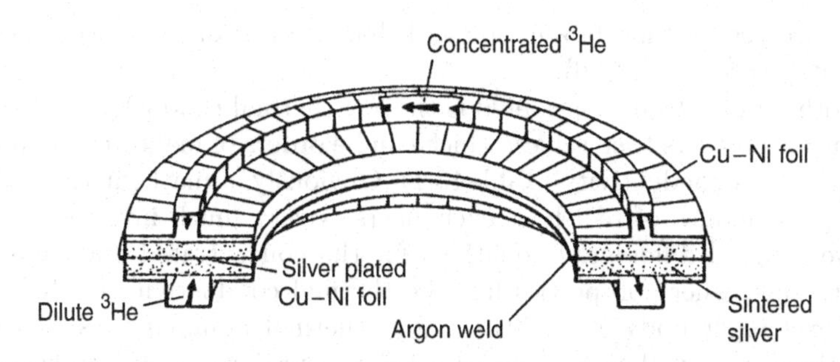

Fig. 7.14. Schematic of a semicontinuous heat exchanger developed by Frossati et al. [7.2, 7.3] of welded Cu–Ni foil filled with submicrometer silver powder sintered to a silver plated Cu–Ni foil

powder (Fig. 7.14) are also taken advantage of in powerful commercial dilution refrigerators. In [7.2, 7.3, 7.11] the two important steps in designing the step heat exchangers are discussed:

(a) Determination of the required total surface area for the heat exchangers for a given minimum temperature at optimum flow rate and at a given heat leak.
(b) Optimization of the size of the flow channels for a given exchanger length to minimize the heat generated by viscosity and axial heat flow.

Calculations of the required surface areas in the heat exchangers can be found in [7.2–7.11, 7.36, 7.37, 7.40]. Assuming for the Kapitza resistance $R_{\mathrm{K}} \propto T^{-i}$ (Sect. 4.3.2) we have for the heat flow $\dot{Q}$ between the two streams in an exchanger with contact area A

$$\dot{Q} = \lambda_{i+1} A (T_1^{i+1} - T_2^{i+1}) . \tag{7.45}$$

Frossati [7.2, 7.11] showed evidence for the following values:

$$\begin{aligned} \lambda_4 &= 63\,\mathrm{W\,m^{-2}\,K^{-4}}, \\ \lambda_3 &= 6.7 \times 10^{-3}\,\mathrm{W\,m^{-2}\,K^{-3}}, \\ \lambda_2 &= 2.1 \times 10^{-4}\,\mathrm{W\,m^{-2}\,K^{-2}}. \end{aligned} \tag{7.46}$$

Typical surface areas required in the last heat exchanger for the following mixing chamber temperatures are then

T_{mc} [mK]	4	8	15	25
A [m^2]	71	34	18	10

assuming $i = 2$, $\dot{n}_3 = 2 \times 10^{-4}\,\mathrm{mol\,s^{-1}}$, $\dot{Q} = 30\,\mathrm{nW}$ [7.11].

As mentioned, the large surface areas necessary in the heat exchangers as well as in the mixing chamber are provided by sintered metal powders.

Various recipes for their fabrication and characterization have been described in the literature (Sect. 13.6).

Another point to be considered is the low thermal conductivity of the sintered metal powders themselves, which can be orders of magnitude lower than the bulk metal conductivity (Table 13.1). Obviously, a mathematical analysis of the performance of real heat exchangers is very much involved. One has to solve coupled differential equations for the concentrated and diluted helium streams, where properties like the thermal conductivity of the liquids, of the exchanger body and of the sinter, thermal boundary resistances between them, viscous heating and osmotic pressure are taken into account to calculate the temperature profile in the exchanger. Figure 7.15 illustrates as an example the effect of the finite thermal conductivities of the liquids in the heat exchangers on the temperatures of the liquids inside the exchanger and

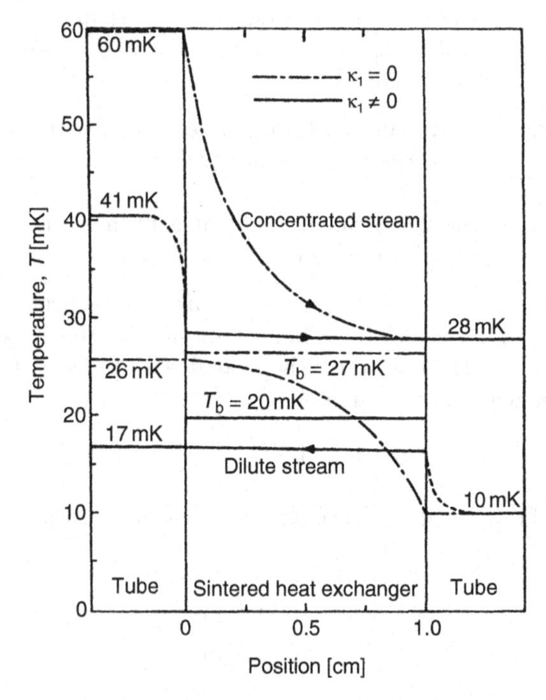

Fig. 7.15. Calculated liquid and body temperature profiles within and near a discrete heat exchanger of length 1 cm for a mixing chamber temperature of 10 mK. One set of curves is calculated neglecting the finite thermal conductivity of the liquids ($\kappa_1 = 0$), and the other set is calculated using the correct liquid thermal conductivities ($\kappa_1 \neq 0$). The liquid volume on the concentrated side is 0.85 cm^3, and the dilute liquid volume is 2.1 times larger. The surface area per unit liquid volume is 400 cm^{-1}. The ^{3}He flow rate is 20 μmol s^{-1}. External heat flow to the heat exchanger and to the mixing chamber is neglected. Obviously a design where the influence of the liquid conductivity is minimized is more favorable [7.8, 7.9]

of the body of the heat exchanger itself. If the finite thermal conductivity of the liquids is taken into account, most of the temperature changes in the two liquid streams occur in the inlet tubes just before the heat exchanger. Even the inclusion of just this one property results in substantial changes in the temperature profiles [7.8, 7.9, 7.39, 7.40].

The selection of the diameters for the tubes connecting the mixing chamber, heat exchangers and the still is important. Here we have to make a compromise for the following reasons. We would like to have a large diameter for the tubes to ensure that the circulation of the rather viscous ^{3}He is large and viscous heating effects are negligible. Yet the diameter of the tubes must be small enough for an osmotic pressure to develop between the mixing chamber and the still, and we also want to have the diameter small so that the refrigerator will not suffer from hydrodynamic instabilities and convection. We have to bear in mind that the ^{3}He concentration decreases toward the still and therefore the density of the mixture in the dilute phase increases when going from the mixing chamber to the still. Typical diameters of the connecting tubes are about 1 mm just below the still up to 10 mm before the mixing chamber of a powerful dilution refrigerator [7.2–7.4, 7.36–7.38].

Viscous heating is given by

$$\dot{Q}_{\text{visc}} = \dot{V}\Delta P = Z\eta V^2 \dot{n}^2 \tag{7.47}$$

for a laminar flow with $Z = 128l/\pi d^4$ in case of a circular cross-section tube of length l and diameter d [cm] and $Z = 12l/ab^3$ in case of a square with width b and height a; $\dot{Q}_{\text{visc}}$ may eventually limit the performance of a dilution refrigerator. Viscous heating is more serious in the dilute than in the concentrated stream, and it increases with decreasing temperature ($\eta \propto T^{-2}$ for a degenerate Fermi liquid, Fig. 2.19). To keep viscous heating of the two streams in a heat exchanger equal, one needs $D_{\text{d}} \approx 1.7 D_{\text{c}}$ [7.9].

Wheatley et al. [7.5–7.7] have given arguments that the practical lower limit for T_{mc} is due to viscous heating and conduction of heat in the flowing dilute phase at the outlet of the mixing chamber; they arrived at

$$T_{\text{mc, min}} = 4d^{-1/3} \quad [\text{mK}] , \tag{7.48}$$

where d is the diameter of the exit tube measured in millimeters. Frossati [7.2, 7.3] has discussed these problems by considering viscous heating and heat conduction contributions for the concentrated and dilute streams. With $\dot{Q}_{\text{visc}} \propto d^{-4}$ (assuming Poiseuille flow) and $\dot{Q}_{\text{cond}} \propto d^2$, the sum of the two is

$$\dot{Q}_{\text{visc}} + \dot{Q}_{\text{cond}} = ad^{-4} + bd^2 , \tag{7.49}$$

yielding an optimal channel size

$$d_{\text{opt}} = (2a/b)^{1/6} . \tag{7.50}$$

He gave the resulting d_{opt} for the various heat exchangers of a powerful dilution refrigerator which has reached $T_{\text{mc}} \simeq 2\,\text{mK}$. Since viscosity and thermal conductivity in both Fermi liquid streams increase with decreasing temperature, their limiting influence becomes more and more severe as lower temperatures are required.

Let me come back to the influence of the thermal boundary resistance. Taking the temperature step due to the Kapitza resistance R_K between the helium liquid and the solid with a surface area A into account [7.2, 7.42] modifies (7.40) for the minimum temperature of a dilution refrigerator to

$$T_{mc}^2 = \frac{0.011\dot{Q}}{\dot{n}_3} + \frac{5.2 R_K \dot{n}_3}{A} \quad \text{if} \quad R_K \propto T^{-3}, \tag{7.51}$$

and

$$T_{mc}^2 = \frac{0.011\dot{Q}}{\dot{n}_3} + (27 R_K \dot{n}_3 T^2)^2 \quad \text{if} \quad R_K \propto T^{-2}. \tag{7.52}$$

Taking the value $A R_K T^2 = 28\,\mathrm{m^2\,K^3\,W^{-1}}$ (Sect. 4.3.2) for (7.52) we arrive at the optimum circulation rate

$$\dot{n}_{3,\,opt} \simeq 3 \times 10^{-3} \dot{Q}^{1/3} A^{2/3} \ [\mathrm{mol\,s^{-1}}], \tag{7.53}$$

and the minimum temperature

$$T_{mc,\,min} = 4.5 \left(\frac{\dot{Q}}{A}\right)^{1/3} \ [\mathrm{mK}]. \tag{7.54}$$

The equations for the performance of a dilution refrigerator have been solved rigorously [7.43]. The results replace (7.51–7.54) and allow to determine the normalized cooling power $\dot{Q}/\dot{n}$ as a function of T in good agreement with experimental data over the whole temperature range, in particular when $\dot{Q}$ does not follow a T^2 dependence any more (Fig. 7.19). These equations also allow to calculate the Kapitza resistance of the heat exchanger as well as the internal heat leak for any circulation rate. One can then perform a diagnosis of the refrigerator and decide whether its performance is limited by the effective surface area of the heat exchanger, the intrinsic heat leak to the mixing chamber, or by the Kapitza resistance of the mixing chamber.

7.4 Examples of ^{3}He–^{4}He Dilution Refrigerators

Figures 7.16 and 7.17 are schematic diagrams of some rather successful "home-made" dilution refrigerators. Figure 7.18 shows the low-temperature part of a commercial dilution refrigerator with a minimum temperature of about 4 mK. The cooling power as a function of temperature of a dilution refrigerator with a dilution unit very similar to the one shown in Fig. 7.18 is depicted in Fig. 7.19. The most powerful ^{3}He–^{4}He dilution refrigerators have been built by Frossati et al. in Leiden [7.2, 7.3] (Fig. 10.22) and by Pickett et al. in Lancaster

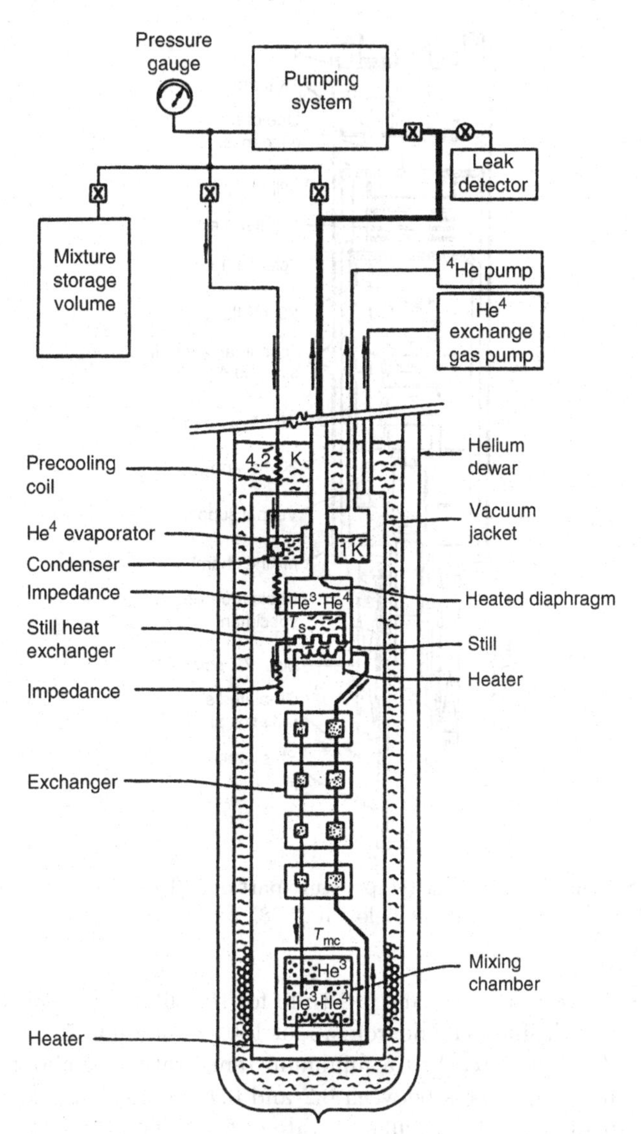

Fig. 7.16. Schematic of a typical ^{3}He–^{4}He dilution refrigerator with four heat exchangers [7.6]

[7.4, 7.38]. The refrigerator in Leiden has a maximum circulation rate of $\dot{n}_3 = 10\,\mathrm{mmol\,s^{-1}}$, a minimum temperature of 1.9 mK at $\dot{n}_3 = 0.85\,\mathrm{mmol\,s^{-1}}$ and a cooling power of 20 μW at 10 mK. It contains the enormous quantity of 1.6 kg Ag, and $A_c = 1{,}000\,\mathrm{m^2}$, $A_d = 1{,}300\,\mathrm{m^2}$ in total in the heat exchangers!

The new refrigerator at Lancaster [7.4] contains a very efficient tubular heat exchanger followed by 15 (!) discrete heat exchangers with diameters of

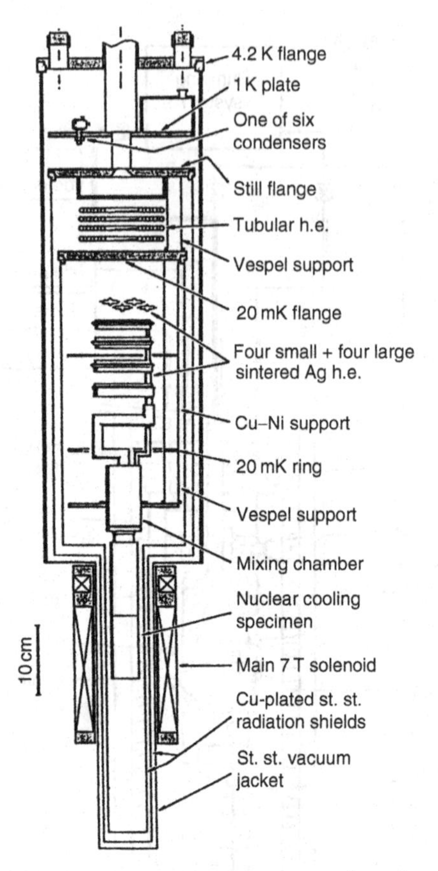

Fig. 7.17. Schematic of the low temperature part of a ^{3}He–^{4}He dilution refrigerator with a nuclear refrigeration stage below it [7.38]

3–6 mm for the concentrated and 4–12 mm for the dilute channels (Fig. 7.20); they are filled with sinters made from about 190 g of nominal 70 nm Ag powder (see Sect. 13.6). It has reached a minimum temperature of about 1.8 mK at rather low circulation rates between 0.2 and 0.4 mmol s^{-1}. Its cooling power is 2 μW at 10 mK and at a circulation rate of 0.27 mmol s^{-1}.

A small ^{3}He–^{4}He dilution refrigerator which can be inserted into a ^{4}He storage dewar (50 mm neck diameter) is shown in Fig. 7.21; it is based on the same principles as the ^{4}He and ^{3}He dipper cryostats discussed in Sects. 5.2.4 and 6.2. It can be cooled from room temperature to the final temperature of about 15 mK within a few hours; a typical circulation rate is 30 μ mol s^{-1}. Such units are available commercially.

For use in high magnetic fields, in particular in high-pulsed magnetic fields, a refrigerator with the dilution unit made from plastics often is required to reduce eddy current heating. The lower end of these refrigerators needs to have a small diameter to fit into the bore of the magnet. Such plastic dilution

Fig. 7.18. Low temperature part of a commercial ^{3}He–^{4}He dilution refrigerator with the still, a continuous heat exchanger, five-step heat exchangers and mixing chamber

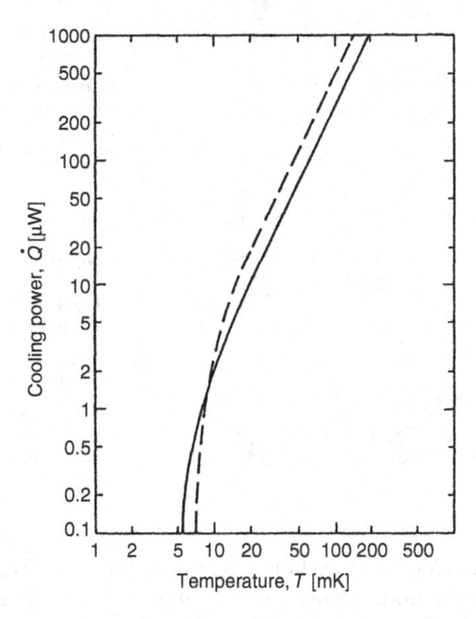

Fig. 7.19. Cooling power as a function of temperature of a commercial ^{3}He–^{4}He dilution refrigerator with a dilution unit very similar to the one shown in Fig. 7.18. Data are for two still heating powers: 8 mW (——) and 22 mW (– – –)

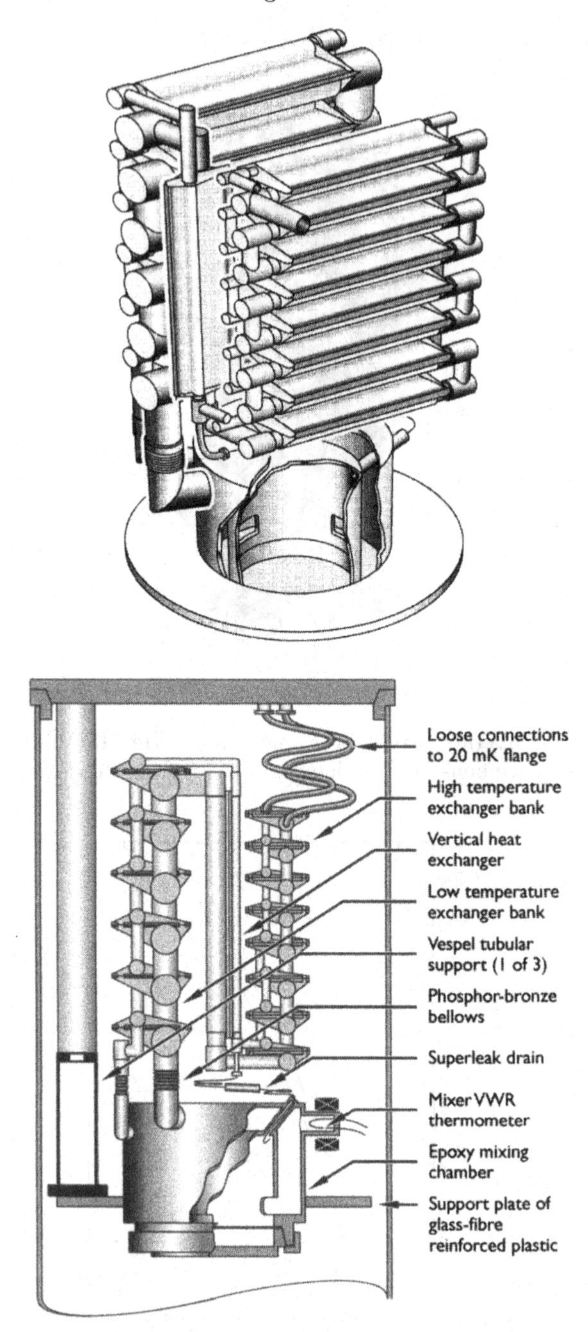

Fig. 7.20. The 15 discrete heat exchangers in the dilution refrigerator of [7.4]. They are filled with Ag sinter and arranged in two stacks above the mixing chamber

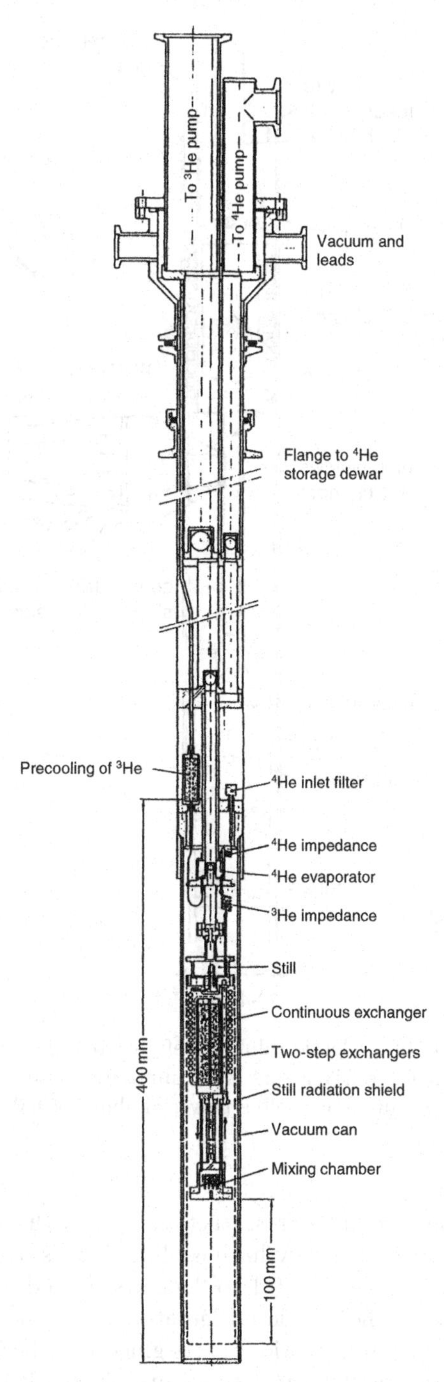

Fig. 7.21. A dipstick ^{3}He–^{4}He dilution refrigerator (courtesy of P. Sekowski, Universität Bayreuth)

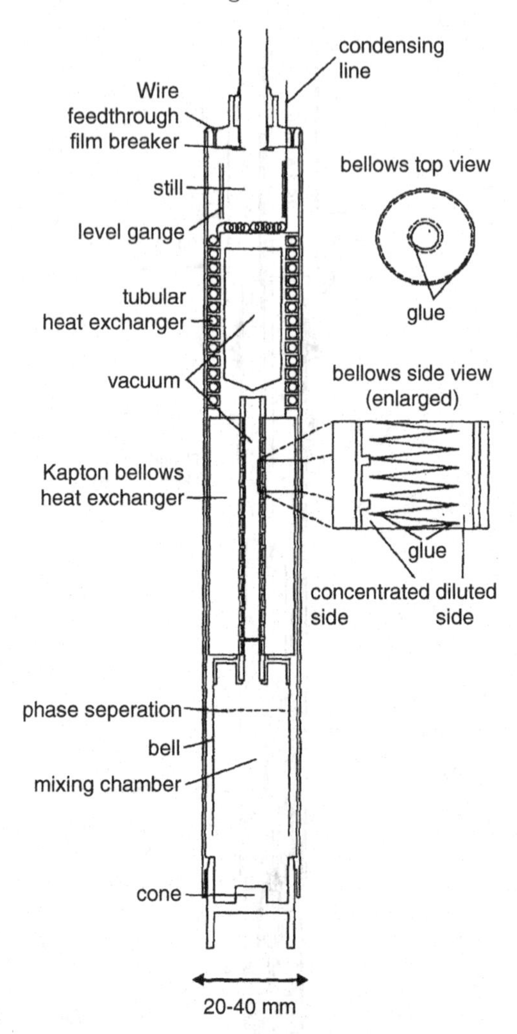

Fig. 7.22. Schematic of a plastic dilution refrigerator. The Kapton bellows heat exchanger is optional for refrigerators with minimum temperatures below 20 mK and flow rates above 0.1 mmol or cooling power of about 30 μW at 30 mK; for details see [7.41]

refrigerators were developed by Frossati et al. [7.41] with outer diameters from 14 to 40 mm (see Fig. 7.22). They have cooling powers of up to 1 (0.01) mW at 100 (20) mK, flow rates from 0.1 to 2 mmol s^{-1}, and a minimum temperature of about 10 mK. The plastic refrigerators are made mainly from easily machinable PVC (polyvinyl chloride) parts glued together with Stycast 1266. Thin-walled CuNi or stainless-steel tubes are glued with Stycast 2850FT – or simply by using a small amount of vacuum grease – to the plastic. In the simpler versions, a tubular heat exchanger of several meters of a coiled Teflon capillary for the concentrated ^{3}He is sitting inside of a spiraling square groove (for the dilute phase) machined into a PVC cylinder. In the more elaborate designs for cooling to below 20 mK and for high circulation rates, a bellow-type heat exchanger made from Kapton foils is added. The refrigerators do not have the usual sintered metal heat exchangers. The total heat exchange areas are between 0.01 and 0.4 m^2 only. The plastic mixing chambers of 2–50 cm^3 for the various refrigerators are closed by a cone with a 5–10° angle sealed by vacuum grease to the mating part [7.3, 7.41]. Another version of a demountable seal for a mixing chamber is described in [7.44]. These joints allow easy access to the mixing chamber for quick exchange of experiments that can be mounted directly onto the bottom cone. Because of their small size, these plastic dilution refrigerators can also be used as a dipstick refrigerator to be inserted directly into a ^{4}He storage dewar (see Fig. 7.21). Reference [7.41] gives detailed instructions for fabrication of the various parts and on the performance of the refrigerators.

Most dilution refrigerators operate according to the principles and designs discussed above. There are a few alternative designs of which I want to describe the most useful ones. Of particular interest are refrigerators in which the circulation of the ^{3}He gas is achieved by internally regenerating adsorption pumps, using mostly activated charcoal [7.45]. Recently, this charcoal adsorption pumping technology already described in Sect. 6.2 for closed ^{3}He cryostats has been introduced to commercially available continuously operating ^{3}He–^{4}He dilution refrigerators [7.46]. Such systems, also for use inside of a ^{4}He storage dewar, have reached a minimum temperature of 15 mK and a cooling power of 40 μW at 100 mK. The cooling time from 4.2 K to 50 mK is 2 h only and removal of the dilution refrigerator from the dewar takes 0.5 h. The key distinguishing feature of the system is the absence of a traditional ^{3}He gas-handling system and mechanical ^{3}He pump. Such a cryogenic ^{3}He cycle can substantially reduce vibrations of a dilution refrigerator because the main source for them is the pumps for circulating the ^{3}He [7.47].

An alternative rather attractive variant is the use of commercially available two-stage closed-cycle refrigerators (Sect. 5.3) to precool the dilution unit and its radiation shields instead of using a cryostat filled with liquid helium. This combination has mostly been advanced by K. Uhlig. In his early designs, he used two-stage Gifford-McMahon refrigerators for precooling with a final temperature of 5.8 K at its second and 35 K at its first stage (for cooling the radiation shield) [7.48]. After varies improvements, in particular to reduce the transfer of vibrations from the closed-cycle refrigerator to the dilution unit by using braided copper straps for thermal connection, he has reached 15 mK at a ^{3}He flow rate of 0.11 mmol s^{-1} and in a later version 8.1 mK at a flow rate of 0.06 mmol s^{-1}. Later on, he introduced a pulse-tube refrigerator for precooling that has no moving cold parts and therefore a much smaller level of vibration [7.49]. The first stage reached 45 K to cool the radiation shield and the second stage could be operated at a temperature as low as about 3 K. At flow rates of 0.3–1 mmol s^{-1}, the achieved minimum temperature of the dilution refrigerator was 9 mK. Another remarkable design of Uhlig's cryogen-free dilution refrigerators is the use of a double-mixing chamber by which he could reduce the minimum temperatures to 7.1 mK for the Gifford-McMahon apparatus and to 4.3 mK for the pulse-tube apparatus. In such a design, the flow of concentrated ^{3}He, after it has entered the first mixing chamber, is divided into two streams. One is diluted in the first chamber. Its cooling capacity is used to precool the second part of the stream, which is then diluted in the second mixing chamber to about a factor of two lower temperatures. The distribution of the flow is determined by the flow impedances of the connecting tubes. A detailed description of the method can be found in [7.48, 7.50]. In all these dilution refrigerators, no separate condenser for the circulating ^{3}He is used. Instead, the incoming ^{3}He is precooled and liquefied by a counterflow heat exchanger that uses the enthalpy of the cold ^{3}He gas pumped out of the still and subsequent Joule-Thompson expansion through a flow restriction. In [7.51], such a combination of Joule-Thompson expansion for precooling the circulating ^{3}He gas and a two-stage hybrid Gifford-McMahon/pulse-tube cooler as the 4.2 K stage of the refrigerator has been described that is being used for neutron-scattering experiments at millikelvin temperatures. In addition, in [7.52] a combined Gifford-McMahon/dilution refrigerator system for precooling a nuclear refrigerator (Chap. 10) to sub-millikelvin temperatures has been presented. Recently, dilution refrigerators with pulse-tube precooling have become commercially available.

The main purpose of all these designs is to avoid (or to minimize) the expensive and bulky room-temperature gas handling and pumping system and to avoid the external supply of liquid nitrogen and helium. The developments demonstrate to which sophistication and the same time simplification the important technology of ^{3}He–^{4}He dilution refrigeration has been developed over the years.

For cooldown and diagnostics of the performance of a dilution refrigerator, resistance thermometers (see Sect. 12.5) should be attached to various flanges, on still and heat exchangers. Besides the usual thermometry techniques to be discused in Chap. 12, a very successful technique to determine the temperature of the liquid in the mixing chamber is a measurement of the T^{-2} dependence of its viscosity (Fig. 2.19) by a wire vibrating in the liquid [7.4, 7.38, 7.53] (Sect. 13.10.2). This wire resonator (for example, 100 μm Ta or NbTi bent to a 3–8-mm diameter semicircle) is set into oscillation by the Lorentz force on the wire when a current at the resonance frequency of a few hundred Hz is passed through it in the presence of a magnetic field provided by a small superconducting solenoid. The width of the resonance curve gives the damping effect of the liquid, which is then converted to temperature by using the known temperature dependence of the viscosity of the liquid which is proportional to T^{-2} for a Fermi liquid at low enough temperatures (Fig. 2.19) and as long as the mean free path of the ^{3}He quasiparticles does not exceed the radius of the wire. Such a vibrating wire can also be employed to locate the position of the phase boundary in the mixing chamber because the viscosity of pure ^{3}He

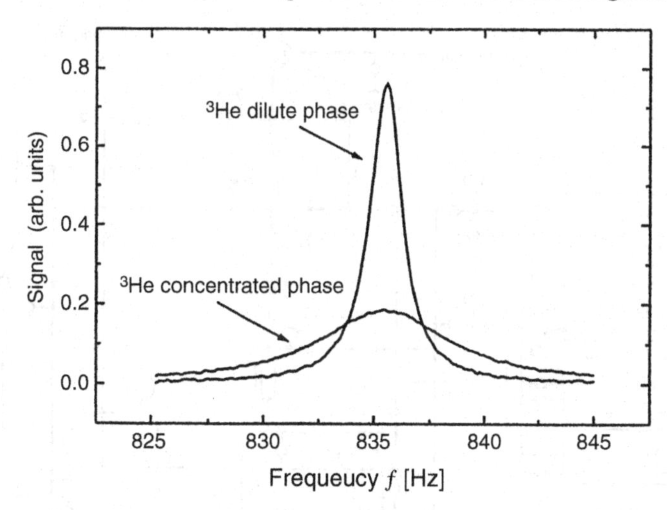

Fig. 7.23. Comparison of the signals from vibrating wires (125 μm Ta in a field of 50 mT) at 10 mK inside the dilute and the concentrated phases of ^{3}He, respectively, in the mixing chamber of the dilution refrigerator of [10.32]. The sensors are used as vibrating wire thermometers (see Sect. 13.10.2) as well as to locate the position of the phase boundary of the liquids inside the mixing chamber

is much larger than that of the mixture (see Figs. 2.19 and 7.23). For more details on vibrating wire viscometers see Sect. 13.10.2.

A dilution refrigerator is a rather complicated apparatus with many parts joined together by welding, hard soldering, soft soldering, or even gluing. To avoid later problems, it should therefore be carefully leak checked at room temperature, at LN_2 temperature, and possibly at LHe temperature in each run. Localizing a leak may be quite time consuming and the refrigerator may have to be separated into various sections to find it, in particular if it is a "cold leak". A procedure to localize a leak which is only penetrated by superfluid helium was described in [7.54]. In addition, the flow rates through the continuous ^{4}He refrigerator and through the dilution unit should be tested at room temperature and at LN_2 temperature.

General considerations for designing a dilution refrigerator are, of course, the space available for experiments and leads, access to experiments and to thermal heat sinks for the leads, mechanical stability, etc.

In addition to the low-temperature part of a dilution refrigerator, one usually needs substantial room-temperature equipment for the pumping and gas handling circuits [7.3, 7.6, 7.9]. A self-explanatory example of such a system is shown in Fig. 7.24. Of course, such a system should not have and should not develop leaks, in order to avoid loss of the expensive ^{3}He gas. This is particularly important for equipment containing moving parts like pumps and valves. One should therefore use valves sealed by bellows and sealed mechanical pumps modified for use with ^{3}He as backing pumps for the usual Roots

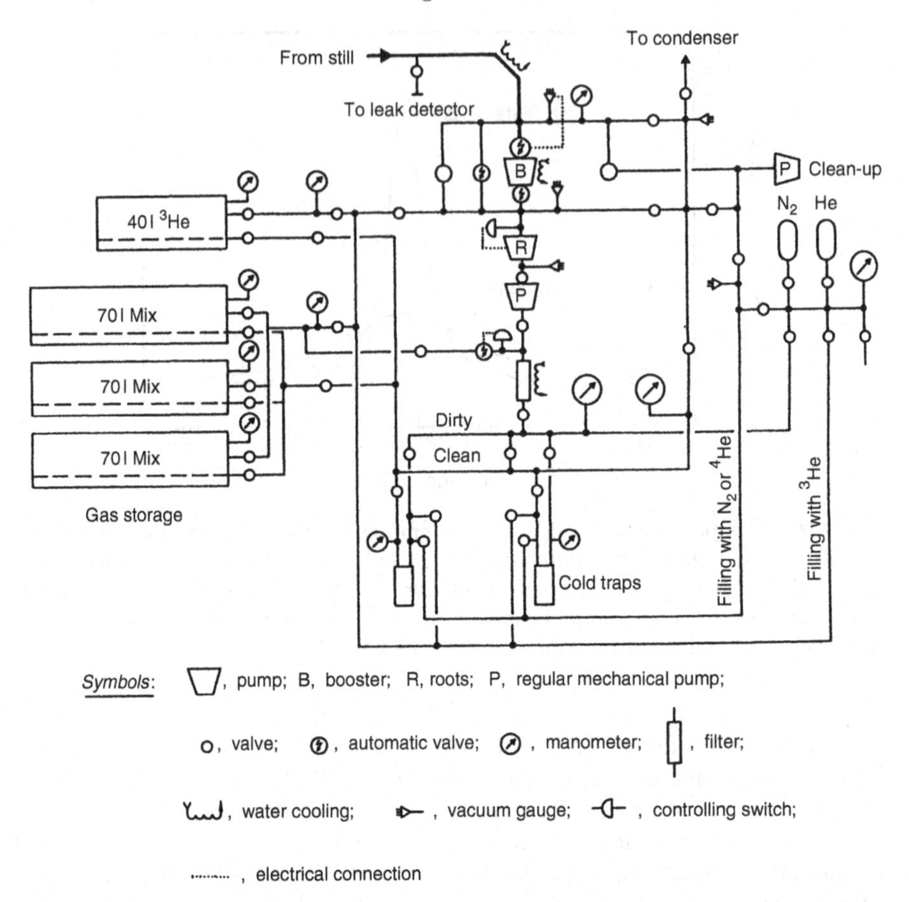

Fig. 7.24. Schematic diagram of the gas handling and pumping system of a ^{3}He–^{4}He dilution refrigerator (courtesy of P. Sekowski, Universität Bayreuth)

or Booster pumps. The system must also contain safety features for the event of cooling water failure or blockages in the refrigerator.

A gas handling system for a dilution refrigerator will always be operated at subatmospheric pressure (to avoid loosing helium gas). If there is a tiny undetected leak, air will get into the circulating gas. In addition, there will be products from cracking of pumping oil. Both the air and the organic gases would condense in the refrigerator and would very soon plug the condenser, capillaries, impedances or heat exchangers of such refrigerators, which are sometimes intended to run for months. To avoid this problem, a "cold trap" at LN_2 temperature (and sometimes possibly a second one at LHe temperature [7.55]) has to be inserted after the pumps to freeze out these gases (Fig. 7.25). With proper precautions, dilution refrigerators have been continuously operated for about a year.

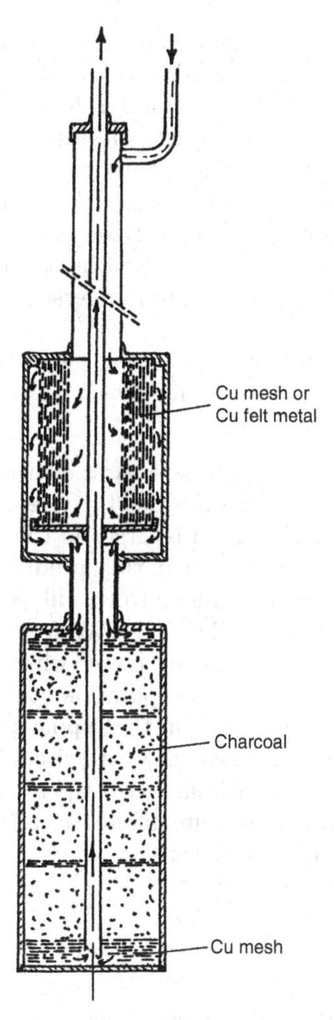

Fig. 7.25. LN_2 cooled activated charcoal/Cu mesh trap (courtesy of R.M. Mueller and P. Sekowski)

Before starting up a dilution refrigerator one has to know the volumes of its various components, to determine the necessary total molar quantity of helium $(n_3 + n_4)$, and the isotopic ratio $n_3/(n_3 + n_4)$ – typically 25% – to position the free surface in the still and the phase separation line in the mixing chamber. Later corrections of n_3 and/or n_4 after the first low temperature trials may be necessary. A capacitance level gauge or a vibrating wire resonator (see Fig. 7.23) in the mixing chamber to monitor the phase separation line is very helpful when adjusting the amount of ^{3}He for maximum performance of the refrigerator [7.3, 7.4, 7.56].

^{3}He–^{4}He dilution refrigerators are among the most important equipment for condensed-matter physics nowadays because many properties of matter have to be investigated at temperatures below 1 K to understand their behavior. To build such a system needs substantial experience and time; to buy it needs appropriate funds. There are several hundred of these refrigerators in operation in research institutes today, and without doubt their number will still increase in the future. Their thermodynamics is well established and problems usually result from design errors rather than from unknown phenomena. Hints of how to start up a dilution refrigerator and how to troubleshoot it are given in [7.57, 7.58].

Let me close this chapter by summarizing the five facts which are the basis for the successful operation of this only known continuous cooling technology for temperatures below 0.3 K.

- ^{3}He–^{4}He mixtures undergo phase separation when cooled below 0.87 K, resulting in two phases between which the ^{3}He atoms can be moved.
- The specific heat of a ^{3}He atom is larger in the dilute phase than in the concentrated phase, which results in the "production of cold" if this atom passes from the concentrated phase to the dilute phase.
- There is a finite solubility of ^{3}He in ^{4}He even for temperatures approaching absolute zero resulting in a cooling power which decreases only with T^2 and not exponentially as in the evaporation cooling process.
- There is a substantial difference in the vapour pressures of ^{3}He and ^{4}He at the same temperature, which allows the circulation of almost pure ^{3}He.
- If the helium mixtures are at different temperatures in a cryogenic apparatus, there is an osmotic pressure difference causing the ^{3}He to flow from the mixing chamber up into the still.

Problems

7.1. At which temperature does the upper, ^{3}He rich phase of a phase-separated ^{3}He–^{4}He mixture contain a ^{4}He concentration of only 10^{-6}?

7.2. At which temperature would the Fermi heat capacity of liquid ^{3}He reach the classical value (7.10), if its heat capacity continued to vary with temperature, as expressed by (7.20)?

7.3. Would the limiting concentration of ^{3}He in a ^{3}He–^{6}He mixture by larger or smaller than in a ^{3}He–^{4}He mixture?

Would the limiting concentration be larger if ^{3}He were replaced by ^{5}He in a mixture with ^{4}He?

7.4. What is the reason for the heat capacity per atom being smaller for ^{3}He in ^{4}He than in pure ^{3}He? How would this result change if ^{4}He were replaced by ^{6}He?

7.5. Calculate the Fermi temperature of liquid ^{3}He–^{4}He mixtures with ^{3}He concentrations of 0.1%, 1%, and 6.6%.

7.6. At which temperature do the osmotic pressures of a saturated mixture calculated by the classical van't Hoff's law and calculated by the equation for a Fermi-degenerate system give the same value?

7.7. The effective mass ratio of pure ^{3}He increases from about 2.8 at saturated vapour pressure to about 4.6 at 20 bar. In the same pressure range, the effective mass ratio for a saturated mixture increases from about 2.4 to 2.8. How would the numerical value for the cooling power of a ^{3}He–^{4}He dilution refrigerator given by (7.30) change if this refrigerator were operated at 20 bar?

7.8. Calculate the optimum circulation rate $\dot{n}_{3,\mathrm{opt}}$ and the minimum temperature $T_{\mathrm{mc,min}}$ taking $AR_{\mathrm{K}}T^3 = 0.3\,\mathrm{m^2K^4\,W^{-1}}$ for (7.51), and $A = 10\,\mathrm{m^2}$.

7.9. Calculate the surface area of the heat exchanger in the mixing chamber of the ^{3}He–^{4}He dilution refrigerator, the performance of which is shown in Fig. 7.19 (for Kapitza resistances see Fig. 4.5 and Sect. 4.3.2).

8

Refrigeration by Solidification of Liquid ^{3}He: Pomeranchuck Cooling

Solidification of matter at the rate $\dot{n}$ usually results in the production of heat according to

$$\dot{Q} = \dot{n}T(S_{\rm sol} - S_{\rm liq}) < 0\,, \tag{8.1}$$

because the liquid's entropy is usually larger than the solid's entropy, as the liquid state is a state of lower order than the solid state.

In 1950 the Russian physicist I. Pomeranchuk predicted that for ^{3}He on the melting curve below about 0.3 K the entropy of the liquid phase is smaller than the entropy of the solid phase (which means the liquid phase is more ordered), and that therefore adiabatic solidification of ^{3}He along the melting line should result in cooling. For quite some time this attractive proposal was not put into practice. Physicists were afraid of the experimental problems which might arise from the required low starting temperature and the possibility of frictional heating during solidification of ^{3}He. However, in 1965 the Russian physicist Anufriev picked up Pomeranchuk's proposal and indeed succeeded in reducing the temperature in his experiment from 50 mK (which he produced by paramagnetic cooling; see Chap. 9) to a final temperature of about 18 mK [8.1]. It is possible that the temperature in his experiment was somewhat lower, but the thermometry was not appropriate for determining lower temperatures conclusively. At the end of the 1960's several American groups, in particular Johnson, Wheatley and their coworkers, started experiments on "Pomeranchuk cooling" and arrived at final temperatures of 2–3 mK (see Fig. 8.6) [8.2, 8.3].

This cooling method was of great importance in the 1970's. Today it is no longer quite as important because above a few millikelvin this "one-shot" method offers essentially no advantage over continuously operating dilution refrigerators, and at lower temperatures nuclear demagnetization (Chap. 10) is much more powerful. I shall discuss the method here anyway for the following reasons:

(a) The method was – as just mentioned – of importance in the 1970's, because it provided the necessary experimental conditions for the detection of the superfluid phase transitions of liquid ^{3}He below 2.5 mK [8.4, 8.5] and for the detection of the nuclear antiferromagnetic phase transition of solid ^{3}He at 1 mK [8.6], both at Cornell University.

(b) This cooling method offers interesting insight into some important properties of matter, in particular liquid and solid ^{3}He, to be discussed in the following.

8.1 Phase and Entropy Diagrams of ^{3}He

The unusual behavior of the melting curve of ^{3}He below 1 K – the basis of Pomeranchuk cooling – is exhibited in Fig. 8.1. Down to about 0.3 K the melting curve shows the usual positive slope, $\mathrm{d}P_\mathrm{m}/\mathrm{d}T > 0$, but below this temperature the slope reverses sign and the pressure increases along the melting line if the temperature is decreased. The melting curve of ^{3}He shows a pronounced minimum at 29.31 bar and 315 mK [8.6–8.10]. Obviously, if we move along the melting curve below this minimum by increasing the pressure and so solidifying more and more ^{3}He, the temperature has to decrease (*Pomeranchuk cooling*).

According to the Clausius–Clapeyron equation

$$\frac{\mathrm{d}P_\mathrm{m}}{\mathrm{d}T} = \frac{(S_\mathrm{liq} - S_\mathrm{sol})_\mathrm{m}}{(V_\mathrm{liq} - V_\mathrm{sol})_\mathrm{m}}, \tag{8.2}$$

(where the differences are taken at melting) and because the molar volume of the liquid phase is always larger (by a constant value of 1.31 cm^3 mol^{-1} at $T < 40$ mK [8.6, 8.7]) than the molar volume of the solid phase, we have

$$\frac{\mathrm{d}P_\mathrm{m}}{\mathrm{d}T} < 0 \text{ if } S_\mathrm{liq} < S_\mathrm{sol}\,. \tag{8.3}$$

This negative slope of the melting curve means that we have the unusual situation of a liquid phase with a smaller entropy than the solid phase (Fig. 8.2). The fact that there is more order in the liquid state results in solidification cooling if we perform an isentropic compression. The resulting

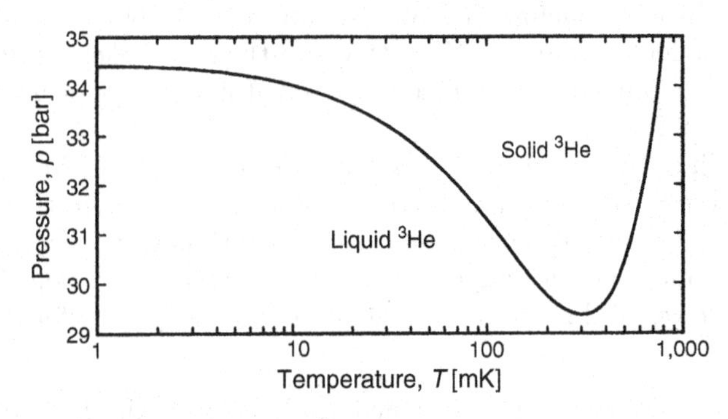

Fig. 8.1. Melting curve of ^{3}He, based on the data of [8.6–8.9]

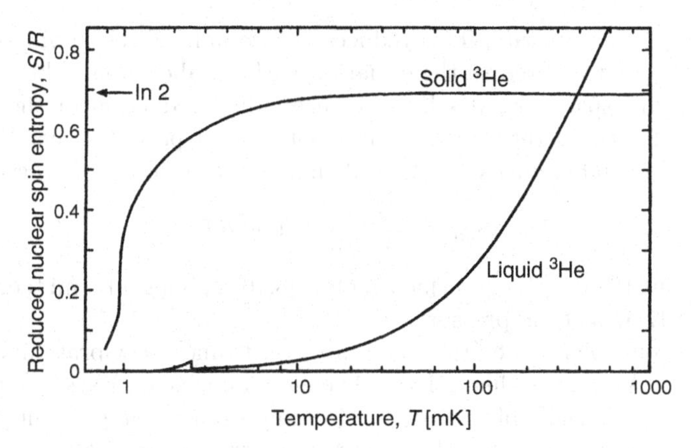

Fig. 8.2. Entropies (divided by the gas constant R) of solid and liquid ^{3}He along the melting curve. The full disorder nuclear spin entropy of solid ^{3}He, $S_s/R = \ln(2I+1) = \ln\ 2$, is marked. The entropy curves cross at the minimum of the melting curve at 315 mK, 29.31 bar. The anomaly in S_{liq} at 2.44 mK is caused by the superfluid transition of liquid ^{3}He. The kink in S_{sol} at 0.90 mK is a result of the nuclear antiferromagnetic ordering of solid ^{3}He

cooling power is

$$\dot{Q} = \dot{n}_{sol} T (S_{sol} - S_{liq})_m \,, \tag{8.4}$$

where $\dot{n}_{sol}$ is the rate of solidification of ^{3}He. The right-hand side of (8.4) is the latent heat of solidification.

8.2 Entropies of Liquid and Solid ^{3}He

Due to its nuclear spin $I = 1/2$ we have a spin disorder entropy of ^{3}He, which in the fully disordered state is given by

$$S = R\ \ln\ 2\,. \tag{8.5}$$

At low enough temperatures, i.e., below the minimum of the melting curve at 315 mK, this disorder entropy is far larger than other entropy contributions of the liquid or solid phases of ^{3}He, which will therefore be neglected in the following. In both states the entropy has to decrease as the temperature decreases because eventually it has to vanish at absolute zero. The decrease results from interactions between the nuclear spins or nuclear moments. The interactions are different in the two states of ^{3}He and the decrease of the spin disorder entropy starts at different temperatures (Fig. 8.2).

Helium-3 atoms are spatially indistinguishable in the liquid state and we have to apply Fermi–Dirac statistics as for conduction electrons in a metal. In liquid ^{3}He the atoms move rather freely and they approach each other quite closely. Therefore, a strong interaction results. Most importantly, the Pauli principle has to be applied to this Fermi liquid. That means that at low enough temperatures we have two particles with opposite spins in each translational energy state. This reduces the volume of the Fermi sphere by a factor of two compared to the situation with parallel spins. This "ordering in momentum space" results in a corresponding reduction of the entropy. Increasing the temperature results in excitation of particles near the Fermi energy, as for conduction electrons, and an increase of the entropy according to

$$S_{\text{liq}} = \frac{\pi^2 R}{2} \frac{T}{T_{\text{F}}} = 4.56\,RT \tag{8.6}$$

(below about 10 mK, along the melting line [8.10]), giving a Fermi temperature of about 1 K at melting pressure.

In the solid phase the ^{3}He atoms are constrained to vibrate about their lattice sites. Again, the thermal kinetic energy and the excitation of phonons, for example, are negligible at millikelvin temperatures; only the nuclear spin contribution matters. Solid ^{3}He at temperatures above 1 mK is a nuclear paramagnet with spin 1/2 and bcc structure. The Pauli principle does not apply to the occupation of energy states because the particles in the solid phase are already distinguished by their spatial parameters. Because of their very weak magnetic dipole and exchange interactions these particles can be described as independent particles for temperatures $T > 10$ mK, resulting in the full spin disorder entropy of $S = R \ln 2$ for 10 mK $< T <$ 1K (Fig. 8.2). Here $C_{\text{sol}} \propto \text{d}S_{\text{sol}}/\text{d}T \approx 0$, or small compared to C_{liq} (compare Fig. 8.3 with Figs. 2.15 and 7.5). The interactions between the nuclear moments in solid ^{3}He become noticeable at temperatures below about 10 mK, so the entropy then starts to decrease in this phase as well. One type of interaction is the direct nuclear dipole–dipole interaction, which was originally believed by

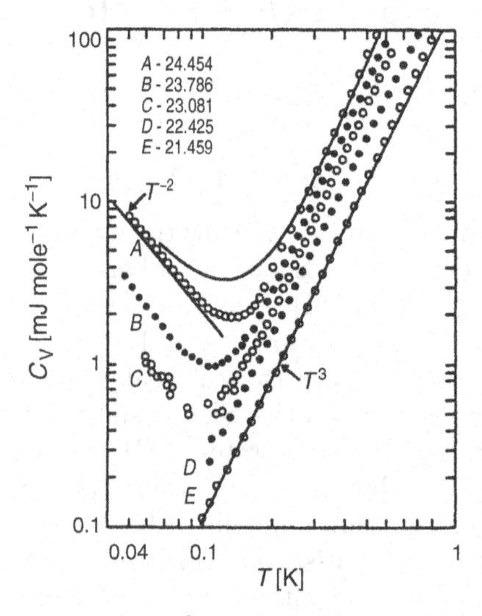

Fig. 8.3. Specific heat of solid bcc ^{3}He versus temperature. The numbers denote the molar volume in cm^3. The high-temperature T^3-behavior is due to phonons, the low-temperature T^{-2}-nature results from the nuclear spin–spin interactions [8.11]

Pomeranchuk to be the only important one; this is a very weak interaction of order 0.1 μK. But solid ^{3}He is a very remarkable material, because here – due to the large zero-point motion – the overlap of the wave functions is so strong that there is direct particle exchange, resulting in an exchange interaction of order 1 mK. Solid ^{3}He is the only system where the term "exchange" is meant in the literal sense of the word: the particles exchange sites.

The nuclear magnetic properties of solid ^{3}He were originally described in terms of a spin $I = 1/2$ Heisenberg magnet with an exchange Hamiltonian [8.12–8.14]

$$\mathrm{H} = -2J \sum_{i<j} I_i I_j \,, \tag{8.7}$$

where J is the exchange coupling constant, which is 0.88 (mK) k_B at melting pressure, resulting in a nuclear antiferromagnetic ordering transition temperature of 0.93 mK [8.6, 8.10][1]. With the general formula for the partition function

$$Z = \mathrm{Tr}\{\exp(-H/k_\mathrm{B}T)\}, \quad \text{and} \quad S = k_\mathrm{B}\frac{\partial(T \ln Z)}{\partial T}\,, \tag{8.8}$$

we obtain for temperatures $T > 3\,\mathrm{mK}$

$$\frac{S_\mathrm{sol}}{R} \simeq \ln 2 - 1.5\left[\frac{J}{k_\mathrm{B}T}\right]^2 + \left[\frac{J}{k_\mathrm{B}T}\right]^3 + \text{higher-order terms}\,, \tag{8.9}$$

for a bcc lattice [8.12–8.15], so that at $T > 10\,\mathrm{mK}$ we have $S_\mathrm{sol}/R = \ln\ 2$ to within 1%.

The details are rather complicated because the particles cannot move "through" each other. Instead, they perform a cyclic motion, leading, in addition to the two-particle exchange, to three-particle and four-particle exchanges, to keep out of each other's way. In more elaborate theories these processes have been taken into account [8.12–8.14]. Anyway, the exchange interaction results in a pronounced decrease of the entropy of solid ^{3}He below about 1 mK due to the antiferromangetic nuclear magnetic ordering [8.6], see Fig. 8.2.

8.3 Pomeranchuk Cooling

As is obvious from Fig. 8.1, for Pomeranchuk cooling we have to start with liquid ^{3}He at $T < 0.32\,\mathrm{K}$ and perform a continuous adiabatic solidification of part of it, meaning that the sum of the entropies should be kept constant as we compress along the melting curve. If the process is isentropic we can calculate the relative amount of solid phase and liquid phase in our experimental chamber from

[1] The transition temperature is 0.90 mK at the new temperature scale PLTS-2000 (Sect. 11.3)

$$n_{\text{total}} S_{\text{liq}}(T_{\text{i}}) = n_{\text{sol}} S_{\text{sol}}(T_{\text{f}}) + (n_{\text{total}} - n_{\text{sol}}) S_{\text{liq}}(T_{\text{f}}) \tag{8.10}$$

and therefore

$$\frac{n_{\text{sol}}}{n_{\text{total}}} = \frac{S_{\text{liq}}(T_{\text{i}}) - S_{\text{liq}}(T_{\text{f}})}{S_{\text{sol}}(T_{\text{f}}) - S_{\text{liq}}(T_{\text{f}})} . \tag{8.11}$$

We see that the lower the starting temperature T_{i}, the smaller the proportion of ^{3}He which we have to solidify to obtain a desired final temperature T_{f}. For example, for $T_{\text{i}} = 25\,\text{mK}$ and $T_{\text{f}} = 3\,\text{mK}$, only 20% of the liquid has to be solidified. It is always advantageous to have a large amount of liquid in the experimental chamber in order to obtain fast thermal equilibrium and to avoid the possibility of crushing solid crystals.

The temperature dependence of the cooling power of this refrigeration process at $5\,\text{mK} \leq T \leq 30\,\text{mK}$ (using (8.6)) is given by

$$\dot{Q} = \dot{n}_{\text{sol}} T (S_{\text{sol}} - S_{\text{liq}}) = \dot{n}_{\text{sol}} T (R \ln\, 2 - 36T) \overset{\propto}{\sim} T \tag{8.12}$$

when n_{sol} moles are solidified per unit time. $\dot{Q}$ decreases proportional to temperature rather than proportional to temperature squared as it did for the dilution refrigerator (Fig. 8.4) because $S_{\text{liq}} \ll S_{\text{sol}} = \text{const.}$ at these temperatures.

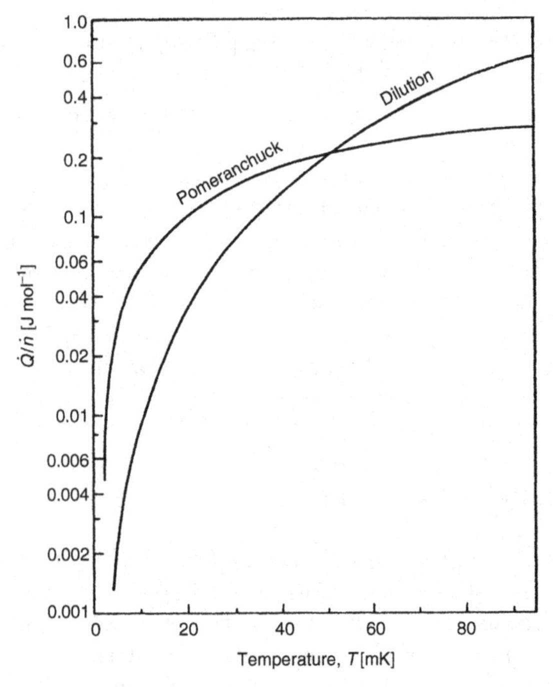

Fig. 8.4. Comparison of the relative cooling powers $\dot{Q}/\dot{n}$ of a Pomeranchuk refrigerator and of ^{3}He–^{4}He dilution refrigerator. In the former case liquid is converted into solid at a rate of $10\,\mu\text{mol}\, s^{-1}$; in the latter case ^{3}He is being removed from the mixing chamber at the same rate [8.16]

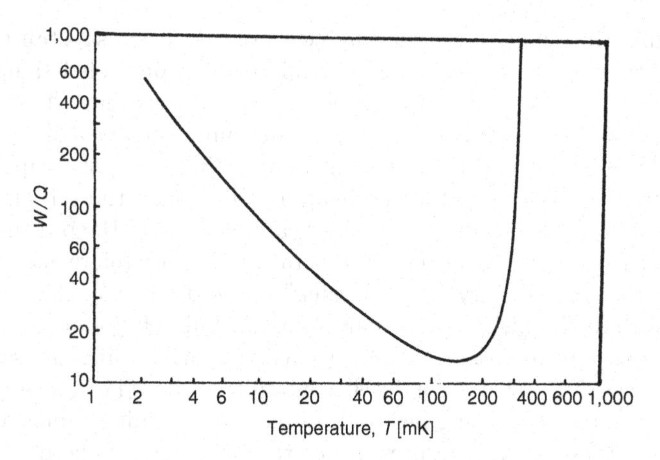

Fig. 8.5. Ratio W/Q of compressional work to heat extraction as a function of temperature [8.16]

Two experimental problems had to be solved before Pomeranchuk's idea could be converted into a refrigeration device. Firstly, when ^{3}He is compressed we perform mechanical work according to

$$\dot{W} = -\dot{n}_{\mathrm{sol}} P_{\mathrm{m}} (V_{\mathrm{sol}} - V_{\mathrm{liq}})\,. \tag{8.13}$$

This has to be compared to the temperature-dependent cooling power (8.12). The ratio of the two is given by

$$\frac{W}{Q} = -\frac{P_{\mathrm{m}}}{T}\frac{\Delta V}{\Delta S} = -\frac{P_{\mathrm{m}}}{T}\frac{\mathrm{d}P_{\mathrm{m}}}{dT}\,. \tag{8.14}$$

This ratio – which is shown in Fig. 8.5 – is always larger than 10 and can reach values above 100 at temperatures below about 8 mK. Work itself would do no harm if performed reversibly, but obviously the system will not cool if even only a small fraction of this mechanical work is irreversibly transformed into heat, for example, by rubbing or crushing of the ^{3}He crystals or by exceeding the elastic limit of the cell wall. Fortunately, only a small fraction of liquid ^{3}He has to be solidified in practice (see earlier). It was a happy surprise that Pomeranchuk cooling worked, an indication that physicists were indeed able to build setups for this refrigeration method where these problems could be avoided (see later).

In the earlier discussions I have assumed thermodynamic equilibrium, $\mu_{\mathrm{liq}} = \mu_{\mathrm{sol}}$, which is justified by the hydrostatic and thermal equilibria established by the liquid phase as long as there is not too much solid ^{3}He in the cell, and if the solid crystals are small enough that thermal conduction and nuclear spin diffusion keep the crystals in equilibrium.

Secondly, the shape of the melting curve of ^{3}He has a substantial impact on how Pomeranchuk cooling can be realized. The precooled sample cell is connected to a room-temperature compressor via a capillary. If we increase the pressure of ^{3}He in this capillary, the minimum of the melting curve at $T = 0.32\,K$ is hit first at some intermediate point along the capillary, and a solid plug will be formed there. This plug will isolate the cell at $T < T_{\min}$ from the room temperature setup and we cannot increase the pressure in it by increasing the gas pressure at room temperature. One therefore has to resort to an indirect pressure increase by "squeezing" the cold cell. How this squeezing is accomplished distinguishes the various Pomeranchuk refrigerators. An indirect hydraulic pressure increase can be achieved by surrounding the sample cell containing the ^{3}He by a second cell containing ^{4}He. The ^{3}He is then first compressed until a solid plug has been formed in the filling capillary. At that moment everything is at the pressure of the minimum of the melting curve (29.3 bar). Any further pressure increase can only be achieved by increasing the pressure in the surrounding ^{4}He chamber, which will compress the walls separating it from the ^{3}He sample cell. The pressure in the ^{3}He sample cell is then given by

$$P(^3\mathrm{He}) = P_{\mathrm{wall}} + P(^4\mathrm{He})\,. \tag{8.15}$$

Of course, it is now also not possible to measure the pressure in the ^{3}He sample cell by connecting it to a room-temperature manometer. We have to measure the pressure in situ at low temperatures, which can be done with a low-temperature capacitive pressure transducer (Sect. 13.1). The ^{4}He is not a heat load on the device because its specific heat and that of the cell walls are very small at these temperatures compared to the specific heat of liquid ^{3}He (Fig. 2.9).

Experimental advances in Pomeranchuk cooling were obtained in three steps. As mentioned at the beginning of this chapter, in 1965, in his original experiment, Anufriev [8.1] reached a temperature of 18 mK (or lower) starting from 50 mK. About four years later, Wheatley, Johnson and co-workers [8.2, 8.3] using a more elaborate design, demonstrated the full power of Pomeranchuk cooling by achieving a minimum temperature of about 2 mK (Fig. 8.6). Very successful Pomeranchuk cells were built at Cornell University [8.4–8.6, 8.17, 8.18] making use of convoluting flexible Be–Cu bellows, as shown in Fig. 8.7. In such a design bellows with different diameters can be combined to work as hydraulic pressure amplifier and to allow for the fact that ^{4}He has a lower freezing pressure than ^{3}He (Fig. 2.4). The Cornell device enabled the scientists there to detect the superfluid transitions of liquid ^{3}He [8.4, 8.5] and the nuclear antiferromagnetic transition of solid ^{3}He [8.6], (Fig. 2.4). More recent applications of Pomeranchuk cooling are investigations of the influence of nuclear polarization on the properties of liquid ^{3}He [8.19–8.21]. Here the advantage is that this cooling method is only weakly influenced by moderate magnetic fields. The cells to be used in changing high magnetic fields are then

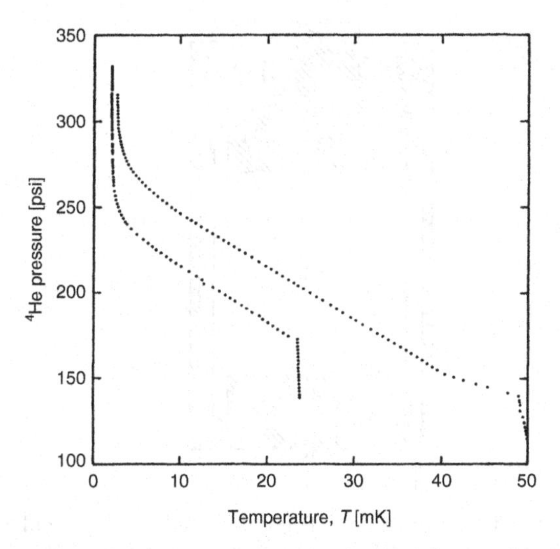

Fig. 8.6. Results of Pomeranchuk cooling experiments [8.2, 8.3]. The starting temperatures were 24 and 50 mK, respectively. The final temperatures were between 2 and 3 mK, respectively. The ^{4}He pressure (in pounds per square inch) is not of any particular quantitative significance, but is simply the experimental parameter which was varied to change the volume of the ^{4}He and ^{3}He cells (see text)

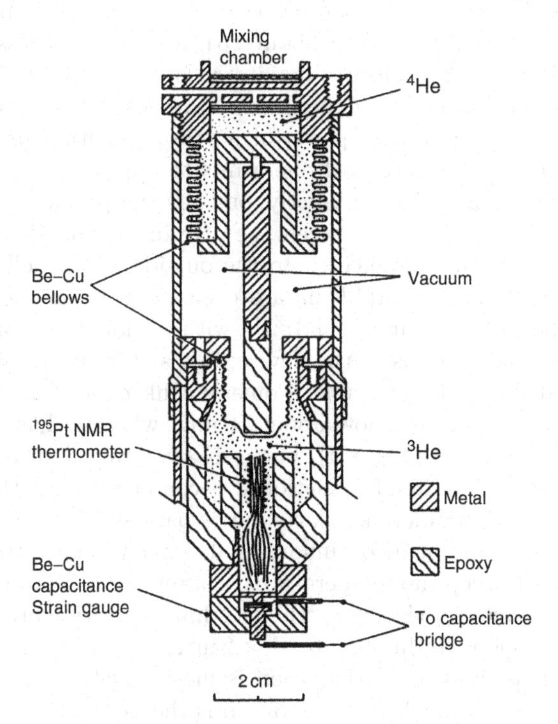

Fig. 8.7. Pomeranchuk refrigeration cell of [8.4, 8.5]

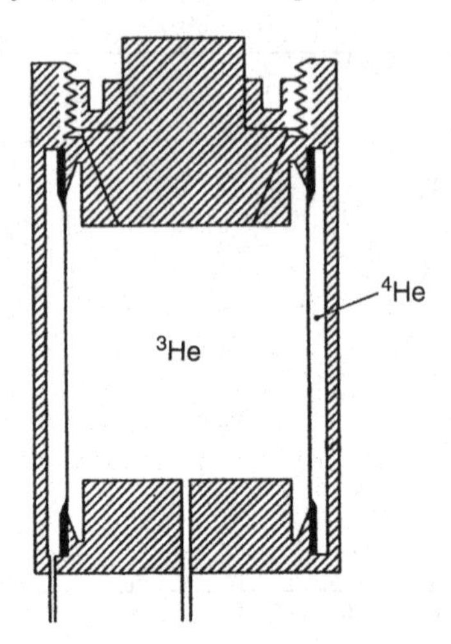

Fig. 8.8. Plastic Pomeranchuk refrigerator cell of Vermeulen and Frossati, University of Leiden, 1989, with a Kapton tube separating the ^{3}He and the ^{4}He spaces

made almost entirely of plastic with a stretched Kapton tube as the pressure transmitter (Fig. 8.8).

In these applications of Pomeranchuk cooling, the cell is precooled by a ^{3}He $-^4$He dilution refrigerator to about 25 mK, and the pressures and volumes in the ^{3}He and ^{4}He cells are monitored by low-temperature capacitive manometers (Sect. 13.1). The experiments have clearly demonstrated that Pomeranchuk cooling is rather powerful if ^{3}He at melting pressure is itself the subject of the investigation. If one wants to apply Pomeranchuk cooling to refrigerate other materials indirectly, one has the problem of the Kapitza thermal boundary resistance between the ^{3}He inside the Pomeranchuk cell and the sample, which is connected to the outside of the cell or floating in the ^{3}He. Thermal contact problems are even more severe because, due to the slope of the melting curve, solid ^{3}He with its low thermal conductivity forms at the warmest places of the device, i.e., at the places where heat has to be absorbed [8.22]. In principle, Pomeranchuk cooling is more powerful than dilution refrigeration below about 50 mK, where it has a higher cooling power at the same $\dot{n}$ (Fig. 8.4), but a disadvantage – in addition to the experimental problems already mentioned – compared with the latter is its discontinuous nature. Pomeranchuk cooling is basically a "one-shot" method. It can be carried out in a semicontinuous mode by continuously compressing at a controlled rate to keep the temperature constant [8.23]. However, it seems to be easier to continuously circulate ^{3}He in a dilution refrigerator. In addition, Pomeranchuck cooling is limited to $T \geq 2$ mK. For the first reasons given earlier, continuous dilution refrigeration is most widely used for $T \geq 5$ mK, and for the last reason nuclear refrigeration is the method of choice for lower temperatures.

Problems

8.1. Why must $dP_m/dT = 0$ at $T = 0$?

8.2. What is the temperature dependence of the cooling power of a Pomeranchuk refrigerator at $3\,\text{mK} \le T \le 10\,\text{mK}$ (see Fig. 2.15 for S_{liq}) and for $10\,\text{mK} \le T \le 316\,\text{mK}$?

8.3. Calculate the crossing point of the cooling power of a ^{3}He–^{4}He dilution refrigerator and a Pomeranchuk refrigerator, as shown in Fig. 8.4 for the ^{3}He rates given in the figure caption.

8.4. Discuss the temperature dependence of the ration W/Q of compressional work to heat extraction as a function of temperature, as depicted in Fig. 8.5.

8.5. Calculate the minimum temperature to which Pomeranchuk cooling would work if the nuclear magnetic ordering of solid ^{3}He did not occur at 1 mK due to strong exchange interaction, but only were caused by nuclear dipole–dipole interaction, see (10.1) (the nuclear magnetic moment of ^{3}He is $\mu = 2.13\,\mu_n$).

8.6. Calculate the solidification rate of ^{3}He in a Pomeranchuk cell necessary to keep a temperature of 8 mK constant at an external heat leak of 1 µW.

9

Refrigeration by Adiabatic Demagnetization of a Paramagnetic Salt

In the year 1922 H. Kamerlingh Onnes reached 0.83 K by pumping on his liquid ^{4}He bath with an enormous battery of pumps. Realizing that there would be no element with a lower boiling point than helium, he predicted that this temperature would remain the minimum temperature achievable by mankind in the laboratory unless somebody discovered a completely new refrigeration technology that did not depend on the latent heat of evaporation. (Of course, he was not aware that there exists a lighter helium isotope, ^{3}He, which pushes this limit down by about a factor of 3.) In 1926, a completely new refrigeration technology was proposed. P. Debye and W.F. Giauque independently made the proposal that lower temperatures could be reached by using the magnetic disorder entropy of electronic magnetic moments in paramagnetic salts, a method later called *adiabatic demagnetization of paramagnetic salts.* Not until seven years later, in 1933, was this proposal converted into a practical realization, when W.F. Giauque and D.P. MacDougall (Berkeley) reached 0.53 K, and a little later W.J. de Haas, E.C. Wiersma and H.A. Kramers (Leiden) reached 0.27 K. In fact, this was the first important low-temperature experiment since the end of the 19th century in which the University in Leiden was not the first to achieve success (for this early work see the relevant references in [9.1–9.3]).

Adiabatic demagnetization of paramagnetic salts was the first method of refrigeration to reach temperatures significantly below 1 K. Today this method can be applied to experiments in the range $2\,\text{mK} \leq T \leq 1\,\text{K}$. For the last decades it has not been used much, because it has been replaced by the ^{3}He–^{4}He dilution refrigerator, which has the substantial advantage of being a continuous refrigeration method. I will discuss adiabatic demagnetization of paramagnetic salts for historical reasons and because it may become of importance again for refrigeration in space flights. In addition, it is the basis for understanding the presently much more important nuclear adiabatic demagnetization to be discussed in Chap. 10.

9.1 The Principle of Magnetic Refrigeration

Let us consider paramagnetic ions with an electronic magnetic moment μ in a solid. We assume the energy ε_m of interaction between the moments themselves as well as with an externally applied magnetic field to be small compared to the thermal energy $k_B T$. This means that we are considering free paramagnetic ions with magnetic moment μ and total angular momentum J with entropy contribution

$$S = R \ln(2J + 1) \tag{9.1}$$

if they are completely disordered in their $2J + 1$ possible orientations with respect to a magnetic field.

As in the case of Pomeranchuk cooling, it is this magnetic disorder entropy which we want to use for refrigeration. At the temperatures of interest in this chapter, this magnetic disorder entropy, which is of the order of joules per mole of refrigerant, is always large compared to all other entropies of the system, for example the lattice and conduction electron entropies, which we will therefore neglect.

If the temperature is decreased, eventually the interactions between the magnetic moments will become comparable to the thermal energy. This will then lead to spontaneous magnetic order, e.g., ferromagnetic or antiferromagnetic orientation of the electronic magnetic moments. As a result the entropy will decrease and approach zero, as required by the third law of thermodynamics. An externally applied magnetic field will interact with the magnetic moments, at least partially orienting them along its axis to create a magnetized state of higher order. Therefore, in the presence of a field the entropy will decrease at a higher temperature than without a field (Fig. 9.1).

The entropy diagram of Fig. 9.1 makes it very easy to understand magnetic refrigeration. We bring the paramagnetic salt in contact with a precooling bath to precool it to a starting temperature T_i. Then a magnetic field B_i is applied to perform an isothermal magnetization at T_i from $B = 0$ to $B = B_i$. During this process the heat of magnetization has to be absorbed by the precooling bath. The next step is thermal isolation of the paramagnetic salt from the surrounding bath which then allows the adiabatic demagnetization to be carried out by reducing the external field from its starting value B_i to a final field B_f, which may be in the millitesla region. The temperature has to decrease accordingly (Fig. 9.1). Finally the cooling agent, the paramagnetic salt, will warm up along the entropy curve at B_f = const. due to an external heat leak until its cooling power has been used up. It is obvious from Fig. 9.1 that a spontaneous magnetic ordering process of the magnetic moments represents the lower limit for magnetic refrigeration. Magnetic refrigeration is a "one-shot" technique, where the demagnetization ends at a low field and then sample and refrigerant warm up.

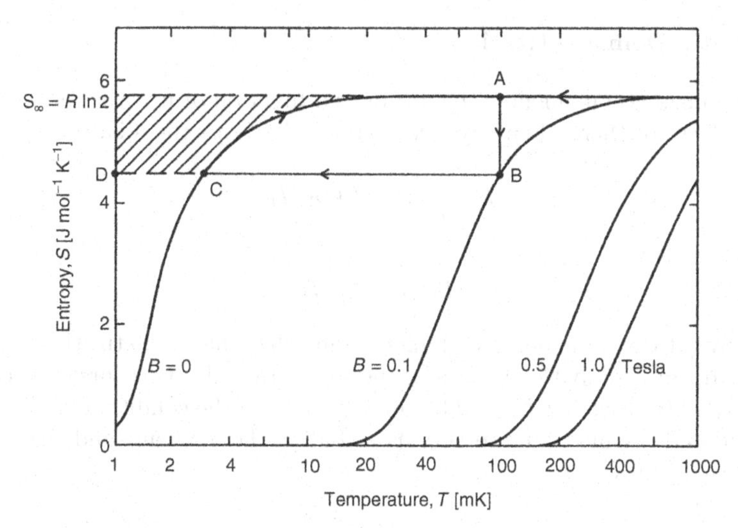

Fig. 9.1. Molar entropy of a single crystal of the paramagnetic salt CMN with angular momentum $J = 1/2$ (Sect. 9.4) as a function of temperature for magnetic fields applied along the cystallographic a axis. For the refrigeration process the salt is first isothermally magnetized (AB), and then after thermal isolation adiabatically demagnetized (BC). Eventually it warms up along the entropy curve at the final demagnetization field, which is zero in the example shown. The heat of magnetization during magnetization is given by the rectangle ABDS$_\infty$. The cooling power of the salt after demagnetization is given by the shaded area [9.4]

9.2 Thermodynamics of Magnetic Refrigeration

In this section I present some simple thermodynamic calculations relevant for the three steps of a magnetic refrigeration process.

Heat of Isothermal Magnetization

The heat of magnetization released when the applied field is increased from zero to B_i and which has to be absorbed by the precooling bath at constant temperature T_i is given by

$$Q(T_\mathrm{i}) = nT_\mathrm{i}[S(0, T_\mathrm{i}) - S(B_\mathrm{i}, T_\mathrm{i})] \tag{9.2}$$

or more exactly

$$Q(T_\mathrm{i}) = nT_\mathrm{i} \int_0^{B_\mathrm{i}} (\partial S/\partial B)_{T_\mathrm{i}} \, \mathrm{d}B = nT_\mathrm{i} \int_0^{B_\mathrm{i}} (\partial M/\partial T)_B \, \mathrm{d}B \,, \tag{9.3}$$

where M is the magnetization (magnetic moment per unit volume). The resulting $Q(T_\mathrm{i})$ is indicated in (Fig. 9.1); it is typically several joules per mole of refrigerant, so it can easily be absorbed by an evaporating helium bath, with its latent heat of several of joules per mole, or by a ^{3}He–^{4}He dilution refrigerator if lower starting temperatures are required.

Adiabatic Demagnetization

For free magnetic moments the entropy is a function of just the ratio of magnetic energy to thermal energy. We therefore have for the adiabatic process

$$S(B_{\mathrm{i}}/T_{\mathrm{i}}) = S(B_{\mathrm{f}}/T_{\mathrm{f}}), \tag{9.4}$$

which results in

$$T_{\mathrm{f}}/B_{\mathrm{f}} = T_{\mathrm{i}}/B_{\mathrm{i}}. \tag{9.5}$$

We arrive at the same result by remembering that the magnetization does not change during adiabatic changes of the magnetic field. Of course, we cannot reach $T_{\mathrm{f}} \to 0$ by letting $B_{\mathrm{f}} \to 0$ because eventually the condition $k_{\mathrm{B}}T \gg \epsilon_{\mathrm{m}}$ is violated and the internal interactions will align the moments and the entropy vanishes (Fig. 9.1).

Warming-up due to External Heating

The cooling power of the salt after demagnetization to B_{f}, or the heat it can absorb as it is warming up, is given by

$$Q(B_{\mathrm{f}}) = n \int_{T_{\mathrm{f}}}^{\infty} T(\partial S/\partial T)_{B_{\mathrm{f}}}\, \mathrm{d}T. \tag{9.6}$$

This hatched area is indicated in Fig. 9.1. It is, of course, substantially smaller than the heat of magnetization due to the fact that the energy is absorbed at a temperature $T < T_{\mathrm{i}}$. Hence the external heat leak should be kept as small as possible.

These results demonstrate that one has to find a compromise between a low final temperature T_{f} [which would require a low final field B_{f}; see (9.5)], and a large cooling power (which requires a large final field; see Fig. 9.1).

For the above analysis we have to know the entropy $S(B,T)$, which can be calculated from data for the specific heat as a function of temperature and field according to

$$S(B,T_1) - S(B,T_2) = \int_{T_2}^{T_1} (C_{\mathrm{B}}/T)\, \mathrm{d}T. \tag{9.7}$$

Another possibility is to use the magnetization

$$M = \frac{\chi B}{\mu_0} \tag{9.8}$$

to calculate the magnetic entropy via the Maxwell relation $(\partial S/\partial B)_T = (\partial M/\partial T)_B$.

9.3 Non-Interacting Magnetic Dipoles in a Magnetic Field

If we have paramagnetic ions carrying a magnetic moment μ and a total angular momentum J, then at temperature T the $(2J+1)$ energy levels with energies ε_{m} will be populated according to

$$P(m) = e^{-\varepsilon_{\mathrm{m}}/k_{\mathrm{B}}T} \sum_{m=-J}^{+J} \mathrm{e}^{-\varepsilon_{\mathrm{m}}/k_{\mathrm{B}}T} , \tag{9.9}$$

and the partition function is given by

$$Z = \left(\sum_{m=-J}^{+J} e^{-\varepsilon_{\mathrm{m}}/k_{\mathrm{B}}T} \right)^{N_0} . \tag{9.10}$$

Once we have calculated the partition function, we can calculate all the required thermodynamic properties, for example

$$\begin{aligned} S &= R\partial(T \ln Z)/\partial T , \\ C_{\mathrm{B}} &= T(\partial S/\partial T)_B , \\ M &= RT(\partial \ln Z/\partial B)_T , \end{aligned} \tag{9.11}$$

and hence the susceptibility $\chi = (\partial M/\partial B)_T$ is known as well. In other words, once we know the energies ε_{m}, everything else can be calculated. Fortunately, if we assume non-interacting magnetic dipoles in an external magnetic field, where $\epsilon_{\mathrm{m}} \ll k_{\mathrm{B}}T$, we have

$$\epsilon_{\mathrm{m}} = -\mu \cdot B = \mu_{\mathrm{B}} g m B , \tag{9.12}$$

with the electronic Landé factor

$$g = 1 + \frac{J(J+1) + S(S+1) - L(L+1)}{2J(J+1)} , \tag{9.13}$$

here, S, J, L refer to the spin, the total and orbital angular momenta, respectively.

In (9.12), $\mu_{\mathrm{B}} = e\hbar/2m_{\mathrm{e}} = 9.27 \times 10^{-24}\,\mathrm{J\,T^{-1}}$ is the Bohr magneton and $m = -J, \ldots, 0, \ldots, +J$ are the magnetic quantum numbers.

Accordingly, the partition function for independent magnetic dipoles is given by

$$Z = \left(\sum_{m=-J}^{+J} e^{-mx} \right)^{N_0} = \left[\frac{\sinh[(J+1/2)x]}{\sinh(x/2)} \right]^{N_0} \tag{9.14}$$

with $x = \mu_B gB/k_B T$. After some manipulation we find

$$S/R = (x/2)\{\coth(x/2) - (2J+1)\coth[x(2J+1)/2]\} + \ln\left[\frac{\sinh[x(2J+1)/2]}{\sinh(x/2)}\right], \quad (9.15a)$$

$$C_B/R = (x/2)^2 \sinh^{-2}(x/2) - [x(2J+1)/2]^2 \sinh^{-2}[x(2J+1)/2], \quad (9.15b)$$

$$M/M_s = [(2J+1)/2J]\coth[x(2J+1)/2] - (1/2J)\coth(x/2), \quad (9.15c)$$

where $M_s = N_0\mu_B gJ$ is the molar saturation magnetization, and the right-hand side of (9.15c) is the polarization P or the Brillouin function $B_J(x)$. This function is tabulated for various J in [9.2, 9.3]. The Brillouin function $B_J(x) = M/M_s$ for various J is shown in Fig. 10.5 as a function of x^{-1} and the dependence of the specific heat C_B on x for various J is depicted in Figs. 3.8, 3.10 and 10.4. These results demonstrate a previously mentioned fact, namely that with the above assumptions the entropy is a function of B/T only. Therefore, in Fig. 9.1 the entropy curves are simply translated if we change the field. In addition, the above results give (9.5), valid for an ideal adiabatic process: S = const. and therefore B/T = const., which means that the starting conditions determine the final state.

The above equations cannot, of course, be used in the limit $B_f \to 0$, because then the assumption of negligible interactions between the moments breaks down. For very small B_f we have to replace B_f in the above equations by an effective field

$$B_{eff} = \sqrt{B_f^2 + b^2} \quad (9.16)$$

acting on the moments, where b is an internal field resulting from the neighbouring moments in the paramagnet. This then leads to

$$T_f = \frac{T_i}{B_i}\sqrt{B_f^2 + b^2}. \quad (9.17)$$

The internal field b (or the magnetic ordering temperature T_c; see below) determines the minimum temperature $T_{f,min}$ that can be reached by demagnetizing to $B_f = 0$. Finally, we find that the magnetization and entropy are kept constant during the demagnetization process or, in other words, that the populations of the various energy levels do not change during the demagnetization (Fig. 10.2). Only their energy difference changes when we decrease the field. Figure 9.1 also demonstrates that the cooling power Q is proportional to the final field B_f. When the demagnetized salt absorbs heat it has to change the population of its energy levels, which means that the spins of the paramagnetic ions have to flip. The ions can absorb more heat if the energy change per spin flip is larger, i.e., if their energy separation, which is proportional to the field B_f, is larger.

Very often the magnetic energy $g\mu_B B$ is substantially smaller than the thermal energy $k_B T$, or $x \ll 1$. In this approximation one does not have to use the full (9.15) but can use the so-called high-temperature approximations. I will give these approximations in Chap. 10, on nuclear demagnetization, where usually $x \ll 1$, because the nuclear moments are rather small.

9.4 Paramagnetic Salts and Magnetic Refrigerators

The behavior of a magnetic refrigerator is mainly determined by the experimental starting conditions (B_i, T_i), the heat leaks (Sect. 10.5), and the properties of the selected paramagnetic salt. Typical starting conditions for paramagnetic refrigeration, $B_i = 0.1$–$1\,T, T_i = 0.1$–$1\,K$, are fairly easy to achieve nowadays. With these starting conditions large entropy reductions are possible (Fig. 9.1).

I shall not discuss here the magnetic and electric interactions of paramagnetic ions in salts; they have been covered in [9.3, 9.5]. In general, the paramagnetic refrigerant, the salt, should have a low magnetic ordering temperature T_c and a large magnetic specific heat in order to achieve a large cooling power, which means a large angular momentum J (Fig. 3.10). The ordering temperature T_c is determined by the interactions between the magnetic moments, which create the internal field b.

Paramagnetic salts suitable for magnetic cooling must contain ions with only partly filled electronic shells, i.e., either 3d transition elements or 4f rare earth elements. The following four paramagnetic substances have often been used for magnetic refrigeration in the so-called "high-temperature range" as well as in the "low-temperature range" (where T_c is the approximate magnetic ordering temperature).

"High"-temperature salts:

$$\begin{aligned} &\mathrm{MAS}: \mathrm{Mn^{2+}SO_4 \cdot (NH_4)_2SO_4 \cdot 6H_2O}; && T_c \simeq 0.17\,\mathrm{K} \\ &\mathrm{FAA}: \mathrm{Fe_2^{3+}(SO_4)_3 \cdot (NH_4)_2SO_4 \cdot 24H_2O}; && T_c \simeq 0.03\,\mathrm{K} \end{aligned}$$

"Low"-temperature salts:

$$\begin{aligned} &\mathrm{CPA}: \mathrm{Cr_2^{3+}(SO_4)_3 \cdot K_2SO_4 \cdot 24H_2O}; && T_c \simeq 0.01\,\mathrm{K} \\ &\mathrm{CMN}: \mathrm{2Ce^{3+}(NO_3)_3 \cdot 3Mg(NO_3)_2 \cdot 24H_2O}; && T_c \simeq 0.002\,\mathrm{K}\,. \end{aligned}$$

Their entropy curves are depicted in Fig. 9.2.

All these salts contain a lot of water of crystallization, which assures a large distance (about 1 nm in CMN) between the magnetic ions and therefore leads to a low magnetic ordering temperature. Details on the properties of these and other suitable salts can be found in [9.1–9.3, 9.5–9.7].

Due to the weak interactions between the rather remote and well-sheltered Ce^{3+} ions in CMN [9.6], which result in a small internal field b and a low ordering temperature T_c (Fig. 9.1), this salt has been quite en vogue as a refrigerant as well as for thermometry (Sect.12.9). Its electronic ground state

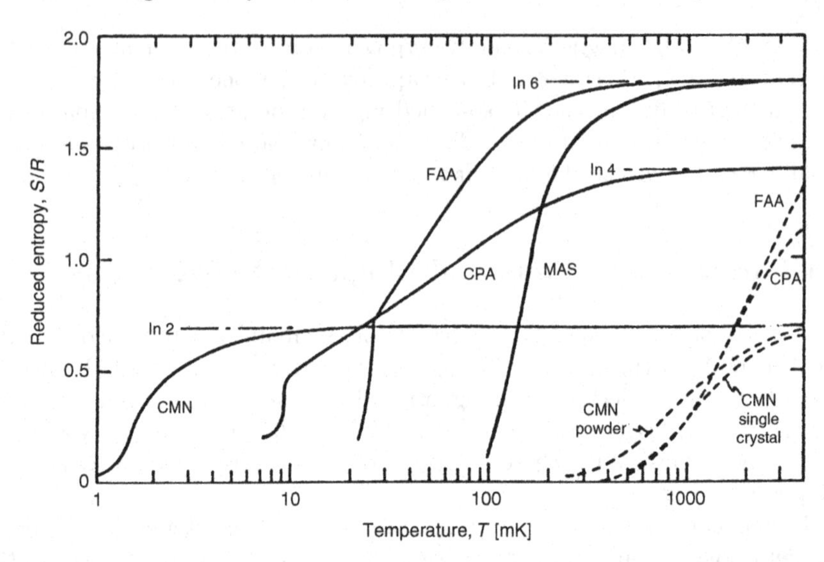

Fig. 9.2. Entropies S (divided by the gas constant R) of four salts suitable for paramagnetic demagnetization as a function of temperature in zero field (——) and in 2 T (– – –). (For the chemical formula of the salts see the text.)

is a doublet with an effective spin of one-half at $T < 1\,\mathrm{K}$. CMN has been extensively used by Wheatley and coworkers [9.7, 9.8] to refrigerate liquid ^{3}He to about 2 mK. This low final temperature is possible due to a magnetic coupling between ^{3}He nuclear moments and Ce electronic moments which strongly reduces the thermal boundary resistance (Sect. 4.3.2). Even lower temperatures can be achieved by using LCMN, a salt where the magnetic Ce ions are partly replaced by non-magnetic La ions [9.9–9.11]. This dilution, of course, reduces the cooling power/volume as well. CMN and LCMN are the salts with the lowest known electronic magnetic ordering temperature. A combination of a moderate dilution refrigerator and a magnetic refrigeration stage of CMN or LCMN at a starting field of about 1 T is a very suitable and reasonably cheap setup for ^{3}He research. For the use of CMN and LCMN in magnetic thermometry, see Sect. 12.9.

Figure 9.3 shows some typical setups for paramagnetic refrigeration used mainly in the 1960s and 1970s before the advent of the dilution refrigerator; they are described in [9.1–9.3, 9.7–9.14] and references therein.

The advantages of paramagnetic refrigeration are that the required starting conditions can be fairly easily achieved with simple dilution refrigerators and superconducting magnets. In addition, the refrigerant can also be used as a thermometer by applying the Curie law $\chi = \lambda/T$ to determine the temperature T by measurements of the susceptibility χ of the salt (Sect. 12.9). However, paramagnetic refrigeration also has some severe drawbacks. The thermal conductivity of dielectric salts is rather poor; a typical value is

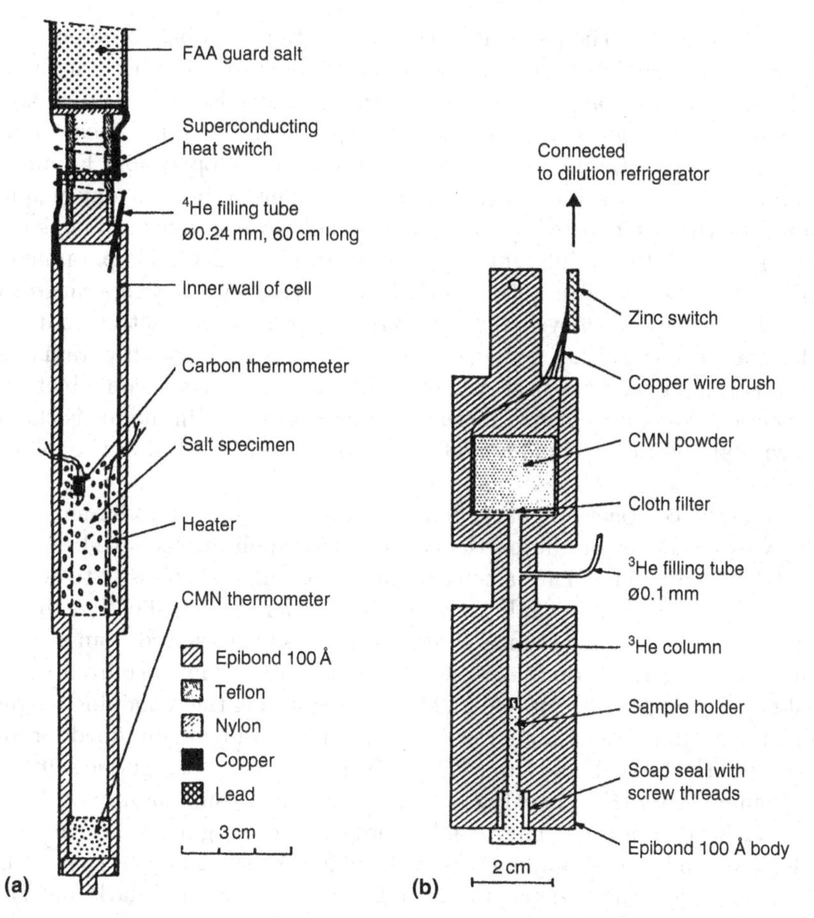

Fig. 9.3. Two paramagnetic refrigerators used by Wheatley and co-workers (**a**): [9.7]; (**b**): [9.12]. These original references and [9.1] should be consulted for details

$10^{-4}\,W\,\mathrm{K}^{-1}\,\mathrm{cm}^{-1}$ at 0.1 K. It is therefore difficult to achieve thermal equilibrium within the salt, leading to temperature gradients and long thermal relaxation times in the millikelvin temperature range. Similarly it is rather difficult to achieve thermal contact to the salt and special construction methods have been developed for paramagnetic refrigeration stages. Probably, the mostly applied one is to compress the powdered salt (typically at 100 bar pressure) together with some "glue" (Apiezon grease or oil or epoxy) and the ends of some fine Cu or Ag wires spread out uniformly in the pill. The other end of the wires should already have been welded to the places where the thermal contact is desired (to a heat switch connecting to the precooling stage and to a metal plate for mounting experiments and thermometers), see Figs. 12.37, and 12.39. In some cases it may be necessary to seal the salt pill to prevent chemical changes (dehydration) leading to deterioration. Consideration of the thermal path from the spins to the lattice of the salt, to the metal wires, and to the sample/thermometer is advised to avoid unnecessary, unpleasant surprises [9.2]; these problems, of course, become more and more serious the lower the temperature range of operation. Finally, the minimum temperature is determined by interactions of the electronic magnetic moments with their surroundings and eventually by spontaneous magnetic ordering, leading to a minimum temperature of about 2 mK for paramagnetic refrigeration (reached in 1953 with CMN). Because these temperatures can nowadays also be achieved by dilution refrigerators, this latter continuous refrigeration method has replaced magnetic refrigeration with paramagnetic salts. For lower temperatures, where dilution refrigerators become inefficient, magnetic refrigeration using nuclear magnetic moments in metals is the only known refrigeration technique and will be discussed in detail in the following chapter.

Recently, compact adiabatic demagnetization refrigerators for small laboratory experiments and, in particular, for space applications have been developed [9.15,9.16]. They have a typical minimum temperature of 30 to 50 mK, a time constant of 1 s at 100 mK, and a cooling power of about 1 μW at this temperature. Single-stage paramagnetic refrigerators are operated from a starting temperature of 1.5 K, double-stage refrigerators can commence from the normal boiling temperature of ^{4}He (4.2 K). By regulating the remaining magnetic field after demagnetization, a stable temperature can be maintained for many hours and the typical hold time at 100 mK is 24 h. As paramagnetic salts ferric ammonium alum (FAA) with $J = 5/2$ or chromic caesium alum (CCA) with $J = 3/2$ have been used. They can be applied for cooling microcalorimeters or bolometers in infrared, millimeter-wave and X-ray astronomy and as test beds for the development of detectors for millimeters, X-rays and dark matter.

Another cause for revival of adiabatic demagnetization using paramagnetic salts may result from the recent introduction of a commercial system, combining a low-vibration 4 K/1 W pulse-tube cooler (see Sect. 5.3) as a precooling stage for an adiabatic demagnetization stage. This setup is offered with a variety of experimental probes, thermometry and magnetic shielding installations for temperatures to 100 mK and as an option with a superconducting magnet supplying 12 T. The device enables the user to perform cryogen-free millikelvin/high-magnetic-field measurements.

Problems

9.1. At which value of x does the high-temperature approximation (10.4) for the magnetic heat capacity deviate by 5% from the exact result given in (9.15)?

9.2. Calculate the cooling power of 1 mole CMN if it is demagnetized from 2 T, 1 K to zero field (Fig. 9.2). Calculate for the same experiment the heat of magnetization which has to be removed if this salt is magnetized isothermally to 2 T. Perform the same calculation for the situation sketched in Fig. 9.1.

9.3. To which temperature does one have to refrigerate a solid containing paramagnetic ions with spin 1/2 and magnetic moments equal to one Bohr magneton in a field of 3 T so that 75% of the atoms are polarized with their spins parallel to the external magnetic field?

10

Refrigeration by Adiabatic Nuclear Demagnetization

There are many elements and compounds that contain no electronic magnetic moment but whose nuclei carry nuclear magnetic moments. These nuclear magnetic moments can be used for magnetic refrigeration in a similar way as the electronic magnetic moments. Nuclear magnetic refrigeration was proposed, and nowadays is applied in several specialized laboratories, to avoid the main disadvantages of electronic paramagnetic refrigeration – the low thermal conductivities and the "high" magnetic ordering temperatures of paramagnetic salts – and to refrigerate into the microkelvin temperature range.

The minimum temperature for magnetic refrigeration is given by spontaneous magnetic ordering. The interaction energy between magnetic moments in the simplest case, for dipole–dipole interaction only, is given by

$$\epsilon_{\mathrm{d}} = \boldsymbol{\mu} \cdot \boldsymbol{b} = \frac{\mu_0 \boldsymbol{\mu}}{4\pi} \cdot \sum_i \left(\frac{\boldsymbol{\mu}_i}{r_i^3} - 3\boldsymbol{r}_i \frac{\boldsymbol{\mu}_i \cdot \boldsymbol{r}_i}{r_i^5} \right), \tag{10.1}$$

where b is the internal dipole field created by neighbours, $\boldsymbol{r}_i$ is the distance between the dipoles with magnetic moments $\boldsymbol{\mu}_i$, and $\mu_0 = 4\pi \times 10^{-7}\,\mathrm{V\,s\,A^{-1}\,m^{-1}}$. This means that the ordering temperature is

$$T_{\mathrm{c}} \propto \frac{\mu^2}{r^3}. \tag{10.2}$$

Because the nuclear magnetic moments are of the order of the nuclear magneton $\mu_{\mathrm{n}} = 5.05 \times 10^{-27}\,\mathrm{J\,T^{-1}}$ rather than of the Bohr magneton $\mu_{\mathrm{B}} = 9.27 \times 10^{-24}\,\mathrm{J\,T^{-1}}$, which we had for the electronic magnetic moments, the ordering temperature due to nuclear dipole–nuclear dipole interaction is much lower, typically of the order of $0.1\,\mu\mathrm{K}$ or less. Therefore, nuclear magnetic refrigeration can be used to much lower temperatures than electronic magnetic refrigeration; it has opened up the microkelvin temperature range to condensed-matter physics.

For electronic magnetic refrigeration we could not use pure metals with their high thermal conductivity because the magnetic ordering temperature of magnetically ordering metals with their conduction electrons polarized by exchange interaction is usually substantially higher than the magnetic ordering temperature of electronic moments in paramagnetic salts. But now for nuclear magnetic refrigeration, where we have the weak nuclear magnetic moments, we can use metals and can take advantage of their high thermal conductivity. In addition, we have the advantage that the density of moments in these metals is substantially higher than the density of electronic moments in the highly diluted paramagnetic salts suitable for magnetic refrigeration, giving a large nuclear entropy density.

But the advantages offered by nuclear magnetic refrigeration due to the small nuclear magnetic moments are counteracted by experimental problems resulting precisely from the small size of these moments. To achieve a reduction of the nuclear magnetic entropy of, at least, a few percent we need rather demanding starting conditions for the magnetic field and for the temperature. If we take, e.g., the "work horse" of nuclear magnetic refrigeration, copper, then even with the starting conditions $B_{\mathrm{i}} = 8\,\mathrm{T}$ and $T_{\mathrm{i}} = 10\,\mathrm{mK}$ we get a reduction of the nuclear magnetic entropy of only about 9% (Fig. 10.1). As shown in Fig. 10.2, even at starting conditions of $B_{\mathrm{i}} = 8\,\mathrm{T}$ and $T_{\mathrm{i}} = 6\,\mathrm{mK}$, the relative population of each of the Zeeman levels of Cu nuclei differs from the population of neighbouring levels by only a factor of two. The cooling power is given by $Q = n\int T\mathrm{d}s$, and because we now also want to work at substantially lower temperature, Q is typically a factor of 1,000 smaller than for electronic magnetic refrigeration. If we take the example of Fig. 10.2 that, starting from $B_{\mathrm{i}} = 8\,\mathrm{T}$ and $T_{\mathrm{i}} = 6\,\mathrm{mK}$, field and temperature are reduced by a factor of thousand to $B_{\mathrm{f}} = 8\,\mathrm{mT}$, $T_{\mathrm{f}} = 6\,\mu\mathrm{K}$ (ideal adiabatic process), each

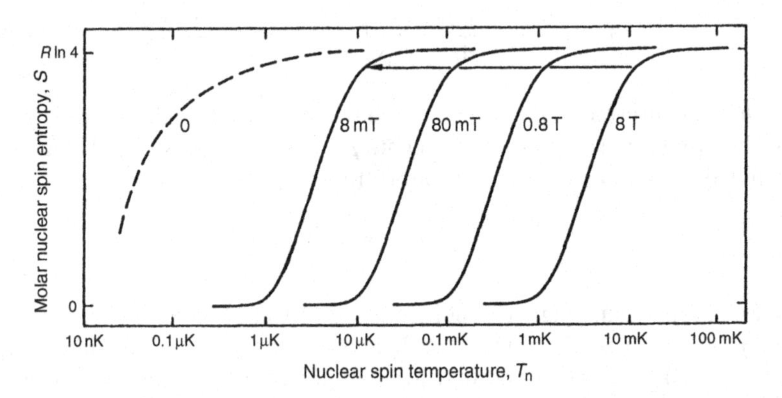

Fig. 10.1. Molar nuclear spin entropy S of Cu nuclei in various magnetic fields as a function of temperature. The arrow indicates an adiabatic demagnetization process from 8 T to 8 mT. The full nuclear spin disorder entropy of copper nuclei with $I = 3/2$ is $R\ln(4)$ as indicated on the vertical axis. The dashed line is the nuclear spin entropy of Cu in zero field according to the data of [10.1]; it indicates the spontaneous antiferromagnetic ordering of the Cu nuclear spins

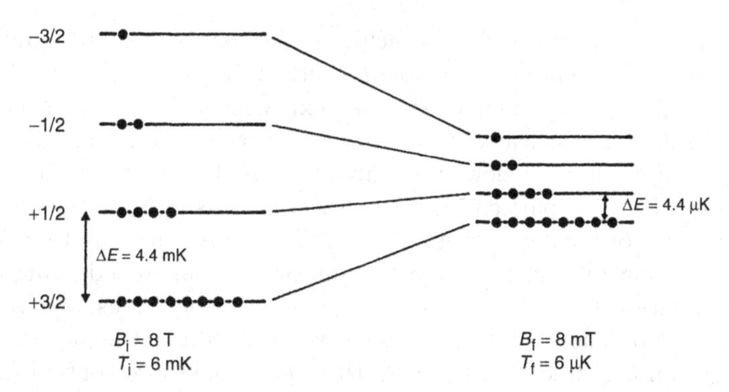

Fig. 10.2. Zeeman levels of Cu nuclei with spin $I = 3/2$ in a starting magnetic field of 8 T and after demagnetization to 8 mT. The relative population of the levels and the indicated energy separation ΔE correspond to a starting nuclear spin temperature of 6 mK and a final nuclear spin temperature of 6 μK. The field to temperature ratio is chosen so that the relative population of neighbouring levels differs by a factor of two

spin flip associated with a transition of a nucleus from one level to the next higher one, gives a cooling power corresponding to a temperature difference of only 4.4 μK for Cu.

A final point to consider is the transfer of the spin temperature. When we demagnetize the nuclear magnetic moments the nuclear spin system may reach a very low temperature, but we have to ask how this very low nuclear spin temperature is transferred to the rest of the system, to the electrons and to lattice vibrations, which determine the "temperature" of our refrigerant – unless we are only interested in the cold nuclear spin system itself. Here again, metals are much more suited than paramagnetic salts because of their much shorter spin–lattice relaxation times. This problem will be discussed later.

Nuclear magnetic refrigeration was first proposed in 1934 by C.J. Gorter, and in 1935 independently by N. Kurti and F.E. Simon. However, due to severe experimental problems and because even the temperature range accessible by electronic magnetic refrigeration had not yet been explored, no experimental realization was tried before the end of World War II. Only in 1956 did a group at Oxford University, led by Kurti, try a practical application of this cooling method [10.2, 10.3]. They succeeded in reducing the *nuclear spin temperature* to the low microkelvin temperature range by using 0.1 mm diameter copper wires as a refrigerant and starting from 12 mK and 3 T. But in their experiment the *temperature of the electrons and of the lattice vibrations* was not reduced below the starting temperature of 12 mK, and the nuclear spins warmed up from the obtained low microkelvin temperature to the starting temperature of 12 mK within a few minutes. Only about a decade later were more trials performed, in particular at Oxford, La Jolla, and Helsinki, to improve the situation. Osgood and Goodkind [10.4], and Symco [10.5] were the first to achieve considerable electron and lattice refrigeration by adiabatic nuclear demagnetization, in the late 1960s. Since the beginning of the 1970s knowledge and experimental equipment have been available to develop nuclear cooling to a refrigeration technology for achieving temperatures clearly below those already available by electronic magnetic refrigeration. Here, in particular, the work of Lounasmaa and his co-workers at Helsinki was of great importance [10.6]. They were the first to combine a ^{3}He–^{4}He dilution refrigerator for precooling with a superconducting magnet for magnetization to obtain the required starting conditions. References to the early work on nuclear refrigeration as well as credit to the pioneers of this technology can be found in [10.7, 10.8]; these two books also contain the basic physics and many of the equations to be used in this chapter. The state of the art of nuclear demagnetization refrigeration was reviewed in 1982 by Andres and Lounasmaa [10.9], and in 1988 by Pickett [10.10].

In principle, the procedure for nuclear magnetic refrigeration is identical to that for electronic magnetic refrigeration. However, in practice there are substantial differences due to the small nuclear moments and very low temperatures involved. Schematic setups of nuclear refrigerators are shown in Fig. 10.3. A superconducting magnet with a maximum field of typically 8 T is required in order to magnetize (polarize) the nuclear moments of typically 10–100 mol Cu. This nuclear refrigeration stage is precooled by a ^{3}He–^{4}He

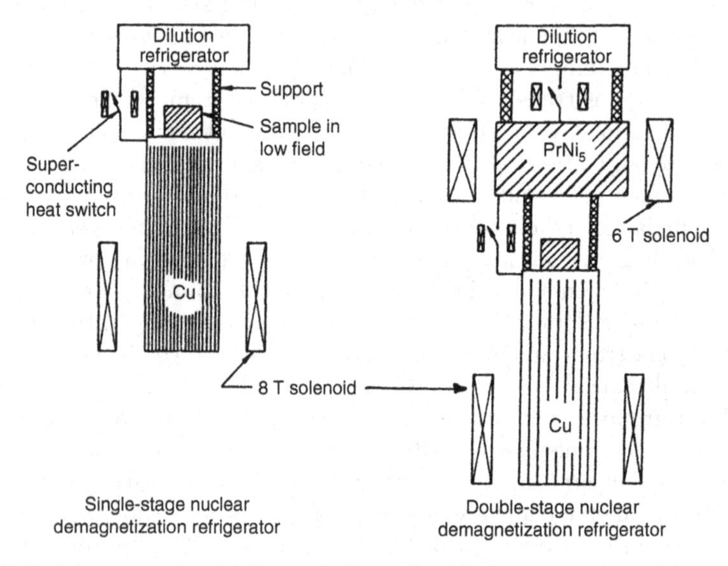

Fig. 10.3. Schematic of a single-stage Cu nuclear demagnetization refrigerator and of a double-stage $PrNi_5$/Cu nulcear demagnetization refrigerator. In the double-stage design the extra $PrNi_5$ stage decreases the starting temperature for demagnetizing the Cu stage. It also decreases the heat leaking by thermal conductance from the parts at higher temperature to the Cu stage, and to experiments and samples

dilution refrigerator – which has to absorb the large heat of nuclear magnetization – to a starting temperature of typically 10–15 mK. A superconducting heat switch (Sect. 4.2.2) is used to make thermal contact between the precooling ^{3}He–^{4}He dilution refrigerator and the Cu nuclear stage. The switch must have a normal-state conductivity compatible with the cooling power of the dilution refrigerator. After the Cu stage has been precooled to the required starting temperature T_i, the superconducting heat switch is opened by reducing the magnetic field on it to zero to thermally isolate the nuclear refrigeration stage. Then demagnetization of the nuclear magnetic moments of the nuclear refrigerant can be started by reducing the main field slowly over several hours or days from the starting field B_i to the required low final value B_f. The experiments and thermometers are usually attached to the top flange of the refrigeration stage outside of the high-field region to avoid disturbances from the demagnetization field. To achieve a lower starting temperature for the demagnetization of the Cu stage, a $PrNi_5$ nuclear demagnetization stage can be used between it and the ^{3}He–^{4}He dilution refrigerator (for details see Sects. 10.7 and 10.8)

10.1 Some Equations Relevant for Nuclear Refrigeration

Again, in principle, we can just use the equations derived in Chap. 9 for electronic magnetic refrigeration. For nuclear refrigeration we have to replace the electronic parameters by the relevant nuclear parameters, i.e., we have to replace $\mu_B\, g$ by $\mu_n\, g_n$. But we have to keep in mind that for nuclear refrigeration the moment μ and the spin $\boldsymbol{I}$ have the same direction, unlike in the electronic case, and therefore in some of the equations the sign is reversed; for example, the energies of the nuclear Zeeman levels in a magnetic field are

$$\epsilon_m = -m\mu_n\, g_n\, B\,, \tag{10.3}$$

where μ_n is the nuclear magneton, $g_n = \mu/I$ is the nuclear g-factor and m runs from $-I$ to $+I$. In addition, for nuclear refrigeration the requirement $\mu\mu_n B \ll k_B\, T\, I$ is usually fulfiled, due to the smallness of nuclear moments, which we consider again as noninteracting. Therefore we can expand the rather cumbersome equations of (9.15) into their high-temperature approximations, so that we have

$$\begin{aligned} S_n &= R\ln(2I+1) - \frac{\lambda_n B^2}{2\mu_0 T_n^2}, \quad C_n = \frac{\lambda_n B^2}{\mu_0 T_n^2} \text{ (Schottky law)}, \\ M_n &= \frac{\lambda_n B}{\mu_0 T_n}, \quad \chi_n = \frac{\lambda_n}{T_n} \text{ (Curie law)}, \end{aligned} \tag{10.4}$$

with $\lambda_n = N_0 I(I+1)\mu_0\mu_n^2\, g_n^2/3k_B$, the molar nuclear Curie constant. The entropy and the specific heat are then functions of $(B/T)^2$ only.

Sometimes the final demagnetization field B_f is so small that the internal field b due to interactions of the magnetic or of the electric nuclear moments is not negligible and we then have to replace B in the above equations by $(B^2 + b^2)^{1/2}$ giving

$$T_\mathrm{f} = \frac{T_\mathrm{i}}{B_\mathrm{i}} \sqrt{B_\mathrm{f}^2 + b^2}. \tag{10.5}$$

As a result, the limiting temperature for nuclear refrigeration is

$$T_\mathrm{f,min} = \frac{b T_\mathrm{i}}{B_\mathrm{i}} . \tag{10.6}$$

For example, for Cu we have $b = 0.36\,\mathrm{mT}$. Because in an adiabatic process $S = \text{const.}$, we should also have $B/T = \text{const.}$, see (10.4). Therefore a plot of the B/T values achieved during magnetic refrigeration is a measure of the adiabaticity or reversibility of the process (see Fig. 10.17).

Recently some very powerful nuclear refrigerators have been built whose behavior can no longer be described in the high-temperature approximation (Sects. 10.7 and 10.8). Figure 10.4 shows that noticeable deviations from the high-temperature approximation occur for the specific heat of Cu in a field

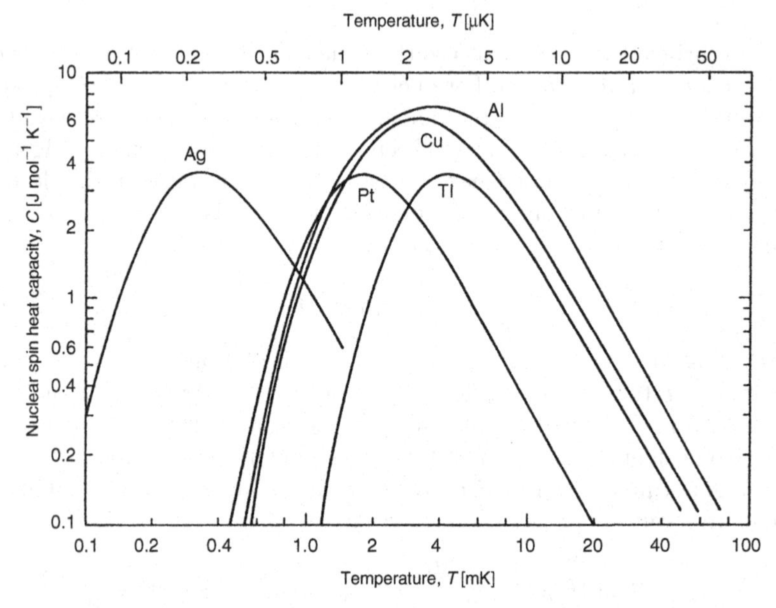

Fig. 10.4. Nuclear heat capacities as a function of temperature for Al ($I = 5/2$), Cu ($I = 3/2$), Tl, Pt and Ag ($I = 1/2$) (however, remember that Pt contains only 33.8% of ^{195}Pt, its only isotope with a nuclear moment). The lower temperature scale corresponds to heat capacity data in a magnetic field of 9 T, the upper temperature scale to data in a field of 7 mT

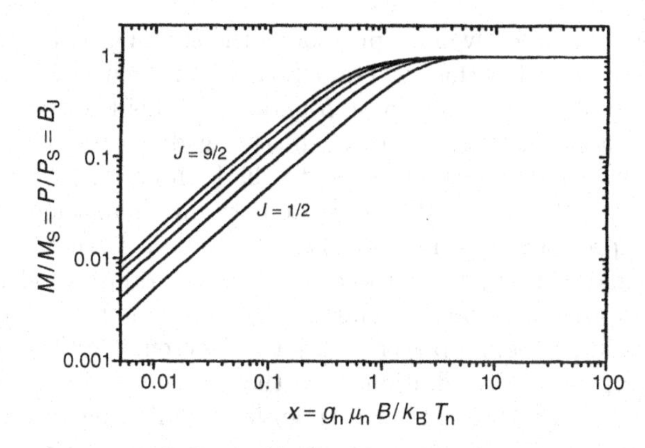

Fig. 10.5. Relative magnetization M/M_s, see (9.15c), or relative polarization P/P_s, or Brillouin function $B_J(x)$ as a function of $x = g_n\mu_n B/k_B T_n$ for nuclei with spins $I = 9/2$–$1/2$ (*from top to bottom*). As examples, we can consider Pt ($I = 1/2$, $M_s = 0.085\,\mathrm{mT}$, $g_n\mu_n/k_B = 0.42\,\mathrm{mK\,T^{-1}}$), Tl ($I = 1/2$, $M_s = 0.36\,\mathrm{mT}$, $g_n\mu_n/k_B = 1.21\,\mathrm{mK\,T^{-1}}$), Cu ($I = 3/2$, $M_s = 1.25\,\mathrm{mT}$, $g_n\mu_n/k_B = 0.56\,\mathrm{mK\,T^{-1}}$), and In ($I = 9/2$, $M_s = 1.33\,\mathrm{mT}$, $g_n\mu_n/k_B = 0.45\,\mathrm{mK\,T^{-1}}$)

of 9 T at $T < 10\,\mathrm{mK}$, for example. This can also been seen, of course, in the nuclear magnetization of Pt, Tl, Cu or In (Fig. 10.5). Quantitatively, a 1% deviation from the Curie law occurs for Cu at $T < 22\,\mathrm{mK}$ ($22\,\mu\mathrm{K}$) or in Pt at $T < 9\,\mathrm{mK}$ ($9\,\mu\mathrm{K}$) in a field of 8 T (8 mT). But in this chapter I will use the high-temperature approximations for our discussion. It is tedious but straightforward to use the full equations (9.15).

10.2 Differences in Experimental Procedure for Nuclear and Electronic Demagnetization

There are two ways to precool and polarize the nuclear moments. Firstly, it can be done *isothermally*, as discussed for electronic magnetic refrigeration (Fig. 9.1). In this case the heat of magnetization is given by

$$Q_a = nT_i\Delta S = -\frac{n\lambda_n B_i^2}{2\mu_0 T_i}\,. \tag{10.7}$$

Another possibility is to switch on the starting field B_i at a high temperature and then precool the polarized nuclear magnetic moments by the ^{3}He–^{4}He dilution refrigerator along the entropy curve at B_i = const to the required starting temperature T_i. In this latter case the heat of magnetization is given by

$$Q_b = \int_{\infty}^{T_i} nC_{n,B}\,\mathrm{d}T = -\frac{n\lambda_n B_i^2}{\mu_0 T_i} = 2Q_a\,. \tag{10.8}$$

In the former case, when we first precool to the starting temperature T_i, the field is slowly increased at this low temperature and the heat of magnetization has to be adsorbed by the dilution refrigerator at this low constant temperature T_i. The Joule heating occurring in the leads to the superconducting magnet (which is quite substantial due to the high currents, which are typically 100 A) has to be adsorbed by the ^{4}He bath for a long time. Even though in the second procedure the heat of magnetization is a factor of two larger, usually one chooses this latter procedure. Now the dilution refrigerator can start doing its work at a higher temperature, where it has a higher cooling power, and its large cooling power makes the difference in the two values for Q rather unimportant. In addition, it is then possible to avoid Joule heating in the ^{4}He bath for an extended time in the following way. The current is put into the superconducting magnet and circulates in a persistent mode though the superconducting magnet and its persistent shunt [10.11], as shown in Fig. 10.6. It is possible even to disconnect the leads and turn off the power supply and yet leave the superconducting magnet at full field. Moreover, the field is very stable, no longer dependent on the stability of the power supply. Such a persistent switch usually consists of a small length of superconducting wire – short-circuiting the solenoid – to which a heater is connected which can drive this piece of superconductor into its resistive state, which is necessary for loading and unloading the current of the main solenoid. The part of the wire to be heated can be embedded in epoxy to provide the necessary thermal resistance to the helium bath. We then have no current between the room temperature power supply and the low-temperature part during the long time of precooling (see Fig. 10.16), thus avoiding ohmic losses. A superconducting solenoid operating in the persistent mode usually produces a field which is more stable and ripple-free than the field of a solenoid connected to a power supply.

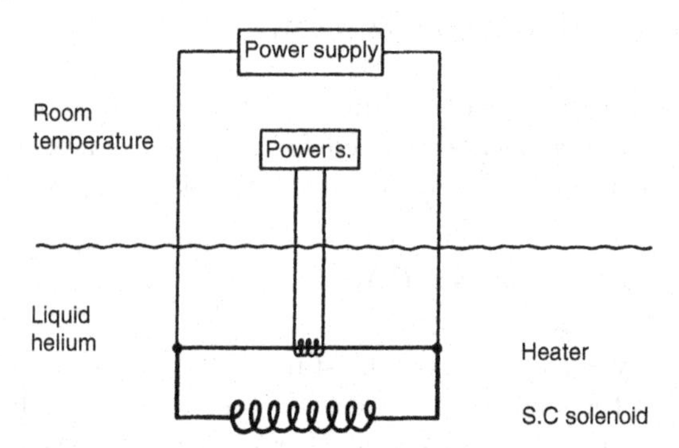

Fig. 10.6. Wiring for a superconducting solenoid to be operated in the persistent mode (see text). The heavy lines indicate the superconducting wire of the magnet and its persistent shunt

10.3 Interaction Between Conduction Electrons and Nuclei

In this section I shall discuss how thermal equilibrium can be established between the various thermal reservoirs of a metal. In nuclear refrigeration we reduce the temperature T_n of the nuclear spin system. Thermal equilibrium among the nuclear magnetic moments is established within the spin–spin relaxation time τ_2. This relaxation time is rather short for metals, with typical values of less than 1 ms (Table 10.1). Hence we can assume thermal equilibrium within the nuclear spin system. For refrigeration we are usually not interested in the final $T_{n,f}$ but much more in the electronic and lattice temperatures of our refrigerant. Fortunately, the nuclear spin system is not isolated from the conduction electrons. If this were the case, it would be practically impossible to refrigerate other materials by nuclear demagnetization. Hence we have to consider how the low temperature of the nuclear spin system can be transferred to electrons and to lattice vibrations. Because the nuclear magnetic moments or spins do interact extremely weakly with the lattice vibrations but reasonably well with the conduction the way heat or cooling is transferred is nuclei $\leftrightarrow$ conduction electrons $\leftrightarrow$ phonons.

10.3.1 Electron–Phonon Coupling

The heat flow between the electron and phonon systems for Cu, for example is given by [10.12, 10.13]

$$\dot{Q}_{e-ph} \cong 2 \times 10^3\, V(T_{ph}^5 - T_e^5) \quad [W], \tag{10.9}$$

where V is the volume [cm^3] of the sample.

This gives a rather small heat flow but fortunately the specific heat of the lattice vibrations decreases with T^3 and is very small in the temperature range of interest to us here. Therefore, the phonon temperature will always follow the electronic temperature within a very short relaxation time. Henceforth we can assume that lattice vibrations and conduction electrons are at the same temperature.

10.3.2 Nucleus–Electron Coupling

The coupling between electrons and nuclei is an electromagnetic interaction acting between the nuclei and those electrons that have a finite charge density at the nucleus. In a first approximation these are only the s-electrons, because electrons with another symmetry have a node of their wave function at the site of the nucleus. In the interaction process a nucleus and a conduction electron undergo a mutual spin flip via the contact interaction.

Table 10.1. Properties of some metals important for nuclear magnetic refrigeration and nuclear magnetic thermometry. In some cases a mean average is given if two isotopes are present. For the two compounds included ($AuIn_2$ and $PrNi_5$) the data are for the element in italics. (The nuclear properties have been taken from [10.22]; the γ and θ_0 values for the noble metals have been taken from [10.23])

isotope/ compound	ρ [g cm^{-3}]	V_m [cm^3 mol^{-1}]	θ_D [K]	γ [mJ mol^{-1} K^{-2}]	T_F [10^4 K]	$\rho_{273\,K}$ [µΩ cm]
^{27}Al	2.70	9.97	428	1.35	13.5	2.65
^{63,65}Cu	8.93	7.11	344	0.691	8.12	1.68
^{93}Nb	8.58	10.9	277	7.79	6.18	12.5
^{107,109}Ag	10.50	10.3	227	0.640	6.36	1.59
^{113,115}In	7.29	15.7	108	1.69	10.0	8.37
^{117,119}Sn	7.31	16.3	200	1.78	11.7	11.0
^{195}Pt	21.47	9.10	239	6.49	10.3	10.6
^{197}Au	19.28	10.2	162	0.689	6.41	2.24
^{203,205}Tl	11.87	17.2	78	1.47	9.46	18.0
Au*In*$_2$	10.3	41.5	187	3.15	9.26	6.3
*Pr*Ni$_5$	8.0	51.0	230	40	–	80

isotope/ compound	structure	T_c [K]	B_c [mT]	isotopic abundance [%]	I
^{27}Al	fcc	1.18	10.5	100	5/2
^{63,65}Cu	fcc	–	–	69.1, 30.9	3/2
^{93}Nb	bcc	9.3	200	100	9/2
^{107,109}Ag	fcc	–	–	51.8, 48.2	1/2
^{113,115}In	tetr	3.41	28	4.3, 95.7	9/2
^{117,119}Sn	tetr	3.72	30.5	7.6, 8.7	1/2
^{195}Pt	fcc	–	–	33.8	1/2
^{197}Au	fcc	–	–	100	3/2
^{203,205}Tl	hcp	2.39	18.1	29.5, 70.5	1/2
Au*In*$_2$	Au:fcc/In:sc	0.21	1.5	4.3, 95.7	9/2
*Pr*Ni$_5$	hex	–	–	100	5/2

We will now define a time τ_1 which the nuclei need to come to thermal equilibrium with the conduction electrons at T_e = const. i.e.,

$$\frac{\mathrm{d}T_n^{-1}}{\mathrm{d}t} = -\frac{(T_n^{-1} - T_e^{-1})}{\tau_1} . \tag{10.10}$$

The time τ_1 is the so-called spin–lattice relaxation time. This name is somewhat misleading for metals because the interaction is not directly between the nuclear spins and the lattice; it is caused by hyperfine interactions between the nuclei and the conduction electrons. Only conduction electrons near the Fermi energy can interact with the nuclei and change their energy due to this interaction because the electron must find an empty energy state at an energy difference corresponding to the small energy exchanged with the nucleus. Electrons which have an energy far below the Fermi energy are not

Table 10.1. — *Continued.* For full caption see previous page

isotope/ compound	Q [barn]	μ [μ_n]	$\gamma/2\pi$[a] [kHz mT^{-1}]	$\gamma_m/2\pi$[b] [kHz mT^{-1}]
^{27}Al	+0.15	3.64	11.09	11.11
$^{63,65}Cu$	−0.211, −0.195	2.22, 2.38	11.29, 12.09	11.31, 12.12
^{93}Nb	0.22	6.14	10.41	10.50
$^{107,109}Ag$	–	−0.113, −0.130	1.723, 1.981	1.732, 1.991
$^{113,115}In$	+0.82, +0.83	5.50, 5.51	9.31, 9.33	9.38, 9.41
$^{117,119}Sn$	–	−0.99, −1.04	15.17, 15.87	15.28, 15.98
^{195}Pt	–	0.597	9.094	8.781
^{197}Au	+0.58	0.143	0.729	0.741
$^{203,205}Tl$	–	1.60, 1.61	24.33, 24.57	24.73, 24.97
Au*In*$_2$	+0.82, +0.83	5.50, 5.51	9.31, 9.33	9.38, 9.41
PrNi_5	−0.06	4.28[c]	13.1[c]	–

isotope/ compound	λ_n/V_m [μK]	λ_n/μ_0 [μJ K T^{-2} mol^{-1}]	κ [K s]	τ_2 [ms]
^{27}Al	0.867	6.88	1.80	0.03
$^{63,65}Cu$	0.570	3.22	1.27, 1.09	0.15
^{93}Nb	1.99	17.2	0.4	0.02
$^{107,109}Ag$	0.0020	0.016	12, 9	10
$^{113,115}In$	1.10	13.8	0.09	0.1
$^{117,119}Sn$	0.015	0.194	0.05	0.18
^{195}Pt	0.0185	0.134	0.030	1.0
^{197}Au	0.0016	0.013	4.6	–
$^{203,205}Tl$	0.210	2.87	0.0044	0.035
Au*In*$_2$	0.83	27.5	0.11	0.7
PrNi_5	33[d]	1418[d]	< 0.001	< 0.01

[a] γ: gyromagnetic ratio of the bare nucleus.
[b] $\gamma_m = \gamma \times$ Knight shift, being the gyromagnetic ratio in the metal.
[c(d)] Values for the nucleus without (with) hyperfine enhancement.

accessible for the interaction. With arguments similar to the ones which led to a linear temperature dependence of the specific heat of conduction electrons (Sect. 3.1.2), we find a linear temperature dependence of the rate τ_1^{-1} at which the nuclei come into equilibrium with the electrons. This then leads to the famous *Korringa law*

$$\tau_1 \, T_e = \kappa \,, \tag{10.11}$$

where κ is a material constant which is a measure of the strength of the hyperfine coupling between electrons and nuclei. Note that (10.11) is valid for $T_e \gg \mu\mu_n B/k_B$; the general expression is [10.14–10.21]

$$\tau_1 = \frac{2\kappa k_B}{\mu\mu_n B} \tanh\left(\frac{\mu\mu_n B}{2k_B T_e}\right). \qquad (10.11')$$

Values of κ are given in Table 10.1 for various metals which are of interest for nuclear magnetic refrigeration. The nuclei exchange energy with each other by performing mutual spin-flips; this self-equilibrium, of course, does not change the macroscopic nuclear magnetization of the sample, i.e., it keeps T_n constant. The rate of this spin–spin relaxation process is $1/\tau_2$. We see from the data in Table 10.1 that usually $\tau_1 \gg \tau_2$ at low temperatures, which means that the nuclei reach thermal equilibrium among themselves much faster than they reach equilibrium with the conduction electrons. It therefore does, indeed, make sense to speak of a "nuclear spin temperature T_n" and of an "electronic or lattice temperature T_e" [10.14–10.17]. We can also see from the data for κ in Table 10.1 that for some metals the relaxation time is only a few seconds, even in the low millikelvin temperature range, but it can reach, for example, about 1 h at 0.3 mK for Cu. Anyway, for metals these are experimentally accessible times. On the other hand, there are no conduction electrons in insulators and the nuclear spin–lattice relaxation time for these materials can be very large, in the range of days or even weeks at millikelvin temperatures. Therefore only metals can be used for nuclear refrigeration.

The Korringa constant is independent of temperature but is a function of the magnetic field if this becomes comparable to the internal field b [10.14–10.17] according to

$$\kappa(B) = \kappa(\infty)\frac{B^2 + b^2}{B^2 + \alpha b^2}, \qquad (10.12)$$

where the coefficient α (about 2–3) depends on the type of internal interactions. For Cu with $b = 0.36\,\text{mT}$ and $\alpha = 2.6$, $\kappa = 0.4\,\text{K s}$ at $B = 0$ and $\kappa = 1.2\,\text{K s}$ at $B \geq 10\,\text{mT}$ [10.1, 10.7]. The discussion of the dynamics to establish thermal equilibrium between nuclei and conduction electrons becomes rather complicated if the high-temperature approximation cannot be used anymore. In such a situation rate equations have to be considered and the behavior depends on the nuclear spin I as well; for details one should consult the original literature [10.14–10.21].

Because for some metals τ_1 can become rather large at low temperatures, it is possible to refrigerate just the nuclear spin system to low microkelvin temperatures and leave conduction electrons and phonons at higher temperatures. This was the case in the original experiment at Oxford in 1956 [10.2, 10.3], and has been used extensively in Helsinki in recent years in studies of nuclear magnetic interactions in Cu, Ag and Rh [10.1, 10.24]. This type of experiment is called *nuclear cooling*. On the other hand, if the refrigeration is performed in such a way that the nuclei pull conduction electrons and phonons to low temperatures as well, the term used is *nuclear refrigeration* and it is the latter which we are considering in this book.

Until now we have discussed the time which "hot" nuclei need to cool to the temperature of "cold" electrons which sit at T_e = const. In the case of nuclear refrigeration it is the other way around: the cold nuclei have to pull the hotter electrons to low temperatures, and both $T_{\rm n}$ and $T_{\rm e}$ will change. In this process we have the heat flow

$$\dot{Q} = nC_{\rm e}\dot{T}_{\rm e} = -nC_{\rm n,B}\dot{T}_{\rm n} \tag{10.13}$$

From the definition for the spin–lattice relaxation time, (10.10, 10.11), for κ we obtain

$$\frac{{\rm d}T_{\rm n}^{-1}}{{\rm d}t} = -\frac{(T_{\rm n}^{-1} - T_{\rm e}^{-1})T_{\rm e}}{\kappa}, \tag{10.14}$$

or

$$\dot{T}_{\rm n} = \frac{(T_{\rm e} - T_{\rm n})T_{\rm n}}{\kappa}. \tag{10.15}$$

Note that if the nuclei experience a change of magnetic field we have

$$\dot{T}_{\rm n} = (T_{\rm e} - T_{\rm n})T_{\rm n}/\kappa + (T_{\rm n}/B)\dot{B}. \tag{10.15$'$}$$

Combining these equations (assuming $\dot{B} = 0$) we find an equation for the rate of change of temperature of the conduction electrons

$$\dot{T}_{\rm e} = -(T_{\rm e} - T_{\rm n})\left(\frac{T_{\rm n}C_{\rm n,B}}{\kappa C_{\rm e}}\right). \tag{10.16}$$

Finally we obtain from (10.15) and (10.16) the rate of change of the difference of the nuclear and electronic temperatures

$$\dot{T}_{\rm n} - \dot{T}_{\rm e} = (T_{\rm e} - T_{\rm n})(1 + C_{\rm n,B}/C_{\rm e})T_{\rm n}/\kappa. \tag{10.17}$$

This latter equation tells us that the hot electrons come into thermal equilibrium with the cold nuclei with the effective time constant

$$\tau_1' = \frac{\tau_1 C_{\rm e}}{C_{\rm n,B} + C_{\rm e}} \approx \frac{\tau_1 C_{\rm e}}{C_{\rm n,B}}, \tag{10.18}$$

where the latter approximation can be used because the specific heat of the conduction electrons is much smaller than the specific heat of the nuclear magnetic moments in an external magnetic field B. For the same reason, the effective time constant τ_1' is much shorter than the spin–lattice relaxation time τ_1. This means that we need only very few nuclear spin flips to remove the small electronic specific heat, i.e., $T_{\rm n}$ stays almost constant, and the electrons can follow the nuclear spin temperature rather quickly.

Everything we have discussed until now was in terms of an adiabatic, reversible process. That means electrons and nuclei are in equilibrium, and there is no external heat load. This is all right because, as mentioned, the phonon and the electronic specific heats are negligible compared to the nuclear specific heat in most cases, and we have neglected external heat loads. However, in general, there will be an external heat flow into the lattice and electron systems, and from there to the nuclei, resulting in a temperature gradient between electrons and nuclei. This more realistic situation will be discussed in Sect. 10.4.

10.4 Influence of an External Heat Load and the Optimum Final Magnetic Field

Let us discuss how much heat the cold nuclei can absorb or how long they will stay below a given temperature. This discussion will lead us to the optimum final demagnetization field, which is essential in determining the cooling power of the "demagnetized" nuclear magnetic moments. We discussed in Sect. 9.2 that for magnetic refrigeration one has to have a finite final field to keep a reasonably large cooling power of the moments and to prevent them from warming up too rapidly.

The cooling power of the nuclei is given by

$$\int \dot{Q} \mathrm{d}t = \int nC_{\mathrm{n,B}}\, \mathrm{d}T\,. \tag{10.19}$$

Using the high-temperature approximation (10.4) for the nuclear specific heat in the final field B_{f} and (10.15) for the rate of change of the nuclear spin temperature, we find for the cooling power of the nuclei (neglecting the very small "load" that the electronic specific heat puts on the nuclei)

$$\dot{Q} = nC_{\mathrm{n,B_f}}\dot{T}_{\mathrm{n}} = (T_{\mathrm{e}} - T_{\mathrm{n}})\frac{n\lambda_{\mathrm{n}}B_{\mathrm{f}}^2}{\mu_0\kappa T_{\mathrm{n}}}\,, \tag{10.20}$$

or

$$\frac{T_{\mathrm{e}}}{T_{\mathrm{n}}} = 1 + \frac{\mu_0\kappa\dot{Q}}{n\lambda_{\mathrm{n}}B_{\mathrm{f}}^2}\,. \tag{10.21}$$

If the high-temperature approximation ($k_{\mathrm{B}}TI \gg \mu\mu_{\mathrm{n}}B$) does not apply, then (10.20) has to be replaced by

$$\dot{Q} = \frac{3n\lambda_{\mathrm{n}}B_{\mathrm{f}}^2}{2\mu_0\kappa}\,\frac{e^{-x_{\mathrm{e}}} - e^{-x_{\mathrm{n}}}}{1 - e^{-x_{\mathrm{e}}}}\,\frac{I(I+1) - \langle m^2\rangle \pm \langle m\rangle}{I(I+1)}\,, \tag{10.20$'$}$$

where $+m$ $(-m)$ corresponds to $m = +I(-I)$ being the lowest energy state, and $x_{\mathrm{e}} = |g_{\mathrm{n}}\mu_{\mathrm{n}}|\,B/k_{\mathrm{B}}T_{\mathrm{e}}$, $x_{\mathrm{n}} = |g_{\mathrm{n}}\mu_{\mathrm{n}}|\,B/k_{\mathrm{B}}T_{\mathrm{n}}$ [10.18–10.20]. We can write (10.21) as

$$T_{\mathrm{e}} = T_{\mathrm{n}}\left(1 + a/B_{\mathrm{f}}^2\right) = \left(\frac{T_{\mathrm{i}}}{B_{\mathrm{i}}}\right)\left(B_{\mathrm{f}} + \frac{a}{B_{\mathrm{f}}}\right), \tag{10.22}$$

where a is a constant. Here we have used the equation $T_{\mathrm{n,f}} = B_{\mathrm{f}}T_{\mathrm{i}}/B_{\mathrm{i}}$, which is not quite correct if the external heat load is not zero. The error introduced by this approximation is small (typically several precent increase in $T_{\mathrm{n,f}}$) and does not change the essential part of our result. The nuclear spin temperature is proportional to the demagnetization field and this proportionality also applies to the electronic temperature as long as the field is not too small. However, for small final demagnetization fields the electronic temperature increases, leading to an increasing difference between nuclear and electronic temperatures due to the heat flow between the two systems, as shown in Fig. 10.7. There is

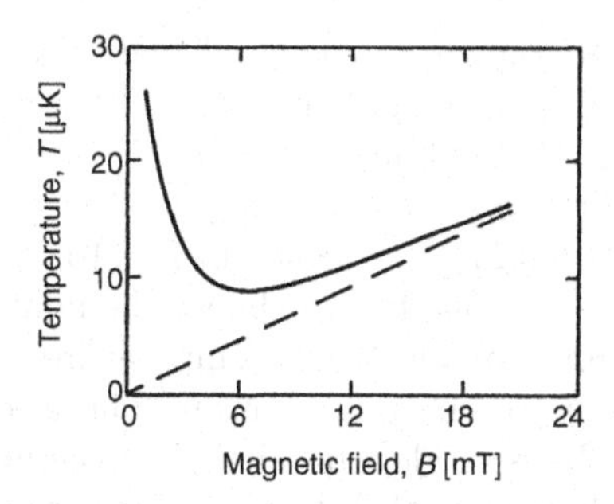

Fig. 10.7. Electronic temperature (——) and nuclear spin temperature (- - -) as a function of magnetic field during demagnetization of Cu nuclei starting from 8 T and 5 mK and assuming a heat leak of 0.1 nW mol^{-1} Cu. The figure shows that for these parameters the lowest electronic temperature of 8 μK would be achieved at a nuclear spin temperature of 4 μK and a magnetic field of 6 mT

an optimum final demagnetization field if we are interested in achieving the minimum electronic temperature. This minimum electronic temperature can be calculated from $\mathrm{d}T_{\mathrm{e}}/\mathrm{d}B_{\mathrm{f}} = 0$, resulting in

$$B_{\mathrm{f,opt}} = \left(\frac{\mu_0 \kappa \dot{Q}}{n\lambda_{\mathrm{n}}} \right)^{1/2} , \tag{10.23}$$

(assuming $b \ll B_{\mathrm{f,opt}}$ as we have done everywhere in this section).

If we use (10.21) for the ratio between electronic and nuclear temperatures, we find that at the optimum final demagnetization field the electrons are a factor of two hotter than the nuclei,

$$T_{\mathrm{e,min}} = 2T_{\mathrm{n}}(\mathrm{at}\, B_{\mathrm{f,opt}}) . \tag{10.24}$$

Equation (10.21) gives $T_{\mathrm{e}}/T_{\mathrm{n}} = 2$ for 10 mol Cu at $\dot{Q} = 1$ nW and $B_{\mathrm{f,opt}} =$ 6 mT (Fig. 10.7).

As a last step I want to calculate how long it takes to warm the nuclei from a temperature $T_{\mathrm{n,1}}$ to a temperature $T_{\mathrm{n,2}}$. This time can be computed from the equation for the rate of change of the nuclear spin temperature,

$$\dot{T}_{\mathrm{n}} = \frac{\dot{Q}}{C_{\mathrm{n},B_{\mathrm{f}}}} . \tag{10.25}$$

The warming time of the nuclei is then given by

$$t = \left(\frac{n\lambda_{\mathrm{n}} B_{\mathrm{f}}^2}{\mu_0 \dot{Q}} \right) (T_{\mathrm{n,1}}^{-1} - T_{\mathrm{n,2}}^{-1}) . \tag{10.26}$$

The product tT_{n} is constant for a constant heat load, and can be used to determine this heat load by plotting tT_{n} or T_{n} as a function of time. If $\dot{Q}$ and the warm-up time t are known, one can calculate the total heat Q that has been absorbed by the nuclear stage. From Q and the specific heat C_{B_f} of the stage, T_{n_f} the starting nuclear temperature for the warm-up of the stage can be calculated. The warm-up rates for both T_e and T_n [and therefore also their difference according to (10.21)] are shown in Fig. 10.8.

Let me stress again that one has to choose the final demagnetization field very carefully if the minimum electronic temperature and a long time for the experiment are the aims of the refrigeration. The electrons and the lattice simply will not cool if B_f is too low because then the nuclear heat capacity is too small, i.e., too little heat is absorbed per nuclear spin-flip.

In the same way, as we have discussed the influence of a constant "heat-leak", we can consider the usual experimental situation that the nuclear spin system with n moles is supposed to refrigerate a load. Let this heat load be n' moles of conduction electrons or of liquid ^{3}He, so that the molar entropy of these Fermi systems can be written as $S' = \gamma' T$. For an adiabatic process from B_i, T_i to B_f, T_f we then have for the total entropies

$$n'\gamma' T_i - \frac{n\lambda_n B_i^2}{2\mu_0 T_i^2} = n'\gamma' T_f - \frac{n\lambda_n B_f^2}{2\mu_0 T_f^2} \,. \tag{10.27}$$

Because the molar entropies of conduction electrons or even of liquid ^{3}He are rather small compared to the entropy change resulting from nuclear refrigeration, they only represent a relatively small load on the nuclear spin system. For example, the volume of liquid ^{3}He can be up to a factor of three larger than the volume of a Cu nuclear refrigerant without deteriorating the performance of the latter significantly [10.10,10.25]. Therefore the final temperature of the

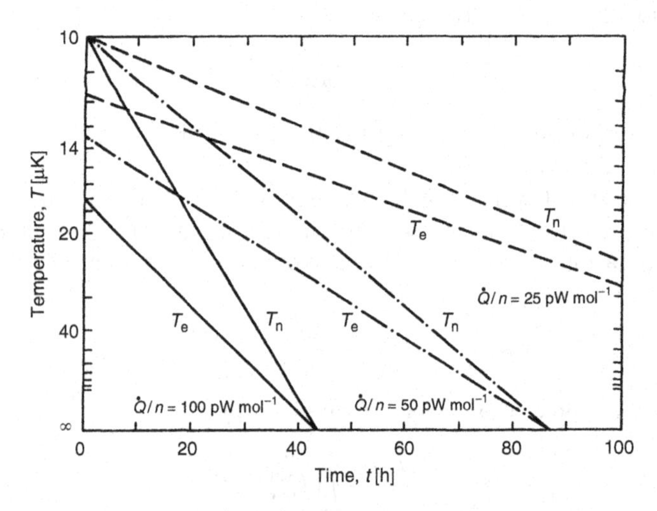

Fig. 10.8. Temperatures, see (10.21), of the nuclear and electronic systems of Cu after demagnetization. The data show the warm-up rates, see (10.26), for the given molar heat leaks $\dot{Q}/n$ to the Cu refrigerant in a final field $B_f = 7\,\mathrm{mT}$ [10.10]

load is determined by $T_i B_f / B_i$ and, above all, by the thermal coupling of it to the refrigerant. A bad thermal coupling will lead to temperature differences between load and refrigerant and to irreversible losses during the refrigeration process.

10.5 Heat Leaks

Because the cooling power of a refrigerator is smaller the lower its temperature regime of operation, more than for all other refrigeration techniques, the success of a nuclear refrigerator depends on the reduction of heat leaks. In addition to limiting the cool-time according to (10.26), minute heat leaks can give rise to very large temperature gradients because of the small thermal conductivities and thermal contact problems at very low temperatures. Whereas in the 1970s total heat leaks of order 10^{-8} W were considered an achievement, the later best nuclear refrigerators reach values of several 10^{-10} W. They operate with heat leaks so low that the heat generated by ionization due to the penetration of the refrigerant by high-energy cosmic rays (about $10^{-14}\,\mathrm{W\,g^{-1}}$) or by residual, natural or artificial, radioactivity can no longer be neglected [10.25–10.31]. The heat leaks in a refrigerator can be divided into "external heat leaks" and "internal heat leaks", which come from the refrigerated material itself and are usually time dependent.

10.5.1 External Heat Leaks

Obvious external heat flow (see Sect. 5.1.2) to the parts at microkelvin temperatures, for example by conduction via residual gas atoms, along mechanical supports, via the heat switch, along electrical leads (typically about 100 in a large nuclear refrigerator, which may include several experiments) or fill capillaries for helium experiments (remember the high thermal conductivity of liquid helium, both ^{3}He and ^{4}He, at low temperatures; see Figs. 2.12, 2.17, and 2.18) have been reduced to a total level of $\dot{Q} < 0.1$ nW [10.25–10.30]. This can be done even in a complicated apparatus by choosing the proper materials and by proper thermal anchoring at intermediate cooling stages. A 10 cm^2 surface at 4 K radiates 10 nW, far too much for a nuclear refrigerator. Therefore, all parts at microkelvin temperatures have to be guarded by a radiation shield connected to a part at lower temperature, such as the mixing chamber or at least one of the heat exchangers of the precooling dilution refrigerator. These shields can be made of brass, which has a conductivity large enough to keep temperature gradients tolerable but small enough to keep eddy current heating (see later) low while the demagnetizing field is changed. Their thermal performance can be improved by glueing highly conducting Cu foils to them. Also, brass is not magnetic (be careful with some stainless steels; see Fig. 3.9 and Sect. 3.4.1), which otherwise could lead to heating effects or could affect the experiments when a magnetic ordering temperature is crossed.

Brass shields seem to behave as well as other sometimes more elaborate designs ("coil-foil", etc.), which are discussed in the literature [10.7, 10.33]. A thermal shield not only reduces radiation but also stops "hot" gas molecules from hitting the inner low-temperature part of a cryostat. In addition, radiation shields and light traps in pumping tubes (made from semicircular baffles soldered to a wire, for example) are required.

The low-temperature parts in every cryostat are connected via transmission lines to room-temperature equipment. These lines couple to RF fields. A very serious consideration is, therefore, RF heating, in particular with regard to thermometers (for the appropriate precautions see Fig. 12.28). Many refrigerators have been built in an RF-shielded room with filtered electrical supplies, which can provide up to 120 dB attenuation at typically 100 kHz–1 GHz [10.30, 10.34]. Then the pumping and gas recovery lines have to be brought in by electrically nonconducting tubes. As stressed in [10.7]: "*To be effective the room must be well-made and maintained. All seams, including those around the door, must be electrically tight. Power lines to the room must be filtered and pumping and service lines electrically grounded as they pass through the walls. A shielded room can be rather easily tested by a portable FM radio which should be absolutely quiet inside. This same instrument can be employed for locating "leaks" although sometimes this is not easy. A well-made room can provide 100 dB attenuation and is obviously very useful to have. It is, however, good practice to build the room first, before the cryostat is installed, and not as an afterthought!*"

But often the noise caused by instrumentation inside of the room can be at least as serious as external noise. It has turned out to be of particular importance to shield the power supplied in most instruments from the rest of the interior; often it may even be necessary to put them into a separate, shielded metal box.

Of course, the leads are usually shielded by the metal cryostat and by the metal tubes through which they run inside of it. They should be twisted pairwise and rigidly fixed. The RF noise they could transmit from their room-temperature ends may have to be cut off by RF filters, as will be discussed in Sect. 12.5.4.

A further very troublesome heat source is mechanical heating from building vibrations, from sound, from mechanical pumps or from thermoacoustic oscillations in the vapour above the helium bath (see "Thermoacoustic Oscillations" in Chap. 5) [10.25–10.30, 10.34–10.38]. Irreversible heating effects can arise from rubbing of adjacent parts, from inelastic bending, and from eddy current heating when electrically conducting parts move in a magnetic field; this last effect will be discussed in Sect. 10.5.2. Most nuclear refrigerators have been built in the basement or at most on the first floor, where building vibrations are smallest. They should be attached to a heavy foundation on damping material and supported by springs. Their mass should be large, so that the resonance frequency is at around 1–2 Hz, which is usually far below all frequencies in the surroundings [10.25–10.30, 10.32, 10.34–10.39]. The damping "springs" can be inflated metal bellows or rubber tubes or commercial pneumatic vibration isolators used in optics laboratories. The compressed air provides the adjustable low spring constant necessary for the isolation [10.37, 10.39]. Usually the most probably paths for transmission of vibrations are pumping and gas handling lines. Mechanical pumps therefore have to be mounted separately and should sit on rubber stoppers. They should be connected to the cryostat by thick-walled rubber hoses and/or flexible soft metallic bellows with compensating bellows T's between them (for the large diameter systems) [10.37, 10.39] or hung in long loose loops. At some intermediate point they should be firmly attached to or partly embedded in a large mass, like a massive concrete wall or a sand-filled box. An example of such a design is shown in Fig. 10.9. A last point of consideration is vibration conducted through the gas phase of the pumping system or of the ^{4}He recovery system, which sees the exhaust of rotary pumps and compressors in the recovery systems. Here, a damping element might be necessary as well. Efficient reduction of vibrational heating in a nuclear refrigerator by using air springs is described in [10.39], which should be consulted besides the references cited above for this problem.

When all these troublesome points have been considered, the effectiveness of the design should be checked with an accelerometer or a seismometer, either of the design described in [10.40] or a commercial one. Of course, the interior of the cryostat should be designed to be as rigid as possible so

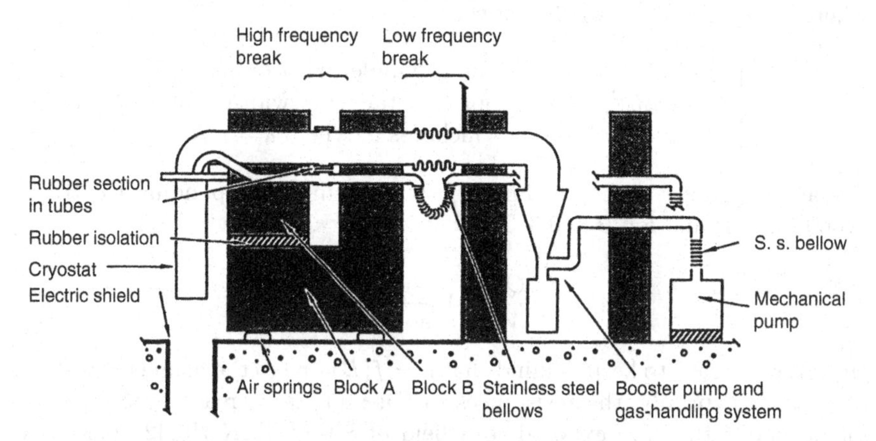

Fig. 10.9. Schematic diagram of a support system for a nuclear refrigeration cryostat. The cryostat is mounted on a concrete block (block A) supported by air springs. On top of block A are placed two smaller blocks (only one shown, block B) resting on thick pads of rubber which carry the wooden beams supporting the cryostat. The pumping tubes are, firstly concreted, into a massive block on the laboratory floor, to remove the vibrations of the pumps, secondly, taken through metal bellows, thirdly, concreted into the main block A, and finally led via a rubber section to be fixed to the sub-blocks B [10.10, 10.36]

that vibrations still transmitted (less than 0.1 μm amplitude and $1\,\mathrm{mm\,s^{-2}}$ acceleration in a reasonably good design) do not do too much harm; later changes – if vibration turns out to be a serious heat source – are very difficult to carry out.

10.5.2 Eddy Current Heating

Because a nuclear refrigerator contains highly conducting metals in changing magnetic fields, we have to consider eddy current heating effects. Obviously one has to operate the superconducting magnet as ripple free as possible; filtering may be necessary. The use of filamentary type superconducting wires has substantially reduced flux jumps but a small decay of the final demagnetization field B_f when the superconducting magnet is operated in the persistent mode has still been observed frequently [10.30,10.41]. However, due to the unavoidable residual vibrations of the apparatus, the highly conducting nuclear stage will always experience a changing field even if only $B \neq 0$.

Eddy current heating is unavoidable when the field is changed at a rate $\dot{B}$ for demagnetization. The heat generated by eddy currents in a body of volume V and with electrical resistance ρ is

$$\dot{Q}_\mathrm{e} = \frac{GV\dot{B}^2}{\rho}, \tag{10.28}$$

where the geometry factor G is given by

$$G = \begin{cases} r^2/8 & \text{for a cylinder with radius } r, \\ (d^2/16)[k^2/(1+k^2)] & \text{for a rectangle of width } w \text{ and} \\ & \text{thickness } d, \text{ where } k = w/d. \end{cases} \tag{10.29}$$

For a Cu cylinder with $r = 1\,\mathrm{mm}$, $\rho = 2\,\mathrm{n\Omega\,cm}$ (corresponding to RRR $\approx$ 1,000 for Cu) and for $\dot{B} = 0.5\ \mathrm{T\,h^{-1}}$ we find

$$\frac{\dot{Q}_\mathrm{e}}{V} \simeq 1\frac{\mathrm{nW}}{\mathrm{mol\ Cu}}. \tag{10.30}$$

However, we have to keep in mind that $\rho = f(B)$ due to the magnetoresistance of metals. Typically, the resistances of Cu and Ag increase by an order of magnitude if they are exposed to a field of 8 T at 4.2 K [10.42–10.44]. The entropy change ΔS_e due to eddy-current heating $\dot{Q}_\mathrm{e}$ should be compared to the reduction of entropy ΔS_B, (10.4), in a field B and at the temperature T,

$$\frac{\Delta S_\mathrm{e}}{\Delta S_\mathrm{B}} = \frac{\mu_0 T r^2}{4\rho\tau\lambda_\mathrm{n}}, \tag{10.31}$$

if the demagnetization is performed exponentially in time with τ being the demagnetization time constant. Under normal experimental conditions this ratio can be kept to a few percent [10.28]. In any case, conducting loops in which varying magnetic flux can give rise to eddy currents should be minimized.

We can ask for the optimum values for the parameters r and $\dot{B}$. Reduction of $\dot{Q}_e$ requires a small demagnetization rate $\dot{B}$. On the other hand, there are other heat sources as well and their influence accumulates if we demagnetize slowly. In practice, rates of $\dot{B} \lessapprox 1\,\mathrm{T\,h^{-1}}$ seem to be appropriate. Modelling of demagnetization experiments with the aim of optimizing the rates $\dot{B}$ can be found in [10.6, 10.8, 10.45–10.48]. Due to a fear of eddy current heating, fine Cu wires were mainly used as a nuclear refrigerant in the early days. Nowadays demagnetization times of many hours are common and for such rates $\dot{B}$ one can show that typical dimensions of 2–3 mm for the refrigerant are optimum values if one wants a large filling factor but small r to reduce $\dot{Q}_e$. This has led to the more appropriate slitted Cu blocks as nuclear refrigerants (see Sect. 10.8)

10.5.3 Internal, Time-Dependent Heat Leaks

Once heat leaks had been reduced to the low nanowatt range, experimentalists found to their surprise that there was a residual, time-dependent heat leak which limits the minimum achievable temperature (Fig. 10.10) [10.26]. Obviously, this heat cannot come from external sources but must be released from the refrigerant or from the refrigerated parts; it depends on how long they have already been kept cold. The origin of this observation is the fact

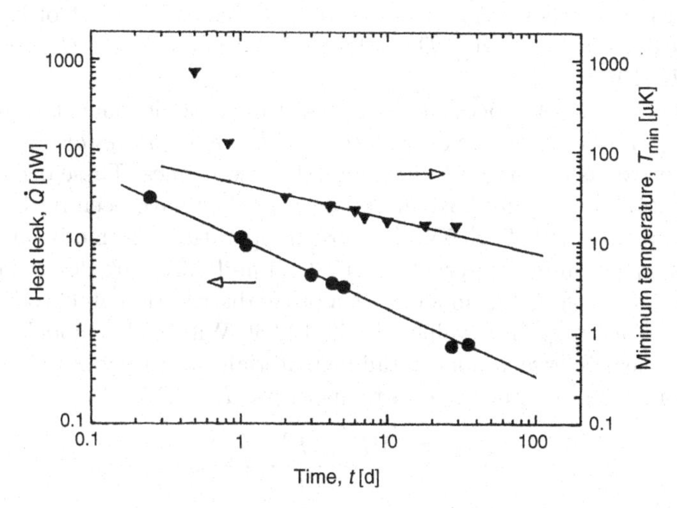

Fig. 10.10. *Left scale*: Low-temperature heat leak $\dot{Q}$ (●) in the nuclear refrigerator shown in Fig. 10.14 as a function of time t after reaching 4 K; the line is $\dot{Q} \propto t^{-3/4}$. *Right scale*: Minimum temperatures (▼) measured by Pt NMR thermometry at the top of the nuclear stage as a function of time after reaching 4 K in this refrigerator; the line is $T_{\min} \propto t^{-3/8}$ [10.30]

that a solid consists of various subsystems which can be at different temperatures due to very long low-temperature relaxation times. The coldest ones, of course, are the "demagnetized" nuclei in the center of the nuclear stage. The electrons there and, in particular, the electrons in the upper flange of the stage are kept by the unavoidable heat leaks at a higher temperature. However, there are other subsystems which may keep themselves at even higher temperatures due to heat releasing processes. These can be gas inclusions, like H_2 (see later), disorder due to lattice defects which may slowly relax (see later), radioactive nuclei which are always present, or other "hot" impurity atoms. The heat-releasing energy sources – except for the radioactive nuclei – are charged up each time the apparatus is warmed up.

The exothermic *ortho–para conversion of* H_2 has been found to be one origin of the time-dependent heat release from refrigerated parts [10.49, 10.50]. Hydrogen at typical concentrations of 10–100 ppm is present in some metals due to their production or purification processes. Many metals, e.g., Cu, Ag, Au, Pt and Rh, cannot dissolve hydrogen in a noticeable amount in their lattice, and H_2 molecules then precipitate in small bubbles with typical diameters of about 0.1 μm. Even though the concentration of H_2 in a metal may be rather small, the heating can be quite severe due to the large rotational energy change of $172[\mathrm{K}]k_B$ for each converted H_2 molecule. The ortho–para conversion of 1 ppm H_2 in 1 kg Cu gives rise to a heat leak of about 5 nW even one week after cooldown (Fig. 2.3). Details are discussed in Sect. 2.2. To get rid of this source of energy release, one has to anneal the metal in vacuum at temperatures close to its melting temperature, which seems to reduce the H_2 concentration to below 0.1 ppm [10.30] (the diffusion constant of H in Cu is $D = D_0 \exp(-E/k_B T)$ with $D_0 = 1.13 \times 10^{-2}\,\mathrm{cm^2\,s^{-1}}$ and $E = 0.40\,\mathrm{eV}$ at $T > 720\,\mathrm{K}$ [10.51]).

Another group of processes giving rise to a time-dependent heat release are *structural relaxations or deexcitations of low energy excitations (like tunneling transitions of protons)* with long relaxation times. These occur mostly in noncrystalline or amorphous materials after they have been refrigerated to low temperatures [10.52–10.55]. The low temperature thermodynamic properties of noncrystalline materials (Sects. 3.1.4 and 3.3.1) are described by the two-level-system tunneling model with a broad distribution of relaxation times which can reach very large values [10.56–10.58]. Within this model, the time-dependent heat release of noncrystalline materials after they have been refrigerated from a temperature T_i to a temperature T_f is given by

$$\dot{Q} = \frac{\pi^2}{24} \frac{V k_B^2 \bar{P}(T_i^2 - T_f^2)}{t}, \tag{10.32}$$

where $\bar{P}$ is the density of two-level tunneling states. Examples of this heat release from quartz glass and from some organic materials are shown in Fig. 10.11. The heat release is typically of order 0.1 (0.01) $\mathrm{nW\,g^{-1}}$, 1 day (week) after these materials have been refrigerated to low temperatures. Because of their heat release and because of their low thermal conductivities

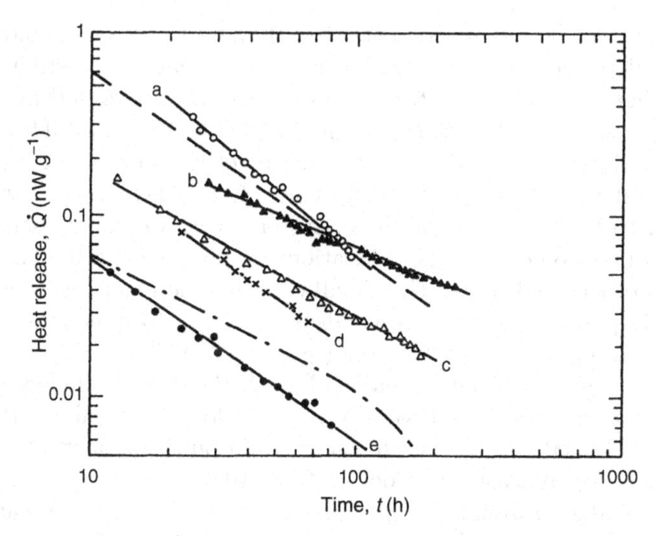

Fig. 10.11. Heat release $\dot{Q}$ for (**a**) Siloxan, (**b**) Stycast 2850FT, (**c**) Stycast 1266, (**d**) Vespel SP22, (**e**) quartz glass (Suprasil W) [10.52];---PMMA (Plexiglass), ·–·–·–: PS (Polystyrol) [10.59] as a function of time after these materials have been cooled to below 1 K. The line through the Suprasil data and the dashed line for the PMMA data correspond to $\dot{Q} \propto t^{-1}$ predicted by the tunneling model for amorphous materials. Deviations from this time dependence according to (10.32) can probably be explained by the complicated nature of some of the investigated materials

(Figs. 3.20–3.23), one should minimize the amount of noncrystalline, particularly dielectric materials (e.g., glues, epoxies, grease, plastics, etc.), which are often not very well-defined materials, in an apparatus which works at microkelvin temperatures. In such a cryostat the minimum achievable temperature is limited by very small heat leaks and such heat leaks can create severe temperature differences. One should also never touch the ultralow temperature parts of a cryostat with bare hands; the transferred "amorphous grease" would give rise to heat release. No heat release (to within 2×10^{-2} nW g^{-1}) has been observed for teflon, graphite, Al_2O_3 [10.52, 10.59], as well as for oak and maple wood [10.60], which are quite useful materials for many cryogenic purposes.

One should also avoid materials with large electronic (resistance wires, Fig. 3.9) or nuclear (Be–Cu) heat capacities and with a low thermal conductivity in a cryostat. They may store energy which is frozen in at higher temperatures and release it as heat over long periods of time after cooldown, leading to a time-dependent heat leak. Simple, pure and well-characterized materials should be used and the others – if unavoidable – should be kept to a minimum. Furthermore, heat may be released by *relaxation of stress*, in particular, *at press contacts* which can be quite numerous in a complicated cryogenic apparatus.

Unfortunately, some experiments [10.1, 10.30, 10.55] seem to indicate that such time-dependent internal-heat release also occurs in crystalline metals; *structural relaxation at grain boundaries or other lattice defects* may possibly be its origin. An example is shown in Fig. 10.10. The typical total heat released from the cold parts of a well-designed nuclear refrigerator during its cold time is of the order of 10 mJ [10.1, 10.26, 10.30]. If this heat release were to be attributed to tunneling transitions of systems with an energy separation of 30 K (3 K) one would need the relaxation of about 10^{-4} (10^{-3}) mol or only some ppm of the cold parts. This small concentration can easily be assigned to lattice imperfections. More research is necessary to understand this heat release and to reduce it to values below some pW mol^{-1}.

Another source of time-dependent heat leaking is *insufficiently pumped exchange gas*. In a nuclear refrigerator one should only use about $10^{-1}(10^{-3})$ mbar of ^{3}He at 300 (10) K as exchange gas and pump on it at a temperature of $T \geq 10$ K for at least 5 h. Cooling from 10 K downward is easy by just circulating cold gas through the dilution refrigerator because heat capacities of solids are already quite small. If too much exchange gas is left in the apparatus, the remaining atoms may desorb from hotter surfaces and condense at the parts at ultralow temperatures, leading to a heat leak which decays with time. ^{4}He should not be used because it has a lower vapour pressure, may interfere with a later mass spectroscopic leak check, and because of its superfluid film flow (Sect. 2.3.5). Hydrogen should not be used as exchange gas because of its ortho–para conversion leading to a large heat release as discussed earlier and in Sect. 2.2.

10.5.4 Heating from Radioactivity and High-Energy Particles

Eventually, when all the heat sources discussed above have been reduced or have decayed to a very small level in a well-designed nuclear refrigerator, the minimum temperature may be limited by ionization heating from radioactivity, emanating from materials in the cryostat or in the laboratory (mostly γ-rays), and from cosmic rays (mostly muons) [10.26]. These heat sources were recently investigated in a nuclear refrigerator [10.31]. Heat leaks of 20 pW originating from background γ–radiation and of 120 pW from cosmic rays were observed and were found to be in good agreement with calculations. By surrounding the cryostat by a 5-cm-thick lead wall, the γ–ray heating was reduced by an order of magnitude and the cosmic ray source decreased by 20%.

10.6 Nuclear Refrigerants

The requirements on a nuclear refrigerator can be divided into requirements on the apparatus and requirements on the material used as a nuclear refrigerant. The requirements on the apparatus are a large ratio for the starting conditions $B_\mathrm{i}/T_\mathrm{i}$ and, most important, a small heat leak to the experiments, to the thermometers, and to the nuclear refrigerant. This last requirement has turned out to be the most important factor for the quality of a nuclear refrigerator.

The requirements on the refrigerant are the following:

- It should be a metal, because we need a small Korringa constant so that the electrons will come into thermal equilibrium with the nuclei in a relatively short time and we need a high thermal conductivity to transport the cold along the refrigerant to experiments and thermometers.
- It should not be a superconductor because if a metal becomes superconducting we lose the conduction electrons for the necessary thermal conductivity (Sect. 3.3.4). We would also loose them as a medium responsible for the hyperfine interaction between nuclei and electrons (τ_1 would become very long). If the material becomes superconducting, its critical field for the superconducting state has to be lower than the final demagnetization field so that we can keep the metal in its normal state.
- The material should be easy to work with mechanically and metallurgically, and should be readily available with high purity.
- It should not show an electronic magnetic ordering transition, because this would produce an internal field in which the nuclear moments would align.
- It should have a small internal field b due to nuclear interactions or a low nuclear magnetic ordering temperature because this limits the minimum achievable temperature (b should be $<1\,\mathrm{mT}$).
- A large fraction of its isotopes should have a nuclear spin $I > 0$ and a large nuclear Curie constant λ_n for a large cooling power.
- The nuclei should either experience a cubic environment (so that electric field gradients vanish) or have a nuclear spin $I = 1/2$ (so that it has no nuclear electric quadrupole moment) to avoid nuclear electric quadrupole interactions.

Let me elaborate on this last consideration. If neither of these last two requirements, cubic symmetry or $I = 1/2$, is fulfilled, the nuclear electric quadrupole moment Q will experience an interaction with the electric field gradient V_{zz}. The nuclei will align and nuclear magnetic refrigeration is not possible anymore. We can consider the following two cases for the nuclear interactions. If the magnetic Zeeman interaction is much larger than the electric quadrupole interaction and if we have an axially symmetric field gradient, we have for the energies of the nuclear hyperfine levels

$$E_\mathrm{m} = -m g_\mathrm{n} \mu_\mathrm{n} B + \frac{e^2 V_{zz} Q}{4I(2I-1)}[3m^2 - I(I+1)]\frac{3\cos^2\theta - 1}{2} \qquad (10.33)$$

with $m = -I, -I+1, \dots, +I$, and θ the angle between B and V_zz.

If the magnetic Zeeman interaction can be neglected compared to the electric quadrupole interaction, we have

$$E_\mathrm{m} = \frac{e^2 V_{zz} Q}{4I(2I-1)}[3m^2 - I(I+1)]\,. \qquad (10.34)$$

In the latter case, the nuclear quadrupole specific heat in the high temperature approximation is given by [10.61]

$$C_{\mathrm{Q}} = \alpha \left(\frac{e^2 V_{zz} Q}{k_{\mathrm{B}} T} \right)^2 - \beta \left(\frac{e^2 V_{zz} Q}{k_{\mathrm{B}} T} \right)^3 + \cdots \qquad (10.35)$$

with

$$\alpha = \frac{R}{80} \frac{(2I+2)(2I+3)}{2I(2I-1)}, \quad \beta = \frac{R}{1120} \frac{(2I-3)(2I+2)(2I+3)(2I+5)}{(2I)^2(2I-1)^2}.$$

There is no material which fulfils all the above requirements. For example, indium has a large nuclear spin, a large nuclear magnetic moment, high thermal conductivity, and a very short Korringa constant (Table 10.1), but it fails to meet the requirement of absence of superconductivity and of nuclear electric quadrupole interaction, limiting its usefulness as a refrigerant to $T \geq 0.3\,\mathrm{mK}$ [10.62, 10.63]. Actually, some of the requirements are even contradictory. A small Korringa constant requires a large electronic density of states at the Fermi energy, as found in transition metals, but this also favors electronic magnetism or superconductivity. And the quest for a large nuclear spin and nuclear moment rules out all noncubic crystals; otherwise there will be quadrupole interactions. If we put all the mentioned restrictive requirements together and look at the periodic system of the elements, there are not many metallic elements left which are suitable for nuclear refrigeration. The relevant parameters for most of them are collected in Table 10.1.

Figure 10.12 shows the reduction of nuclear spin entropies of Cu, In, Nb and $PrNi_5$ in a field of 5 T; here not an equal number of moles but equal volumes are compared, which may be more relevant for practical applications. From this figure it is obvious that $PrNi_5$ is the most appropriate nuclear refrigerant if the starting temperatures $T_{\mathrm{i}} > 10\,\mathrm{mK}$, moderate starting fields B_{i} and moderate final temperatures $T_{\mathrm{f}} \geq 0.4\,\mathrm{mK}$ are required. The favorable properties of $PrNi_5$ and other hyperfine-enhanced van Vleck paramagnets will be discussed in Sect. 10.7. The next best candidates would be Al, In and Nb, according to their rather large nuclear moments. However, superconducting transitions with the critical fields $B_{\mathrm{c}} > 10\,\mathrm{mT}$ are the main obstacles to using In (also large quadrupole interaction [10.62, 10.63]), Nb (almost type-II superconductor), and Tl (poisonous and very strong oxidation [10.41, 10.42]). Some of them may still be suitable nuclear refrigerants for some special applications. The disadvantages of In can be avoided by using it in the compound $AuIn_2$ (cubic; B_{c} only 1.5 mT) [10.64]. However, recent experiments [10.65] with this compound have shown a spontaneous ferromagnetic ordering transition of the In nuclei at 35 μK (Fig. 3.14). In addition, $AuIn_2$ is very brittle. The possible suitability of some metal hydrides with their high proton moment density remains to be demonstrated [10.66].

In the end, it turns out that at present Cu seems to be the most suitable "workhorse" for nuclear refrigeration. It has two isotopes with $I = 3/2$, and

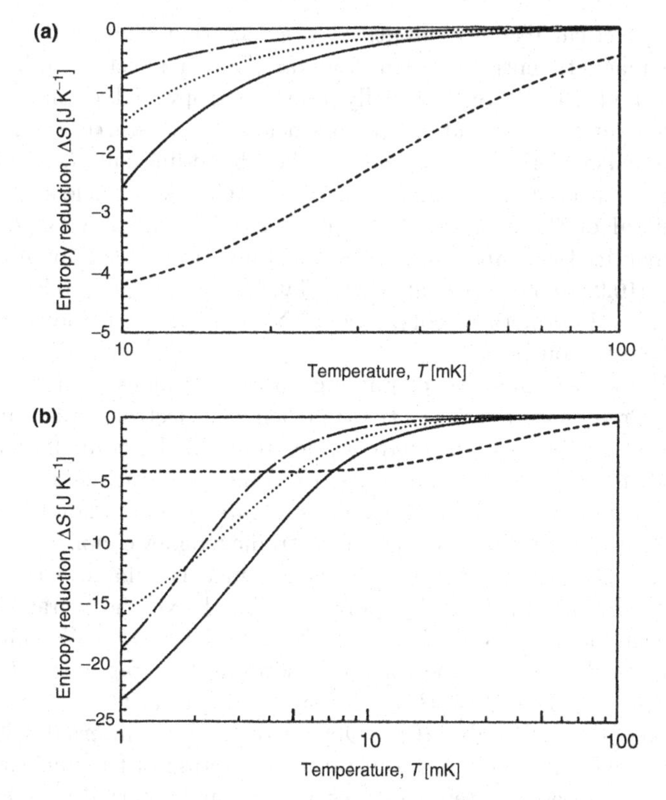

Fig. 10.12. Nuclear spin entropy reduction ΔS of 1.4 mol Nb (——), 1.0 mol In (·······), 2.0 mol Cu (·–·–·), and 0.3 mol $PrNi_5$ (– – –) in a field of 5 T and at 10 mK $\leq T \leq$ 100 mK (**a**) and at 1 mK $\leq T \leq$ 100 mK (**b**). The amount of each material has been chosen to give the same volume of refrigerant

rather similar and reasonably large nuclear magnetic moments and nuclear Curie constants. Above all, Cu can be obtained in high purity at a reasonable price, has good metallurgical and handling properties, is not superconducting for $T > 10\,\mu$K (if at all), is cubic, and has a very good thermal conductivity. Its nuclear ordering temperature is $T_{n,c} < 0.1\,\mu$K and the internal field is small (0.36 mT) [10.1]. Platinum would be a good candidate, too, but only 33.8% of its isotopes have a nuclear moment and above all the moment is small. In addition, Pt is very expensive. However, Pt has recently been demagnetized to 1.5 μK [10.67], and it is the main candidate for nuclear magnetic resonance thermometry because of its small τ_1, rather large τ_2, and because it has $I = 1/2$ (Sect. 12.10).

10.7 Hyperfine Enhanced Nuclear Refrigeration

Starting conditions for nuclear refrigeration of $B_i/T_i \approx 1{,}000$ (T/K), corresponding to $B_i \approx 8$ T and $T_i \approx 8$ mK, for example, represent in most cases an upper practical technical limit. This means that the relative reduction of nuclear magnetic entropy of Cu, for example, is restricted to $\Delta S/S \leq 0.1$ (Figs. 10.1 and 10.12). Substantially larger entropy reductions or nuclear polarizations at less demanding starting fields B_i and starting temperatures T_i can be achieved if one uses the so-called hyperfine enhanced van Vleck paramagnets instead of a simple metal like Cu as a nuclear refrigerant. The potential of these materials for nuclear refrigeration was pointed out by Altshuler in 1966, and then experimentally established by Andres and coworkers [10.9, 10.68–10.70] and later by Mueller et al. [10.28, 10.71]. In particular, the intermetallic compound $\mathrm{PrNi_5}$ turned out to be very suitable for the requested purpose.

Van Vleck paramagnets containing rare earth ions such as $\mathrm{Pr^{3+}}$ in hexagonal $\mathrm{PrNi_5}$ have a temperature-independent electronic susceptibility at low temperatures because they have an electronic singlet nonmagnetic ground state of their 4f electron shells.[1] An external magnetic field changes the electronic configuration of the 4f electrons by mixing higher non-singlet states into the ground state. This induces an electronic magnetic moment generating a hyperfine field B_int at the $^{141}\mathrm{Pr}$ nucleus which is enhanced compared to the externally applied field B. For polycrystalline $\mathrm{PrNi_5}$ the average hyperfine enhancement factor is $K = B_\mathrm{int}/B = 11.2$ [10.70–10.72]. We can still use the equations deduced in the foregoing sections for magnetic refrigeration but we have to replace B by $B(1+K)$, and the nuclear Curie constant λ_n is enhanced by $(1+K)^2$ (Table 10.1). Obviously, the large internal field seen by the $^{141}\mathrm{Pr}$ nuclei will result in a substantial reduction of the nuclear spin entropy even at rather high temperatures and in small external magnetic fields; for example, $\Delta S/S \cong 70\%$ at $T_\mathrm{i} = 25\,\mathrm{mK}$ and $B_\mathrm{i} = 6\,\mathrm{T}$ for $\mathrm{PrNi_5}$ (Figs. 10.12, 10.13 and 10.23). The cooling capacity/volume of $\mathrm{PrNi_5}$ is rather large. Hence a moderate dilution refrigerator and a simple superconducting magnet are sufficient to achieve the necessary starting conditions for nuclear refrigeration with $\mathrm{PrNi_5}$; it is the *poor man's workhorse* for nuclear refrigeration. Hyperfine enhanced nuclear refrigeration is an intermediate step between paramagnetic electronic refrigeration and nuclear refrigeration with simple metals, taking advantage of some of the useful features of these two methods.

The minimum temperature that can be achieved is, of course, again given by the nuclear magnetic ordering temperature $T_\mathrm{n,c}$. Due to the strong hyperfine interaction in rare earth van Vleck compounds, $T_\mathrm{n,c}$ is in the millikelvin range for relevant Pr compounds; for example, $T_\mathrm{n,c} = 0.40\ (2.6)\,\mathrm{mK}$ for $\mathrm{PrNi_5}$ ($\mathrm{PrCu_6}$) [10.71, 10.73]. The internal field will then be quite large; $b = 65\,\mathrm{mT}$ in $\mathrm{PrNi_5}$, for example. The minimum temperature which has been achieved with such compounds is $0.19\,\mathrm{mK}$ for $\mathrm{PrNi_5}$ [10.28]. Recent experiments on $\mathrm{Pr_{1-x}Y_xNi_5}$ have shown that the nuclear spin ordering temperature, and

[1] The Ni ions in $\mathrm{PrNi_5}$ "behave like Cu" because the holes in its electronic d-shell which are responsible for the paramagnetic susceptibility of Ni in other compounds are filled up by the outer ($6\mathrm{s}^2\ 5\mathrm{d}^1$) electrons of Pr.

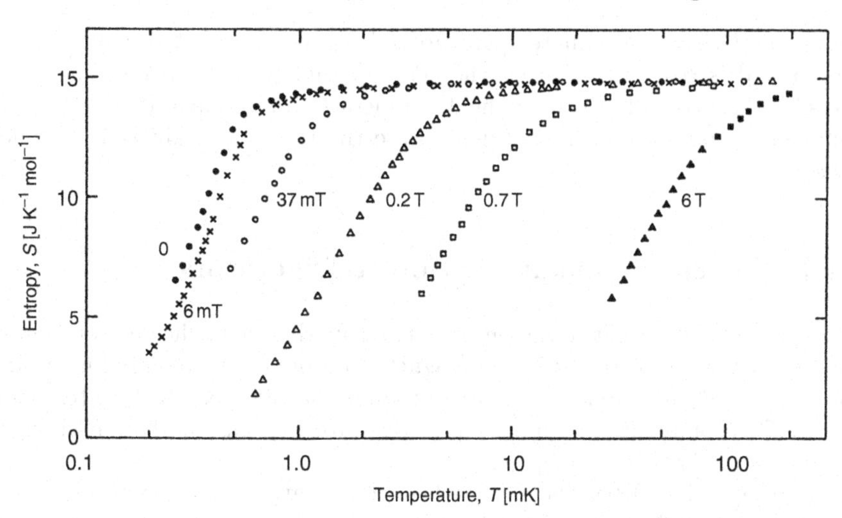

Fig. 10.13. Nuclear spin entropy of $PrNi_5$ as a function of temperature in the indicated fields [10.71]

therefore the accessible temperature range, can be substantially reduced by diluting Pr by a few percent of the nonmagnetic Y, still keeping a rather high cooling power/volume [10.72]. This is quite different to the dilution of Ce by La in LCMN, where substantial reductions of T_c can only be achieved by a very high degree of dilution (see references quoted in Sect. 9.4). Of course, at the conditions accessible with these compounds, the high-temperature approximations utilized in this chapter cannot be used anymore; we have to take the full equations, such as (9.15).

Some practical aspects to be considered in the use of $PrNi_5$ as a nuclear refrigerant are the following. It has a very small, and still unknown, nuclear spin–lattice relaxation time, but due to its small thermal conductivity ((0.5–1) W Km^{-1} [10.28, 10.74, 10.75], something like brass) usually a long thermal relaxation time. It is very brittle, and the use of H_2 gas for thermal coupling in the cooldown process should be avoided because $PrNi_5$ adsorbs H_2 strongly and may then crack. The compound can be soldered with Cd (use ZnCl flux) to Cu wires for thermal contact; the Cd should be kept in its normal conducting state by a high enough final demagnetization field. A complete coating of $PrNi_5$ with Cd keeps possibly cracked parts together. $PrNi_5$ is usually used in the form of rods of 6–8 mm diameter and typically 10 cm length, and because of the low thermal conductivity of $PrNi_5$ it is advisable to solder high-purity Cu wires along the entire length for thermal anchoring [10.28, 10.29]. Good $PrNi_5$ samples should have RRR $\geq$ 20; lower values may indicate the presence of excess Ni or of other Pr–Ni compounds, which may cause trouble in the demagnetization, limiting the cooling power and the minimum temperature.[2] A good test of the quality of the compound is to check whether the electronic magnetization is a linear, reversible function of magnetic field. Minicracks developing with time and number of cooldowns can also deteriorate the thermal conductivity of such stages.

10.8 Nuclear Demagnetization Refrigerators

The progress of nuclear refrigeration is particularly due to the development of filamentary wire superconducting magnets to produce the required fields and of ^{3}He–^{4}He dilution refrigeration to produce the required starting temperatures. This combination in a nuclear refrigerator was first used at Helsinki, as described in [10.6, 10.7, 10.9].

In the early 1970s, the Cu nuclear refrigeration stage was almost exclusively a bundle of insulated fine Cu wires to reduce eddy current heating [10.7, 10.9]. Only later it was realized that this problem is not as serious as anticipated and today favorite designs consist of an assembly of welded Cu rods [10.28,10.29], see Figs. 10.24 and 10.25, or sheets [10.76,10.77], or of a slit Cu block [10.1,10.30,10.32], see Figs. 10.14, 10.15 and 10.20, to improve structural stability and filling factor, to reduce the number of joints, and to avoid the wire insulation. One can show that with typical experimental parameters [$\mathrm{RRR_{Cu}} \approx 10^2$–$10^3$, $T_\mathrm{i} \approx 10\,\mathrm{mK}$, demagnetization rate $\approx 1\,T\,\mathrm{h}^{-1}$] a thickness of the Cu rods or sheets of about 3 mm is the optimum when maximum filling factor and eddy current losses are compared (Sect. 10.5.2).

A review on some of the successful nuclear refrigerators built before 1981 has been published by Andres and Lounasmaa [10.9]. Since then the technology has been advanced, in particular by the Cu nuclear demagnetization refrigerators built at Lancaster [10.25] and Bayreuth [10.30].

In the nuclear demagnetization refrigerator at Bayreuth [10.30] (Figs. 10.14, and 10.15), the ^{3}He–^{4}He dilution refrigerator – whose cooling power is shown in Fig. 7.19 ($T_\mathrm{min} = 4.5\,\mathrm{mK}$, $\dot{Q} = 2.5\,\mu\mathrm{W}$ at 10 mK) – is able to precool the 104 mol (6.6 kg) of the 275 mol (17.5 kg) Cu stage (RRR $\approx 1,000$) which are in a field of 8 T to a temperature of 13 (10) mK within 1 (4) day(s) (Fig. 10.16). The temperature of the magnetized Cu stage during the precooling phase decreases with time according to $T \propto t^{-1/3}$. This is the expected behavior at least as long as the cooling power of the dilution refrigerator varies as T^2 [Fig. 7.19 and (7.38)], the nuclear specific heat of the magnetized nuclear stage varies as T^{-2}, (10.4), and they are connected by a metallic link of thermal conductivity $\kappa \propto T$. Demagnetization of the 104 mol Cu from 10 mK/8 T ($\Delta S/S \approx 9\%$) to 4 mT results in a *measured* minimum electronic

[2] The nominal composition should contain some excess Pr, as in $\mathrm{PrNi_{4.98}}$, because some of it will be oxidized during the arc-melting production process from the constituent elements or may already be present as oxide.

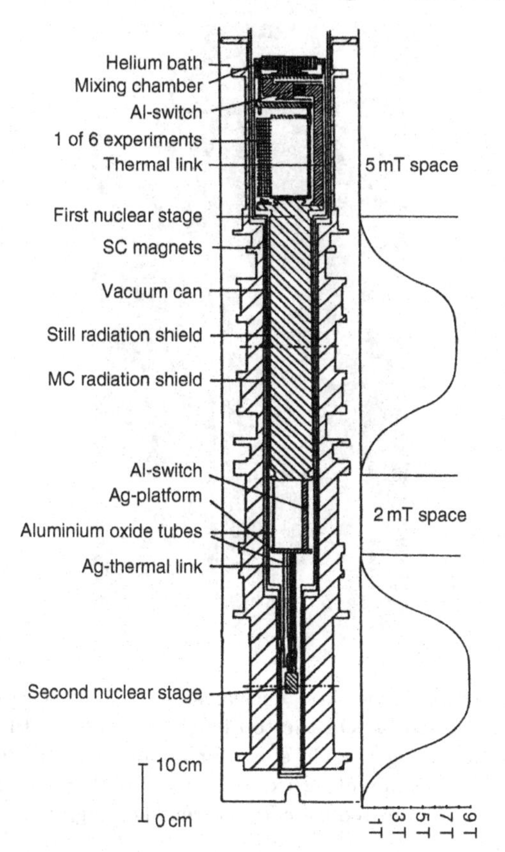

Fig. 10.14. Schematic of the low-temperature part of the nuclear refrigerator at the University of Bayreuth with the two superconducting magnets and their field profiles. The first nuclear stage contains 17.5 kg or 275 mol Cu, of which 6.6 kg or 104 mol Cu are effectively in a field of 8 T. The second nuclear stage contains 0.13 kg or 2 mol Cu in a field of 9 T [10.30]. The superconducting Al heat switches and the used Pt NMR thermometers in the low-field experimental spaces are of our design discussed in Sects. 4.2.2 and 12.10.3, respectively

temperature of 15 μK in the low-field experimental region of the refrigerator containing nine sites for experiments and thermometers (Fig. 10.17), and a *calculated* minimum electronic temperature of 5 μK at the center of the stage. This powerful Cu nuclear stage has also been used to precool a 2 mol Cu stage in 9 T (see Fig. 10.14) to 3.5 mK. The demagnetization of the latter stage gave *measured* electronic temperatures between 10 and 12 μK [10.30].

With this refrigerator, the lowest ever achieved equilibrium temperature with $T_{\text{nuclei}} = T_{\text{electrons}} = T_{\text{phonon}}$ has been obtained for a 0.16 mol (32 g) Pt sample [10.67]. To this end, the Pt sample in a calorimetric setup in a field of 0.37 T, similar to the one shown in Fig. 3.13, was precooled by the first Cu nuclear cooling stage to 103 μK. Its nuclei were then demagnetized to

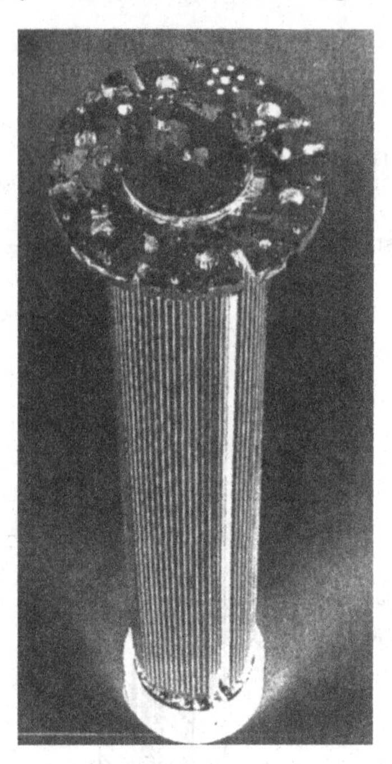

Fig. 10.15. Photograph of the first nuclear stage (17.5 kg Cu) of the nuclear refrigerator shown in Fig. 10.14. On the top flange one can see the large crystallites which result from annealing the Cu stage for about 4 days at 950° in 1 μbar O_2. The various holes are for mounting experiments and thermometers. The stage is slit, except at the flanges, to reduce eddy current heating. The total length of the stage is 525 mm [10.30]

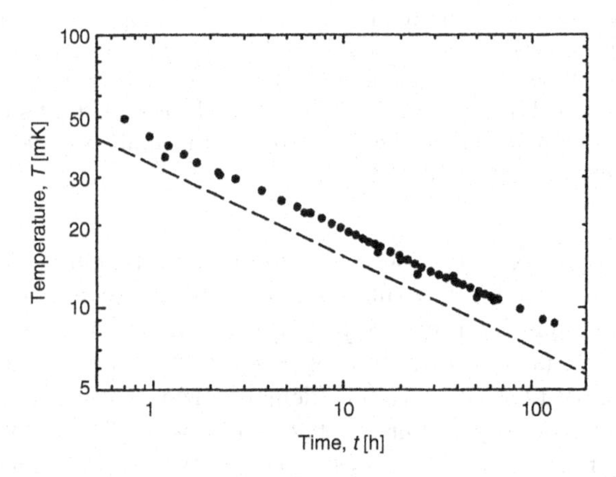

Fig. 10.16. Precooling of 104 mol Cu magnetized in 8 T in the nuclear refrigerator shown in Fig. 10.14 by a ^{3}He–^{4}He dilution refrigerator with the cooling capacity indicated in Fig. 7.19 as a function of time. The dashed line represents $T \propto t^{-1/3}$ for comparison [10.30]

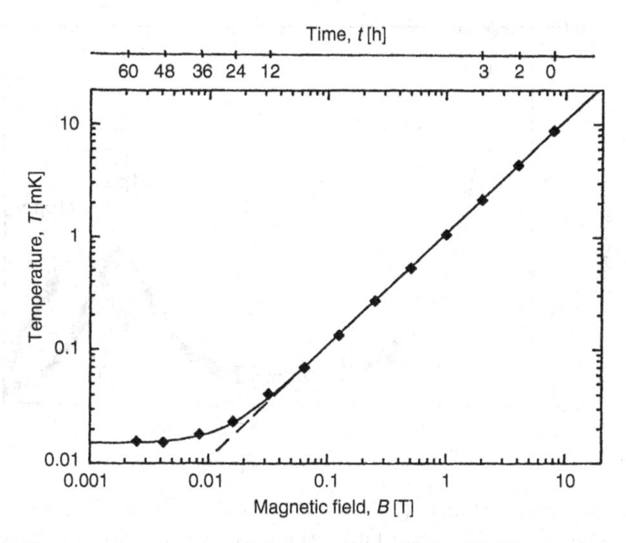

Fig. 10.17. Temperature measured by a pulsed Pt NMR thermometer in the experimental region of the first Cu nuclear refrigeration stage of the Bayreuth nuclear refrigerator (Fig. 10.14) as a function of the applied magnetic field. The demagnetization behaves in an ideally adiabatic way, B/T = const., and there are no temperature differences between demagnetized Cu nuclei and Pt thermometer to within 3% to at least 70 μK. The demagnetization rate was adjusted so that nuclei and electrons in the Cu stage stay in thermal equilibrium as long as possible [10.30]

2.5 mT to cool themselves as well as the electrons and phonons of the sample to about 1.5 μK; further reduction of the field to a value <0.05 mT reduced the temperature of the nuclei only to 0.3 μK (Fig. 10.18).

Recently, a new nuclear refrigerator was described in [10.48]. It consists of two concentric slitted cylinders of Cu as nuclear stages connected to each other by high-purity Al screws acting as heat switches; a third nuclear stage can be inserted into them. Starting from 8.3 mK/8.5 T, a minimum temperature of 23 μK measured by a massive Pt NMR thermometer was reached. The paper also discusses an optimized thermodynamic description of nuclear refrigeration.

If it is intended to refrigerate ^{3}He by demagnetizing a metallic nuclear refrigerant, the contact area between ^{3}He and the refrigerant has to be large to overcome the thermal boundary resistance problem (Sect. 4.3.2). In this situation a metallic refrigerant of Cu (or $PrNi_5$ [10.78]) in the form of compressed powders or sinters is most appropriate. This design has similarities to that for refrigerating and making thermal contact to liquid helium using CMN powder (Sect. 9.4). The Lancaster group [10.25] has developed very efficient Cu sinter nuclear refrigeration stages for this purpose. They use a nested rather than a series design for the nuclear stages, using the outer one as a thermal guard. The stages contain a mixture of Ag and Cu sinters on 1 mm

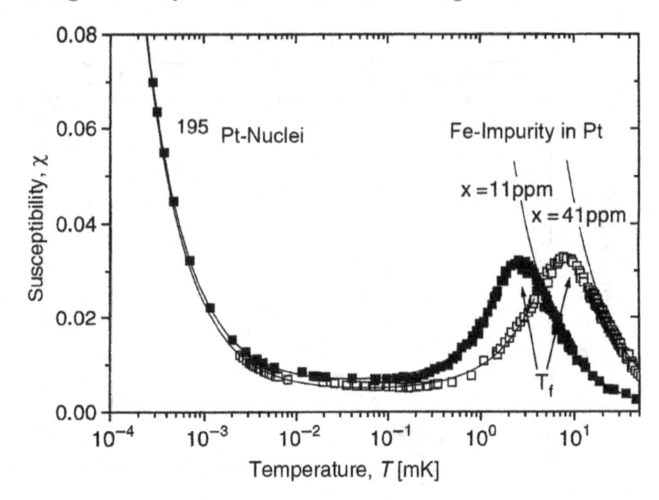

Fig. 10.18. AC susceptibility χ of two Pt samples which contained $x = 11$ and 41 ppm Fe, respectively. At $T > 0.1\,\mathrm{mK}$, the data are the *electronic* susceptibilities χ due to the spin glass behavior of the giant electronic magnetic moments of $8\mu_{\mathrm{B}}$ caused by the Fe impurities in the Pt samples. They show Curie behavior at T > T_f; the latter are the indicated spin freezing temperatures of the samples which depend on the Fe concentration. The increase of χ at $T < 0.1\,\mathrm{mK}$ is the *nuclear* susceptibility of the nuclear magnetic moments of ^{195}Pt, which again shows Curie behavior to the lowest obtained temperatures. In these experiments, a minimum electronic or equilibrium temperature of $T_{\mathrm{e}} = 1.5\,\mu\mathrm{K}$ was obtained. Afterward, the Pt nuclei alone continued to cool to a minimum temperature of T_{n}=0.3 μK [10.67]

Cu plates in epoxy cells acting both as a refrigerant and as heat exchangers for thermal contact to the liquid helium to be refrigerated and investigated in the cells (Fig. 10.19). Of course, the cells are demagnetized together, and the liquid sample and the thermometers are exposed to the changing demagnetizing field. The electrical conduction of the sinter is low enough that eddy current heating during demagnetization is negligible. As shown by (10.27), only a comparatively small amount of Cu is needed to refrigerate some cm^3 of liquid ^{3}He. According to [10.10, 10.79] this cryostat has refrigerated liquid ^{3}He to about 100 μK. Using such a design without a metal sinter on the Cu plates, i.e., using just three Cu plates in series – separated by superconducting heat switches – as refrigerant, the group achieved 7 μK, measured with a ^{195}Pt NMR thermometer on the inner Cu plate of about 9 g (0.134 mol) only [10.10].

In this book I shall not discuss the methods of *nuclear cooling* by which only the nuclear spin temperature is reduced, leaving the lattice and electrons at a higher temperature. This can be achieved by the method of dynamic polarization and demagnetization in the rotating frame for insulators, as developed by Abragam, Goldman, and their coworkers [10.14–10.16], or by the usual "brute force" demagnetization procedure discussed here for metals

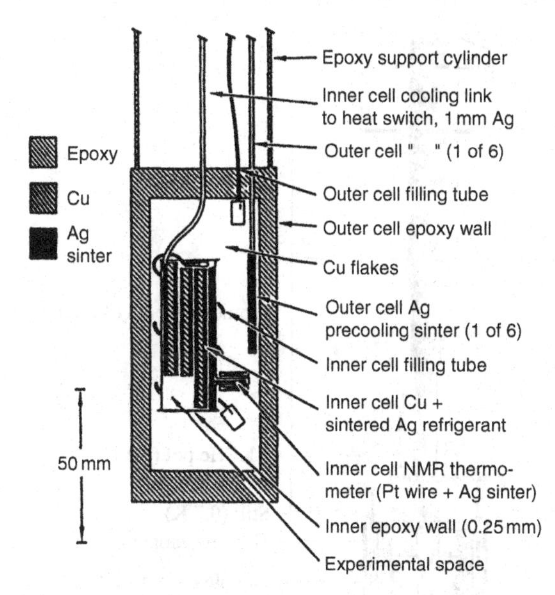

Fig. 10.19. Nuclear refrigeration stage used at the University of Lancaster for refrigerating liquid ^{3}He (see text) [10.10, 10.25]

if the nuclear spin–electron coupling is weak enough (large Korringa constant) and the demagnetization is performed within a few minutes instead of many hours [10.1,10.24]. Nuclear cooling is important for the study of nuclear cooperative phenomena but not suitable for refrigerating other samples. Two-stage Cu–Cu nuclear demagnetization refrigerators intended for studies of nuclear magnetic ordering phenomena have been built in Helsinki and are described in [10.1, 10.6, 10.24, 10.32] (Fig. 10.20).

For the most recent design of a two-stage cascade Cu nuclear refrigerator [10.32], the group in Helsinki has put all the experience from their former nuclear refrigerators [10.1, 10.6, 10.7, 10.9] as well as the substantial progress of other groups, in particular from [10.30] together. In this refrigerator (Fig. 10.20), the first nuclear stage contains 170 mol Cu with slits spark cut into a Cu block, of which 98 mol are effectively in a field of 9 T. It is precooled by a powerful commercial dilution refrigerator (10 μW at 10 mK) to 10 mK within 50 h. Demagnetization of this stage results in a measured minimum electronic temperature of 67 μK in a field of 10 mT, probably limited by a heat leak of still 6 nW 2 weeks after cooldown; temperatures below 100 μK can be kept for 2 months. These temperatures were measured again by a Pt NMR thermometer of standard design (Sect. 12.10.3). A smaller second nuclear stage in a field of 7 T can be precooled by the first stage. Special care has been taken for electric filtering of the leads, magnetic shielding, and vibration isolation. The group in Helsinki has used their refrigerators to cool and to investigate the nuclear spin systems of Cu, Ag [10.1] and Rh [10.24] to

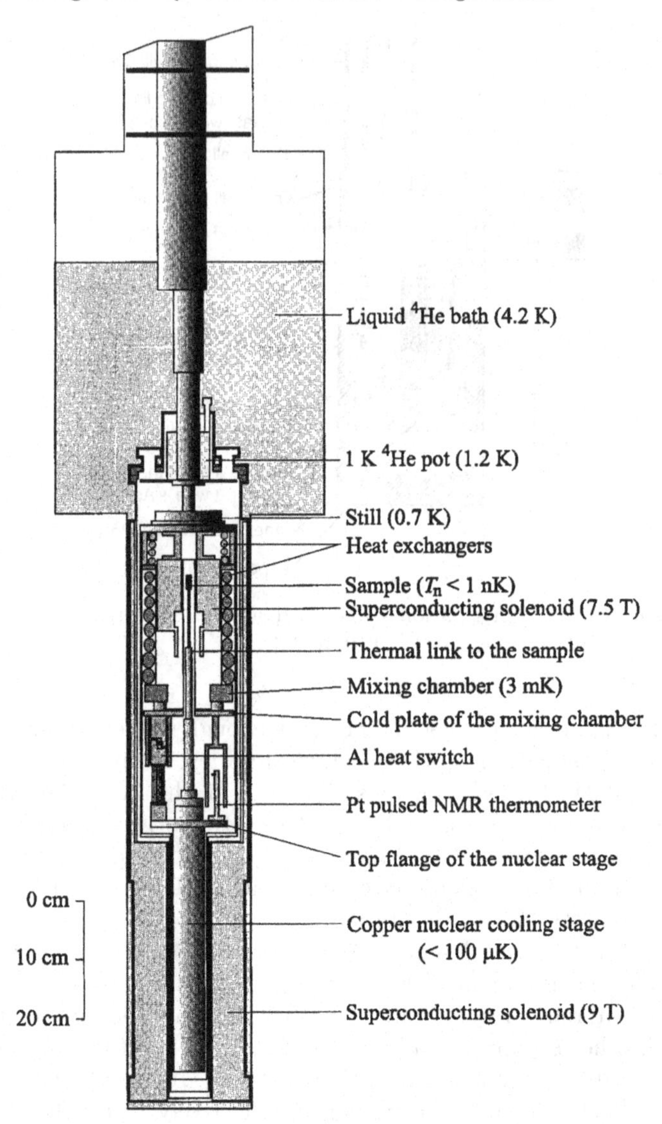

Fig. 10.20. Double-stage nuclear demagnetization refrigerator at the Technical University of Helsinki (see text) [10.32]

sub-nanokelvin *nuclear* spin temperatures while the electrons and phonons of their samples were kept at around 0.1 mK. This large temperature difference is possible due to the long nuclear spin – lattice relaxation times in these metals at very low temperatures (see Table 10.1).

If moderately low final temperatures of the order of 0.4 mK are required, these temperatures can be achieved using $PrNi_5$ with its large hyperfine enhancement of fields as a refrigerant with rather modest superconducting magnets and dilution refrigerators for the starting conditions, see Sect. 10.7. Simple $PrNi_5$ nuclear refrigerators have been described by Andres [10.9,10.70] and by Greywall [10.80] (Fig. 10.21). A compact 0.9 mol $PrNi_5$ nuclear refrigeration stage (31 bars of 6 cm length and 7 mm diameter) has been described in [10.81]. This stage, together with its 4 T, 5 cm bore demagnetization solenoid, is mounted in the free volume of the heat exchangers of a dilution refrigerator used for precooling (Fig. 10.22). This design allows the construction of a very compact nuclear refrigerator which is less susceptible to vibrations than usual cryostats where the nuclear stage is mounted below the mixing chamber of the dilution refrigerator. This apparatus has been modified by mounting a 6 T magnet to the still of the ^3He–^4He dilution refigerator and a 61 mol Cu nuclear stage inside of the heat exchangers of the dilution refrigerator [10.82]. The torus-shaped mixing chamber [10.34] allows easy access to the nuclear stage and to the experiments on its bottom.

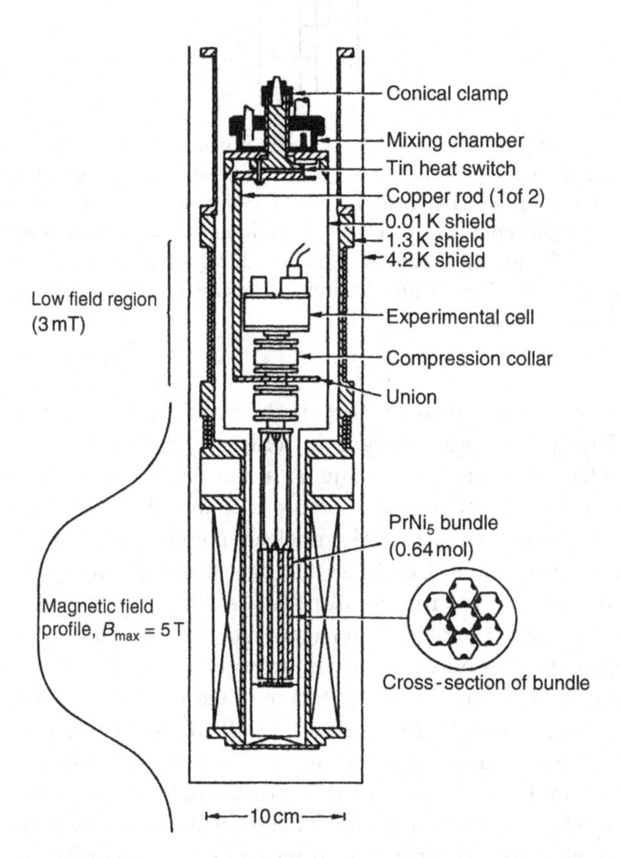

Fig. 10.21. Low-temperature part of a $PrNi_5$ nuclear refrigerator. The refrigerant consists of 0.64 mol $PrNi_5$ in the form of seven 8 mm dimeter hexagonal rods of 95 mm length. This refrigerator has cooled ^3He samples to below 0.3 mK [10.80]

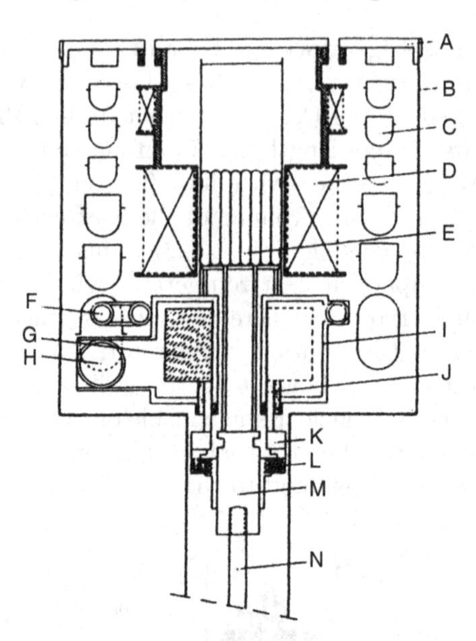

Fig. 10.22. Lower part of the dilution refrigerator of [10.34, 10.81] with a $PrNi_5$ nuclear demagnetization stage. (A, cold plate; B, cold plate thermal radiation shield; C, heat exchanger of dilution refrigerator; D, demagnetization magnet; E, $PrNi_5$ stage; F, input tube connection; G, heat exchanger for nuclear stage; H, output tube connection; I, mixing chamber (epoxy); J, leg of heat exchanger; K, copper ring for thermal anchoring; L, heat switch; M, silver post; N, cold finger extending to the high field region)

If very low temperatures or very high cooling powers are required, the compound $PrNi_5$ can bridge the gap in temperature where the cooling power of dilution refrigerators for precooling a Cu nuclear stage becomes very small, which is at around 10–15 mK, but where one would like to have a powerful precooling refrigerator to reduce the nuclear magnetic entropy of copper as a nuclear refrigeration stage (Fig. 10.23). In this situation one can build a rather powerful two-stage nuclear refrigerator using a dilution refrigerator as the first precooling stage, a $PrNi_5$ nuclear refrigeration stage as the second precooling stage, bridging the gap between the dilution refrigerator and the copper, and then the final copper nuclear refrigeration stage to reach the final minimum temperature (Fig. 10.3) [10.26, 10.28, 10.29, 10.83]. With such a combination substantially higher entropy reductions can be achieved. The large hyperfine enhanced nuclear heat capacity of the intermediate $PrNi_5$ stage can also be used as a very efficient thermal guard for the Cu stage and experiments, as well as thermal anchor for everything leading to lower temperatures. The design of such a two-stage nuclear refrigerator built in Jülich [10.26–10.28] is shown in Figs. 10.24 and 10.25. It requires two superconducting magnets (6 T for the $PrNi_5$ and 8 T for the Cu) and two superconducting Al heat switches. The

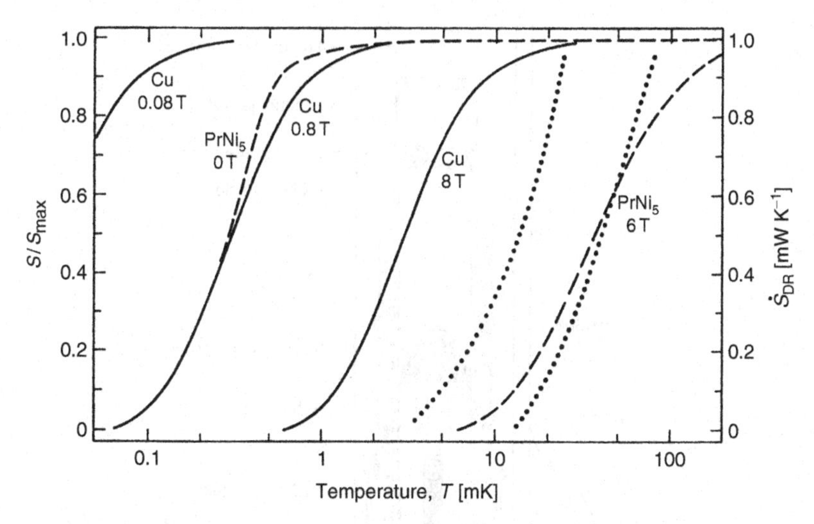

Fig. 10.23. Reduced nuclear spin entropy of Cu (——) and of $PrNi_5$ (- - -) in the indicated magnetic fields compared to the rate of entropy reduction of typical ^3He–^4He dilution refrigerators (·······). The reduction of the nuclear spin entropy of $PrNi_5$ occurs at a higher temperature than that for Cu because the ^{141}Pr nuclei see an effective field which is a factor of 12 enhanced compared to the externally applied field [10.26, 10.28]

Jülich refrigerator contains 4.3 mol $PrNi_5$ (60 rods, 6.4 mm diameter, 120 mm length each soldered with Cd to six 1 mm diameter Cu wires) and effectively 10 mol Cu (96 rods of $2 \times 3\,mm^2$ and 245 mm length of which about 120 mm are inside of the field region of 8 T), welded at one end and electrically isolated but bolted together at the other. The $PrNi_5$ stage (starting from 25 mK, 6 T corresponding to $\Delta S/S \approx 0.7$) precools the Cu stage in 8 T to typically 5 mK, reducing the Cu nuclear spin entropy by about 26%. Figure 10.26 shows the development of the temperature of the experiments in this refrigerator after a demagnetization from 3.8 mK/8 T to 0.01 T. The minimum temperature which the $PrNi_5$ stage in this refrigerator has reached is 0.19 (0.31) mK starting from 10 (23) mK and 6 T, and the Cu stage of it has refrigerated experiments to 38 μK [10.26, 10.28]. The low heat leak in this refrigerator results from its rigid construction (reducing eddy current heating) and the very small quantity of "ill-defined" materials. The refrigerator can stay below 50 μK (300 μK) for about 2 weeks (at least 1 month). These were great achievements in the late 1970s when this refrigerator was built. Similar two-stage nuclear refrigerators have been built in Tokyo (11 mol $PrNi_5$, 19 mol Cu; $T_{min} = 27\,\mu K$) [10.29] and in Nagoya (2.6 mol $PrNi_5$, 8 mol Cu) [10.83]. With the advent of more powerful dilution refrigerators [10.34, 10.36, 10.84] the advantage or necessity of an intermediate $PrNi_5$ precooling stage for the final Cu nuclear refrigeration may have disappeared [10.1, 10.30, 10.32].

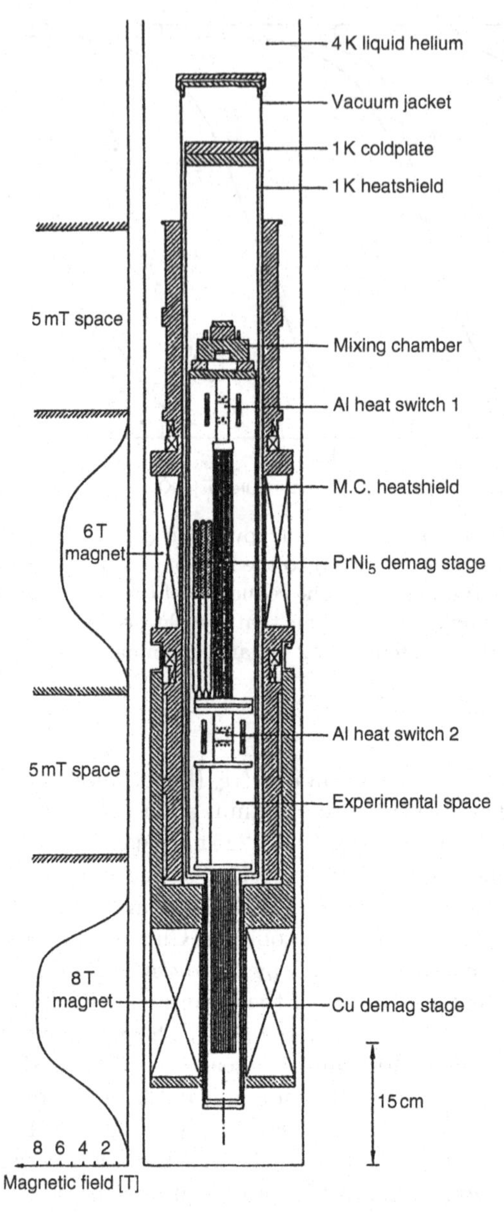

Fig. 10.24. Schematic of the low-temperature part of the $PrNi_5$/Cu nuclear refrigerator at the KFA Jülich and field profiles of its two superconducting magnets (see text) [10.28]. The superconducting Al heat switches and the Pt NMR thermometer are of our design discussed in Sects. 4.2.2 and 12.10.3, respectively

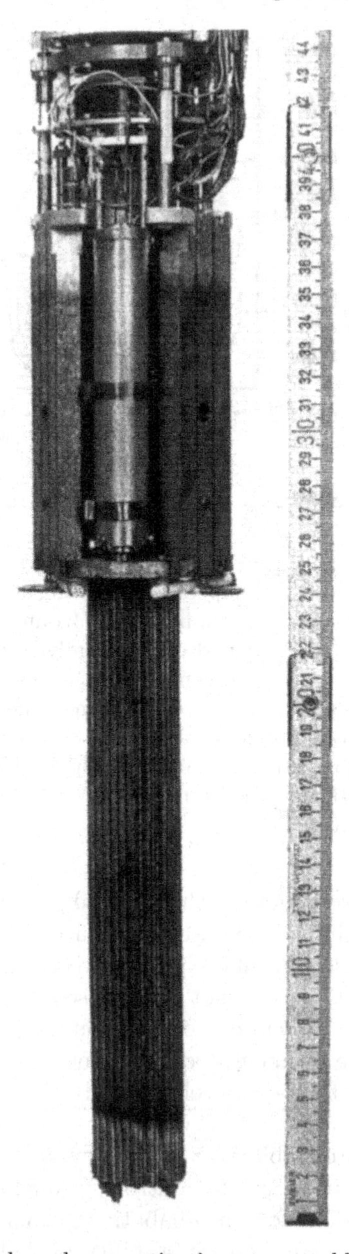

Fig. 10.25. Copper nuclear demagnetization stage welded to the bottom of the experimental region of the nuclear refrigerator shown in Fig. 10.24. At the top the lower end of the second superconducting heat switch can be seen. In the experimental region, a Nb tube contains a Pt NMR thermometer (see Sect. 12.10.3). The thermal connection between the heat switch and the Cu nuclear stage is made by three massive Cu legs with two grooves reducing eddy current heating in each of them. At the bottom of the experimental region the centring device can be seen [10.28]

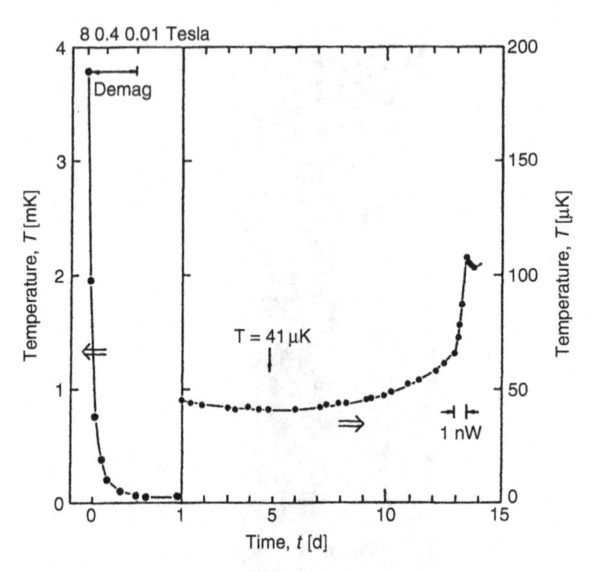

Fig. 10.26. *Left*: The decrease of temperature, starting at 3.8 mK, of a sample in the double stage nuclear refrigerator shown in Figs. 10.24 and 10.25 when the magnetic field on the Cu refrigerant is decreased exponentially in time for 10 h from 8 T to 0.01 T. *Right*: The development of the sample temperature after demagnetization on an expanded temperature scale and compressed time scale. It takes about 4 days for the apparatus to relax to its minimum measured electronic temperature of 41 μK. The sample was kept below 50 μK for 10 days. On the 13th day after starting the demagnetization an additional heat input of 1 nW was supplied to the sample to accelerate the warm-up [10.26]

In all these nuclear refrigerators the thermal path between refrigerant and samples/thermometers in the experimental space should be kept as small as possible to keep temperature differences between them small. Most experiments and thermometers should not be exposed to the large and changing demagnetization fields. Therefore the superconducting main magnet for demagnetization has to be compensated at its end(s) to reduce the field in the experimental space to a tolerable level, typically ≤ 5 mT when the coil is fully energized.

The best temperature stability in magnetic refrigerators is achieved if the magnetic field itself is used as the control parameter. An automated temperature regulation system for an adiabatic paramagnetic demagnetization refrigerator achieving a temperature stability of 2 μK at 100 mK has been described in [10.85].

Figure 10.27 shows the development of the minimum equilibrium temperatures achieved by nuclear refrigeration of nuclei in metals and available for refrigerating samples in the field-free region of the refrigerators. Table 10.2 summarizes the minimum temperatures to which various materials have been refrigerated.

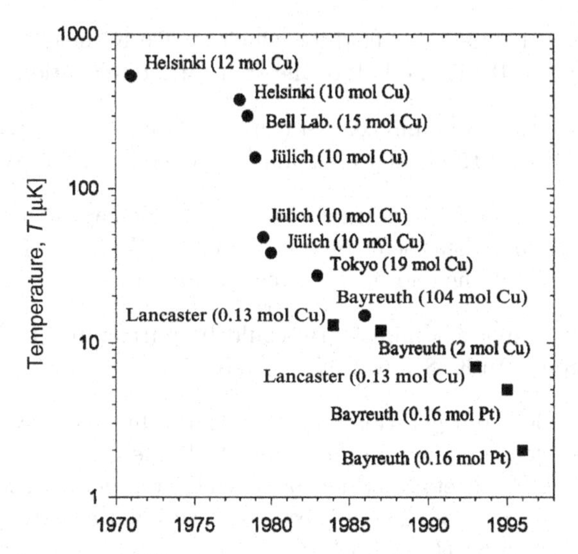

Fig. 10.27. Minimum equilibrium temperatures ($T_{\text{nuclei}} = T_{\text{electrons}} = T_{\text{phonons}}$) measured by the indicated laboratories at their nuclear refrigeration stages consisting of the indicated number of moles of Cu (●). In addition, equilibrium temperatures measured at demagnetized Cu or Pt experimental setups connected to nuclear refrigeration stages are indicated (■)

Table 10.2. Minimum temperatures to which the given materials have been cooled by nuclear magnetic refrigeration

Refrigerated material	Minimum temperature	Reference
Liquid ^{3}He; liquid ^{3}He–^{4}He	0.10 mK	[10.78, 10.86–10.88]
Solid ^{3}He	34 μK	[10.89]
Massive Cu nuclear refr. stage	10 μK	[10.30]
Metals	1.5 μK	[10.67]
Nuclear moments in metals	0.28 nK	[10.24]

Problems

10.1. At which nuclear spin temperature T_{n} would Cu nuclei order magnetically, if we assume that this order occurs roughly when the nuclear magnetic interaction energy μb_{i} becomes comparable to the thermal energy $k_{\text{B}}T_{\text{n}}$? The internal field created by the neighbour nuclei $b_{\text{i}} \approx 0.36\,\text{mT}$.

10.2. Calculate the cooling power and the heat of magnetization if 10 mol Cu are magnetized at 15 mK in 8 T and then adiabatically demagnetized to 8 mT.

10.3. How long would it take a ^{3}He–^{4}He dilution refrigerator with a circulation rate of 1 mmol ^{3}He/s to precool 20 mol of Cu in 8 T to 15 mK?

10.4. To which temperature does one have to cool copper to achieve a 50% entropy reduction in a field of 8 T?

10.5. Calculate the magnetic field for which at 10 mK the effective time constant given by (10.18) is one half of the spin–lattice relaxation time for Cu.

10.6. Calculate the minumim temperatures and optimum magnetic fields, as shown in Fig. 10.7, if the heat leak could be reduced to 0.03 nW mol^{-1} Cu.

10.7. Calculate the warm-up time of Cu nuclei demagnetized to 10 µK in a field of 4 mT at a heat leak of 0.1 nW mol^{-1}. What is the temperature difference $T_e - T_n$ at the end of the demagnetization?

10.8. How many moles of liquid ^{3}He could be refrigerated by 1 mol Cu demagnetized from 10 mK, 8 T to 8 mT, see (10.27).

10.9. A Cu nuclear refrigeration stage has to be slit to reduce eddy current heating during demagnetization (Sect. 10.5.2). If one makes the slits too wide or the remaining Cu sheets too thin, too much Cu refrigerant is lost. Calculate the optimum thickness of the Cu sheets for a residual resistivity ratio of Cu of 10^3 and $T = 10$ mK, $\dot{B} = 0.5$ T h^{-1}.

10.10. How many moles of two-level tunneling systems with an energy separation of 10 K are necessary to explain a total heat release from a nuclear stage of 20 mJ?

10.11. Calculate at which time after cool-down the heat release from 10 g Stycast 1266 is equal to the heat release from ortho–para conversion of 1 µmol H_2 (Sect. 2.2).

10.12. How large has the electric field gradient V_{zz}, which Cu nuclei may see, to be so that the nuclear quadrupole specific heat (10.35) equals the nuclear magnetic specific heat in a field of 10 mT (use the high-temperature approximation $C_n \propto T^{-2}$).

10.13. What is the reason for the nonmonotonic shift of the entropy curves in Fig. 10.13 with the field?

10.14. Make a rough estimate of the hyperfine enhancement for the magnetic field which Pr nuclei see in $PrNi_5$ from the entropy data shown for this compound in 6 T and for Cu in 8 T in Figs. 10.13 and 10.23. Determine from the 0 T result for PrNi, displayed in the same figure, the internal magnetic field.

11

Temperature Scales and Temperature Fixed Points

In low-temperature physics and technology the measurement of a temperature is very often as difficult and, of course, as important as actually reaching that temperature. Therefore, low-temperature physicists have to spend a substantial part of their time considering thermometry. Before we can talk about thermometry we have to talk about temperature; we have to define a temperature scale. Actually, temperature is one of the most important parameters in physics and technology. It is one of the basic units in all systems of units. In spite of this fact – as we will see in the following – the absolute temperature is not very accurately known; our knowledge of temperature is much less precise than our knowledge of time, length, mass, voltage, etc. But let me start with a recollection of the definition of the temperature scale and some temperature fixed points.

11.1 Thermodynamic Temperature

The definition of the temperature scale is obtained from the Carnot cycle, which is based on the second law of thermodynamics. This reversible cycle gives

$$\oint T^{-1}\mathrm{d}Q = 0\,, \tag{11.1}$$

which is equivalent to

$$T/Q = \text{const.} \tag{11.2}$$

Hence, this or any other process gives the temperature only in terms of ratios or to within a multiplicative constant; the absolute values in a temperature scale have to be fixed by definition.

The most important proposal for the definition of a temperature scale, in general everyday practical use, was made in 1742 by A. Celsius, who proposed that the range of temperature between that at which water boils and that at which ice melts should be divided into 100°:

$$1742\,\text{Celsius}: T(\text{H}_2\text{O boiling}) - T(\text{H}_2\text{O melting}) = 100^\circ\,. \tag{11.3}$$

In 1887 the name "centigrade" and in 1948 the name "degree Celsius" were internationally accepted for the unit of this scale. Of course, we know that we have to decide at which pressure this definition should be valid; the accepted value is 1 bar.

Physicists know that the definition of a temperature scale in which the temperature of melting ice is set at 0°C is not very appropriate for physics and for many processes in nature, because we have an absolute zero in temperature. That an absolute zero exists had actually already been realized 40 years before Celsius' proposal, in 1702, by the French scientist Amontons. He found from his experiments with gases that there must be a lower limit for the temperature, and he estimated this lower limit to be at ≈−240°C. This was an outstanding achievement for that time.

Almost 150 years passed before Lord Kelvin made the proposal in 1848 to take absolute zero, T_0, as the starting point of a thermodynamic temperature scale. It was not until 100 years later, in 1954, that this and the proposal that one should count from T_0 in steps of

$$1^\circ = 1\,\text{K} = T_{\text{triple}}(\text{H}_2\text{O})/273.16 \tag{11.4}$$

was internationally adopted. Thus the unit of the thermodynamic temperature is defined as 1/273.16 of the temperature of the triple point of water which is taken as the second fixed point of the Kelvin scale. A degree in this scale is the same size as in the Celsius scale. The following values relate the Celsius to the Kelvin scale

$$\begin{aligned} 1848\,\text{Kelvin}: T(\text{H}_2\text{O melting}) &= 0^\circ\text{C} = 273.15\,\text{K},\\ T_0 &= 0\,\text{K} = -273.15^\circ\text{C}\,. \end{aligned} \tag{11.5}$$

So, we have a point for the zero of our temperature scale, and in principle we can use an experiment which is related to a Carnot cycle to establish the temperature scale, but this is rarely done by experimentalists. A handier procedure is to establish a number of fixed temperature points and to use them to calibrate the chosen thermometric method, which makes use of some temperature dependent property of a suitable material in the desired temperature range.

Presently, the Kelvin is defined by the temperature of the triple point of water which is a material property with a present standard uncertainty 3×10^{-7}. It would be a great advantage if this unit could be related to fundamental constants, as it has been possible for most of the other units. For the Kelvin, the corresponding fundamental constant is the Boltzmann constant k_B, because temperature always appears as thermal energy $k_\text{B}\text{T}$ in fundamental laws of physics. Unfortunately, the present standard uncertainty for k_B is

2×10^{-6}, which is even worse than the uncertainty for the water triple point. This value is based on one measurement applying one method; this is unsatisfactory for such an important fundamental constant. Therefore, substantial efforts at several national standards laboratories are under way to determine k_B by a variety of methods. The aim is an adoption of a reliable value for k_B with an uncertainty of 1 to 2×10^{-6} in 2011 by the Comite' International des Poids et Messures; this "modest" target value already indicates the involved difficulties.

11.2 The International Temperature Scale ITS-90

In 1968 international agreement was reached on the definition of an offical temperature scale, at least for temperature above about 14 K. This temperature scale, the IPTS-68 (amended in 1975) [11.1] is given by fixed temperature points which are defined by equilibrium phase transitions of pure substances. It superseded the earlier scales IPTS-48, ITS-48 and ITS-27. There was no internationally adopted and binding temperature scale for temperatures lower than 14 K, only recommendations for temperatures between 0.5 and 30 K. One was the "provisional 0.5–30 K temperature scale (EPT-76)" [11.2, 11.3] (see Sect. 11.4.1), which again is given by fixed points which are defined by equilibrium phase transition temperatures of pure substances, in particular the transition temperatures into the superconducting states of five metals at $0.5\,\mathrm{K} \leq T < 10\,\mathrm{K}$ (Table 11.7). In addition, it was recommended to use the vapour pressures of ^{4}He and ^{3}He at $0.5\,\mathrm{K} \leq T \leq T_{\mathrm{crit}}$ whose temperature dependences are given with a very high accuracy in Tables 11.3 and 11.4.

Unfortunately, already at the time of the international agreement on IPTS-68 it was realized that there are substantial errors (up to the order of several 10^{-4}) in this temperature scale, deviations from the thermodynamic temperature which are too large for an international standard. It was also obvious that the deviations of the helium vapour pressure scale from IPTS-68 and EPT-76 were a few millikelvins. Even though EPT-76 was established substantially later than IPTS-68 and had a much higher accuracy, the need for a new international scale was obvious. After substantial efforts in various laboratories, the *International Temperature Scale of 1990*, ITS-90, came into effect on January 1, 1990 [11.4–11.7]. It was adopted by the Comité International des Poids et Messures at its meeting in September 1989 and has been the official international temperature scale since 1 January 1990. There are significant differences between the new ITS-90 and the earlier scales; for example, 0.6 mK at 10 K between ITS-90 and EPT-76, and 9 mK at 20 K between ITS-90 and IPTS-68.

The ITS-90 extends from 0.65 K to the highest temperatures practicably measurable in terms of the Planck radiation law using monochromatic radiation. The defining fixed points of the ITS-90 are mostly phase transition

Table 11.1. Defining fixed points of the temperature scale ITS-90 [11.4–11.7]

material[a]	equilibrium state[b]	temperature (K)	uncertainty [mK]
He	VP	3–5	
e–H_2	TP	13.8033	0.5
e–H_2 (or He)	VP (or CVGT)	≈17	
e–H_2 (or He)	VP (or CVGT)	≈20.3	
Ne	TP	24.5561	0.5
O_2	TP	54.3584	1.0
Ar	TP	83.8058	1.5
Hg	TP	234.3156	1.5
H_2O	TP	273.16	0
Ga	MP	302.9146	1
In	FP	429.7485	3
Sn	FP	505.078	
Zn	FP	692.677	
Al	FP	933.473	
Ag	FP	1,234.93	
Au	FP	1,337.33	
Cu	FP	1,357.77	

[a] All substances (except ^{3}He) are of natural isotopic composition. e–H_2 is hydrogen at the equilibrium concentration of the *ortho-* and *para-* molecular forms.

[b] The symbols have the following meanings: *VP*, vapour pressure point; *TP*, triple point; *CVGT*, gas thermometer point; *MP, FP*, melting point, freezing point (at a pressure of 1.01325 bar).

temperatures of pure substances, given in Table 11.1. In its overlapping ranges the ITS-90 is defined in the following ways [11.4–11.7]:

(a) 0.65–5.0 K: Three vapour pressure/temperature relations of ^{3}He (0.65–3.2 K) and of ^{4}He (1.25–5.0 K) given by

$$T = \sum_{i=0}^{9} A_i \left(\frac{\ln P - B}{C} \right)^i , \tag{11.6}$$

with the constants A_i, B and C given in Table 11.2 (T in Kelvin, P in Pascal, $1\,\text{Pa} = 10^{-5}$ bar). The equation is, in fact, valid to 0.5 K.

Tables 11.3 and 11.4 give the new vapour pressure temperature values for ^{3}He and ^{4}He in the range from 0.6 to 5.2 K calculated from (11.6). These tables supersede the former $T_{58}(T_{62})$ scale for the ^{4}He (^{3}He) vapour pressure.

(b) 3.0–24.5561 K (triple point of Ne): Constant-volume helium gas thermometer (Sect. 12.1) calibrated at points number 1, 2 and 5 of Table 11.1, and with the P/T relation given in [11.4–11.7].

(c) 13.8033 (triple point of H_2)–1,234.93 K (freezing point of Ag): Electrical resistance of platinum (Sect. 12.5.1), calibrated at defining fixed points

Table 11.2. Values of the coefficients A_i, and of the constants B and C for the ^{3}He and ^{4}He vapour pressure equations (11.6) and the temperature range for which each equation is valid [11.4–11.7]

coeff. or constant	^{3}He (0.65–3.2 K)	^{4}He (1.25–2.1768 K)	^{4}He (2.1768–5.0 K)
A_0	1.053 477	1.392 408	3.146 631
A_1	0.980 106	0.527 153	1.357 655
A_2	0.676 380	0.166 756	0.413 923
A_3	0.372 692	0.050 988	0.091 159
A_4	0.151 656	0.026 514	0.016 349
A_5	−0.002 263	0.001 975	0.001 826
A_6	0.006 596	−0.017 976	−0.004 325
A_7	0.088 966	0.005 409	−0.004 973
A_8	−0.004 770	0.013 259	0
A_9	−0.054 943	0	0
B	7.3	5.6	10.3
C	4.3	2.9	1.9

and with various Pt resistance thermometers designed for the particular T ranges. For the determination of temperatures, various relations of temperature to resistance ratio $W(T_{90}) = R(T_{90})/R(273.16\,\mathrm{K})$ are given.

(d) Above 1,234.93 K (freezing point of Ag): Planck's radiation law.

The ITS-90 contains detailed instructions about how to calibrate a thermometer relative to it in the various temperature ranges, as well as differences to earlier scales [11.4–11.7]. It can be used down to 0.5 K with a thermodynamic inaccuracy of about 1 mK, and of about 0.5 mK between 1 and 5 K. Uncertainties at higher temperatures are 1 mK (at 13.8 K), 5 mK (at 933 K), and 10 mK (at 1,235 K).

There is still substantial work going on to improve the ITS-90 and to check its thermodynamic consistency. For example in [11.8], reports can be found on the recent work on new sealed fixed point devices aiming to improve the accuracy of the cryogenic triple points of ITS-90 to 0.1 mK or of the influence of the deuterium content on the triple point of hydrogen. Because it is known that ITS-90 deviates by about 1 mK from the thermodynamic temperatures at $T < 1$ K [11.6–11.18], very careful new ^{3}He vapor-pressure measurements have been performed at PTB from 0.65 to 1.2 K [11.9]. In these investigations, different vapor-pressure cells and different pressure-sensing tubes between the sample and the manometers at room temperature have been used. This allowed reducing the errors from heat input to the samples, and in particular from thermo-molecular pressure differences as well as from aerostatic head corrections (see Sect. 12.2). Pressures were measured at room temperature with calibrated commercial differential capacitive diaphragm gauges and temperatures were measured with calibrated rhodium-iron resistance thermometers

Table 11.3. Helium-3 vapour pressure (kPa) according to ITS-90 [11.4, 11.5]

T (K)	0.00	0.01	0.02	0.03	0.04	0.05	0.06	0.07	0.08	0.09
0.6	0.071	0.079	0.087	0.096	0.105	0.116	0.127	0.139	0.152	0.166
0.7	0.180	0.195	0.211	0.229	0.247	0.267	0.287	0.308	0.330	0.353
0.8	0.378	0.404	0.431	0.459	0.489	0.520	0.552	0.586	0.621	0.657
0.9	0.695	0.734	0.775	0.817	0.861	0.907	0.954	1.003	1.054	1.106
1.0	1.160	1.216	1.274	1.333	1.395	1.459	1.523	1.590	1.660	1.731
1.1	1.804	1.880	1.957	2.037	2.118	2.202	2.288	2.376	2.466	2.559
1.2	2.654	2.752	2.851	2.954	3.059	3.165	3.275	3.387	3.501	3.618
1.3	3.738	3.860	3.985	4.112	4.242	4.375	4.511	4.649	4.790	4.934
1.4	5.081	5.231	5.383	5.538	5.697	5.858	6.022	6.189	6.360	6.533
1.5	6.709	6.889	7.071	7.257	7.446	7.638	7.834	8.033	8.235	8.440
1.6	8.649	8.861	9.076	9.295	9.517	9.742	9.972	10.20	10.44	10.68
1.7	10.92	11.17	11.42	11.68	11.93	12.19	12.46	12.73	13.00	13.28
1.8	13.56	13.84	14.13	14.42	14.72	15.02	15.32	15.63	15.94	16.26
1.9	16.58	16.99	17.23	17.56	17.90	18.24	18.58	18.93	19.28	19.64
2.0	20.00	20.37	20.74	21.11	21.49	21.87	22.26	22.65	23.05	23.45
2.1	23.85	24.26	24.68	25.10	25.52	25.95	26.38	26.82	27.26	27.71
2.2	28.16	28.61	29.08	29.54	30.01	30.49	30.97	31.45	31.94	32.44
2.3	32.94	33.44	33.95	34.47	34.99	35.51	36.04	36.58	37.12	37.67
2.4	38.22	38.77	39.33	39.90	40.47	41.05	41.63	42.22	42.82	43.41
2.5	44.02	44.63	45.24	45.87	46.49	47.12	47.76	48.40	49.05	49.71
2.6	50.37	51.04	51.71	52.38	53.07	53.76	54.45	55.15	55.86	56.58
2.7	57.29	58.02	58.75	59.49	60.23	60.98	61.74	62.50	63.27	64.04
2.8	64.82	65.61	66.41	67.21	68.01	68.83	69.65	70.47	71.30	72.14
2.9	72.99	73.84	74.70	75.57	76.44	77.32	78.21	79.10	80.00	80.91
3.0	81.83	82.75	83.68	84.61	85.96	86.51	87.46	88.43	89.40	90.38
3.1	91.37	92.36	93.37	94.38	95.39	96.42	97.45	98.49	99.54	100.60
3.2	101.66	102.73	103.82	104.90	106.00	107.10	108.22	109.34	110.47	111.61

(see Sect. 12.5.1). The data with a standard uncertainty of about 0.1 mK confirm the above-mentioned deviation and the requirement to amend ITS-90 to bring it in agreement with the new low-temperature scale PLTS-2000 (see below).

11.3 The New Provisional Low-Temperature Scale PLTS-2000

The International Temperature Scale ITS-90 has its lower end at 0.65 K. However, important research is being performed at substantially lower temperatures, requiring an internationally accepted, thermodynamically consistent temperature scale down to at least 1 mK. For this purpose, a new low-temperature scale, the PLTS-2000 [11.10] has been introduced in January 2000 and was formally accepted by the Comite' International des Points et

Table 11.4. Helium-4 vapour pressure (kPa) according to ITS-90 [11.4, 11.5]

T (K)	0.00	0.01	0.02	0.03	0.04	0.05	0.06	0.07	0.08	0.09
1.2	0.082	0.087	0.093	0.100	0.107	0.115	0.123	0.131	0.139	0.148
1.3	0.158	0.168	0.178	0.189	0.201	0.213	0.226	0.239	0.252	0.267
1.4	0.282	0.298	0.314	0.331	0.348	0.367	0.387	0.407	0.428	0.449
1.5	0.472	0.495	0.519	0.544	0.570	0.597	0.625	0.654	0.684	0.715
1.6	0.747	0.780	0.814	0.849	0.885	0.922	0.961	1.001	1.042	1.084
1.7	1.128	1.173	1.219	1.266	1.315	1.365	1.417	1.470	1.525	1.581
1.8	1.638	1.697	1.758	1.820	1.883	1.948	2.015	2.084	2.154	2.226
1.9	2.299	2.374	2.451	2.530	2.610	2.692	2.776	2.862	2.949	3.039
2.0	3.130	3.223	3.317	3.414	3.512	3.613	3.715	3.818	3.925	4.032
2.1	4.141	4.253	4.366	4.481	4.597	4.716	4.836	4.958	5.082	5.207
2.2	5.335	5.465	5.597	5.731	5.867	6.005	6.146	6.288	6.433	6.580
2.3	6.730	6.882	7.036	7.192	7.351	7.512	7.675	7.841	8.009	8.180
2.4	8.354	8.529	8.708	8.889	9.072	9.258	9.447	9.638	9.832	10.03
2.5	10.23	10.43	10.64	10.84	11.05	11.27	11.48	11.70	11.92	12.15
2.6	12.37	12.60	12.84	13.07	13.31	13.55	13.80	14.05	14.30	14.55
2.7	14.81	15.07	15.33	15.60	15.87	16.14	16.42	16.70	16.98	17.26
2.8	17.55	17.84	18.14	18.44	18.74	19.05	19.36	19.67	19.98	20.30
2.9	20.63	20.95	21.28	21.61	21.95	22.29	22.64	22.98	23.33	23.69
3.0	24.05	24.41	24.77	25.14	25.52	25.89	26.27	26.66	27.05	27.44
3.1	27.84	28.24	28.64	29.05	29.46	29.87	30.29	30.72	31.14	31.58
3.2	32.01	32.45	32.89	33.34	33.79	34.25	34.71	35.17	35.64	36.11
3.3	36.59	37.07	37.56	38.05	38.54	39.04	39.54	40.05	40.56	41.08
3.4	41.60	42.12	42.65	43.18	43.72	44.26	44.81	45.36	45.92	46.48
3.5	47.05	47.62	48.19	48.77	49.35	49.94	50.54	51.13	51.74	52.35
3.6	52.96	53.57	54.20	54.82	55.46	56.09	56.73	57.38	58.03	58.69
3.7	59.35	60.02	60.69	61.37	62.05	62.73	63.43	64.12	64.83	65.53
3.8	66.25	66.96	67.69	68.41	69.15	69.89	70.63	71.38	72.14	72.90
3.9	73.66	74.43	75.21	75.99	76.78	77.57	78.37	79.17	79.98	80.80
4.0	81.62	82.44	83.27	84.11	84.95	85.80	86.66	87.52	88.38	89.26
4.1	90.13	91.02	91.91	92.80	93.70	94.61	95.52	96.44	97.37	98.30
4.2	99.23	100.18	101.13	102.08	103.04	104.01	104.98	105.96	106.95	107.94
4.3	108.94	109.94	110.95	111.97	113.00	114.03	115.06	116.11	117.15	118.21
4.4	119.27	120.34	121.42	122.50	123.51	124.68	125.79	126.89	128.01	129.13
4.5	130.26	131.40	132.54	133.69	134.84	136.01	137.18	138.36	139.54	140.73
4.6	141.93	143.13	144.35	145.57	146.79	148.03	149.27	150.52	151.77	153.04
4.7	154.31	155.58	156.87	158.16	159.46	160.77	162.09	163.41	164.74	166.08
4.8	167.42	168.78	170.14	171.51	172.89	174.27	175.66	177.07	178.47	179.89
4.9	181.32	182.75	184.19	185.64	187.10	188.56	190.04	191.52	193.01	194.51
5.0	196.08	197.53	199.06	200.59	202.13	203.68	205.24	206.81	208.39	209.97

Mesures in October 2000. It is based on thermodynamic properties of liquid and solid ^{3}He, in particular its melting curve.

In the 1990s, several low-temperature laboratories had advanced the precision and reliability by which the P–T relation of a mixture of solid and liquid ^{3}He could be measured over more than three decades in temperature, from about 1 mK to about 1 K (Sect. 12.3). The melting pressure of ^{3}He was chosen because it is a thermodynamic property usable over a very wide temperature range, complications from impurities and imperfections in the substance are essentially absent, thermodynamic consistency checks are possible (see below), and because of the relative ease and reliability by which it can be unequivocally realized in laboratories around the world. In addition, liquid and solid ^{3}He due to its outstanding superfluid and nuclear magnetic properties provide four unique fixed points along the melting curve (see Fig. 2.4): the minimum of the melting pressure at $P_{min} = 2.93113$ MPa, allowing a calibration of the used pressure transducer; the two unique transitions to the superfluid A-phase and from the A- to the superfluid B-phase of liquid ^{3}He as well as the nuclear antiferromagnetic transition of solid ^{3}He offer temperature calibration points. The values for these temperature-pressure fixed-points are given in Table 11.5.

Because of the melting pressure minimum, a plug of solid ^{3}He will form in the sensing line between a pressure transducer at room temperature and the sample at low temperatures at a point where $T(P_{min})$ is reached first, isolating the pressure cell. Therefore, the pressure has to be measured in situ at low temperatures, usually by a low-temperature capacitive transducer as described in Sect. 13.1. The transducer has to be calibrated at a temperature above the one of the melting curve minimum (or at another fixed point of liquid and solid ^{3}He mentioned above). The plug, of course, leads to a very good stability of the low-temperature setup. More information on helium melting pressure thermometry can be found in Sect. 12.3.

The use of the ^{3}He melting pressure as a temperature standard was already proposed by Scribner and Adams in 1970 [11.11]. It was in full use at Cornell University in the late 1970s [11.12], and later substantially refined by Greywall [11.13]. The work leading eventually to the ^{3}He melting pressure scale PLTS-2000 [11.10] was mostly performed in the nineties by the following three laboratories:

Physikalisch-Technische Bundesanstalt (PTB), Berlin; scale PTB-96 [11.6, 11.14–11.16]:

The scale of this laboratory is based in particular on noise thermometry using a SQUID as detector (Sect. 12.7). The data have been obtained on a 20-μΩ resistor in five measurements over 11 h with a gate time τ=20 ms from 0.88 mK to 1 K. The obtained noise temperatures were substantiated by measurements of the susceptibility of a single crystal of the paramagnetic salt

Table 11.5. Fixed points of liquid and solid ^{3}He according to the PLTS-2000 scale [11.10]

minimum of melting curve	2.93113 MPa	315.24 mK
superfluid A-transition at P_{melt}	3.43407 MPa	2.444 mK
superfluid B-transition at P_{melt}	3.43609 MPa	1.896 mK
nuclear magn. transition of solid ^{3}He at P_{melt}	3.43934 MPa	0.902 mK

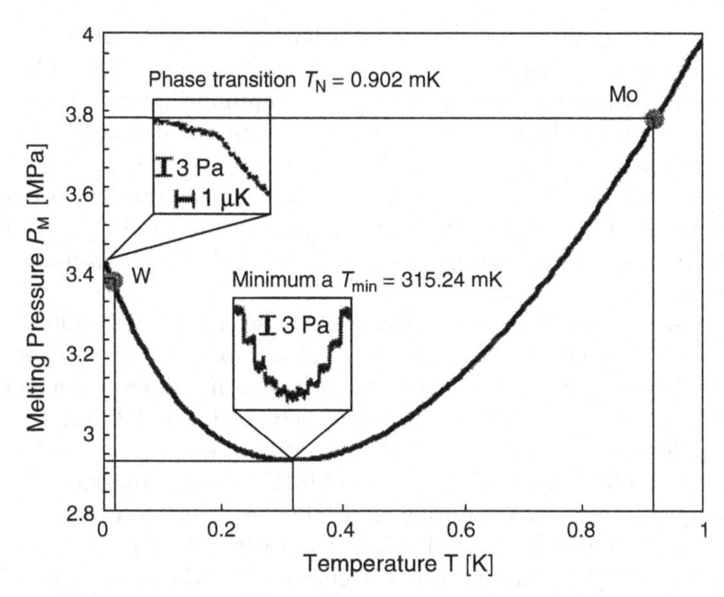

Fig. 11.1. Melting pressure of ^{3}He as a function of temperature. The insets demonstrate the high resolution of the pressure measurements performed by a capacitive strain gauge similar to the one shown in Fig. 13.2. In the inset for the pressure minimum, the pressure is plotted vs. time. T_N is the temperature of the nuclear antiferromagnetic transition of solid ^{3}He, whereas W and Mo indicate the superconducting transitions of these metals taken as reference values. Reprinted from [11.6], publ. by IOP Publishing. See also [11.16]

CMN (Sect. 12.9) from 23 mK to 1.2 K, as well as by pulsed nuclear magnetic resonance on ^{195}Pt at 250 kHz/28.5 mT in a pure platinum sample with the rather long spin-lattice relaxation time of 0.9 ms (see Sect. 12.10.3) from 0.9–50 mK. For the nuclear magnetic moments of platinum it is know that they follow a Curie law into the low microkelvin range (see Fig. 10.18). The ^{3}He pressure was measured by a capacitive transducer (Sect. 13.1) at low temperatures with a resolution of 0.1 Pa at 3 MPa (Fig. 11.1). It was calibrated by a room temperature quartz oscillator pressure transducer and a pressure balance carrying a calibration against the national pressure standard of PTB. The standard uncertainties are 40 Pa for the pressure at the melting curve minimum, and 1(0.04)% in temperature at 1 mK(1 K). A detailed description of the design, construction, installation, filling to the correct gas pressure, the room temperature part, calibration, and use of the PTB pressure transducers can be found in [11.16].

National Institute of Standards and Technology (NIST), Boulder, Colorado; scales NIST-92/-98 [11.17, 11.18]:

The temperature scale of this laboratory, too, is based on resistive SQUID noise thermometry used as a primary method from 6–738 mK. It was checked from 1.25–3.0 K against the Curie-Weiss law of the susceptibility of CMN, which was calibrated against the ITS-90, as well as against a ^{60}Co nuclear orientation thermometer from 7–22 mK. The reproducibility and accuracy of the data was 0.1%. The PTB and NIST temperature scales agreed within experimental error after the original NIST temperatures were reduced by 0.15% when an error in the calculation of the noise temperature was discovered. The dominant uncertainty in the NIST scale is the uncertainty of 15 ppm in the effective area of the piston pressure gauge used for establishing the ^{3}He pressure scale.

University of Florida, Gainesville, Florida; scale UF-95 [11.19]:

This group used the anisotropic emission of the γ-rays from ^{60}Co (Sect. 12.11) as a primary thermometer with counting times of 8 h at each of 63 T–P points along the ^{3}He melting curve between 0.5 and 25 mK. This thermometer did serve to calibrate a ^{195}Pt wire NMR thermometer (Sect. 12.10.3) operated at 3.427 MHz with a precision of 0.2% achieved by signal averaging over the above-mentioned counting time; this latter temperature sensor was used only below 7 mK. The stated accuracy of their scale at T >0.93 mK is 1%. A quartz pressure transducer with an accuracy of 0.04% was used to calibrate the capacitive pressure transducer at low temperatures. A small correction of the minimum pressure value toward that of the NIST value improved the pressure scale.

The remaining differences between the scales of the three laboratories and their likely origins were examined in a collaboration between the involved laboratories. The consistency checks relied in particular on the application of thermodynamics to the melting pressure-temperature relation. This analysis followed essentially the reasoning of Greywall [11.20] who had calculated the melting pressure by using the Clausius-Clapeyron equation and his data for the heat capacities of liquid and solid ^{3}He (see below). Eventually, agreement on slight adjustments of the various scales as well as on a compromise equation for the melting pressure of ^{3}He was reached, and the PLTS-2000 scale was subsequently formulated. This equation is (for P in MPa)

$$P = \sum_{i=-3}^{+9} a_i T^i \tag{11.7}$$

at 0.9 mK < T <1 K, and with the coefficients a_i given in Table 11.6 [11.10]. In addition, the mentioned fixed points of liquid and solid ^{3}He at low millikelvin temperatures (Table 11.5) are incorporated in the PLTS-2000. The standard uncertainty of the PLTS-2000 in thermodynamic terms is 0.5 mK at $T > 0.5$ K, decreasing linearly to 0.2 mK at 100 mK. Eventually, it is about 0.3% at 25 mK and 2% at 0.9 mK. The standard uncertainty in pressure is about 60 Pa. The consistency with the ITS-90 temperatures is 0.3 mK at 1 K.

Work related to improvements of the PLTS-2000 scale has still been going on since its acceptance. At PTB, the thermodynamic consistency of the scale

Table 11.6. Coefficients a_i of (11.7) for the ^{3}He melting curve according to the Provisional Temperature Scale PLTS-2000 (in the units MPa; K) [11.10]

$a_{-3} =$	$-1.385\ 544\ 2 \times 10^{-12}$
$a_{-2} =$	$4.555\ 702\ 6 \times 10^{-9}$
$a_{-1} =$	$-6.443\ 086\ 9 \times 10^{-6}$
$a_0 =$	$3.446\ 743\ 4 \times 10^{0}$
$a_1 =$	$-4.417\ 643\ 8 \times 10^{0}$
$a_2 =$	$1.541\ 743\ 7 \times 10^{1}$
$a_3 =$	$-3.578\ 985\ 3 \times 10^{1}$
$a_4 =$	$7.149\ 912\ 5 \times 10^{1}$
$a_5 =$	$-1.041\ 437\ 9 \times 10^{2}$
$a_6 =$	$1.051\ 853\ 8 \times 10^{2}$
$a_7 =$	$-6.944\ 376\ 7 \times 10^{1}$
$a_8 =$	$2.683\ 308\ 7 \times 10^{1}$
$a_9 =$	$-4.587\ 570\ 9 \times 10^{0}$

has been checked with new measurements using noise and platinum NMR thermometry [11.21], supported again by calculations using Greywall's specific heat data [11.20]. These investigations indicated that the deviation of PLTS-2000 from thermodynamic consistency with the specific heat data is of order 1%.

Substantial efforts have been undertaken within an EU project to disseminate PLTS-2000 in Europe [11.22]. This has led to instrumentation and several thermometers for this purpose: a current sensing noise thermometer (Sect. 12.7), a CMN susceptibility thermometer (Sect. 12.9), new ^{195}Pt and ^{3}He NMR thermometry devices (Sect. 12.10), a Coulomb blockade thermometer (Sect. 12.6), a ^{3}He–^{4}He second-sound acoustic thermometer (T obtained from the strong temperature dependence of the velocity of second sound in a 1% mixture) [11.23], the superconducting fixed-point device SRD 1000 (Sect. 11.4.3), as well as ruthenium oxide resistance thermometers (Sect. 12.5.3). They are being checked as secondary devices and fixed points against the noise and the melting curve thermometers.

I shall now reproduce the arguments given by Greywall [11.20] to demonstrate the thermodynamic consistency of his temperature scale and of the input data. Because in the beginning of the 1980's there were differences of up to 40% in the specific heat of liquid ^{3}He measured at various laboratories, he performed accurate measurements of this specific heat in the temperature range 7 mK–2.5 K. From these data he was able to calculate the entropy of liquid ^{3}He along the melting curve

$$S_{\mathrm{liq,m}} = \int_0^{T_\mathrm{m}} \left(\frac{C_{\mathrm{v,liq}}}{T'} \right)_\mathrm{m} \mathrm{d}T' \,. \tag{11.8}$$

With the Clausius-Clapeyron equation

$$S_{\mathrm{sol,m}} = S_{\mathrm{liq,m}} - (V_{\mathrm{liq,m}} - V_{\mathrm{sol,m}}) \left(\frac{\mathrm{d}P}{\mathrm{d}T} \right)_{\mathrm{m}}, \tag{11.9}$$

and the known values of the slope of the melting curve [11.13] and the difference of the molar volumes of ^{3}He in the liquid and solid phases along the melting curve ($1.31\,\mathrm{m^3\,mol^{-1}}$) [11.12, 11.24], he was able to calculate the entropy of solid ^{3}He along the melting curve. This calculation gave the values (Fig. 11.2)

$$S_{\mathrm{sol,m}}/R = 0.688(0.693) \text{ at } T = 0.02(0.32)\,\mathrm{K}. \tag{11.10}$$

These values are in excellent agreement with the value which is expected between about 10 mK (which is substantially above the nuclear magnetic ordering temperature of 0.93 mK) and 0.32 K (which is below a temperature where other excitations, e.g., phonons, start to contribute), which would be

$$S_{\mathrm{sol,m}}/R = \ln(2) = 0.693 \tag{11.11}$$

for the fully disordered ^{3}He nuclear spins.

These are the same arguments used by the above-mentioned three groups to check the thermodynamic consistency of their new scale. Greywall's temperature scale [11.13] was the most accurate and most widely used temperature scale for the millikelvin temperature range until the PLTS-2000 scale was established. It is in good agreement (to within about 2%) with the PLTS-2000 scale at T<40 mK. For his measurements, he used a calibrated superconducting fixed-point device SRM 768 (Sect. 11.4.2) to calibrate an LCMN susceptibility thermometer (Sect. 12.9), ensuring that on this temperature scale the

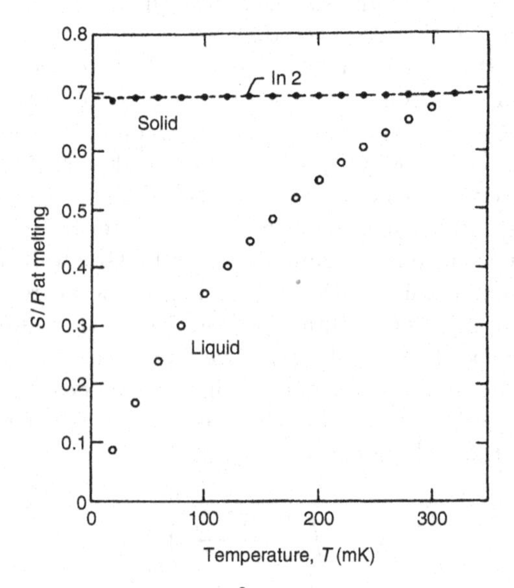

Fig. 11.2. Entropies of liquid and solid ^{3}He along the melting curve calculated from measured specific heat data of liquid ^{3}He [11.20]

specific heat of liquid ^{3}He, which he measured, would follow at low temperatures the linear law predicted by the Landau theory for a Fermi liquid at $T \ll T_{\mathrm{Fermi}}$ (see Fig. 2.15).

11.4 Practical but not Officially Accepted Low-Temperature Fixed Points

There are various very useful temperature fixed points – besides the unique fixed points of liquid and solid ^{3}He at low millikelvin temperatures on the ^{3}He melting curve (Table 11.5) - for low-temperature thermometry which are in wide use but which are not officially accepted or officially recommended.

11.4.1 Fixed Points of EPT-76

Even though the low-temperature scale EPT-76 [11.2, 11.3] is not an official scale anymore, it contains a number of fixed points which are very useful for low-temperature thermometry. These fixed points are given in Table 11.7. They differ from ITS-90 by less than 1 mK at $T < 13$ K, which is close to the uncertainty of EPT-76 at these temperatures. In detailed comparisons between the Curie–Weiss behavior of the susceptibility of powdered CMN (Sect. 12.9) and EPT-76 as well as ITS-90 (which have the same thermodynamic basis below 4 K), consistency was found between 3 and 1.2 K; below 1.2 K the differences increase monotonically with decreasing temperature and reach 1.3 mK at 0.5 K.

11.4.2 The NBS Superconducting Fixed-Point Device

Superconducting transitions are very suitable fixed points for low-temperature thermometry, owing to their ease of detection, their reproducibility, their accuracy and fast response time, if the precautions discussed below are taken into account.

The American National Bureau of Standard (NBS; now National Institute of Standards and Technology, NIST) did offer a device (NBS-SRM 767 a) containing the five superconductors with transitions between 0.5 and 7.2 K of the EPT-76 temperature scale, as well as Nb with $T_c = 9.3$ K [11.25].

Table 11.7. Some fixed points (in K), mostly from the superseded low-temperature scale EPT-76 [11.2, 11.3]

T_b (H_2)[a]	T_{tr} (H_2)	T_c (Pb)	T_b (^{4}He)[a]	T_c (In)	T_λ (^{4}He)[b]	T_c (Al)	T_c (Zn)	T_c (Cd)
20.27	13.80	7.20	4.21	3.415	2.1768	1.180	0.851	0.520

[a] At a pressure of 1 bar.

[b] At saturated vapour pressure.

However, thermometry in the Kelvin range is not too great a problem and is widely performed with the helium vapour pressure scale.

For a number of years the NBS (NIST) also offered a very useful superconducting fixed-point device for the millikelvin temperature range (NBS-SRM 768) [11.26]. The temperatures for the superconducting transitions of five metals contained in this device together with other relevant information are given in Table 11.8. The temperature scale of this device is based on a scale called "NBS-CTS-1" established at 0.01–0.5 K by the American NBS by ^{60}Co γ-ray anisotropy (Sect. 12.11), by Josephson noise thermometry (Sect. 12.7), and the paramagnetic susceptibility of CMN (Sect. 12.9) [11.27]. Each device is individually calibrated. The reproducibility and traceability to the NBS temperature scale is about 0.1–0.2 mK. This device is very useful for calibrating other millikelvin thermometers and it is a simple standard, widely used in low-temperature laboratories working in the millikelvin temperature range.

The design of this device, a self-contained assembly of coils and samples, is shown in Fig. 11.3. Thermal contact between the samples and the Cu body is provided by Cu wires welded to the body and connected to the samples by varnish. The device contains two primary coils to generate a magnetic field ($B = 0.016 \cdot I$) and two secondary coils to measure the AC susceptibility of the samples; they are wound directly on the primary coils. The coils are connected in series opposition, so that only four leads are required. When a sample enters the superconducting state it expels the magnetic field (Meissner effect) generated by the primary coil, which is observed as a change in the mutual inductance of the secondary coils. A typical electronic setup to measure

Table 11.8. Properties of the superconductors in the NBS superconducting fixed point device SRM 768 [11.26]

material	T_c (mK)	B_c (mT)	RRR
W	15.5–15.6	0.12	10^3
Be	22.6–22.8	0.114	79
$Ir_{0.8}Ru_{0.2}$	99–100	(type II)	2.5
$AuAl_2$	159–161	1.21	50
$AuIn_2$	203–206	1.45	50

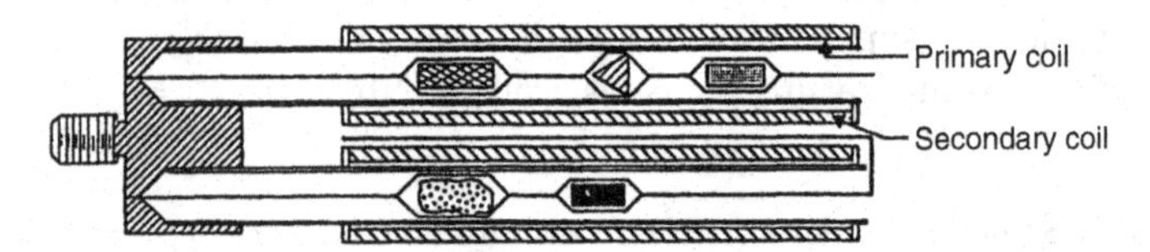

Fig. 11.3. Schematic of the superconducting fixed-point device SRM 768 of the NBS (now NIST) containing five superconducting samples as well as the primary and secondary coils for measuring the transition temperatures of these metals in a Cu holder [11.26]

the superconducting transitions is shown in Fig. 11.4; the generated signals are 0.1–1 μV.

The NBS (NIST) has put substantial effort into the selection of samples with reproducible, sharp superconducting transitions and into investigations of external influences on these transitions, in particular by a magnetic field. The influence of an external magnetic field on the transition of Be is shown in Fig. 11.5. An external magnetic field shifts the superconducting transition to lower temperatures. Even more important, in an external magnetic field the superconducting transition is a first-order phase transition. Therefore supercooling effects can occur for pure elements, as is shown for pure Be. In the device this effect is reduced by spot-welding small pieces of Al to the W and Be samples so that Al, with $T_c \simeq 1\,\mathrm{K}$, serves as a nucleation center to induce superconductivity in the samples by the proximity effect. Because the

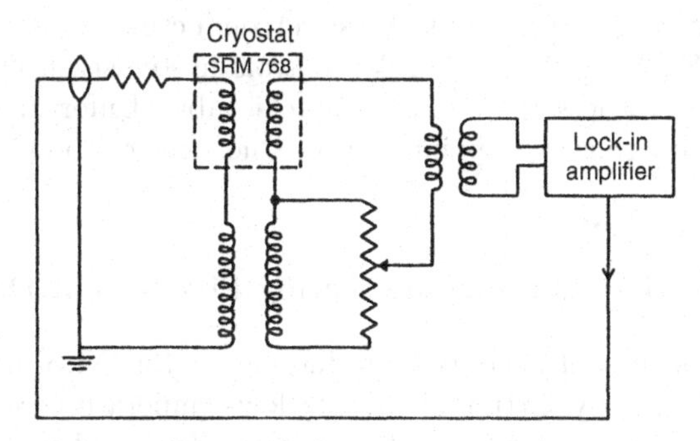

Fig. 11.4. Mutual inductance bridge for detecting the superconducting transitions of a superconducting fixed-point device like the SRM 768 shown in Fig. 11.3

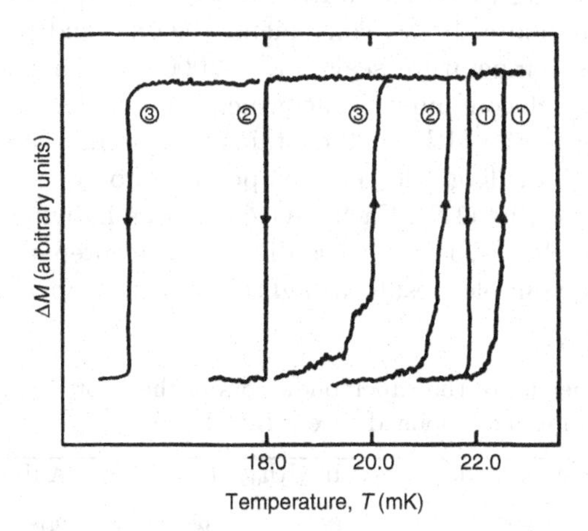

Fig. 11.5. Superconducting transitions of pure Be indicating the influence of a magnetic field. The data in cooling and warming are taken in fields of 0.5 μT (1), 9.5 μT (2) and 19 μT (3) [11.26]

earth's magnetic field (about 50 μT) is substantially larger than fields which have a substantial influence on some of the transitions of the metals in this device, one has to shield carefully (to less than 10^{-7} T) against the earth's magnetic field. This shielding is of course even more important in an experimental setup where large fields are applied, as in a magnetic refrigerator. I shall discuss magnetic field shielding in Sect. 13.5.

The experimental conditions recommended by the NBS (NIST) for use of their device are:

(a) ambient magnetic field: ≤ 1 μT
(b) peak-to-peak field in primary coil: ≤ 2.3 μT for tungsten, ≤ 0.40 μT for the other superconductors (Joule and eddy current heating: 1.8 and 0.08 nW, respectively).

The reproducibility and the shifts in T_c in a 1 μT field are about 0.1 mK. The sweep rate used to trace out the superconducting transitions should be less than 0.1 mK min^{-1} (for Be and W) to avoid hysteresis effects; then widths of the transitions of less than 1 mK are achievable. Unfortunately this very useful superconducting fixed point device is no longer produced by the NBS (NIST).

11.4.3 The SRD 1000 Superconducting Fixed-Point Device

Since NIST has discontinued the production of the NBS-SRM 768 fixed-point device, it is very fortunate for the low-temperature community that the Dutch company Hightech Developments Leiden, Leiden, NL in cooperation with Leiden University and PTB Berlin has developed and is offering the new Superconducting Fixed-Point Device SRD 1000 [11.28, 11.29]. Its main purpose is to provide a direct, convenient, reliable, and practical means to transfer the new temperature scale PLTS-2000 (see Sect. 11.3) to the low-temperature research community. The principle of the device is identical to the SRM 768. However, the SRD 1000 contains ten carefully prepared superconducting samples providing reference temperatures between 15 and 1,200 mK (see Table 11.9 and Fig. 11.6). Each device has been individually calibrated at PTB against the PLTS-2000 because differences in preparation and in inhomogeneities of the samples result in slightly different transition temperatures.

Table 11.9. Properties of the superconductors in the pilot production of the SRD 1000 superconducting fixed-point device [11.28, 11.29]

material	W	Be	$Ir_{80}Rh_{20}$	$Ir_{92}Rh_{08}$	Ir	$AuAl_2$	$AuIn_2$	Cd	Zn	Al
nominal T_c(mK)	15	21	30	65	98	145	208	520	850	1,180
90/10% width (mK) <	0.2	0.3	0.5	0.5	0.5	0.5	1	4	3	4

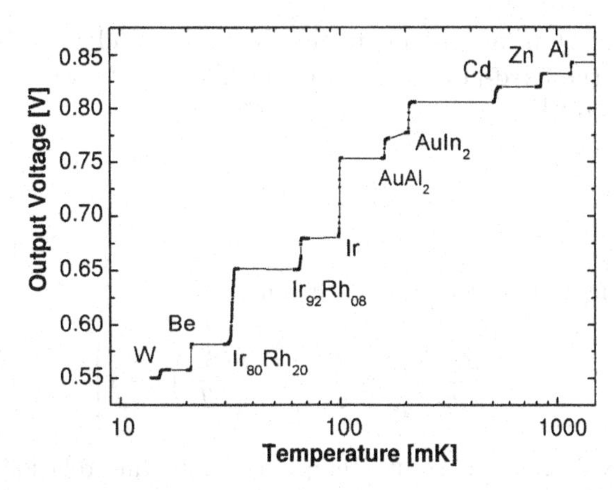

Fig. 11.6. Inductively measured superconducting transitions of the ten samples in the superconducting fixed-point device SRD 1000 of [11.28–11.30]

The change in magnetization of each sample is detected by planar microcoils, realized by a thin-film niobium structure on a silicate wafer. The device is provided with the low-temperature sensor (10 mm diam., 50 mm length), a degaussing system for the magnetic shields (see Sect. 13.5), a preamplifier, and a phase-sensitive mutual inductance measurement system giving a DC output signal with a staircase pattern (Fig. 11.6).

Prototypes of the device have been evaluated by several European institutes for metrology [11.30]. The results confirm that (1) the uncertainties in the transition temperatures are about 0.1–0.3%, or less than 0.2 mK for the samples with the lower transition temperatures and between 1 and 4 mK for Cd, Zn, and Al; (2) the field dependence of the transition temperatures is about −0.1 mK/μT. Magnetic shielding by a Cryoperm magnetic shield and a superconducting niobium shield degaussed in situ before the measurement (see Sect. 13.5) is very effective (attenuation factor higher than 500), so that depressions of T_c due to residual and measuring fields (AC-field with 0.3 μT amplitude) can be neglected.

11.4.4 The Superfluid Transition of Liquid ^{4}He

The superfluid transition of pure ^{4}He at saturated vapour pressure provides a very sharp fixed-point reference for thermometry at 2.1768 K (see Fig. 2.10). Self-contained, sealed fixed-point devices, based on the rapid change of thermal conductivity at the superfluid transition of ^{4}He, have been described in [11.31–11.33]. They can have nanokelvin resolution, are virtually immune to drift, and almost unaffected by magnetic fields ($\mathrm{d}T_\lambda/\mathrm{d}B \approx 0.3\,\mu\mathrm{K/T}$). However, the transition depends weakly on pressure ($\mathrm{d}T_\lambda/\mathrm{d}P = -8.78\,\mathrm{mK/bar}$), and therefore – via gravity – on sample height ($\mathrm{d}T_\lambda/\mathrm{d}h = 1.27\,\mu\mathrm{K/cm}$), somewhat on ^{3}He impurity concentration x ($\mathrm{d}T_\lambda/\mathrm{d}x = -1.45\,\mu\mathrm{K/ppm}$), and, above all, on heat flux Q, primarily due to the thermal boundary resistances of the device [11.31]; the fixed-point temperature should be taken by extrapolation to $Q = 0$ at 2.1768 K.

Problems

11.1. The Joule–Kelvin coefficient is given by

$$\mu \equiv \left(\frac{\partial T}{\partial p}\right)_H = \frac{V}{C_p}\left[\frac{T}{V}\left(\frac{\partial V}{\partial T}\right)_p - 1\right] . \tag{11.12}$$

Since it involves the absolute temperature T, this relation can be used to determine the absolute temperature T. Consider any readily measurable arbitrary temperature parameter ϑ (e.g., the height of a mercury column). All that is known is that ϑ is some (unkown) function of T, i.e., $\vartheta = \vartheta(T)$.

(a) Express (11.12) in terms of the various directly measurable quantities involving the temperature parameter ϑ instead of the absolute temperature T, i.e., in terms of $\mu' \equiv (\partial\vartheta/\partial p)_H$, $C_p' \equiv (\partial Q/\mathrm{d}\partial)_p$, $\alpha' \equiv \mathrm{v}^{-1}(\partial V/\partial\vartheta)_p$, and the derivative $\mathrm{d}\partial/\mathrm{d}T$.
(b) Show that by integrating the resulting expression, one can find T for any given value of ϑ if one knows that $\vartheta = \vartheta_0$ when $T = T_0$ (e.g., if one knows the value of $\vartheta = \vartheta_0$ at the triple point where $T_0 = 273.16\,\mathrm{K}$).

11.2. Perform a qualitative calculation of the deviation of the actual vapour pressure curve of liquid ^{4}He, as given by (11.6) and Table 11.4, to the approximate equation deduced in Sect. 2.3.2.

11.3. Calculate the heat of magnetization at the superconducting transitions of the metals mentioned in Table 11.8, if the transitions occur in a field of 1 µT, see also Problem 4.2. How much are the superconducting transitions of these metals shifted by a field of 1 µT?

12

Low-Temperature Thermometry

A thermometer is a device by which we can measure a property of matter which is connected via a physical law to the concept of "temperature", where the latter is defined by thermodynamics. Therefore knowledge about the obtained temperature depends both on the quality of the measurement and on the theory on which this physical law is based. In general, thermometers can be divided into two groups.

For *primary thermometers* our theoretical knowledge about the measured property of matter is good enough to calculate the thermodynamic temperature from it without any calibration. Examples of primary thermometers are gas thermometers (using the relation between pressure, volume and temperature for gases) (Sect. 12.1), measurements of the velocity of sound in a gas [$v = (\kappa\rho)^{-1/2}$], the Coulomb blockade of electrons in tunnel junctions (see Sect. 12.6), thermal noise of an electrical resistor (using the Nyquist equation, see Sect. 12.7), and the angular anisotropy of gamma rays emitted from radioactive nuclei (using the Boltzmann population of the various hyperfine levels, see Sect. 12.11). Usually primary thermometry is difficult and mostly left to specialized national standards laboratories. Therefore the *secondary thermometers* are the "workhorses" of thermometry in research laboratories. For these thermometers, theory is not at a stage such that we can calculate the temperature directly from the measured property. Secondary thermometers have to be calibrated with a primary thermometer and/or at fixed points, as discussed in Chap. 11. Secondary thermometers are often much more sensitive than primary thermometers, and usually they are much more convenient to use. Examples will be discussed in this chapter.

There are various requirements which a useful thermometer has to fulfil:

- The thermometer should have a wide operating temperature range and should be insensitive to environmental changes, such as magnetic fields
- The property x to be measured must be easily, quickly, reproducibly, and exactly accessible to an experiment

- The temperature dependence of the measured property, $x(T)$, should be expressible by a reasonably simple law
- The sensitivity $(\Delta x/x)/(\Delta T/T)$ should be high
- The thermometer should reach equilibrium in a "short" time, both within itself and with its surroundings whose temperature it is supposed to measure. Therefore it should have a small heat capacity, good thermal conductivity and good thermal contact to its surroundings. In particular, the thermal contact problem is ever present for thermometry at $T \leq 1\,\mathrm{K}$.
- The relevant measurement should introduce a minimum of heat to avoid heating of the surroundings of the thermometer and, of course, above all, heating of itself; this becomes more important the lower the temperature.

There are many possible choices for a thermometer, and one of the first considerations has to be to select the one most appropriate for a particular experiment and temperature range. The recent achievements in refrigeration have required corresponding progress in thermometry. The substantial advancements in this field are reflected by the enormous number of papers on low-temperature thermometry published in journals such as the *Journal of Low Temperature Physics, Review of Scientific Instruments* and *Cryogenics.* It is impossible to give sufficient credit to all this work. Additional information can be found in [12.1–12.11].

The oldest methods of thermometry use thermometers based on the expansion of gases, liquids or solids when the temperature is changed. I discuss one of them here in Sect. 12.1 of this chapter.

12.1 Gas Thermometry

The gas thermometer [12.1, 12.2, 12.8–12.11, 12.14], in principle, is a primary thermometer based on the relation for an ideal gas

$$PV = nRT\,, \tag{12.1}$$

assuming that there are no interactions between the gas molecules and that their own volume is negligible. Real gases, of course, deviate from the behavior of an ideal gas, and therefore corrections to this equation ("virial coefficients") have to be applied. The measurement can be performed with a fixed amount of gas either at constant pressure or at constant volume. Figure 12.1 schematically illustrates the setups for these two procedures.

One of the requirements for gas thermometry is that the dead volumes (valves, tubes, manometers) are small and hopefully constant, because one has to correct for them. In addition, the thermal and elastic volume changes of the various components, and absorption and desorption of gas from the container walls have to be taken into account and, last but not least, the deviations from the ideal gas behavior, which are particularly important at low temperatures, must be known. In general, high-precision gas thermometry

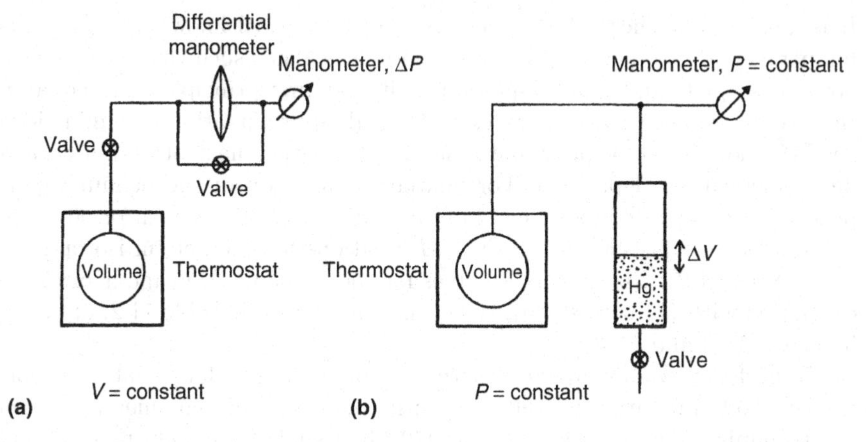

Fig. 12.1. Schematic setups for gas thermometry at (**a**) **V** = const. and (**b**) **P** = const.

is very tedious and therefore the gas thermometer is mostly used by national laboratories for calibration purposes or to establish the temperature scale. As discussed in Sect. 11.2, it is one of the devices used to establish the ITS-90, and that is the main reason for mentioning it here.

Other versions of gas thermometers are the acoustic gas thermometer (using the relation for the velocity of sound $v_s = (RTC_P/C_VM)^{1/2}$) or the dielectric constant thermometer (using the relation for the dielectric constant $\varepsilon = \varepsilon_0 + \alpha_0 \, n \, V^{-1}$, with ε_0 the exactly known dielectric constant and α_0 the static electric dipole polarizability of atoms). Again, these are the equations for the ideal gas, which have to be corrected by virial expansions for real gases [12.10].

12.2 Helium Vapour Pressure Thermometry

Another method relying on the behavior of the gas phase and on the Clausius–Clapeyron equation (2.8) is the determination of temperature by measurement of the vapour pressure above a cryogenic liquid. It is less complicated and more accurate than gas thermometry because the vapour pressure depends strongly on the temperature, see (2.10). Here one has to use the liquid which is appropriate for the relevant temperature range. Vapour pressure data for H_2, Ne, N_2 and O_2 can be found in [12.1], but, of course, the temperatures there have to be converted to the new ITS-90 scale (Sect. 11.2). Most frequently used for low-temperature physics is the measurement of the vapour pressure of the liquid helium isotopes which defines the ITS-90 scale from 0.65 to 5.0 K (Sects. 2.3.2 and 11.2). Helium vapour pressure thermometry is of particular importance for the calibration of resistance thermometers in the range from 0.5 to 4.2 K (Sect. 12.5). A vapour pressure thermometer has to be calibrated.

It is a secondary thermometer because our theoretical knowledge about the temperature dependence of the vapour pressure is not such that we can write down a complete and simple equation for it from first principles. The equation that we derived in Sect. 2.3.2 was a very rough approximation, assuming ideal gas behavior for the vapour and assuming the latent heat of evaporation to be temperature independent. Fortunately, calibration of the helium vapour pressure has been done very carefully by various laboratories in recent years to establish the ITS-90, where the *P-T* relations for the helium isotopes are given (Sect. 11.2). The new *P-T* values for the ^{3}He and ^{4}He vapour pressures calculated with (11.6) and with the coefficients given in Table 11.2, are listed in Tables 11.3 and 11.4.

To perform vapour pressure thermometry one needs data relating vapour pressure to temperature as well as a sensitive, calibrated manometer, and one has to minimize experimental errors [12.14]. Usually it is not sufficient just to measure the vapour pressure above the boiling cryoliquid because liquids have a low thermal conductivity (except superfluid helium), and therefore substantial temperature gradients may develop within them. For example, liquid ^{3}He at SVP shows a pronounced maximum in its density at 0.5 K. This anomaly and the low thermal conductivity of liquid ^{3}He (Figs. 2.17 and 2.18) at those temperatures can lead to a severe temperature (density) gradient in ^{3}He baths at $T < 0.5\,\mathrm{K}$, with the coldest part at the bottom, because in this case gravitationally driven convection does not occur. As a result, the temperature at the top of the liquid which determines the vapour pressure may be substantially different from the temperature lower down in the bath where the experiment is situated. The problem can be reduced by using only a thin layer of ^{3}He and well-conducting container walls. These problems are absent when superfluid ^{4}He with its very large thermal conductivity is used, but here problems may arise from the superfluid helium film (Sect. 2.3.5) creeping to warmer parts of the apparatus, where it evaporates. Also, one should never use the pumping tube to connect the cryogenic liquid to the manometer, because there is usually a pressure gradient along it. The connecting capillary should also not be connected to a point at a temperature lower than the one to be measured because here part of the vapour would recondense. This problem can be eliminated by using a vacuum jacket around the capillary. Figure 12.2 shows an improved method of vapour pressure thermometry using the vapour above the liquid in a small cell connected to the experimental whose temperature should be determined. Even doing it in this way may introduce errors if the capillary – connecting the vapour pressure bulb at low temperatures with the manometer at room temperatures – has a small diameter and if the vapour pressure is low. A so-called "thermomolecular pressure difference" between the low-temperature bulb and the room-temperature manometer will develop when the mean free path λ of the gas particles becomes comparable to the radius r of the connecting capillary [12.4, 12.12–12.14]. The equation for this difference is

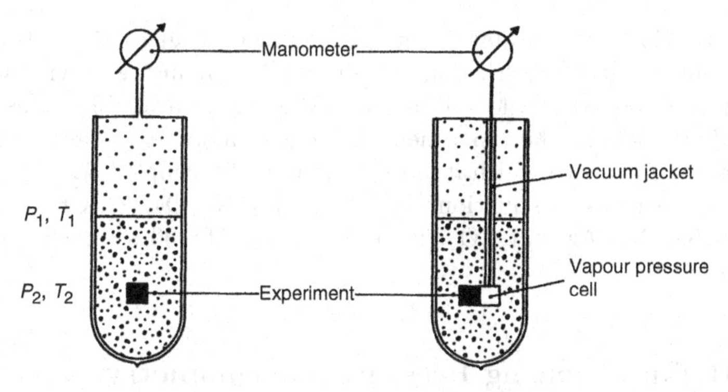

Fig. 12.2. Schematic setups for vapour pressure thermometry. *Left*: The vapour pressure above an evaporating cryoliquid is measured. Substantial temperature differences between the top of the liquid and the experiment can result due to the hydrostatic pressure head and temperature gradients in the liquid, which is usually not a good conductor (except for superfluid helium). *Right*: A small vapour pressure cell is connected to the experiment to avoid these problems. The capillary from this cell to the manometer at room temperature is vacuum jacketed to avoid changes due to a change in height of the cryoliquid

Table 12.1. Thermomolecular pressure correction for ^{3}He in a tube of radius r and with $T_{\mathrm{warm}} = 300\,\mathrm{K}$ and $T_{\mathrm{cold}} = 2\,\mathrm{K}$ at its ends [12.4, 12.12, 12.13]

rP_{warm} (mm mbar)	0.1	0.3	0.7	1.5	4
$P_{\mathrm{cold}}/P_{\mathrm{warm}}$	0.59	0.82	0.92	0.97	0.995

$$\frac{\mathrm{d}P}{\mathrm{d}T} = \frac{P}{2T} f(r/\lambda)\,, \tag{12.2}$$

where f is a function of the ratio r/λ. Table 12.1 lists some values for this correction for ^{3}He vapour with $T_{\mathrm{cold}} = 2\,\mathrm{K}$ and $T_{\mathrm{warm}} = 300\,\mathrm{K}$. The equation for the vapor pressure – if measured at room temperature – is then given by

$$P = P_{\mathrm{measured}} + \quad \Delta P_{\mathrm{hydrost}} + \Delta P_{\mathrm{therm.mol.}} \tag{12.3}$$

All these problems can be avoided if the pressure is measured not with a manometer at room temperature but with a cold manometer connected in situ directly to the refrigerated experiment. This low-temperature manometric vapour pressure thermometry can be done with very high precision with capacitive manometers, which will be discussed in Sect. 13.1. A capacitive vapour pressure thermometer with a resolution of about 10^{-9} over the limited temperature range $1.6\,\mathrm{K} \leq T \leq 2.2\,\mathrm{K}$ that uses a low temperature capacitive pressure transducer has been described in [12.15]. The design is similar to the earlier one of Greywall and Busch [12.16] to be discussed in Sect. 13.1, which has been used for vapour pressure thermometry below 1 K, but the new design has a better long-term stability and resolution. Of course, it is only possible to take advantage of such a high-sensitivity thermometric device when an adequately designed stable thermal environment is available. This can be obtained by shielding the experiment with one or more temperature-regulated platforms or shields to which it is weakly thermally coupled [12.17].

The vapor pressure equations for liquid e-H_2, Ne, O_2, Ar, CH_4, CO_2, and for solid Ne, N_2, Ar and CO_2, in terms of the ITS-90 scale can be found in [12.1, 12.18].

12.3 Helium Melting Pressure Thermometry

The physics of liquid and solid ^{3}He determining the shape of its melting curve has already been discussed in Chap. 8; the melting curve is described by the Clausius–Clapeyron equation (8.2). The melting pressure of ^{3}He exhibits a pronounced temperature dependence (Fig. 8.1), which has been used as a rather precise thermometric standard, in particularly to establish the new temperature scale PLTS-2000 [12.19] (see Sect. 11.4) in the temperature range $0.9\,\mathrm{mK} < T < 1.0\,\mathrm{K}$ [12.19–12.24]. In addition, as also discussed in Sect. 11.3, the temperature of the minimum of the melting curve, the superfluid transitions of liquid ^{3}He, and the nuclear antiferromagnetic ordering transition of solid ^{3}He on the melting curve provide well-defined temperature fixed points independent of the pressure measurement, which can easily be detected using a melting curve thermometer. Alternatively, the pressures at which these transitions occur (Table 11.5) can be used to check the pressure calibration.

As a result of the earlier considerations, several groups investigating the properties of ^{3}He along the melting curve have used the melting pressure for thermometry at $T < 0.3\,\mathrm{K}$ in a similar way as helium vapour pressure is used for thermometry at higher temperatures. The pressure is measured as a function of temperature in situ capacitively with a capacitive manometer, like the ones discussed in Sect. 13.1, filled with a mixture of liquid and solid ^{3}He. With a typical pressure resolution of $10\,\mu\mathrm{bar}$, the precision of temperature measurement is 3×10^{-4} at 1 mK, 3×10^{-5} at 10 mK and 5×10^{-6} at 100 mK (Fig. 11.1), [12.25]. The standard uncertainties of the temperature scale PLTS-2000, which is based on ^{3}He melting pressure thermometry are, of course, substantially worse; they are given in Sect. 11.3.

The ^{4}He impurities shift the minimum of the ^{3}He melting curve and change its slope. However, these effects are negligible at low ^{4}He impurity concentrations due to preferential adsorption of the heavier isotope on walls and due to phase separation at low temperatures (7.2). At higher ^{4}He concentrations, the effects can become noticeable. They are $\Delta P_{\mathrm{min}}/P_{\mathrm{min}} \approx 0.1$ bar for $x_4 = 2\%$, and the temperature errors in melting curve thermometry would be in the several percent range for such concentrations [12.26]. In addition, high magnetic fields depress the ^{3}He melting curve [12.27–12.29] because the entropy of the solid phase is field dependent due to partial alignment of the nuclear moments of ^{3}He by the field. The equation for the field depression is

$$\Delta P = -a_2(T)B^2 - a_4(T)B^4, \tag{12.4}$$

with the temperature dependent functions a_i given in [12.29]. Typical values are $\Delta P \approx 0.25(0.15)$ bar at 3.5 (10) mK in a field of 10 T. A typical resulting temperature error in ^{3}He melting curve thermometry would be $\Delta T/T \approx 2\%$ at 25 mK and 5 T [12.29], which is much smaller than typical impacts of magnetic fields on resistance thermometers, for example (see Sect. 12.5).

The advantages of a ^{3}He melting pressure thermometer are the high resolution (about 1 μK) and reproducibility (several ppm), essentially zero power dissipation, insensitivity to RF radiation and (almost) to magnetic fields. A drawback is the rather large specific heat of liquid ^{3}He in the thermometer, the need of ^{3}He, of a fill capillary, and of a gas-handling system. A further advantage of the ^{3}He melting curve as a temperature scale is the fact that it can easily be transferred between laboratories without requiring the exchange of calibrated devices. It has therefore become for the millikelvin temperature range what the helium vapour pressure curves had already become for the Kelvin temperature range (Sects. 11.2, 12.2). Very careful new determinations of the ^{3}He melting pressures were discussed in Sect. 11.3. In particular, the detailed discussion of possible error sources for the pressure calibration (hydrostatic head, thermal expansion, etc.) in [12.20, 12.22] is illuminating for everyone interested in careful pressure measurements.

A high-resolution ^{4}He melting-pressure gauge for use near 2 K has been described in [12.30]; it is stable to a few 10^{-9} K over several hours.

12.4 Thermoelectricity

If the ends of a metallic wire are at different temperatures a voltage will develop along the wire. This voltage ΔU is the absolute thermoelectric force; the corresponding thermoelectric power is defined as

$$S = \frac{\Delta U}{\Delta T}. \tag{12.5}$$

Generally the thermoelectric power between two different metals is measured with a reference junction at a reference temperature, usually at 0°C to avoid influences from the contacts at the voltmeter (Fig. 12.3).

The advantages of thermometry based on thermoelectric power are: it is a local measurement using a point sensor; the device and the measurement are simple (but see below); the device is rather insensitive to magnetic fields (except for magnetic alloys, see below); the device has a small specific heat; the measurement occurs essentially without heat input; the results are reproducible, because in general there is no change even after the device has been warmed up and cooled down repeatedly.

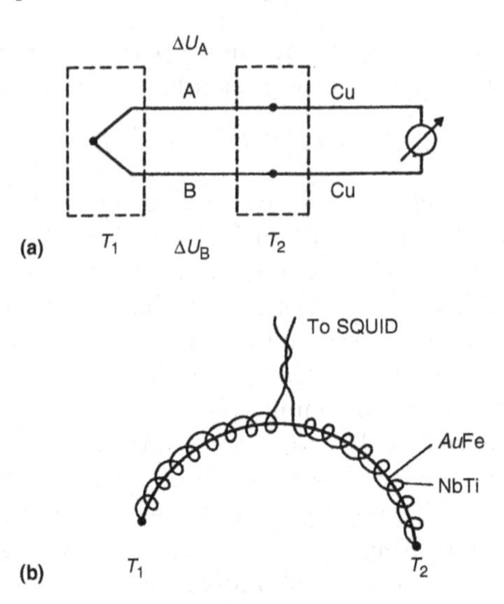

Fig. 12.3. Wiring for thermoelectric thermometry. (**a**) The general setup, where the thermoelectric power is created at the welded junction between wires A and B, which is at a temperature T_1. These two wires are connected to the Cu leads which eventually lead to the measuring instrument. (**b**) Wiring suitable for thermoelectric thermometry at very low temperatures. Here the thermopower of a wire of *Au*Fe, for example, is compared to that of a superconducting wire of NbTi, which does not create any thermopower. At low temperatures the small thermopowers have to be measured with a SQUID

Unfortunately the thermoelectric power vanishes for $T \rightarrow 0$. Therefore, thermometry based on thermoelectric measurements with one of the usual combinations of thermocouple wires – for example, Constantan/Cu – becomes insensitive below 10 K (Fig. 12.4). However, new materials and very sensitive voltage measuring devices have been developed allowing thermoelectric thermometry to be extended even to the millikelvin temperature range. The newly developed suitable alloys are highly diluted magnetic systems like Au + a few up to several 100 ppm Fe [12.1, 12.2, 12.31–12.34], or Pd + a few up to several 100 ppm Fe [12.35–12.38]; usually commercially "pure" Au or Pd contain enough Fe. The first is a so-called Kondo alloy, where the magnetic moment of Fe in the Au matrix is strongly temperature dependent. Values for its thermoelectric power can be found in [12.1, 12.2, 12.34], see Fig. 12.4. It has been investigated in detail in [12.34] at 10 mK (here, $S = 10\,\mathrm{nV\,K^{-1}}$) to 7 K showing that the optimum Fe concentration is 300 ppm (Fig. 12.5). Of course, these still small voltages cannot be measured relative to a reference point at 0°C, because there the thermopower is very large so that the reference temperature would have to be extremely well regulated. For these low-temperature thermocouples the reference junction is usually kept in a liquid ^{4}He bath at 4.2 K. The

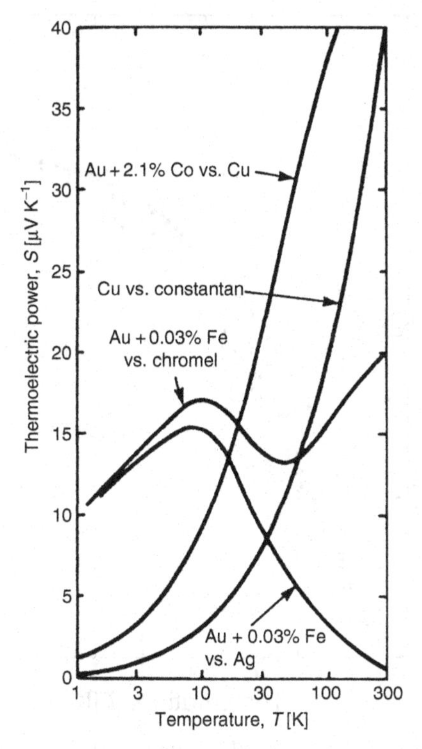

Fig. 12.4. Thermoelectric powers of some metal pairs [12.1]

second alloy, whose thermopower is shown in Fig. 12.6, is a so-called giant moment spin glass. Here the Fe impurity polarizes the highly paramagnetic Pd matrix, giving rise to a giant magnetic moment of up to $14\,\mu_B$. These giant moments freeze in at low temperatures with statistical orientations ("spin glass freezing") [12.37, 12.38]. Of course, the properties of these dilute magnetic systems depend strongly even on small magnetic fields [12.37, 12.38].

Spurious thermoelectric voltages may develop in a wire due to chemical inhomogeneity or physical strain if it is exposed to a temperature gradient; these effects in the leads to room temperature may be larger than the desired low-temperature signal from the junction. The leads have therefore to be carefully selected and fixed when they have to be brought out to room temperature. To avoid nuisance thermopowers one should join the various metals without using a third metal. Welding is the most appropriate joining procedure for a thermocouple. Very often thermopowers are a problem if one has to measure small voltages, because thermopower develops at soldering joints, at switches, and other contacts between different metals. Therefore sensitive measurements, for example in bridges, mostly have

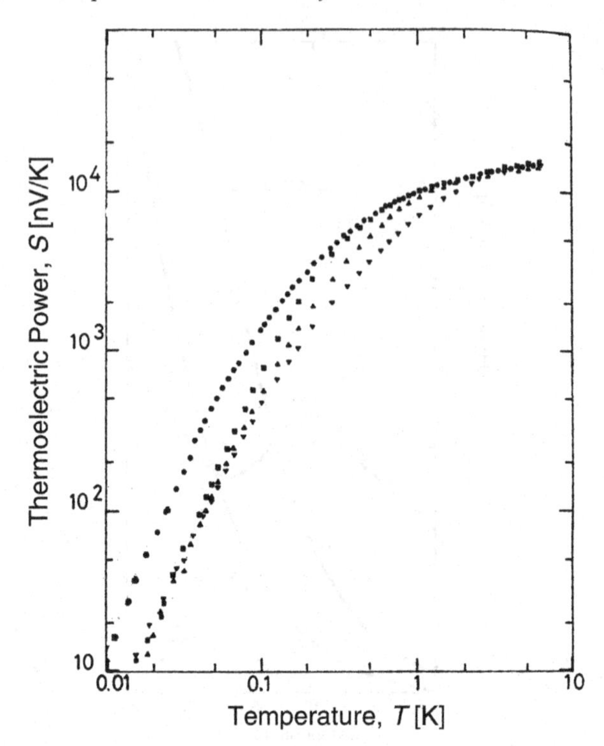

Fig. 12.5. Thermoelectric power of $Au\mathrm{Fe}_x$ alloys with Fe concentrations $x = 56$ ppm (▲), 123 ppm (■), 300 ppm (●), and 700 ppm (▼) as a function of temperature [12.34]

to be AC measurements. Because no voltage drop can be developed along a superconducting wire, a superconductor, for example NbTi, with $S = 0$ is the appropriate reference wire in a low-temperature thermocouple (Fig. 12.3b). The high-sensitivity device for measuring very low thermoelectric voltages at low temperatures is the so-called Superconducting Quantum Interference Device (SQUID), allowing a resolution of 1 μK at 1 K with a Au + 0.03 at.% Fe thermocouple, which has a thermopower of about $-9\,\mu\mathrm{V\,K}^{-1}$ at 1 K [12.1, 12.2, 12.10, 12.31–12.33].

Generally, a practical low-temperature limit for the use of thermocouples is 10 K, with some effort and the use of magnetic alloys the temperature range can be extended to about 1 K, and the practical limit for thermoelectric thermometry is about 0.1 K. It has been shown that, in principle, thermopower can be used for thermometry to about 10 mK (Fig. 12.6). This has not been applied in routine temperature measurements because there are more suitable methods available, which will be discussed in the following sections. However, thermoelectric thermometry can be attractive in situations where temperature differences have to be measured.

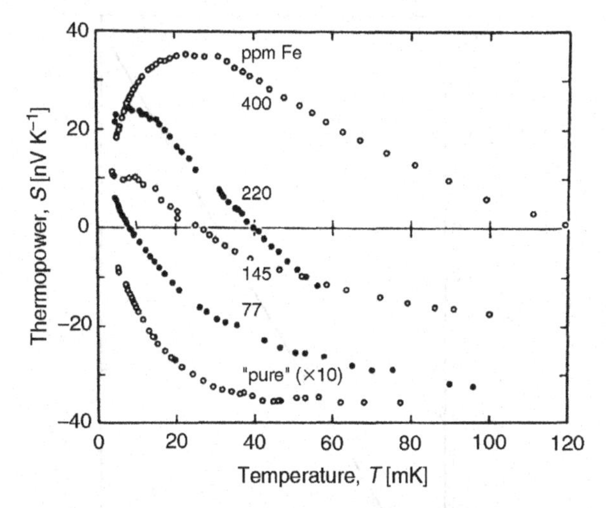

Fig. 12.6. Thermoelectric power of Pd doped with the indicated amounts of Fe (in ppm) as a function of temperature at millikelvin temperatures. The values for the "pure" sample have been multiplied by 10 for clarity [12.36]

12.5 Resistance Thermometry

Resistance thermometry is based on the temperature dependence of the electrical resistance of metals, semiconductors or complicated compounds (see below). It is probably the simplest and most widely used method of low-temperature thermometry. The devices are readily available and the measurements are easy. As we will see below, the problems at very low temperatures are thermal conductivity, thermal contact, and self-heating of the device due to the measuring current and RF absorption. In addition, the resistivity usually has no simple, a priori known temperature dependence, therefore a resistance thermometer is a secondary thermometer. Many different kinds of resistance thermometers are commercially available, often even calibrated for a particular temperature range.

12.5.1 Metals

The pure metal most commonly used for resistance thermometry is platinum [12.1, 12.9–12.11]. It is one of the standards for interpolation between fixed points of the ITS-90 (Sect. 11.2). Platinum is chemically resistant; it can be obtained with high purity (diminishing the temperature-independent residual resistivity part); it is ductile, so it can be drawn into fine wires; and its resistance has a rather large temperature coefficient. The temperature-dependent resistance of a commercial Pt-100 thermometer shown in Fig. 12.7 illustrates the linear R–T dependence over an appreciable temperature range. A Pt-100 thermometer is a platinum resistor with a resistance of exactly

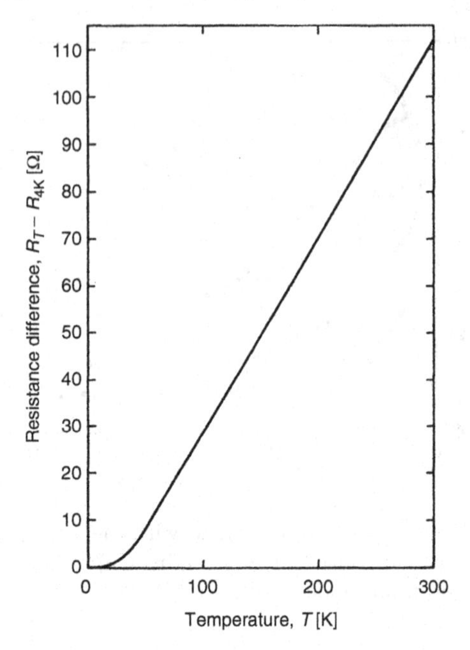

Fig. 12.7. Difference of the resistances at temperature T and 4 K of platinum as a function of temperature. The resistance difference varies linearly with temperature between room temperature and about 50 K

Table 12.2. Temperature–resistance values of a Pt-100 resistance thermometer

T (K)	10	15	20	25	30	40	50	60	77(!)	100
R(Ω)	0.09	0.19	0.44	0.94	1.73	4.18	7.54	11.45	18.65	28.63
T (K)	125	150	175	200	225	250	273.2(!)	300		
R(Ω)	39.33	49.85	60.23	70.50	80.66	90.72	100.00	110.63		

(For more data points see [12.1])

100 Ω at 0°C. Values at other temperatures are given in Table 12.2. Pt-100 thermometers or Pt thermometers with other resistance values are available commercially with calibration tables. They are well annealed and either fused in quartz glass or encapsulated in a quartz or metal tube which is filled with helium gas for thermal coupling of the Pt wire to its surroundings. It is rather important that the Pt wire is supported strain-free so it does not change its properties on repeated cooling and warming. Cheaper Pt thermometers are manufactured by lithographic techniques as metal films.

As can be seen from Figs. 12.7 and 12.8, sensitive thermometry with a Pt resistor or with other pure metals is only possible down to about 10 K; at lower temperatures we are in the temperature-independent residual resistivity range. As in the case of the thermopower, one can increase the sensitivity

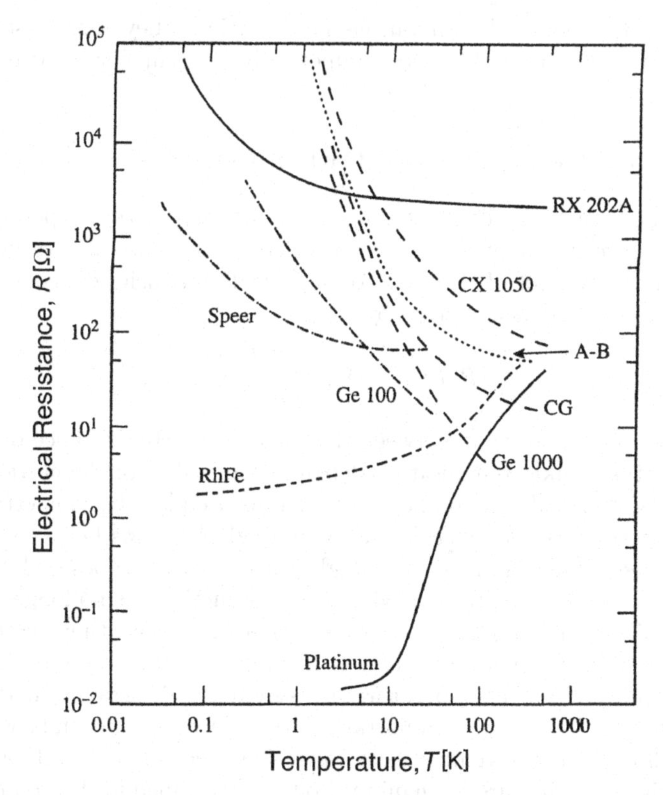

Fig. 12.8. Temperature dependence of the electrical resistances of some typical low-temperature thermometers. A–B denotes an Allen–Bradley carbon resistor, Speer is also a carbon resistor and CG is a carbon-in-glass thermometer; Ge 100 and Ge 1000 are two commercial germanium thermometers; CX 1050 is a CERNOX and RX 202A is a RuO_2 commercial thermometer (see below). The lower two curves are for the two indicated metals. Reprinted from [12.1], copyright (2002), Oxford Univ. Press

of the resistance of a metal to temperature changes at low temperatures by introducing magnetic impurities. This is shown for Cu with Fe impurities in Fig. 3.25. Like *Au*–Fe, the alloy *Cu*–Fe is a Kondo alloy, where the Fe moment is strongly temperature dependent at low temperatures, leading to an *increase* of the resistance with decreasing temperatures below about 15 K. The magnetic alloy most widely used for resistance thermometry at low temperatures is the commercially available alloy *Rh*–0.5% Fe [12.8,12.10,12.11]. Above 30 K its resistance behaves similarly to Pt, whereas below 10 K its resistance remains sensitive to temperature changes due to magnetic scattering of conduction electrons at the Fe impurity atoms; it then decreases almost linearly with decreasing temperature at $0.1\,\mathrm{K} < T < 1\,\mathrm{K}$ (Fig. 12.8). *Rh*–Fe resistance thermometers are also in use to realize temperature scales (Sect. 11.3.1) and to transfer them between various laboratories. They show a stability of 10 mK/year and can be obtained commercially as incapsulated wires or thin film thermometers.

12.5.2 Doped-Germanium and Carbon Resistors

The resistance of a semiconductor does not *decrease* with temperature as it does for a pure metal; it *increases* with decreasing temperature because of the decreasing number of carriers. For an "ideal" intrinsic semiconductor the temperature dependence is given by

$$R(T) = \alpha \, \exp\left(\frac{\Delta E}{2k_{\mathrm{B}}T}\right) , \tag{12.6}$$

where ΔE is the energy gap between the valence and conductance bands, and α contains the weakly temperature-dependent mobility of the carriers. However, the resistance of semiconducting materials used for thermometry usually does not agree with this relation, and empirical equations between R and T have to be used. One can reach very high resistances and, above all, very high sensitivities at low temperatures where the conductivity is no longer intrinsic ($\Delta E \gg k_{\mathrm{B}}T$) but results from impurities donating or accepting electrons. Actually, very often the resistance of a semiconductor becomes too high below 1 K to be suitable for a temperature measurement. Resistance thermometry with semiconductors is an important and widely used secondary thermometry technique in the temperature range between about 10 mK and 10 K. Two favorite materials are germanium, specifically doped for low-temperature thermometry, and carbon in the form of commercial carbon resistors from the electronics industry.

Doped Germanium

Germanium used as a low-temperature thermometer is specially doped with 10^{15}–10^{19} atoms cm^{-3} of As ("n-type") or Ga ("p-type"). Below 100 K the conductance is due to holes or electrons which the dopant delivers. Such a thermometer is particularly suited for the temperature range $0.3\,\mathrm{K} \leq T \leq 40\,\mathrm{K}$. At lower temperatures the resistance often becomes too large ($> 1\,\mathrm{M\Omega}$), but with special doping these thermometers have been used to 30 mK. The temperature dependences of various Ge thermometers are shown in Figs. 12.8 and 12.9.

A great advantage of the germanium thermometers is the stability of their resistance values. Even after repeated cyclings between room temperature and low temperatures, or after an extended time on the shelf, the deviations at 4 K are typically only about 1 mK or, more generally, about 0.1%. Therefore these thermometers can be bought calibrated with a computer fit for the resistance–temperature relation to 50 mK (but this increases their price by at least an order of magnitude!). Such a computer fit has to be performed, because, unlike

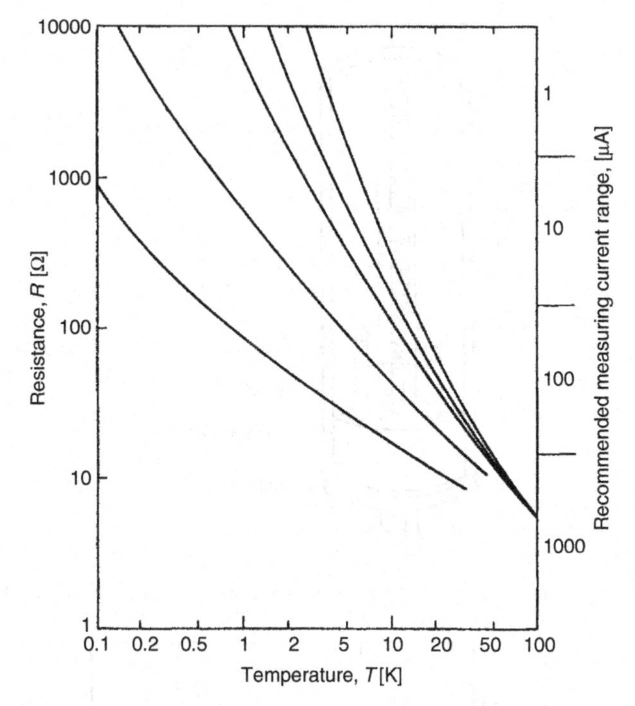

Fig. 12.9. Typical temperature dependence of the resistances of some commercial germanium thermometers. The right-hand vertical scale shows the measuring current recommended for these thermometers to avoid overheating them

an "ideal" semiconductor, the T–R relation for a heavily doped germanium thermometer is not simple, but given, for example, by

$$\ln R = \sum_{n=0}^{m} \alpha_n (\ln T)^n \text{ or } \ln T = \sum_{n=0}^{m} \beta_n (\ln R)^n . \tag{12.7}$$

To assure the reproducibility and stability of the device, the encapsulated single crystal Ge chip has to be supported strain-free on its four gold leads (Fig. 12.10). Again, for an encapsulated Ge thermometer the thermal coupling is provided by helium exchange gas as well as by its leads. This coupling and the thermal conductivity of the chip are weak. Therefore one has to keep the measuring current and hence the Joule heating low. Typical, safe values for the allowed measuring current and power are 10 (1; <0.1) μA and 10^{-7} (10^{-10}; <10^{-12}) W at 10 (2; <1) K. Another point of importance is the strong and orientation-dependent magnetoresistance of germanium thermometers, which increases with temperature sensitivity (Fig. 12.11) [12.39, 12.40]; at low temperatures they should not be used in the Tesla field range.

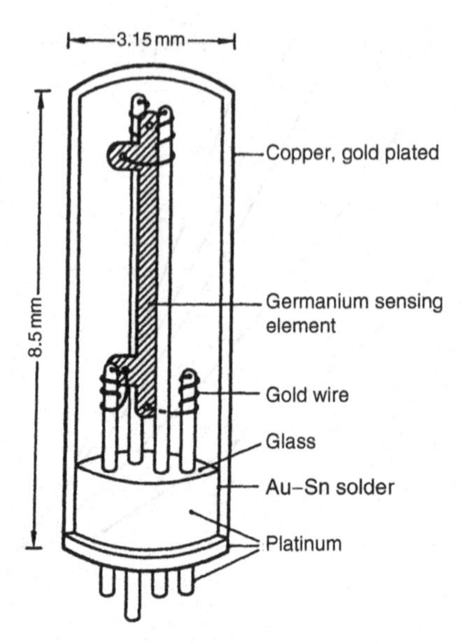

Fig. 12.10. Typical construction of a commercial germanium thermometer

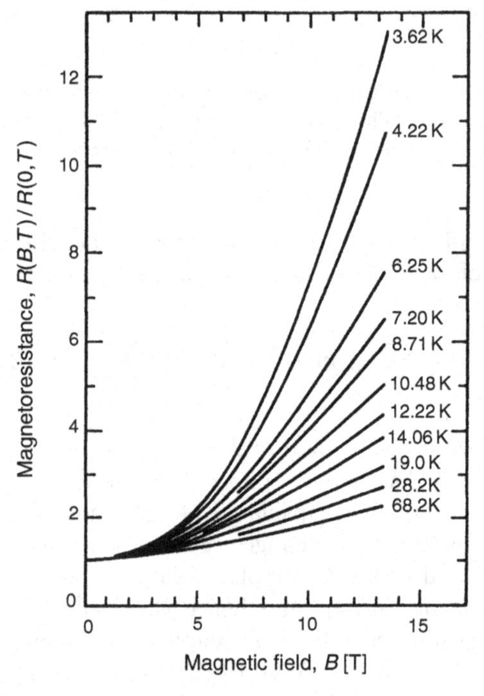

Fig. 12.11. Magnetic field dependence of the electrical resistance of a germanium thermometer ($R_{4\mathrm{K}} = 856\,\Omega$) at the indicated temperatures [12.39]

Carbon Resistance Thermometers

In contrast to Ge thermometers, carbon composition resistors applied for low-temperature thermometry are not specially manufactured for thermometry but are taken from the mass production of the electronics industry. Therefore their price is less than 1 € each, which is two orders of magnitude lower than the price of the specially produced Ge thermometers. Due to this low cost and their very suitable *R–T* behavior, carbon resistance thermometers were for a long time the most widely used secondary thermometers in the upper millikelvin and Kelvin temperature ranges.

Pure carbon is not a semiconductor. The negative *R–T* characteristic of commercial carbon resistors results from their production process, which consists of pressing and sintering fine carbon particles together with some glue. The resistance is probably mostly determined by the contact resistance between the particles and by composition. Particle size and production method have an essential influence on the behavior of carbon resistors. As a result, the resistance of carbon resistors changes from sample to sample and often is not very reproducible after thermal cycling due to the stresses involved. Furthermore, they show ageing and sometimes even changes during a low-temperature experiment. The typical drift in resistance of a carbon thermometer is of the order of 10^{-3} per time decade [12.41, 12.42]; no time dependence has been observed for germanium-resistance thermometers. One therefore should repeatedly cycle a carbon resistor between room temperature and liquid-nitrogen temperature before using it as a thermometer. And one has to avoid overheating it while soldering on its short leads. Anyway, the calibration should be repeated in each run if temperature has to be measured to within a few percent or even better.

For low-temperature thermometry almost exclusively the following three commercial types of carbon thermometers with the given properties were in use:

Allen–Bradley 1/8 W; small, cheap; good at $T > 1$ K; too large R (and dR/dT) at lower T [12.4, 12.43].

Matsushita 1/8 W; types ERC 18 GK and ERC 18 SG; small; useful at $T \geq 10$ mK [12.44–12.46].
Speer 1/2 W; type 1002; larger; particularly useful in the millikelvin temperature range to about 10 mK because they have a less steep R–T dependence [12.47–12.49].

Typical temperature dependences of the resistance of these thermometers are shown in Figs. 12.12–12.15. With such thermometers, a resolution of order 10 μK at $T \leq 1$ K is possible. The particular type most suited for the desired purposes – most importantly the temperature range of interest – has to be chosen.

Application of carbon thermometers for thermometry *below* 1 K presents some problems. These are their rather high heat capacity (Fig. 12.16) and

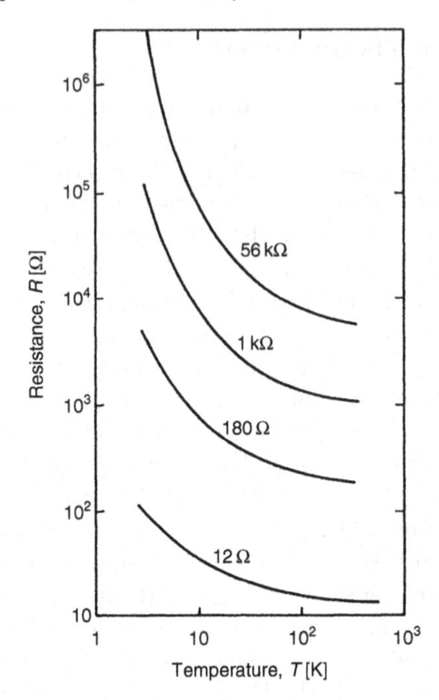

Fig. 12.12. Temperature dependence of the resistances of four Allen–Bradley carbon resistors (1/8 W) with the indicated room-temperature resistances [12.43]

their low thermal conductivity. These two intrinsic properties, as well as the difficulty of thermal coupling of the thermometers to their environment while avoiding electrical contact can lead to very long thermal relaxation times [12.49] and, even worse, to thermal decoupling from their environment. This may lead to saturation of the resistance at low temperatures, mostly due to pick-up of RF noise (Fig. 12.15). Typical values for the thermal resistance of Speer resistors are $R \simeq 10^4 T^{-3}(\mathrm{K\,W}^1)$ [12.48]; in this reference it was shown that the bottleneck for Speer carbon resistors is the thermal resistance in the carbon itself rather than the Kapitza boundary resistance.

The most difficult problem for carbon thermometry in the millikelvin temperature range is therefore establishing thermal equilibrium within the thermometer and between the thermometer and its surroundings. The problem of coupling these thermometers thermally to the area whose temperature is supposed to be measured has the consequence that joule heating from the measuring current or due to pick-up of RF waves has to be kept very small. As a general rule the heat input to the thermometers should be at most

$$\dot{Q}(\mathrm{W}) < 10^{-6}\, T^4\ (\mathrm{K}^4)\,, \tag{12.8}$$

if overheating of the resistor is to result in a temperature error of not more than about 1%. This limit gives, for example, $\dot{Q} < 10^{-14}\,\mathrm{W}\,(10^{-10}\mathrm{W})$ for

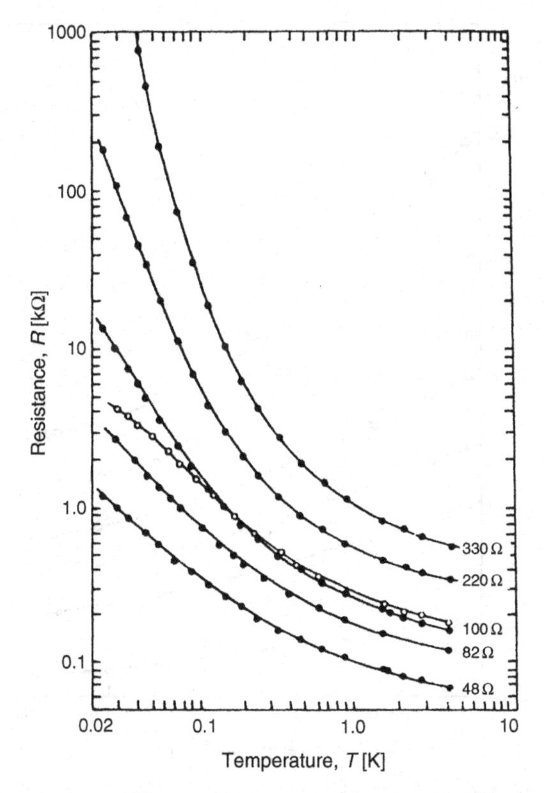

Fig. 12.13. Temperature dependence of the resistances of several Matsushita carbon resistors (grade ERC-18 SG, 1/8 W) with the indicated room temperature resistances (•). For comparison the temperature dependence of a Speer carbon resistor (grade 1002, 1/2 W, 100 Ω) is shown as open points [12.45]

a well-shielded and well-coupled piece of carbon at 10 mK (100 mK) [12.43, 12.48, 12.50, 12.51].

For the Kelvin temperature range it is adequate to put the carbon thermometer into a tightly fitting hole filled with vacuum grease (Fig. 12.17a) [12.57] or to glue it into a Cu foil for thermal contact (Fig. 12.17b). However, these thermometers should not be used in their original form if temperatures below 1 K are to be measured. For the millikelvin temperature range one should use carbon slices about 0.1 mm thick to achieve adequate thermal response and contact [12.50, 12.51, 12.58]. For this purpose first the phenolic cover should be removed from the resistor. Then one side of the carbon core should be ground down with fine sandpaper. The resulting semicircular piece of carbon can be glued to a Cu sheet (separated by a thin cigarette or lens paper) for mechanical strength and for better thermal contact. Then the second side of the carbon can be ground down to a remaining thickness of 0.05–0.1 mm (or a resistance of 1–2 kΩ if Speer resistors are used), see Fig. 12.17c. Finally a

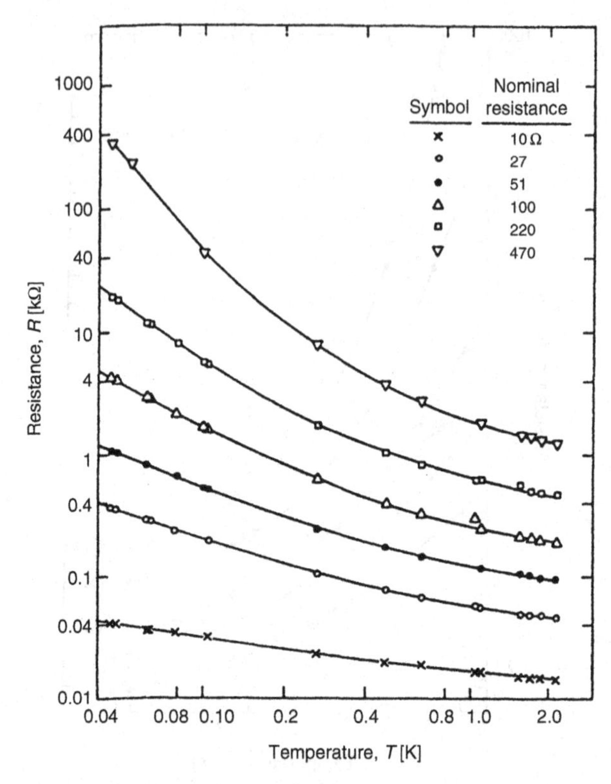

Fig. 12.14. Temperature dependence of the resistances of various Speer carbon resistors (grade 1002, 1/2 W) with the indicated room temperature resistances [12.47]. For more recent data for the 220 R and 470 R units at 50 mK to 300 K see [12.49]

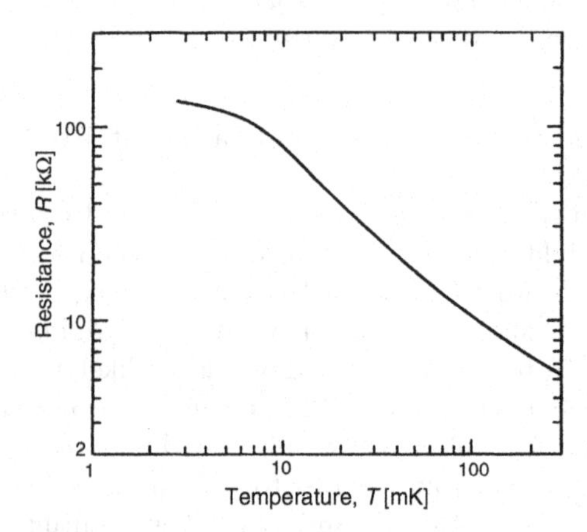

Fig. 12.15. Temperature dependence of a ground-down Speer carbon resistor (Fig. 12.17c). The resistance seems to saturate below about 7 mK due to a heat input of about 10^{-14} W from the measuring current and from the RF-noise pick-up [12.50]

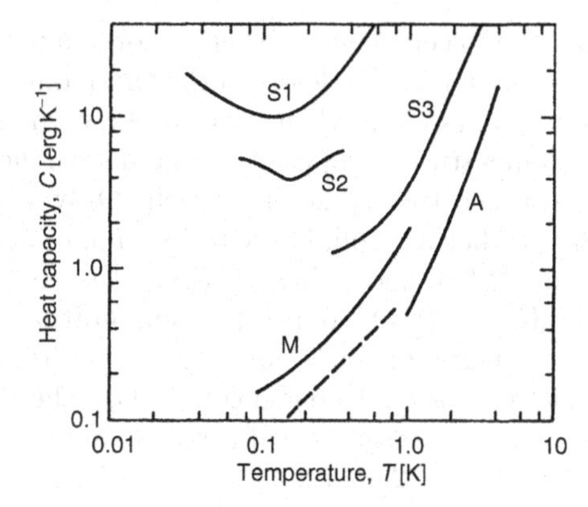

Fig. 12.16. Heat capacity of modified carbon resistors. S1 Speer 220 Ω (about 30 mg) [12.52]; S2: Speer 220 Ω (25 mg) [12.53]; S3 Speer 220 Ω (17 mg) [12.54]; A Allen–Bradley 116 Ω (11 mg C and 2 mg Cu) [12.55]; M: Matsushita, 100 Ω (about 5.5 mg C and Cu each) [12.56]. The dashed curve is an estimate of the contribution of the copper leads to the heat capacity of the Matsushita resistor

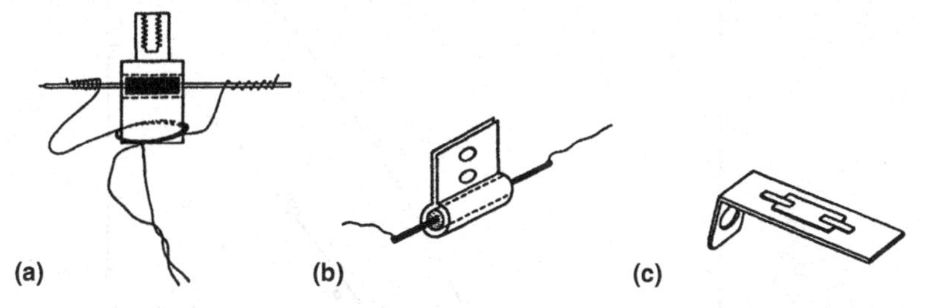

Fig. 12.17. Recommended designs for the use of a carbon resistor as low-temperature thermometers. For the Kelvin temperature range it is sufficient to put it in a hole of a Cu holder which can be filled with grease for thermal coupling [12.52] (**a**) or to glue the unmodified carbon thermometer into a Cu foil (**b**). The leads to the thermometer should be thermally heat sunk to the Cu holder. For the millikelvin temperature range one should use only a 0.1 mm thick slice of a carbon resistor to achieve adequate thermal contact (**c**). For this purpose one side of the resistor should be ground down. The remaining semicircular piece should then be glued with a thin layer of epoxy to a metal heat sink for protection against breakage (put a cigarette paper in between for electrical isolation). Then the second side of the resistor can be ground down to give the required thickness of the thermometer. It should then be covered with a copper foil for shielding against RF pick-up. The leads, of course, should again be heat sunk to the place whose temperature the carbon slice is supposed to measure. From there on the leads should be superconducting to reduce heat transmitted along them

thin Cu foil should be glued to the upper side for electrical and thermal shielding of the thermometer. The varnish or epoxy will penetrate the carbon, giving it a better mechanical stability and preventing microcracks. Such thermometers have been found to have much more reproducible characteristics than in their original shape; they keep their calibration for years and through many cooldowns to within about 1%. The leads to the thermometer have to be well heat sunk at various points along their way in the cryostat and eventually to the area whose temperature is supposed to be determined; this, of course, applies to other resistance thermometer as well. Only with such a design, the sensitivity and the thermal coupling can be maintained even in the low millikelvin range (Fig. 12.15), and a thermal time constant of some minutes is possible even at 10 mK [12.51]. Of course, the temperature dependence of the resistance of a carbon thermometer is not simple; one has to fit an equation like (12.7) to obtain their $R(T)$ dependence. Carbon thermometers show an often non-monotonic magnetoresistance of typically only a few percentage per tesla [12.8, 12.10, 12.40, 12.44, 12.46, 12.59], see Figs. 12.18–12.20.

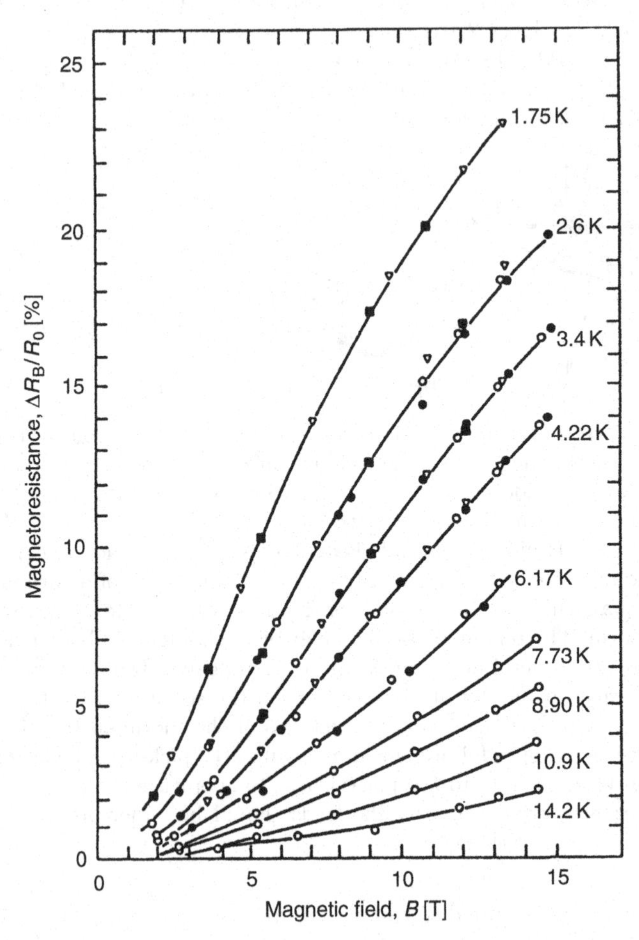

Fig. 12.18. Magnetoresistances of 47 Ω, 1/4 W Allen–Bradley carbon thermometers [12.40]

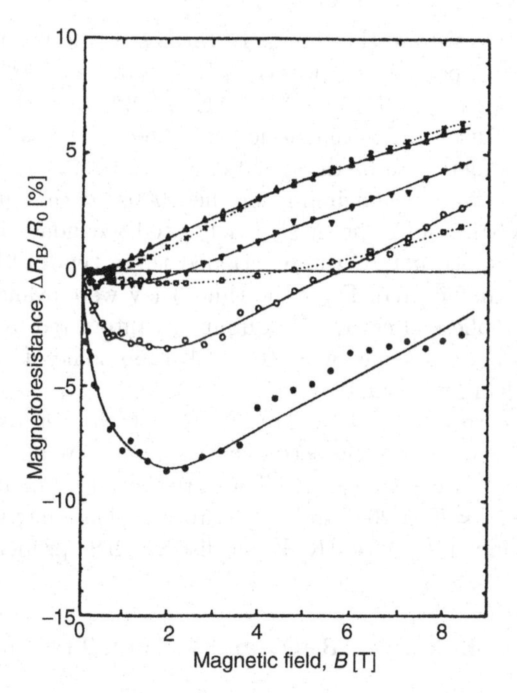

Fig. 12.19. Magnetoresistances of a Matsushita carbon resistor (68 Ω, 1/8 W) at 0.05 K (●), 0.1 K (○), 0.25 K (▽), 0.5 K (△), 1.0 K (x), and 4.2 K (□) [12.46]

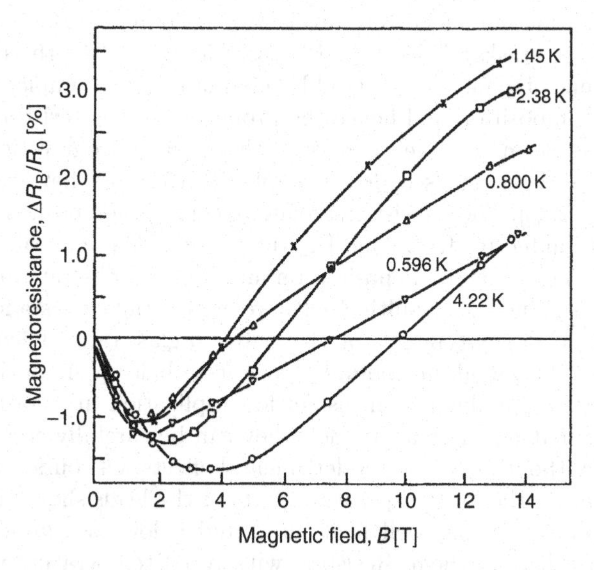

Fig. 12.20. Magnetoresistances of a 220 Ω, 1/2 W, grade 1002 Speer carbon resistor [12.40]. For more recent data for the 470 R unit see [12.49]

For the Speer resistors, the change of the resistance in magnetic fields of less than 10 T was reported to be less than 10% at 0.2–4 K [12.59] and for fields up to 18 T, to be less tan 12% at 0.1–1.6 K [12.49]. In general, they are not recommended for use in high magnetic fields; here the new oxide compound resistance thermometers to be discussed in Sect. 12.5.3 are surely preferable.

For decades, from the beginning of the 1960s to the end of the 1990s, these carbon resistors were the most widely used secondary low-temperature thermometers. Unfortunately, the mentioned three types of carbon resistors are no longer manufactured. For some time, they were available from stocks kept by some suppliers of cryogenic equipment. Interested users may still try there or with some colleagues who may still keep some of these very useful resistors in stock [12.9, 12.49, 12.60].

Because the carbon-composition resistors are no longer available, it is very fortunate that new (and even better suited) commercial, cheap resistance thermometers have become available. They are the thick-film chip RuO_2-based resistors for 30 mK $< T <$ 20 K and the ceramic zirconium oxynitride resistors for 0.1 (better only 1 K) to 400 K. I will discuss their properties in the next section.

12.5.3 Oxide Compounds: RuO_2 and Cernox Thermometers

Thick-Film Chip Resistors Based on RuO_2

The thick-film chip resistors based on RuO_2 [12.61–12.74] can be obtained from the relevant manufacturers of electronic components (Dale Electronics, Norfolk, Nebraska, USA; NV Philips, Eindhoven, Netherlands; ALPS Electric Comp., Japan; Siegert GmbH, Germany, for example; however, be careful: the composition and hence the properties of the resistors from some manufacturers seem to have changed with time) or from various suppliers of cryogenic equipment (see, for example [12.75]), or they can be home-made [12.76–12.78]. These resistors are metal-ceramic composites consisting of a mixture of conductive RuO_2 and Bi_2RuO_2 embedded in a lead silicate glass ($PbO–B_2O_3–SiO_2$) matrix, deposited on an alumina substrate and heated to above its glass point. The resulting negative non-metallic resistance characteristic (Fig. 12.21) depends mostly on the metal-to-glass ratio. Their advantages are reproducibility, weak magnetoresistance (see below), small size and mass (a few milligrams, making them useful for application in calorimeters), and low cost. For making thermal contact, they can be carefully epoxied or glued to a sample without indication of detrimental effects. Of course, if used in the millikelvin range, the low-temperature parts of the leads should again preferably be superconducting, well heat-sunk, and a low-pass filter (Fig. 12.28) should be installed on them. Resistors with room temperature values in the range of 0.5 to a few kΩ seem to be quite suitable for low-temperature thermometry with a sensitivity comparable to that of Ge resistance thermometers. Their resistance can be well fitted by the empirical equation [12.70]

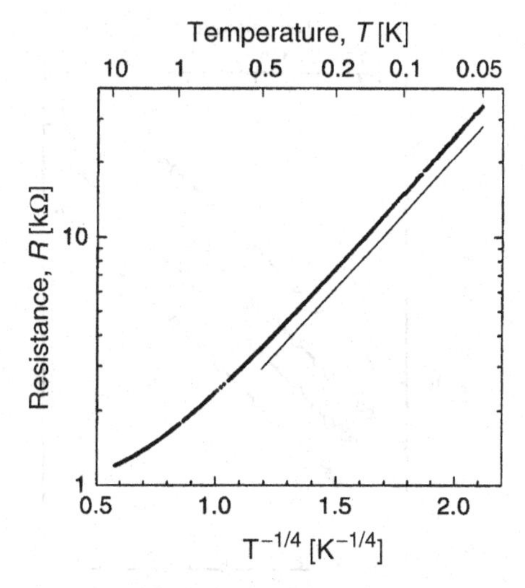

Fig. 12.21. Resistance of a commercial RuO_2-based resistor versus $T^{-1/4}$. This temperature dependence fits the data well between 0.05 and 0.5 K as indicated by the straight line. Reprinted from [12.72], copyright (2001), with permission from Elsevier. For commercial RuO_2 based resistors with different characteristics, see [12.75]

$$\ln R = \sum_{n=0}^{2} A_n \, (\ln T)^n \, . \tag{12.9}$$

In a limited temperature range, the expression

$$R = R_o \exp(T_o/T)^{1/4} \tag{12.10}$$

with the constants R_o and T_o depending on composition fits the data well [12.61–12.63, 12.66, 12.70, 12.71, 12.76–12.78], (Fig. 12.21). This dependence corresponds to an electronic conduction mechanism due to variable-range hopping in three dimensions between localized states near the Fermi energy with a constant electronic density of states. The temperature range can be extended by using somewhat larger exponents [12.61, 12.63, 12.65, 12.66, 12.70, 12.71, 12.76–12.78], (Fig. 12.22), indicating a crossover to an exponent 1/2 due to possible influences from Coulomb interaction, which leads to a gap in the density of electronic states at the Fermi energy.

Several researchers have investigated the magnetoresistance of these thermometers [12.61–12.67, 12.70–12.74]. The results do not agree in detail and they seem to depend somewhat on the batch. However, for most of them, the magnetoresistance is negative for fields up to about 1 T, then goes through a broad minimum, and is positive at higher fields (Fig. 12.23). The magnetoresistance becomes larger and the size of the minimum becomes more pronounced with decreasing temperature. The resistance change from the zero field value to the resistance minimum at about 1 T (at 8 T) is typically less than 1(5)%

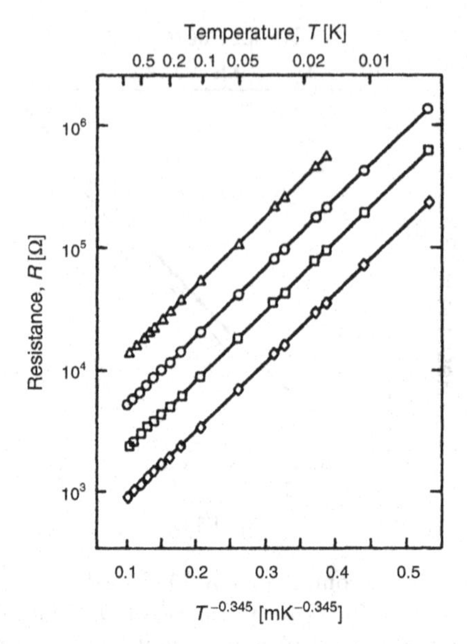

Fig. 12.22. Resistances vs. $T^{-0.345}$ of four different RuO_2 resistors with approximate room temperature values of 0.5 kΩ (◊), 1 kΩ (□), 2 kΩ (○), and 4.7 kΩ (△), respectively. The *upper* horizontal scale shows the temperature in Kelvin [12.65]

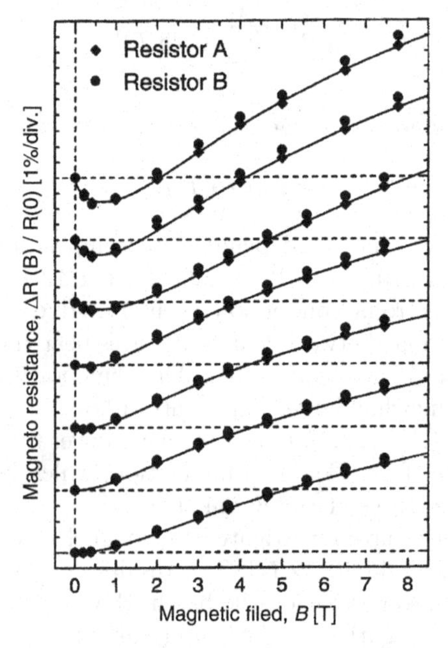

Fig. 12.23. Relative change of the resistance of a two commercial RuO_2-based resistors as a function of magnetic field at 0.05, 0.06, 0.09, 0.12, 0.16, 0.20, and 0.24 K (from top to bottom). The origin of each data set is offset for clarity. Reprinted from [12.72], copyright (2001), with permission from Elsevier

of the zero field value at 0.5 K; at 30 mK these changes are about 4(10)% of the zero field value [12.70–12.72]. However, larger values have also been reported [12.73,12.74]. The magnetoresistance at temperatures $T > 1$ K fortunately is less than 1% for $B < 8$ T [12.70,12.71], which is important for many applications. A dependence of the magnetoresistance proportional to $B^{1/2}$ has been discussed in [12.72,12.73]. In [12.63] and [12.73], the magnetoresistance has been investigated to 20 and 32 T, respectively.

Schottky-type anomalies (Sect. 3.1.5) in the heat capacity of commercial thick-film chip resistors have been observed in measurements at $0.01\,\mathrm{K} < T < 2\,\mathrm{K}$ with a peak at 0.4 K [12.69]. The observed maximum corresponds to $C_{max}/\mathrm{mass} = 30\,\mu\mathrm{J\,K^{-1}\,mg^{-1}}$ (Fig. 12.24). The origin may be magnetic impurities in the alumina substrate, which could be removed or thinned for calorimetry experiments, for example [12.64,12.69].

These resistors change their values during about the first 60 thermal cycles, but become very stable afterward according to [12.63, 12.73]. Hence, they should be calibrated only after such a number of thermal cycles have been performed. When used in the millikelvin temperature range, the applied power should be 10^{-13} W or less to avoid overheating and if 1% accuracy is the aim [12.68,12.70,12.72].

In [12.76–12.78], it is reported how such thermometers can be home-made from commercially available RuO_2 paste using standard screen-printing techniques followed by drying and firing. Both the R–T characteristics and the sensitivities can be adjusted by changing the printing geometry and composition. They have been investigated to 50 mK [12.76]. In [12.77,12.78], a detailed investigation of the dependence of the properties of these home-made resistors on their composition, in particular the crossover from the above-mentioned

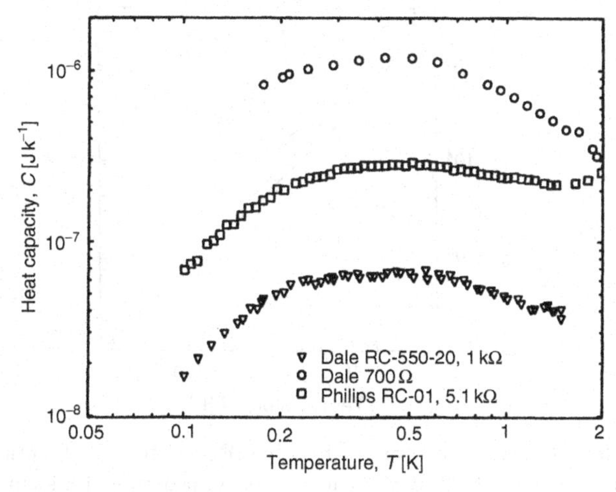

Fig. 12.24. Heat capacity of thick-film chip resistors based on RuO_2 [12.69]

exponent 1/4 to an exponent 1/2 by changing temperature and/or composition, has been presented. The properties are in general comparable to those of the commercial resistors.

Thin-Film Ceramic Zirconium Oxynitride (CERNOX) Resistors

The "CERNOX" thermometers [12.75, 12.79–12.81] are manufactured commercially by reactive sputter deposition of zirconium in an atmosphere containing Ar, N_2, and O_2 onto a sapphire substrate of about 0.3 mm thickness. The resulting zirconium oxynitride film has a thickness of about 0.3 μm. The negative temperature characteristic (Fig. 12.25) can be varied by adjusting the partial pressures of the gases. The electrical conductivity changes from metallic to semiconducting behavior with increasing O_2 content because oxygen will be incorporated into the ZrN lattice, which is enlarged by this incorporation. The thermometers are available with different packaging options and with temperature sensitivities $(T/R)(\mathrm{d}R/\mathrm{d}T)$ between about -0.6 and -3, although units having the range -1.2 to -1.9 with resistance values of 1–10 kΩ at 4.2 K are recommended for thermometry [12.79]. Their properties make them good, fast thermometers from 0.1 K to even above room temperature, with only a small change in sensitivity. These sensors are robust, small (mass less than 3 mg, making them ideal for high-field – see below – heat-capacity measurements), ease of thermal contact, sensitive, have a fast response time (a time constant of less than 1 ms at 1.7 K was reported for a bare CERNOX resistor in [12.81]; however, the supplier gives a conservative value of 1.5 ms at 4.2 K for the device), low sensitivity to magnetic fields, and can be bought with calibration [12.75].

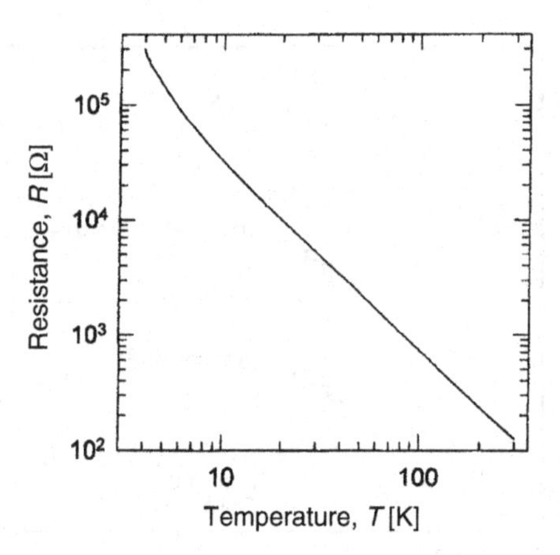

Fig. 12.25. Resistance of a commercial CERNOX resistor as a function of temperature: Reprinted from [12.81], copyright (1998), with permission form Elsevier. For commercial CERNOX resistors with different characteristics, see [12.75]

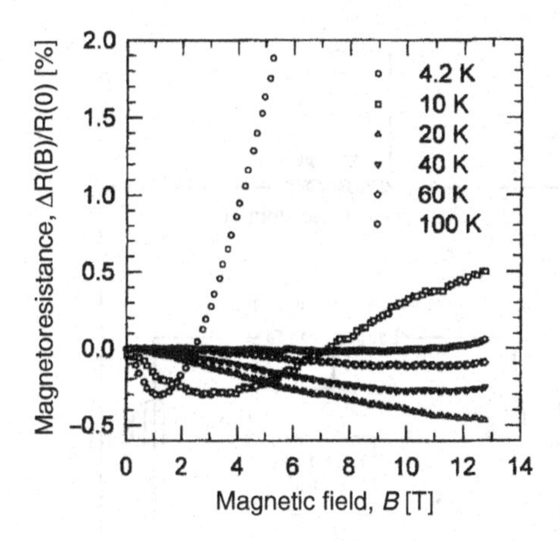

Fig. 12.26. Relative change of the resistance of a commercial CERNOX resistor (same whose data are shown in Fig. 12.25) as a function of magnetic field at the indicated temperatures. Reprinted from [12.81], copyright (1998), with permission from Elsevier

The weak dependence of the resistance of these sensors on magnetic field above 1 K has been investigated by a number of researchers [12.79, 12.81–12.83]. Similar to the above-discussed RuO_2 resistors, (as well as to the Matsushita and Speer carbon thermometers, see Figs. 12.19 and 12.20) there seem to be competing negative (dominating at low fields) and positive (dominating at higher fields) contributions to their magnetoresistance with a rather small net effect (Fig. 12.26). The change of resistance in fields up to 13 T is between 1 and 3% at 4.2 K and less than 0.5% at 10 K and above [12.79, 12.81, 12.82]. At 32 T, changes of less than 3% for temperatures above 4.2 K have been reported [12.79]. The magnetoresistance seems to depend on the individual sensitivities of the sensors. Above liquid nitrogen temperatures the magnetoresistance becomes so small that most users will not bother correcting for it, many may do so even at lower temperatures. *These resistors are probably the best choice for use in high magnetic fields at temperatures above 1 K.* However, below 1 K and at fields of 17 T, anomalously large resistance decreases have been observed in [12.83].

Procedures to correct for magnetic field dependences of resistance thermometers at low temperatures to better than 1% for fields up to 8 and 31 T, respectively, have been discussed in [12.84, 12.85].

Good overviews on the large variety of calibrated and uncalibrated commercial resistance thermometers including information on their sensitivity, stability, field dependence, and packing options as well as on the related electronics are given in [12.75].

12.5.4 Resistance Bridges

In low-temperature resistance thermometry one has to measure very small voltages, often in the nanovolt range, because low-temperature resistance

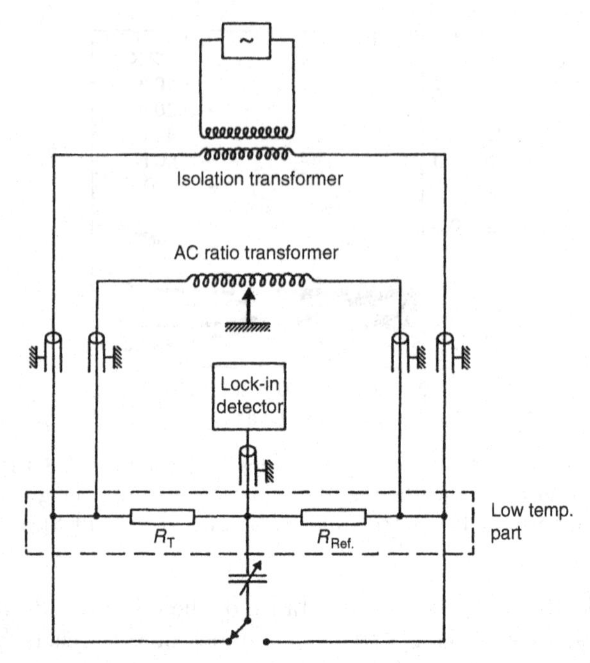

Fig. 12.27. Schematic diagram of a bridge for measuring the resistance of low-temperature thermometers. The adjustable capacitor is used to compensate a capacitive unbalance in the bridge [12.17]

thermometry has to be performed at very low powers (see earlier). Therefore, in general a direct current–voltage measurement is not applicable. The solution is to resort to a suitable bridge design for the electronic equipment [12.6, 12.17, 12.43]. In general, the so-called five-wire method is applied (Fig. 12.27) in order to separate current and voltage leads. In this way, the resistance of the leads is not measured along with the resistance of the thermometer. The measurement should be performed with AC current to avoid problems arising from thermoelectricity and to obtain higher sensitivities. The measurements are performed at low frequency, e.g., 40 Hz, to keep capacitive effects leading to out-of-phase signals at a low level. The simple design of the usual Wheatstone bridge has to be improved for adequate low temperature resistance thermometry. Some of the possible problems and possibilities for improvements are the following. One has to be very careful about grounding; ground loops have to be avoided to keep nuisance currents in the thermometer small. All leads and the thermometer have to be shielded from RF, because the thermometer often has a rather high resistance and, together with its leads, it may be well matched to absorb sizable amounts of RF energy floating around in the laboratory. Figure 12.27 shows a bridge design which avoids some of the problems that may arise with a standard Wheatstone bridge. In particular, it reduces the influence of the resistances r_l of the leads on the cold sensor because the equation for the resistance R_T of the thermometer is given by

$$R_\mathrm{T} + r_1 = R_\mathrm{Ref} + r_2 , \tag{12.11}$$

whereas for a simple Wheatstone bridge r_1 and r_2 both add to R_T. This design also decouples the AC-power supply from the bridge and it uses an inductive voltage divider ("ratio transformer") to null the voltage across the bridge. A ratio transformer has a very small error per reading, is nearly unaffected by age, temperature and voltage, and it has a low impedance. The null detector is a phase-sensitive lock-in amplifier with a high voltage sensitivity. Finally, the reference resistor, which may be a stable metal film resistor, is kept at low temperatures to keep its resistance constant and the lead resistances in the two arms of the bridge as equal as possible to reduce the noise signal and improve the stability. A variable capacitor is included to null out the out-of-phase signal. Such a bridge can be homemade, with the lock-in amplifier and the ratio transformer as the most expensive components, or one can buy an AC resistance bridge commercially. Some modern resistance bridges for thermometry do not use a reference resistor but compensate the voltage drop across the resistor using feedback electronics. In this way, it is possible to realize fully automatic bridges, that can be controlled and read out by a computer.

The leads into the cryostat should be twisted pairwise, rigidly fixed and well shielded to avoid induced currents due to movements in electromagnetic fields which are always present. In the low-temperature part they should be superconducting if the working range is $T < 1\,\mathrm{K}$ to keep the heat flow to the thermometer small, and they have to be thermally anchored, preferably at a temperature as close as possible or slightly below the temperature the thermometer is exposed to. This thermal anchoring of the leads can be performed by glueing them on a Cu rod to which a layer of lens or cigarette paper has been glued with GE 7031 varnish (diluted by a toluene–methanol mixture). The Cu rod should be well annealed to keep its thermal conductivity high.

Equipment for very low temperature experiments is very often installed in a shielded room. The main advantage of such a setup is not so much that the shielded room keeps RF power away from the refrigerator, but that it keeps RF power away from the leads and from the thermometers to avoid heating of the temperature sensors. A final step to keep nuisance RF heating to the low-temperature sensor at a tolerable level is to instal a low-pass filter at low temperatures just in front of the low-temperature resistive elements. A design of such a filter is shown in Fig. 12.28 [12.6, 12.43, 12.48, 12.51]. It should be kept in a cold Cu housing for shielding.

A very successful design of a thermometer made from a Speer grade 1002–100 Ω-1/4 W resistor with appropriate thermal anchoring and shielding of it as well as of its leads has been described in [12.51]. The thermometer is ground down to 0.15 mm and fixed with Stycast 1266 epoxy; it showed a response time of 5 min at its lowest useful temperature of 5 mK, where it still did not show any saturation effects, which is attributed to the cold filter, the careful electromagnetic shielding of thermometer and leads, and the small measuring power of 4 fW.

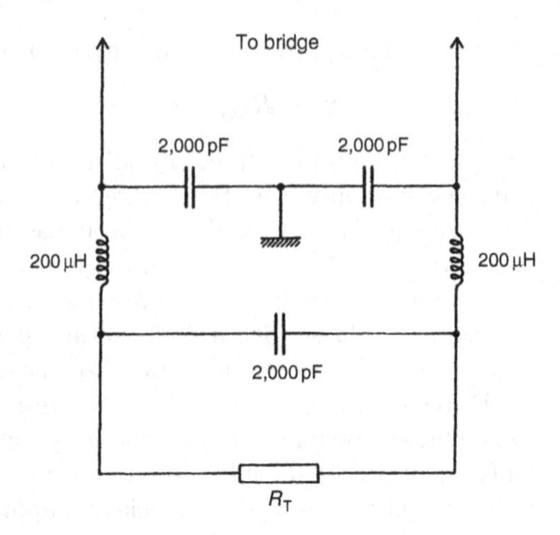

Fig. 12.28. Low pass filter between a resistor R_T used as a low-temperature thermometer and the bridge by which its resistance is measured. The cutoff frequency for the shown values of capacitors and inductances is about 0.1 MHz. The filter should be put into a Cu case to which all elements are well heat sunk. The inductances can be homemade and for the capacitors one could use commercial mica or, better, styroflex capacitors

12.6 Coulomb Blockade Thermometry

Modern micro- and nano-lithography, in particular electron-beam lithography, has opened the way to new concepts in thermometry with very small, very fast and even primary sensors. Because of their small size of order 1 μm^2, they measure the temperature locally – for example on the tip of a scanning tunneling microscope – they can be integrated into other micro-devices, and they can be fabricated reproducibly in large numbers in well-developed, controlled processes. However, one has to take into account that they allow only very small measuring powers to avoid self-heating and that the devices measure the *electron* temperature. The latter is particularly important at low temperatures because the heat transport between electrons and phonons decreases with T^5 ([12.86] and references in [12.87], p. 223), so that the electrons easily decouple thermally from the underlying lattice.

On one hand, important and well-developed micro-sensors are superconductor-insulator-normal metal (SIN) tunnel junctions. Their I-V characteristic depends on the temperature of the normal electrode only, yielding the Fermi distribution of the electrons as long as $T/T_c < 0.4$. They have a very steep tunneling characteristic at temperatures near to the superconducting transition temperature, which has made them attractive X-ray detectors, infrared bolometers, as well as thermometers for on-chip micro-coolers (see Sect. 13.9). SIS junctions on the other hand have very high energy resolution and are therefore used as photon and particle detectors but not much as thermometric devices [12.87].

However, the most promising micro-thermometric sensor is the Coulomb blockade thermometer (CBT) using single-electron tunneling; their development has been mostly advanced by the work of Pekola and coworkers [12.87–12.91]. In this device, tunnel junctions are connected in series by metallic islands. If the resistance of the junctions is higher than the quantum resistance $R_Q = h/4e^2$, a well-defined number of electrons are located on the islands. At finite temperature, single electrons (or holes) can occasionally tunnel through the barrier of the junctions due to their thermal energy. Each extra electron increases the Coulomb charging energy E_C of the islands. For small junctions with small capacitances, E_C can exceed the ambient thermal energy and the Coulomb repulsion prevents the tunneling of additional electrons. This "Coulomb blockade" leads to a drastic decrease of the tunneling current at $k_\mathrm{B}T < E_\mathrm{C}$ and low bias voltages V so that e $V < E_\mathrm{C}$.

A Coulomb blockade thermometer, however, is operated in the so-called weak Coulomb blockade regime ($k_\mathrm{B}T \approx E_\mathrm{C}$) where single-electron tunneling effects play a role and temperature influences the electric transport properties. In this regime, the I-V characteristic does not show a sharp Coulomb blockade gap but is smeared over the range e $V \sim k_\mathrm{B}T$. One then has a competition between thermal energy $k_\mathrm{B}T$, electrostatic energy e V at bias voltage V, and charging energy $E_\mathrm{C} = e^2/2C^*$ due to extra or missing individual electrons on the metallic electrodes, where C^* is the effective capacitance of the system. It is convenient not to measure the nonlinear I–V characteristic directly but to measure the differential conductance $G = \mathrm{d}I/\mathrm{d}V$ as a function of the bias voltage V at a frequency of typically a few tens of Hz. The result is a temperature dependent bell shaped dip in conductance around zero bias (Fig. 12.29) which is due to the Coulomb blockade; it increases with decreasing temperature. In lowest order of $E_\mathrm{C}/k_\mathrm{B}T$ one finds for the differential conductance of a symmetric linear array of N uniform junctions in series [12.87–12.89]

$$G(V)/G_\mathrm{T} = 1 - (E_\mathrm{C}/k_\mathrm{B}T)g(eV/Nk_\mathrm{B}T), \tag{12.12}$$

With $E_\mathrm{C} = e^2\ (N-1)/NC^*$; G_T is the asymptotic conductance at high voltage, and $g(x) = (x\sinh x - 4\sinh^2 x/2)/(8\sinh^4 x/2)$. The full width at half minimum of the conductance dip has in lowest order of $E_\mathrm{C}/k_\mathrm{B}T$ the universal value

$$V_{1/2} = 5.439Nk_\mathrm{B}T/e, \tag{12.13}$$

which does not dependent on any device parameter (except the number N of junctions), and allows to determine T without calibration. Its measurement to determine T requires measuring a full conductance curve by sweeping the bias voltage, which will take a few minutes. In a faster mode, one can measure

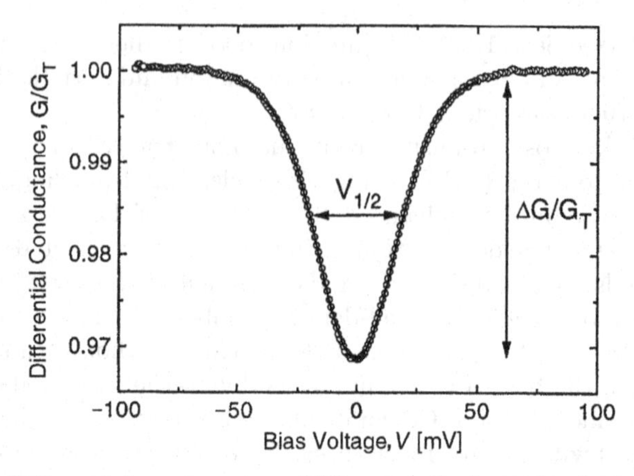

Fig. 12.29. Differential conductance G/G_{T} of a Coulomb blockade thermometer with $N = 20$ tunnel junctions at $T = 4.2\,\mathrm{K}$. The full line is the theoretical prediction (12.12). The two temperature dependent parameters $V_{1/2}$ prop. T (12.13) and $\Delta G/G_{\mathrm{T}}$ prop. T^{-1} (12.14) are indicated [12.88]

the depth of the dip, which is proportional to inverse temperature, and again in lowest order in $E_{\mathrm{C}}/k_{\mathrm{B}}T$, is given by

$$\Delta G/G_{\mathrm{T}} = E_{\mathrm{C}}/6k_{\mathrm{B}}T. \tag{12.14}$$

Equations (12.13) and (12.14) are the central results for a rather simple – requiring only a low-frequency resistance measurement – primary ($V_{1/2}$ with no adjustable parameter) or secondary ($\Delta G/G_{\mathrm{T}}$ with E_C as device parameter) thermometer. Introducing higher-order terms in $E_{\mathrm{C}}/k_{\mathrm{B}}T$ allows extending the range of operation of a CBT – which usually has a dynamic range of two decades in temperature – to lower T [12.87, 12.88].

A Coulomb blockade thermometer is a series of connected tunnel junctions biased at $+V/2$ and $-V/2$ at its two ends. Experiments have shown an excellent agreement between experimental data and theory, for example for the linear dependence of $V_{1/2}$ on temperature T (Fig. 12.30) or on the number N of tunnel junctions (Fig. 12.31) [12.87, 12.88]. Of course, one may have to take into account possible non-uniformities of the junctions. However, this influence is weak with changes in the I-V parameters of less than 1% if the areas of the junctions or their resistance vary by 10%, for example [12.87–12.89]. At low temperatures, a CBT can have an accuracy of 1% at 50 mK to 4 K, and 3% at 20 mK. The temperature ranges are limited by the finite height of the barrier as well as by the vanishing of the dip at high T (around 30–50 K). At low temperatures, the decoupling of the electrons from the lattice, the influence of background charges and possibly the approach to the full Coulomb blockade are the limiting factors [12.87–12.89].

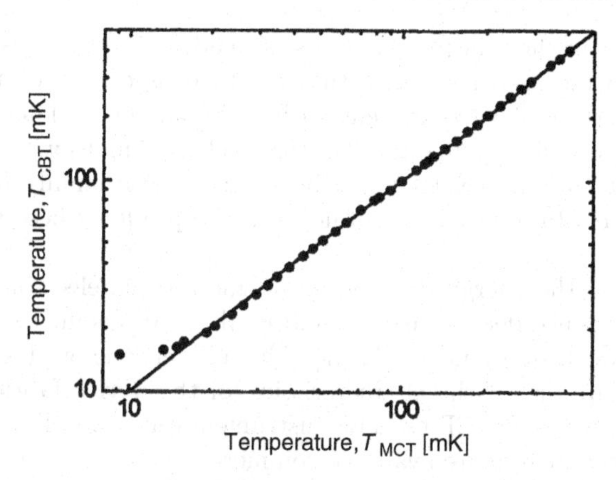

Fig. 12.30. Temperature T_{CBT} determined with a Coulomb blockade thermometer compared to the temperature determined with a ^{3}He melting curve thermometer. Saturation of the CBT values below 20 mK indicates the decoupling of the electrons from the phonons in the CBT [12.88, 12.89]

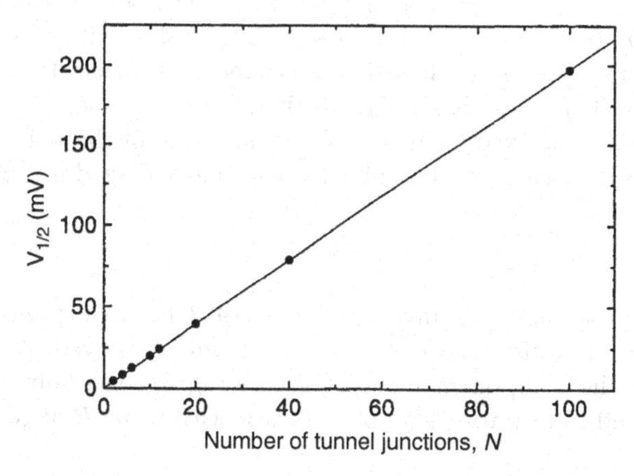

Fig. 12.31. Dependence of the full width $V_{1/2}$ at half minimum of the differential conductance of a Coulomb blockade thermometer (see Fig. 12.29) on the number of tunnel junctions of this thermometer at $T = 4.2$ K [12.88]

The problem of electron–phonon coupling at low temperatures seems to limit this thermometry device to $T > 10$ mK at present due to overheating the electrons by the measuring power (Fig. 12.30) (which is, of course, zero at zero bias) as well as by parasitic heat leaks. Actually, these devices are well suited to measure quantitatively the electron-phonon coupling at low temperatures [12.87, 12.89]. A CBT is insensitive to magnetic fields, because its operation is based on electrostatic properties. In [12.90], a sensor consisting of 100 tunnel junctions (100 nm normal conducting Al, Al_2O_3, 200 nm Cu) in series with ten of them in parallel (total size $30\,\mu\mathrm{m}^2$, total resistance $300\,\mathrm{k}\Omega$) was shown to be insensitive to within 3% to magnetic fields up to 27 T at $T = 50\,\mathrm{mK}$. An earlier experiment with a similar device had shown no observable magnetic field dependence of the CBT reading to within 1% in fields up to 23 T at 0.42–1.46 K [12.90]. The precision was mainly limited by the high noise level of the resistive coil necessary to produce these high magnetic fields.

A review of the energy distribution of mesoscopic electron systems, and how it can be controlled, measured, and used in various micro- or nanostructured devices can be found in [12.87]. This reference as well as [12.91] gives detailed descriptions of the characteristics, of the device fabrication and on the instrumentation for CBTs. This instrumentation as well as the Coulomb blockade thermometers are available commercially.

12.7 Noise Thermometry

The conduction electrons in a metal perform random thermal movements (*Brownian motion*), which result in statistical voltage fluctuations of a resistive element. Therefore all resistive elements of an electronic circuit are noise sources. This noise is statistical, therefore we cannot make statements about its value at a fixed time, and the mean value of the noise voltage vanishes. However, we can calculate an effective time averaged mean square noise voltage

$$U_{\mathrm{rms}} = \sqrt{\langle u^2 \rangle_t}\,. \tag{12.15}$$

This noise voltage was investigated in 1928 by J.B. Johnson (*Johnson noise*) and H. Nyquist (*Nyquist theorem*). Nyquist arrived at the following equation for the component of the time averaged noise voltage within the frequency band from ν to $\nu + \mathrm{d}\nu$ of a resistor with value R at temperature T:

$$\langle u^2(\nu) \rangle_{\mathrm{t}} = 4k_{\mathrm{B}} T R \,\mathrm{d}\nu\,, \tag{12.16}$$

valid for $\nu \ll k_{\mathrm{B}}T/h$.

If we measure within a frequency band of width $\Delta\nu$ we obtain

$$\langle u^2 \rangle = \int_{\nu}^{\nu+\Delta\nu} 4k_{\mathrm{B}}\, TR \,\mathrm{d}\nu = 4k_{\mathrm{B}}\, TR\Delta\nu\,. \tag{12.17}$$

This relation between noise voltage and resistance as well as temperature can be obtained by the following reasoning [12.92, 12.93]. If we connect two resistors – which act as noise sources providing random AC voltages of uncorrelated frequencies and phases – via lossless leads (Fig. 12.32), the power transported from R_1 to R_2 and vice versa is

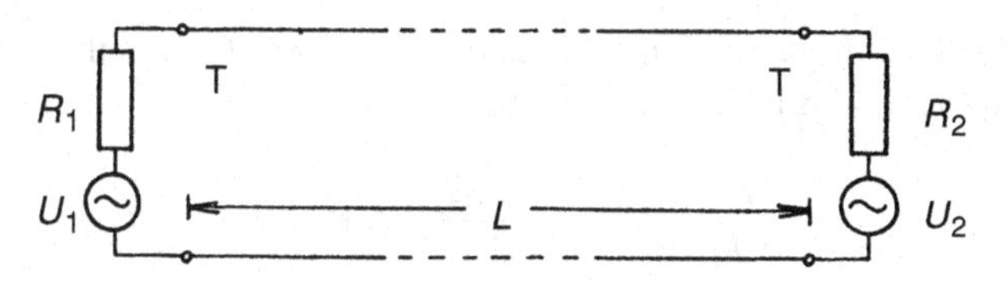

Fig. 12.32. Wiring to obtain the Nyquist theorem (see text)

$$\dot{Q}_{12} = R_2 I_1^2 = \frac{R_2 U_1^2}{(R_1+R_2)^2}$$

and (12.18)

$$\dot{Q}_{21} = R_1 I_2^2 = \frac{R_1 U_2^2}{(R_1+R_2)^2},$$

respectively. In thermal equilibrium $\langle \dot{Q}_{12} \rangle = \langle \dot{Q}_{21} \rangle$, and we have

$$\frac{\langle U_1^2 \rangle}{R_1} = \frac{\langle U_2^2 \rangle}{R_2}, \tag{12.19}$$

the R dependence of the noise voltage we were looking for.

To obtain the temperature dependence of the noise voltage we set the two resistances equal to each other, and join them by a connection with vanishing ohmic resistance but with an impedance $Z = R$. If we short the leads at time t, then the total transported energy at this time is trapped, and the enclosed wave train can be written as a sum of the eigenfrequencies of the transforming leads with length L,

$$\nu_n = \frac{nv}{2L} \tag{12.20}$$

with $n = 1, 2, \ldots$ and v being the transport velocity in the leads.

For large n, we can write

$$\mathrm{d}\nu = \frac{v\ \mathrm{d}n}{2L} \tag{12.21}$$

In thermal equilibrium at temperature T each mode has an electric and a magnetic degree of freedom with a mean energy $k_\mathrm{B}T/2$. Therefore the total energy in the interval $\mathrm{d}\nu$ is given by

$$\langle \mathrm{d}Q \rangle = 2\ \mathrm{d}n k_\mathrm{B} T/2 = 2 k_\mathrm{B} T L\ \mathrm{d}\nu / v\,, \tag{12.22}$$

which is just the energy Q transported from R_1 to R_2, or vice versa, during the time $t = L/v$,

$$\langle \mathrm{d}\dot{Q}_{12} \rangle = \langle \mathrm{d}\dot{Q}_{21} \rangle = \langle \mathrm{d}Q \rangle / 2t = k_\mathrm{B} T\ \mathrm{d}\nu\,. \tag{12.23}$$

Therefore we have, with (12.18) and $R_1 = R_2$,

$$\langle u^2(\nu) \rangle = 4 \dot{Q} R = 4 k_\mathrm{B} T R\ \mathrm{d}\nu\,. \tag{12.24}$$

If we perform a quantum mechanical calculation, the result is the replacement of $k_BT/2$ (valid at $h\nu \ll k_BT$) by

$$\frac{h\nu/2}{\exp(h\nu/k_BT)-1} \Rightarrow \frac{k_BT}{2} \quad \text{for} \quad h\nu \ll k_BT\,, \tag{12.25}$$

where the last requirement, $\nu/T \ll 20\,\mathrm{GHz\ K^{-1}}$, is always fulfilled in practice. Equations (12.16, 12.17) are also obtained for this limit.

The experimental problems in using noise voltages for thermometry result mainly from the rather small size of the effect [12.3, 12.4, 12.10, 12.11, 12.92–12.97]. If we take, for example, $R = 1\,\mathrm{k\Omega}$ and $\Delta\nu = 1\,\mathrm{kHz}$, then

$$\text{at } 4\,\mathrm{K}: \quad U_{\mathrm{rms}} \simeq 10^{-8}\,\mathrm{V},\ \dot{Q} \simeq 10^{-19}\,\mathrm{W}, \tag{12.26}$$

and

$$\text{at } 10\,\mathrm{mK}: \ U_{\mathrm{rms}} \simeq 10^{-9}\,\mathrm{V}, \dot{Q} \simeq 10^{-22}\,\mathrm{W}. \tag{12.27}$$

At least the latter values are usually not measurable with semiconductor amplifiers and one has to use a SQUID with its extremely low intrinsic noise as the amplifying element. Two types of absolute noise thermometer have been used, which both measure the noise voltage by a resistor at the temperature T using a SQUID as a detector and amplifier.

In one method, the "resistive SQUID method", the resistor is directly connected to the Josephson tunneling junction(s) of the SQUID [12.10, 12.20, 12.94, 12.95]. Making use of the AC Josephson effect with a voltage across the junction, the SQUID serves as a very accurate voltage-to-frequency converter. The equation relating the voltage V to the Josephson frequency, $\nu_J = (2\,e/h)\,V$, contains only fundamental constants; the frequency is 484 MHz for a voltage of $1\,\mu\mathrm{V}$. Thus, the noise voltage is measured by a frequency counter. The detector bandwidth $\Delta\nu$ is equal to $1/2\ \tau$ (τ: gate time of the frequency counter; in [12.95] for example, a time of 20 ms was chosen). Most importantly, the result is independent of circuit parameters like bandwidth or amplification factor, making it in principle a primary temperature technique. However, the oscillations are extremely small and special amplification and detection electronics is required. Further problems of this method, for example modification of the Johnson noise of the resistor by the Josephson junction and the impact of noise from external parts of the circuit, are discussed in [12.20, 12.94, 12.95]. Nonetheless, nowadays uncertainties as small as a few 0.1% at temperatures above 10 mK and 1% to about 1 mK can be obtained.

In the alternate technique, the "current sensing noise thermometer", the resistor is inductively coupled to the SQUID, which serves as a low-noise amplifier of the current generated by the thermal noise in the input coil (Fig. 12.33) [12.96, 12.97]. Analogous to (12.24), the mean square noise current flowing in the SQUID input coil per unit bandwidth $\Delta\nu$, arising from the thermal noise in the sensor, is given by

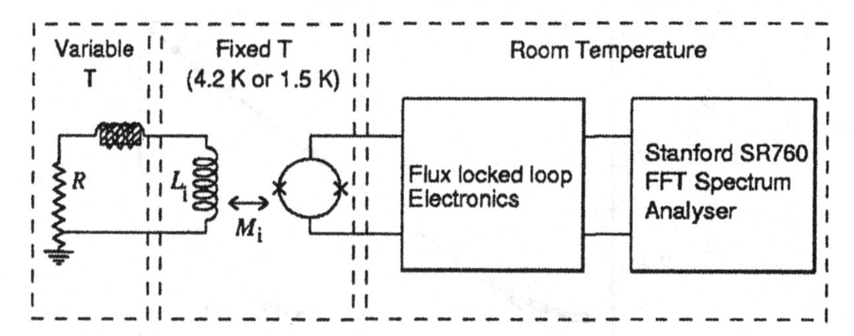

Fig. 12.33. Schematic diagram of the current sensing noise thermometer of [12.97]. One end of the noise resistor R is grounded to the thermal reservoir whose temperature is to be measured. The SQUID with its input coil L_i is held at a fixed temperature of 1.5 or 4.2 K for stability. A superconducting fixed-point device (see Sect. 11.4.3) can be incorporated into the SQUID input circuit, providing temperature fixed-points for calibration. Reprinted from [12.97], copyright (2003), with permission from Elsevier

$$\langle I^2 \rangle = [4k_BT/R]\Delta\nu/[1 + (2\pi\nu\tau')^2], \tag{12.28}$$

with the time constant $\tau' = L/R$ determining the bandwidth, and L the total inductance of the input circuit. A rather small value R for the resistor, typically 0.1–1 mΩ, has to be chosen to match the circuit to the SQUID. In order to use this method in an absolute mode as a primary thermometer, a number of the parameters of the relevant circuit have to be measured independently.

In [12.96], all the required parameters of the circuit were measured or calculated, and the device was indeed used as a primary thermometer with an accuracy of 3% from 4 mK to 4 K. By the use of the SQUID as a low-noise amplifier, the device noise temperature was kept at 50 μK. Whereas in these early circuits for noise thermometry bulk resistive RF-SQUIDs with one Josephson junction were used as detectors and amplifiers, they were replaced in the more recent designs [12.97] of current noise-sensing thermometers by thin-film resistive DC-SQUIDs with two Josephson junctions. Their lower noise and higher sensitivity result in a higher accuracy and a shorter measuring time. The noise of the SQUID can be parameterized by the coupled energy equivalent sensitivity ε_c (for the definition see [12.97]) of the minimum detectable current in its input coil; it is around $500h$ (h: Planck's constant) for present commercial SQUIDs. This parameter ε_c determines the noise temperature

$$T_N \approx \varepsilon_c/2k_B\tau' \tag{12.28a}$$

and the precision $\Delta T/T$ (see below).

The SQUID is used in a flux-locked loop mode. This design results in a significant improvement in speed and the possibility to use the circuit to measure much lower temperatures. Using a 0.34-mΩ Cu foil resistor, the equivalent amplifier noise temperature in the circuit of [12.97] was 8 μK in the frequency range of 0.5–1 kHz. Using their setup as a primary thermometer

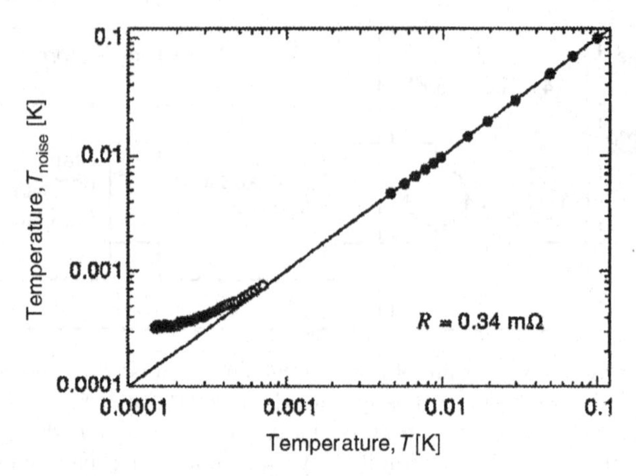

Fig. 12.34. Temperatures obtained from the current sensing noise thermometer of [12.97] vs. temperatures from a ^{3}He melting pressure thermometer (•) and a platinum NMR thermometer (○). At $T < 4.5$ mK, the noise thermometer reads hotter because of a heat leak into it. Reprinted from [12.97], copyright (2003), with permission from Elsevier

(by measuring the resistance value and SQUID gain) and fitting the measured frequency-dependent noise spectra to (12.28), the authors have achieved 1% agreement of the measured temperature with the temperature from the ITS-90 scale (Sect. 11.2) at 4.2 K. In a secondary mode, the device was calibrated at 100 mK and temperatures were measured from 4.2 K to below 1 mK. They agreed down to 4.5 mK to within 1% to temperatures obtained from a ^{3}He melting curve thermometer using the PLTS-2000 scale (Sect. 11.3). At lower temperatures, the temperatures from the noise thermometer were higher than those from a platinum NMR thermometer (Fig. 12.34), indicating a heat leak into the noise sensor. About 3.3 min of measuring time were required for 1.5% accuracy in temperature. With a higher resistance value, for example 5 mΩ, the noise temperature would increase to 30 μK, but 1% accuracy could be achieved in 10 s only.

Because of the stringent experimental requirements, noise thermometry will be used mostly by standards laboratories for calibration purposes while the experimenter will resort to some simpler means of thermometry. Indeed, noise thermometry has played an essential role in establishing the new temperature scale PLTS-2000 (Sect. 11.3). Very detailed investigations performed in this context have shown that the two main error sources of current noise thermometers seem to be parasitic noise from the measuring equipment, which adds to the noise from the thermometer, as well as heat leaking into the noise thermometer. The first problem can easily be taken into account in the analysis if it is temperature independent [12.10, 12.95].

To reduce the impact of parasitic heat input, a SQUID noise thermometer without any leads to the sample was recently used between 6 mK and

4 K [12.98, 12.99]. These authors used a superconducting pick-up coil wound around a high-purity Au sample as noise source. The coil is connected to the input coil of a DC-SQUID held at 1.2 K to form a superconducting flux transformer. The thermal motion of the conduction electrons in the Au sample cause fluctuations of the magnetic flux in the pick-up coil which are sensed by the SQUID. A resolution of 1% was achieved within 13 s.

In principle, a noise thermometer is a primary thermometer. In practice there are many severe experimental problems besides the small voltages, so that this method is not in wide use in research laboratories. Some of these problems are the following. All resistive elements of the electronic setup for noise thermometry are noise sources themselves. This situation has been improved by using a SQUID as a low-temperature amplifier. The remaining equipment is at temperatures between 4 K and 300 K and noise from the resonance circuit, for example, may affect the performance of the SQUID as well. Another important parameter is the effective bandwidth of the electronics and, last but not least, one has to determine the amplifying factor or gain of the total system. If noise thermometry is used as secondary thermometry, the system can be calibrated by putting the resistor at a known temperature. Another possibility is to build two identical electronic circuits and calibrate the two against each other, where one of them has the sensor as the noise source at unknown temperature and the other one has its sensor at a known temperature.

If we detect noise voltages, we are detecting statistical events, which have to be time averaged and, as for all statistical events, the accuracy depends on the measuring time t. If $t\Delta\nu \gg 1$, then the standard deviation for the temperature measurement via noise thermometry is given by

$$\Delta T/T = (\alpha\tau/t)^{1/2} . \tag{12.29}$$

with α a constant of order unity and depending on the method used. This means that the accuracy does not depend on temperature, which is an advantage if we want to apply this method at low temperatures.

12.8 Capacitance Thermometry

Capacitance measurements are as simple as resistance measurements and they usually have the advantage of negligible heating of the investigated samples. Actually these measurements can be performed with very high precision, in general much more accurately than a resistance measurement with comparable electronic effort. Therefore, if one has a material whose dielectric constant changes with temperature this will be a very attractive thermometric parameter to be measured in a capacitor. Indeed, some dielectric materials, in particular amorphous or glassy materials, show changes of their dielectric constant down to the lowest investigated temperatures [12.99–12.101]. Investigations by Frossati and coworkers [12.102, 12.103] of the low-frequency

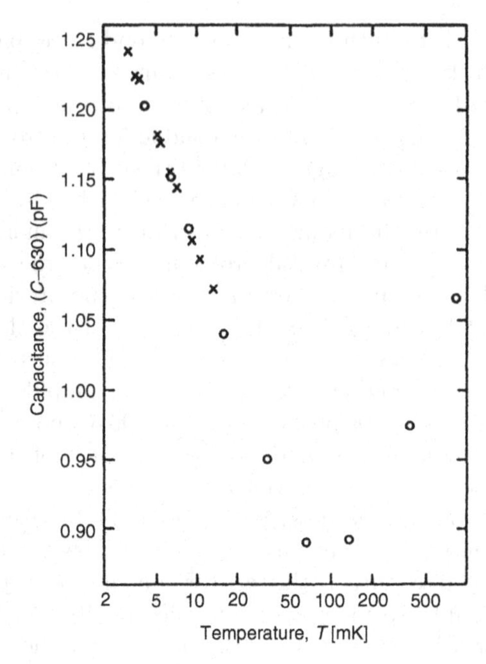

Fig. 12.35. Capacitance (−630 pF) of a capacitive glass thermometer (vitreous silica containing 0.12% OH^- ions) as a function of temperature. The data points were obtained at excitation frequencies of 1.0 and 4.7 kHz, at zero magnetic field (×) and at $B = 9\,\mathrm{T}$ (○) [12.103]

capacitance of a glass thermometer made from SiO_2 which contained about 1,200 ppm OH^- gave very encouraging results in the temperature range from 4 to 100 mK (Figs. 12.35 and 12.36). In particular, they showed that such a thermometer is field independent for fields up to 9 T (within 5%), and it has a time constant of order 1 s to the lowest investigated temperatures [the specific heat is typically 10 T (erg/g K), see Sect. 3.1.4]. In addition, heating effects are negligible, typically $< 10^{-12}$ W, because of the very small dielectric losses of vitreous silica, and because it is almost not affected by stray RF fields which limit the use of carbon resistors to $T > 5$ mK, for example (Sect. 12.5.2). The measurements can be performed with a commercial capacitance bridge or a bridge, as shown in Fig. 3.32 (with the inductances replaced by capacitors), see also the end of Sect. 13.1.

The dielectric constant ϵ of glasses (for example, Suprasil, Homosil, Spectrosil, BK 7, Kapton, smoky quartz [12.100–12.106]) as a function of temperature passes through a minimum whose position (at about 70 mK for 1 kHz) depends on frequency ν as $T_{\min} \underset{\sim}{\propto} \nu^{1/3}$ [12.102, 12.103]. At lower temperatures ϵ is independent of frequency for $h\nu \ll k_B T$, with a slope $\mathrm{d}\ln\epsilon/\mathrm{d}\ln T \approx -10^{-4}$ (Fig. 12.35) and saturating at a few millikelvin (but see [12.106]). This temperature dependence of ε is, like the thermal and acoustic properties of glasses at low temperatures, explained by the tunneling

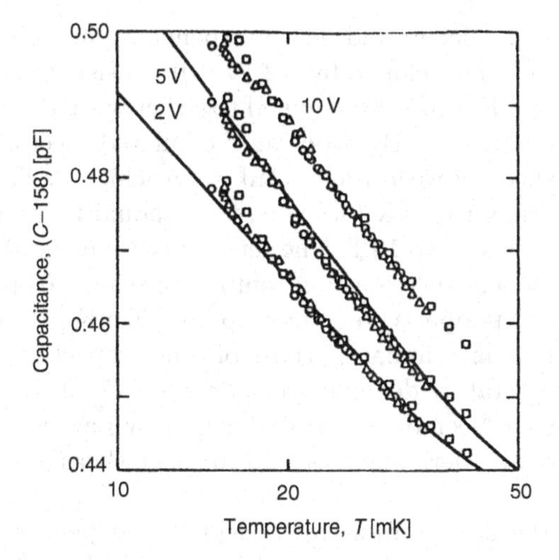

Fig. 12.36. Capacitance (−158 pF) of a capacitive glass thermometer as a function of temperature at the three indicated measuring voltages but all at the same frequency (4.7 kHz). The measurements were performed in magnetic fields of 0.0 T (○), 0.25 T (△), and 6.0 T (□) [12.102]

model [12.99–12.101]. The sensitivity $\mathrm{d}\ln\varepsilon/\mathrm{d}\ln T$ of Suprasil and Homosil glasses increases with their OH^- content for concentrations between 200 and 1,000 ppm, indicating that the OH^- dipoles dominate the dielectric behavior [12.104]. It was found that the dielectric constant depends also on the excitation voltage [12.102, 12.103, 12.105] (Fig. 12.36). Therefore these thermometers should be operated at constant voltage. The latter nuisance effect is influenced by the design of the glass thermometer (construction of electrodes, contacts, leads, etc.). Typical shapes of these thermometers are plates or tubes of 10–40 mm length and width, and some 0.1 mm thickness, covered with fired or sputtered noble metal electrodes, giving typical capacitances of 0.1–1 nF. The measurements are performed at some kilohertz and with voltages of 0.1 V to some volts applied to the electrodes, giving a sensitivity of about 10^{-4} pF at 100 pF and a temperature resolution of 0.1 μK at some millikelvins, even though the absolute changes of ϵ with T are small (Figs. 12.35 and 12.36). Unfortunately, the thermometers usually have to be recalibrated in each cooldown, mostly due to changing contributions from the leads. The recent result [12.107] that the velocity of sound and the internal friction of metallic and dielectric glasses change with temperature to at least 0.1 mK may make these thermometers suitable even for the microkelvin temperature range.

A capacitance thermometer based on the incommensurate crystal $(Pb_{0.45}Sn_{0.55})_2P_2Se$ has shown an order of magnitude higher sensitivity $\mathrm{d}\ln C/\mathrm{d}\ln T$ of about $5*10^{-3}$ and a negligible field dependence for fields up to 20 T [12.108].

A rather simple, efficient and field-independent parallel plate capacitance thermometer has been developed by [12.109]. It consists of 17.8-μm-thick copper foils and 7.6-μm Kapton sheets glued together with a thin layer of Stycast 1266 and rolled into a tube. By choosing the foil and sheet dimensions appropriately, capacitances between 1 and 52 nF were achieved. The capacitors were investigated at 20 mK to 1.6 K (actually, they should be usable to room temperature) and at fields up to 18 T. The sensitivity is about 10^{-4} and the upper limit of the field dependence is 1% in temperature and 70 ppm in capacitance in the investigated temperature range up to 400 mK. Their frequency and voltage dependences are similar to those of other capacitance thermometers.

A suitable temperature dependence of ϵ down to 7 mK and the absence of a field dependence had been demonstrated earlier by Lawless and coworkers for $SrTiO_3$ glass–ceramic thermometers [12.110]; such thermometers are available commercially.

Capacitance thermometers with the negligible dependence of their dielectric constant on magnetic fields seem to be a very good choice for thermometry in high magnetic fields. However, the thermal coupling of a dielectric material whose capacitance is supposed to be measured and hysteresis effects have to be investigated; actually, most of the capacitance thermometers have to be recalibrated in each cooldown. In addition, the time-dependent heat release of non-crystalline solids (Sect. 10.5.3) has to be taken into account.

12.9 Magnetic Thermometry with Electronic Paramagnets

The magnetic thermometric method to be discussed in this section has been for a long time the conventional way to measure temperatures below 1 K. Its basis is the T^{-1} dependence of the magnetization

$$M = \frac{\lambda B}{\mu_0 T} \tag{12.30}$$

or of the susceptibility

$$\chi = \mu_0 M/B = \lambda/T \tag{12.31}$$

of a paramagnet with the Curie constant

$$\lambda = \frac{N_0 J(J+1)\mu_0 \mu_B^2 g^2}{3k_B}\,. \tag{12.32}$$

These equations are valid in the high-temperature approximation $\mu\mu_B B \ll k_B T J$, where the susceptibility follows the Curie law. The above relations were deduced in Chaps. 9, 10 for non-interacting magnetic moments in an external magnetic field.

Due to the simple relation between the measured parameters χ or M and temperature, magnetic thermometry, in principle, is a primary method. However, again there are problems and deviations, making magnetic thermometry in practice a secondary method. The moments in paramagnetic dielectrics experience a local magnetic field B_{loc} which is different from the externally applied field B_{ex} and which has the following three components [12.3, 12.4, 12.111]:

$$B_{\mathrm{loc}} = B_{\mathrm{ex}} + B_{\mathrm{d}} + B_{\mathrm{w}} \,. \tag{12.33}$$

The demagnetization field B_{d} resulting from the magnetization of the paramagnet in a field is given for an ellipsoid magnetized in the direction of its symmetry axis by

$$B_{\mathrm{d}} = -f\mu_0 M/V \,. \tag{12.34}$$

The geometry factor f can be calculated for various ratios of length to diameter of an ellipsoidal sample

l/d :	1	1.5	2	3	4	5	6	∞
f :	0.333	0.233	0.174	0.108	0.075	0.056	0.043	0

These effects are of great importance, of course, for superconductors (which are ideal diamagnets) as well as for ferromagnets; they can be quite appreciable for paramagnets as well.

The Weiss field B_{w} results from the neighbouring partially aligned dipoles and is

$$B_{\mathrm{w}} = \alpha\mu_0 M/V \,. \tag{12.35}$$

The parameter α depends on the symmetry of the crystal; for a cubic crystal $\alpha = 1/3$. We then have

$$B_{\mathrm{loc}} = B_{\mathrm{ex}} - (f - \alpha)\mu_0 M/V \,. \tag{12.36}$$

The local susceptibility

$$\chi_{\mathrm{loc}} = \mu_0 M/B_{\mathrm{loc}} = \lambda/T \tag{12.37}$$

follows the Curie law. Actually we are measuring

$$\chi = \mu_0 M/B_{\mathrm{ex}} \,, \tag{12.38}$$

which leads to

$$\chi = \chi_0 + \frac{\lambda}{T - (\alpha - f)\lambda/V} = \chi_0 + \frac{\lambda}{T - \Delta} \,, \tag{12.39}$$

where a (hopefully temperature independent) background contribution χ_0 resulting from other parts of the experimental setup – besides the thermometric sample – has been included. The latter relation (without the first term) is known as the Curie–Weiss law. The Weiss constant Δ depends on the shape of the sample, the symmetry of the crystal, and the interactions between the moments. It can be small or may even vanish for a sphere of a crystal with cubic symmetry, where $B_{loc} = B_{ex}$. In addition, of course, it is small for small Curie constants λ. Because of the usually unknown parameters in (12.39), we have to calibrate our thermometer at several temperatures, making it a secondary thermometer. This calibration can be performed against the ^{3}He vapour pressure at $T \geq 0.4$ K, against a superconducting fixed point device or the ^{3}He melting pressure at $T \leq 0.3$ K, etc.

Substances suitable for electronic magnetic thermometry are paramagnets containing elements with partly filled 3d or 4f electronic shells. As mentioned in Chap. 9, the paramagnetic salt with the lowest ordering temperature is CMN with $T_c \simeq 2$ mK. This value can be further reduced by partly replacing the magnetic Ce^{3+} ions in CMN by non-magnetic La^{3+} ions, possibly to $T_c \leq 0.2$ mK if only 5% or less cerium remains [12.112–12.115]. These salts can be used over the widest temperature range known and indeed it is the most widely applied paramagnetic thermometer [12.3, 12.4, 12.7, 12.8, 12.10, 12.20, 12.25, 12.96, 12.111–12.121]. It has also played an important role in establishing the new low-temperature scale PLTS-2000 (see Sect. 11.3). A CMN single crystal is a nearly perfect paramagnet if the temperature is not too low, say for $T > 5$ mK. The use of a single-crystal sphere is, of course, inconvenient. Therefore, in most cases powders compressed to a cylindrical shape have been used. However, its thermal conductivity is low and it is difficult to make thermal contact to it. Hence, usually a mixture of CMN powder plus grease (or another suitable liquid) has been compressed together with a brush of fine metal wires (Fig. 12.37). Thermal contact is then made via these wires. Of course, if the temperature of liquid ^{3}He is to be measured, CMN powder can be immersed in the liquid itself (Fig. 12.37), taking advantage of the low thermal boundary resistance between these two materials (Sect. 4.3.2). Many groups have used such a thermometer in the shape of a cylinder with its diameter equal to its length to keep deviations from the Curie law small and to make their temperature scales comparable [12.20, 12.116, 12.117, 12.121]. It has been

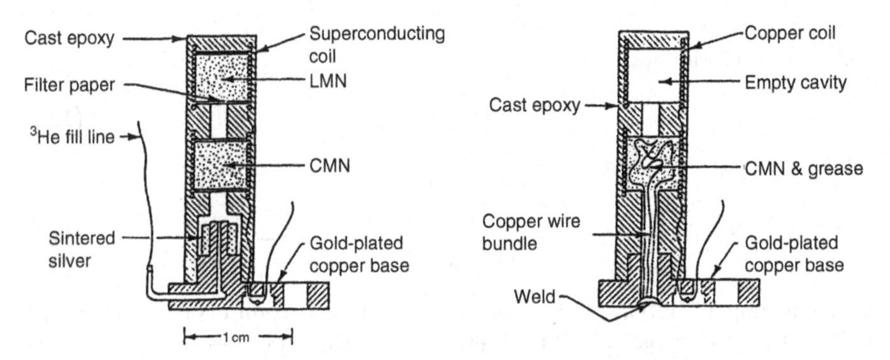

Fig. 12.37. Schematic of setups for paramagnetic thermometry with paramagnets such as CMN; for details see text [12.25]

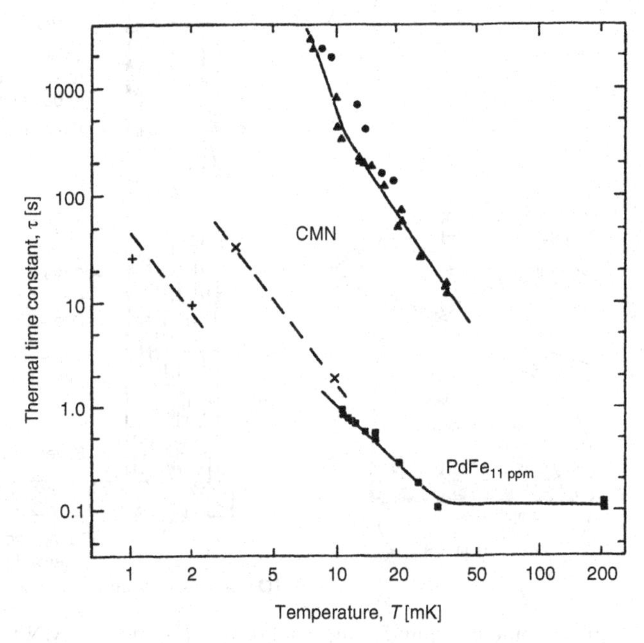

Fig. 12.38. Thermal time constant of the indicated paramagnetic thermometers. The data (+, ×, •, ▲) for CMN are from [12.25, 12.118, 12.120], respectively. The PdFe data are from [12.120]

found that the Weiss constant $|\Delta|$ is in the range of several 0.1 mK and that such a CMN thermometer can follow the thermodynamic temperature scale to 3 mK to with in about 0.2 mK, but a more typical limit is 5–8 mK. The thermal time constant of such thermometers still is of order 10 (10^3) s at 40 (10) mK (Fig. 12.38) [12.25, 12.118, 12.120]. This has recently been improved by a design described in [12.118], see Fig. 12.39. This thermometer follows a Curie–Weiss law between 7 and 250 mK with $\Delta = -0.004$ mK. To obtain a resolution of 10^{-4} at 8 mK, a drive level in a bridge of the design shown in Fig. 3.32 was necessary which gave a heating of 10 pW (probably mostly eddy currents in the Ag). Most important, this CMN thermometer has a time constant of only 10 s at 2 mK (Fig. 12.38). With a modified fitting equation, it has been used to 1 mK, and, with the Ce^{3+} ions partly replaced by La^{3+} ions, even to 0.4 mK. With the improvement of measuring techniques the use of salts with larger Curie constants than CMN – and correspondingly higher ordering temperatures [12.3, 12.4, 12.7] – does not seem to be warranted anymore, so CMN seems to remain unrivalled among the salts. This has been confirmed by the recent results of [12.20], where resolution and reproducibility of about 10^{-4} have been obtained between 0.5 and 3 K for a CMN powder thermometer.

To avoid the thermal equilibrium and chemical stability problems, in particular dehydration, (at 25°C the water vapour pressure of CMN corresponds to 25% humidity) encountered when paramagnetic salts are employed for

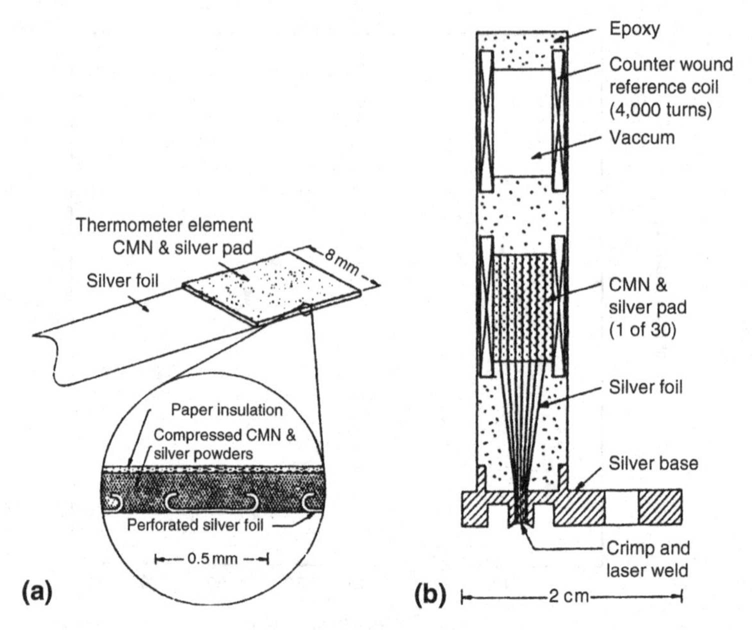

Fig. 12.39. Single element (**a**) and completed design (**b**) of the CMN thermometer of [12.118]

thermometry, various dilute paramagnetic *alloys* have been used for magnetic thermometry. These metallic alloys have a reasonably good thermal conductivity, and it is easy to make thermal contact to them. An example is Cu with some ppm of Mn [12.122]. With such a sample, 1% accuracy in a field of 0.1 mT at 5 mK can be obtained using a SQUID as the detecting element.

At the University of Bayreuth, we have very successfully used *Pd*Fe alloys with Fe concentrations of the order of 5–30 at ppm [12.38, 12.120, 12.123]. If one buys "pure" Pd of 4 N (or 5 N) purity it usually contains just the right concentration of Fe impurities. In these alloys, the Fe and its Pd surroundings have a "giant" magnetic moment of $14\,\mu_B$, giving a rather large signal. Spin glass freezing of these moments occurs at T_f (mK) $\simeq 0.1 x_{Fe}$ (ppm) (where x_{Fe} is the concentration of Fe in ppm) [12.38], and the susceptibility follows a T^{-1} dependence to quite low temperatures (Figs. 12.40 and 12.57); the data indicate that $|\Delta| < 0.2$ mK. The accuracy of the calibration is better than 1%, and the sensitivity is about 10^{-4} at 10 mK and 10^{-3} at 100 mK, comparable to CMN thermometers. The very simple design of such a thermometer is shown in Fig. 12.41. Due to its good thermal conductivity and ease of making thermal contact, such a thermometer has a rather short time constant of at most 1 (0.1) s at temperatures above 10 (30) mK (Fig. 12.38). Such a PdFe thermometer calibrated with an NBS superconducting fixed-point device (Sect. 11.4.2) is our standard thermometric method for the temperature range between about 5 and 500 mK. The only drawback of a PdFe thermometer is

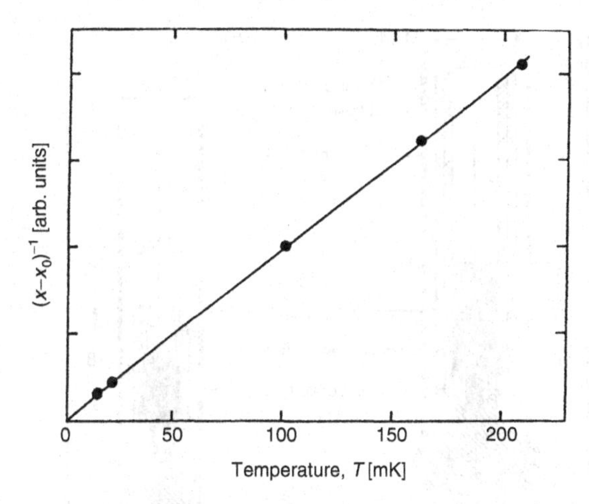

Fig. 12.40. Reciprocal of the susceptibility χ minus a background susceptibility χ_0 of a *Pd*Fe sample containing 15 ppm Fe as a function of temperature. The temperatures for calibration of the susceptibility thermometer are obtained from a superconducting fixed-point device such as the one shown in Fig. 11.1 [12.120]

its sensitivity to magnetic fields (it has to be shielded from stray fields, but the earth's field is okay, therefore one can use a superconducting shield) [12.38].

Another paramagnetic metallic spin glass very suitable for susceptibility thermometry is *Au*Er^+ [12.124]. Because of the weaker exchange interaction between Er ions in Au compared to Fe in Pd, Er concentrations of several 100 ppm can be used without the magnetization showing significant deviations from the Curie–Weiss law. For example, the AC susceptibility χ of an Au sample containing 660 pm Er follows this law to about 1 K [12.124]. The ratio of the temperature at the maximum of χ to the content of Er^+ ions (1.4 μK/ppm) is about three orders of magnitude lower than for Fe^{3+} impurities in Cu (1 mK/ppm). In the setup shown in Fig. 3.28, this thermometer has been used for thermal conductivity measurements to 6 mK.

High-resolution magnetic thermometers using paramagnetic salts, in particular for the investigation of critical phenomena like heat capacity and thermal conductivity at the very sharp lambda transition of liquid ^{4}He, have been developed by the groups of Lipa et al. [12.125–12.127], (Fig. 12.42) and then Ahlers et al. [12.128]. Both groups have used copper ammonium bromide ($Cu(NH_4)_2Br_4\ 2H_2O$) either directly in liquid helium or grown onto a matrix of Cu wires for thermal contact. The salt is used at temperatures closely above its magnetic ordering transition at 1.83 K, where the susceptibility changes strongly with temperature. This change induces a persistent current in a superconducting coil wound around the sample. The current is transferred by a superconducting flux transformer to an RF-SQUID and amplified for readout. Typical sensitivities are between 0.3 and 1 μK/Φ_o(Φ_o: flux quantum) in a field of 10 mT near the Curie temperature of the salt. With a resolution of

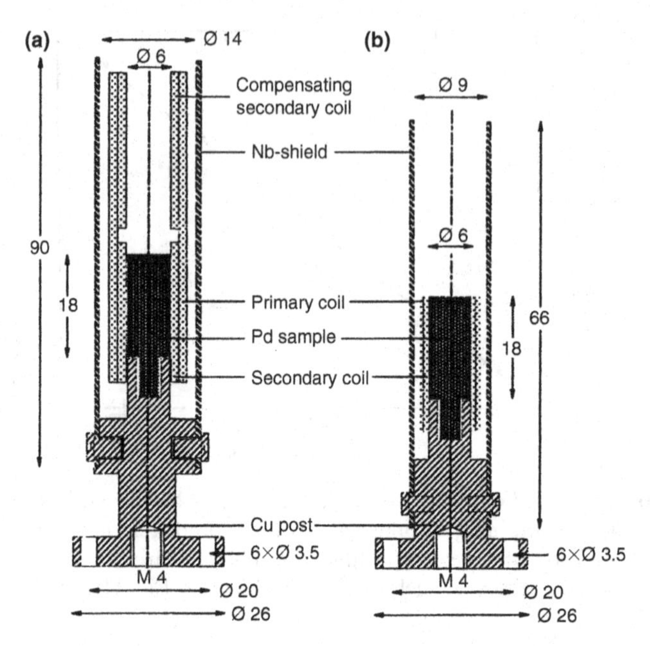

Fig. 12.41. Design of *Pd*Fe susceptibility thermometers. (**a**) With compensated secondary coils; (**b**) without compensated secondary coils. Dimensions are given in mm. The sample is a palladium rod containing 10–30 ppm Fe. Slits should be cut into the Pd rod to reduce eddy-current heating. If the susceptibility is measured with a SQUID then the primary coil may contain two layers of 0.1 mm Cu-cladded NbTi wire and the secondary coil ten layers of 25 µm Cu wire. If the susceptibility is measured with a mutual inductance bridge the number of windings should be increased, e.g., to 400 and 10^4, respectively. The two coils can be wound directly onto each other. The sample is protected against magnetic fields by a superconducting Nb shield

10^{-4} Φ_o of the SQUID this then results in sub-nanokelvin temperature resolution (Fig. 2.10b), a noise level of 3×10^{-10} K/$\sqrt{\text{Hz}}$, and drift rates of less than 0.1 nK h^{-1} [12.125–12.127]. Of course, to achieve these results requires a careful design with a very stable control of the temperature of four thermal control stages in series with the calorimeter as well as an effective magnetic shielding of thermometers and leads to the SQUID.

Using salts with Curie temperatures in other temperature ranges, this type of thermometer may be used there. For example, the miniature susceptibility thermometer of [12.129] using $\text{La}_x\text{Gd}_{1-x}\text{Cl}_3$ with different x to adjust the Curie temperature to the desired temperature range ($T_c = 2.2$ K for $x = 0$) has achieved a resolution of 0.2 nK in a 1-Hz bandwidth, a drift rate of less than 0.2 nK h^{-1} at 2.2 K, and a time constant of 30 ms.

A high-resolution magnetic thermometer using the giant-magnetic-moment alloys PdMn_x (or PdFe_x) has been described in [12.130]; again, the Curie temperature can be adjusted by changing x, for example to 1.5–3 K

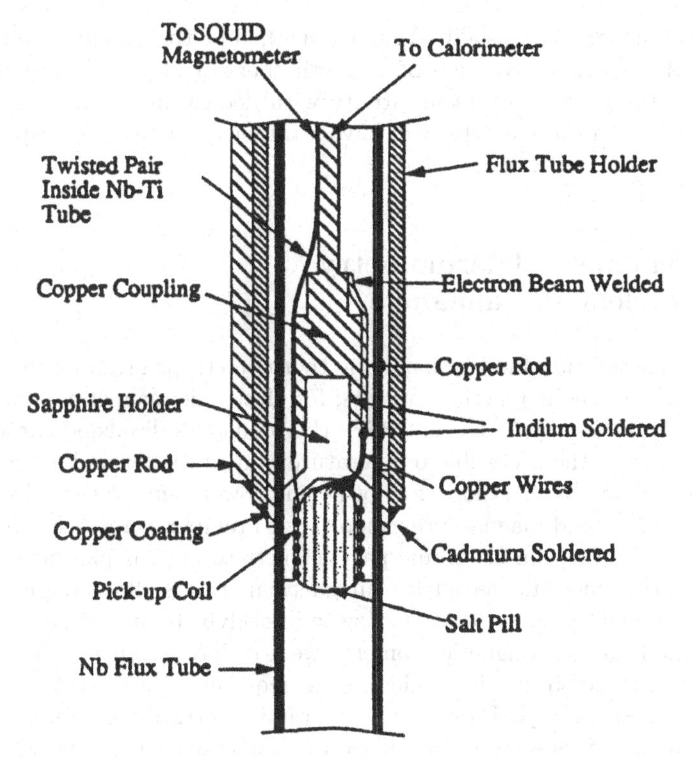

Fig. 12.42. High-resolution paramagnetic susceptibility thermometer used in [12.125–12.127] for measurements of the heat capacity of liquid ^{4}He very close to its superfluid transition temperature (see Fig. 2.10)

for $x = 0.6$–0.9 at.% Mn. The advantages of this metal compared to the salts are its inertness and that thermal contact can be easily established. The authors also use a SQUID magnetometer for detection of the signal and have achieved comparable resolution, noise level, and drift rates. For experiments with very limited space, the group has used sputtered films of $PdFe_{0.68\,at.\%}$ with somewhat less resolution [12.131].

The lower limit for a paramagnetic thermometer is given by the deviation of χ from a Curie–Weiss law. The upper limit of the thermometer's temperature range is governed by the loss in sensitivity due to $\chi \propto T^{-1}$, which is usually at about 1 K, depending on the material and on the electronic equipment.

Examples of typical electronic setups for measuring electronic susceptibilities of paramagnets suitable for thermometry are shown and discussed in Sect. 3.4.2.

Magnetic thermometry has the advantages that it is based on a simple and quick measurement, one can have a rather large sensitivity of typically 10^{-4}, there is essentially no heating due to the measurement, one has a simple

T–χ relation (therefore calibration is easy), and the thermometer can be homemade. Another advantage of magnetic thermometry for low-temperature physics is the increase of its sensitivity with decreasing temperature. If well designed, such a thermometer will have a rather short thermal response time of a few seconds at 10 mK.

12.10 Magnetic Thermometry with Nuclear Paramagnets

The low-temperature limit of magnetic thermometry is given by the ordering temperature of the magnetic moments; for electronic magnetic moments this is about 1 mK for the CMN and PdFe thermometers discussed earlier. With the extension of the accessible temperature range to lower and lower temperatures thermometric methods for lower and lower temperatures have to be invented. To extend magnetic thermometry to temperatures below a few mK one has to switch from electronic paramagnets to nuclear paramagnets. The change to the much smaller nuclear magnetic moments allows the use of magnetic thermometry to at least the low microkelvin temperature range. Due to the small nuclear magnetic moments we now have a much smaller susceptibility and therefore much smaller signal, requiring more sensitive detection methods; these are SQUID or resonance techniques [12.132–12.136], which will be described in this section. In addition to the extension of the temperature range, we can now switch to pure metals with their good thermal conductivity and contact and fast nuclear spin–lattice relaxation, resulting in fast thermal response times – as we did when going from electronic magnetic refrigeration to nuclear magnetic refrigeration.

The parameter to be measured is the nuclear susceptibility

$$\chi_n = \lambda_n / T_n \quad \text{(Curie law)} \tag{12.40}$$

with the Curie constant

$$\lambda_n = \frac{N_0 I(I+1)\mu_0 \mu_n^2 g_n^2}{3k_B} \, . \tag{12.41}$$

These equations are valid in the "high-temperature limit" $g_n \mu_n B \ll k_B T$, where the nuclear magnetic dipoles can be treated as non-interacting.

In many metals this nuclear Curie law should be valid to at least a few microkelvin (or possibly even 0.1 μK). This seems to be true for Cu and Pt to within about 1% in fields of less than 1 mT (see Sect. 10.1 and Figs. 10.4 and 10.5) [12.137]. However, recently deviations have been observed for the intermetallic compound $AuIn_2$ [12.138] and for Tl [12.139], for example, already at about 100 μK in fields of some millitesla, therefore some caution may be appropriate when new materials are introduced for thermometry.

Requirements for the validity of the nuclear Curie law are the absence of changing internal fields due to nuclear magnetic or electronic magnetic ordering in the relevant temperature range, and the absence of nuclear electric quadrupole interactions (which means cubic lattice symmetry or, better, $I = 1/2$; see Sect. 10.6).[1] In addition, we need a short τ_1 (because we are measuring nuclear spin temperatures T_n but are usually interested in the electronic temperature T_e or in the equilibrium temperature of the material) and the absence of a superconducting transition. Nuclear magnetic thermometry dominates the low- and the sub-millikelvin temperature range but at higher temperatures the signals can become rather weak and may be dominated by contributions from electronic magnetic impurities. Therefore an adequate measuring technique – mostly a resonance method – is of utmost importance.

12.10.1 Non-Resonant, Integral Detection of Nuclear Magnetization

Of course, we can directly measure M_n or χ_n statically as we do in electronic magnetic thermometry. However, now due to the small nuclear moment necessarily we need a SQUID as detector with its high sensitivity to measure nuclear magnetization. In the experiment whose results are shown in Fig. 12.43 [12.141], the magnetization of a 6 N pure Cu sample in a field of 0.25 mT was investigated with a sensitivity of 10^{-3} flux quanta ϕ_0 ($\phi_0 = 2 \times 10^{-8}$ mT cm^2), resulting in $\Delta T/T = 5 \times 10^{-4}$ or 5 μK at 10 mK in a field of 1 mT. The result gave agreement with the calculated nuclear Curie constant of Cu and $M_n \propto 1/T$ in the investigated temperature range even up to 0.9 K! However, in general, one measures a combination of nuclear and electronic magnetizations $M_n(T) + M(T)$ in a direct, non-resonant measurement, and then 1 ppm Fe or Mn in Cu would give the same signal as the 100% Cu nuclei. As shown by Hirschkoff et al. [12.142, 12.143], electronic contributions from the wire of the primary or secondary coils or from the insulation already may cause serious problems. These problems can mostly be eliminated by measuring in a magnetic field large enough to saturate the large electronic moments at the temperatures involved (also at the calibration points!), which, for example, is $B/T \geq 1\,\mathrm{T\,K^{-1}}$ for 3d elements in the cubic intermetallic compound $AuIn_2$ (Fig. 12.44) [12.144]. In this latter work a simpler but much less sensitive commercial fluxgate magnetometer was used instead of a SQUID in the temperature range 15–500 mK. If the electronic moments are saturated, they only give a temperature-independent background to the signal. Of course, there is the possibility of other background contributions from the sample holder, coils, etc. which may give temperature-independent as well as temperature-dependent contributions. The direct method has experimental problems but, unlike the following methods, it does not lead to heating of the sample due to the measuring process.

[1] This, of course, does not apply if "nuclear quadrupole resonance" is applied for thermometry, as proposed for very low temperatures in [12.140].

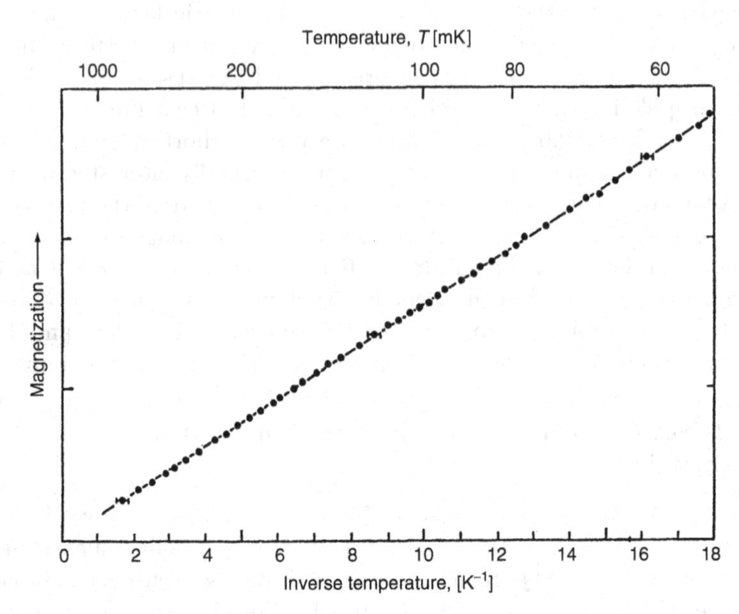

Fig. 12.43. Temperature dependence of nuclear magnetization of copper as a function of inverse temperature (the temperature scale is shown on the top horizontal axis). The static nuclear magnetization of Cu in a field of 0.25 mT was measured with a SQUID. The temperatures were deduced from the ^{3}He melting pressures [12.141]

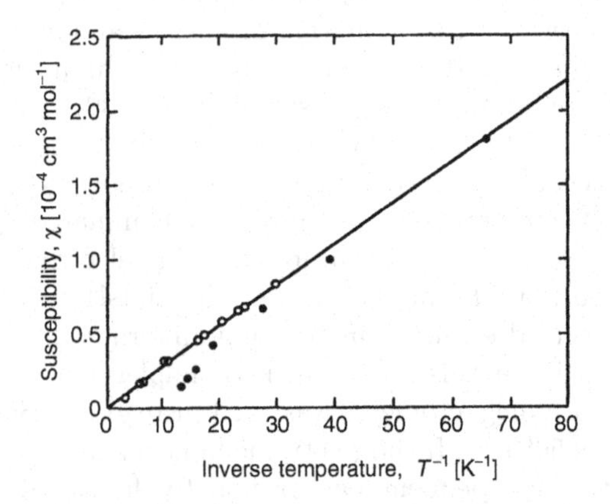

Fig. 12.44. Molar nuclear susceptibility of $AuIn_2$ in different magnetic fields plotted against T^{-1}. The measuring fields are 20 mT (●), 193 and 640 mT (○), respectively. The last two magnetic fields are strong enough to saturate the electronic moments, so that only the changes in nuclear susceptibility are measured (data from [12.144])

12.10.2 Selective Excitation but Non-Resonant Detection of Nuclear Magnetization

A variation of the method discussed earlier, avoiding the problem of background contributions, is the resonant, selective destruction of the nuclear polarization in an external field B_z by irradiating the sample with an RF field at resonance. The frequency of the RF field has to be in resonance with the Zeeman splitting

$$h\nu = \mu\mu_{\mathrm{n}}B_z/I \,. \tag{12.42}$$

It has to be perpendicular to the polarizing static field B_z, so it can induce transitions from the lower to the upper nuclear levels. In this way the static magnetization of the nuclei is changed and this change can be detected with a SQUID (Fig. 12.45) [12.145, 12.146]. The method combines the high sensitivity of the SQUID with the selectivity of resonant excitation; electronic contributions or contributions from other nuclei are not detected because they are not in resonance. Some results for "SQUID NMR" on Cu are shown in Fig. 12.46. The destroyed nuclear magnetization recovers with the spin–lattice relaxation time τ_1. This method is often applied when the more conventional NMR techniques to be discussed later are inappropriate; for example, for very broad resonances, experiments at very low frequencies [12.147,12.148] or when heating effects are a problem. A disadvantage of this method is that a change of magnetization of all the materials within the SQUID sensing volume will lead to changes in the baseline of the signal (Fig. 12.46).

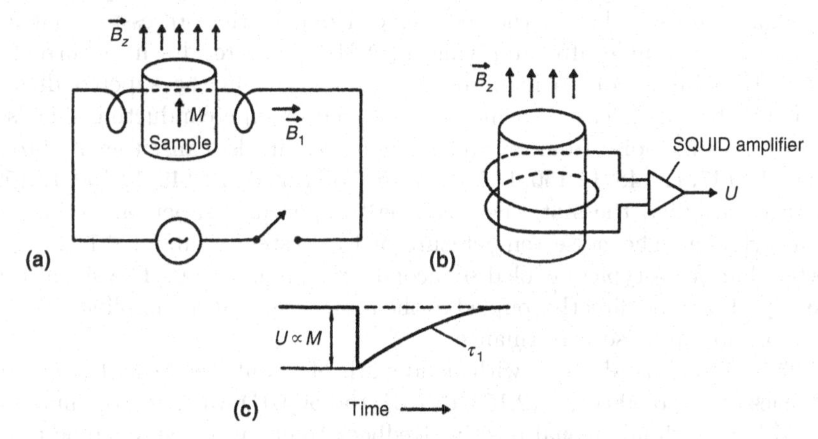

Fig. 12.45. Schematic for the technique of SQUID NMR thermometry. (**a**) The magnetization of the sample induced by the static field B_z is destroyed by applying a RF pulse of strength B_1 at the resonance frequency. (**b**) The resulting change of the magnetization is detected by a SQUID amplifier. (**c**) The voltage U detected by the amplifier – which is proportional to the magnetization M – changes with time when the magnetization recovers with the spin lattice relaxation time τ_1

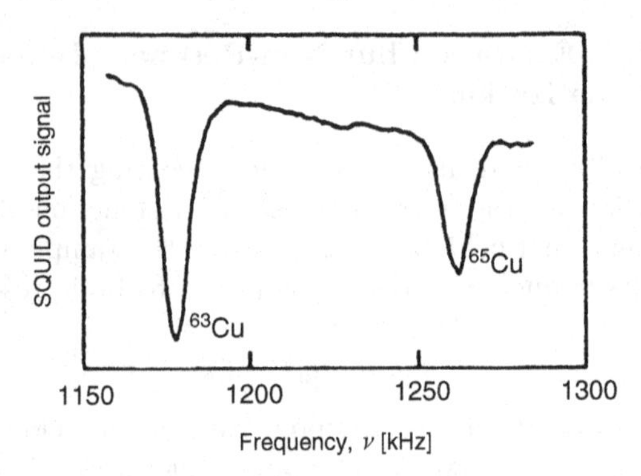

Fig. 12.46. A SQUID NMR spectrum of Cu taken at a sweep rate of 15 kHz s^{-1} at 0.65 K in $B_0 = 104\,\mathrm{mT}$. The vertical signal represents the total nuclear magnetization of the specimen which is reduced when the frequency is swept through the resonance of the two Cu isotopes [12.145]

The method of low-frequency pulsed SQUID NMR [12.147–12.149] has been substantially improved by using very sensitive commercial DC-SQUIDs as first-stage, low-noise preamplifier. They have a coupled energy equivalent sensitivity of the minimum detectable current in the input coil of the SQUID of a few $100h$ at a bandwidth of 3.4 MHz, (h: Planck's constant) [12.150–12.152]. In these spectrometers, the superconducting NMR receiver coil around the sample together with the input coil of the SQUID form a flux transformer circuit; their inductances should be similar to optimize the coupled flux. A changing magnetic flux in the receiver coil due to the precessing magnetization of the sample after applying an NMR pulse results in a current in the SQUID input coil to maintain a constant flux in the superconducting loop. The RF input circuit, which is contained in superconducting shields in order to reduce noise from extraneous magnetic fields, can either be broadband [12.147, 12.148, 12.150, 12.151, 12.153] or tuned [12.149, 12.150, 12.152]. For field stability, the static field can be trapped in a superconducting niobium cylinder. The noise temperature of the system is about 0.1 K, much better than for a typical cooled semiconductor preamplifier. The signal from the SQUID can be directly coupled to the room temperature amplifier without deterioration in noise performance.

In the broadband mode with bandwidth of about 3 MHz at Larmor frequencies of up to 500 kHz [12.150, 12.151], the SQUID was operated in a flux-locked loop with additional positive feedback to enhance the dynamic range, to improve gain stability, and giving a short recovery time of order 10 μs; this allows investigations of samples with short spin-spin relaxation times. A substantial advantage is the possibility of changing the NMR frequency without changing the detection electronics. Experimental results obtained with such a system are shown in Fig. 12.47. Similar broadband, low-noise SQUID NMR setups have been described in [12.153].

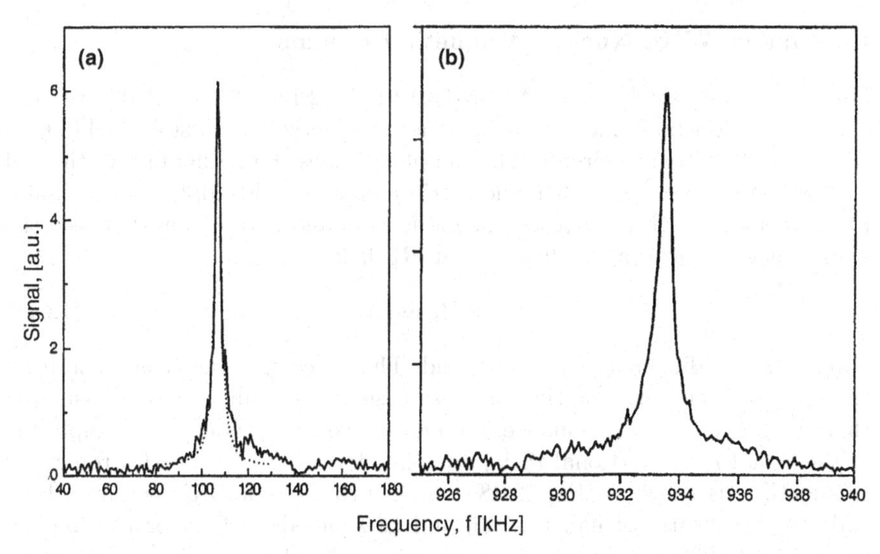

Fig. 12.47. Fast-Fourier transforms of free-induction decay signals obtained by SQUID NMR in: (**a**) a broadband circuit for 6×10^{19} ^{195}Pt nuclear spins in a single crystal of UPt_3 in the superconducting mixed phase at 40 mK [12.151], as well as in: (**b**) a tuned circuit for 5×10^{18} ^{3}He nuclear spins at 4.2 K. Reprinted from [12.151] and [12.152], respectively, copyright (2000), with permission from Elsevier

As discussed in [12.150, 12.152], a tuned input circuit gives a higher signal-to-noise ratio at higher frequencies, above a few 100 kHz. In this mode, the pick-up coil and the SQUID input coil form part of a resonant circuit at around 1 MHz, with a noise of order 0.1 K. Here the receiver coil is made from normal conducting wire. Such a system has been used in particular to investigate very small numbers of spins, for example 10^{18} spins in thin ^{3}He films, or when the RF-penetration is limited to the skin depth in well-conducting metallic samples (Fig. 12.47).

12.10.3 Resonant Excitation and Resonant Detection of Nuclear Magnetization

The most commonly used method in nuclear magnetic thermometry are Nuclear Magnetic Resonance (NMR) techniques which avoid electronic or any other non-resonant contributions. There are two ways to do nuclear magnetic resonance: in the continuous wave mode or with pulses. Excellent books describe these very important methods which are not only of relevance for thermometry but have found many applications in physics, chemistry, biology, and medicine. I cannot discuss these methods in as much detail as their importance would require but refer to relevant books and review articles (see, e.g., [12.3, 12.4, 12.7, 12.132–12.136]).

Continuous Wave Nuclear Magnetic Resonance

For continuous wave NMR (CW NMR) [12.132–12.135,12.154,12.155] we consider a sample whose nuclear moments are magnetized by a static field B_z in the coil of a resonance circuit. The sample changes the inductance of the coil by the factor $(1 + \chi_n)$. When the static magnetic field applied to the sample is swept through the nuclear magnetic resonance, transitions between the nuclear energy levels are induced by an RF field

$$B_y = B_1 \sin(\omega t) \tag{12.43}$$

applied perpendicular to the static field. The necessary energy is taken from the resonance circuit, resulting in a decrease of its quality factor. Assuming that the line shape at resonance is temperature independent, the amplitude of the signal is proportional to χ_n and therefore to T_n^{-1}. The electronics for this method is shown in Fig. 12.48. One has to use small RF power to keep eddy-current heating small [12.156] and to avoid disturbing or heating the nuclear spin system from thermal equilibrium; therefore the signals are rather small. To increase the sensitivity, very often the frequency or the magnetic field is modulated at audio frequencies and the signal is detected with lock-in techniques.

It can be shown that the signal at resonance, $\omega = \omega_0$, in CW NMR is proportional to the imaginary part of the dynamic nuclear susceptibility

$$\chi_n(\omega_0) = \chi_n' - i\chi_n'' , \tag{12.44}$$

with

$$\chi_n''(\omega_0) = \frac{\omega_0 \tau_2^*}{2} = \frac{\chi(0) B_z}{\Delta B_z} \propto 1/T_n , \tag{12.45}$$

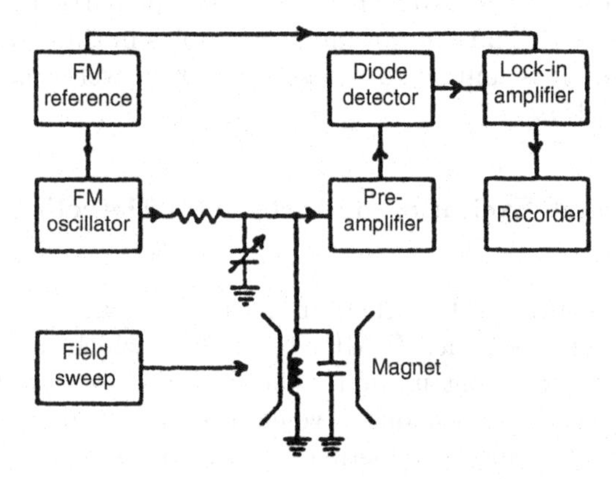

Fig. 12.48. Block diagram of the electronics used by Corruccini et al. [12.155] for CW NMR. The cw oscillator is frequency modulated at 100 Hz as the magnetic field is swept through the resonance of the sample inside of the coil of the resonance circuit

in the case that the signal is not saturated; here ΔB_z is the resonance width at half maximum. Hence the dynamic susceptibility is enhanced with respect to the static one by the ratio $B_z/\Delta B_z$ which can be of order 10^2–10^3 for Cu or Pt, for example.

Pulsed Nuclear Magnetic Resonance

In the today mostly used pulsed nuclear magnetic resonance technique [12.132–12.136, 12.157–12.161], again a static field B_z keeps a nuclear magnetization M_n in the z-direction. But now we apply perpendicular, in the y-direction, a short pulse of a sinusoidal field $B_y = B_1 \sin(\omega t)$ at the resonance frequency given by (12.42). This RF pulse tips the nuclear magnetization by an angle

$$\theta = \pi B_1 / B_z \tag{12.46}$$

away from the z-direction. The magnetization M_n then precesses around B_z at the resonance frequency ω_0; the resulting transverse component $M_\mathrm{n} \sin\theta$ rotates in the xy-plane and is detected in a receiver coil (Fig. 12.49). The magnetization along B_z has been reduced from M_n to $M_\mathrm{n} \cos\theta$. The coil for

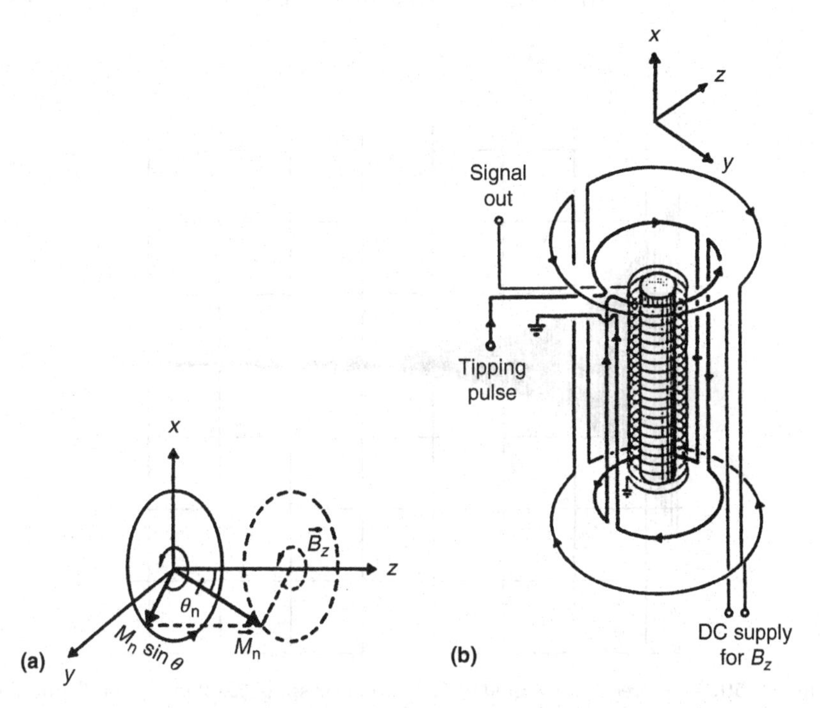

Fig. 12.49. Schematic of pulsed nuclear magnetic resonance thermometry. (**a**) Principle of the method. (**b**) Schematic of the coil arrangement [12.3, 12.4, 12.159, 12.171]

the tipping pulse and for detection of the signal can be identical (Figs. 12.55 and 12.56) because the tipping and detection do not occur simultaneously. The precession signal in the receiver coil decays with a characteristic time given by the transverse spin–spin relaxation time τ_2 due to the dephasing of the rotating spins. An always present inhomogeneity δB_z of the static field will speed up the dephasing. The effective decay time [12.132–12.136] is then given by

$$\frac{1}{\tau_2^*} = \frac{1 - P^2}{\tau_2} + \frac{\mu\mu_{\mathrm{n}}}{\hbar I}\delta B_z \,, \tag{12.47}$$

where P is the polarization, hence

$$M_{xy}(t) = M_{\mathrm{n}}(0)e^{-t/\tau_2^*} \,. \tag{12.48}$$

Such a free induction decay signal is shown in Fig. 12.50. The pulse length ($\tau_{\mathrm{p}} = \theta\hbar/\mu\mu_{\mathrm{n}}B_1$ at resonance) should, of course, be short compared to τ_2^*. This gives a lower limit on the tipping angle or on the magnitude of the pulse for a given tipping angle.

The alternating voltage induced in a detection coil in the x-direction around the sample is then given by

$$U(t) = \alpha\omega M_{\mathrm{z}} \sin\theta \sin(\omega t) \exp(-t/\tau_2^*) \,, \tag{12.49}$$

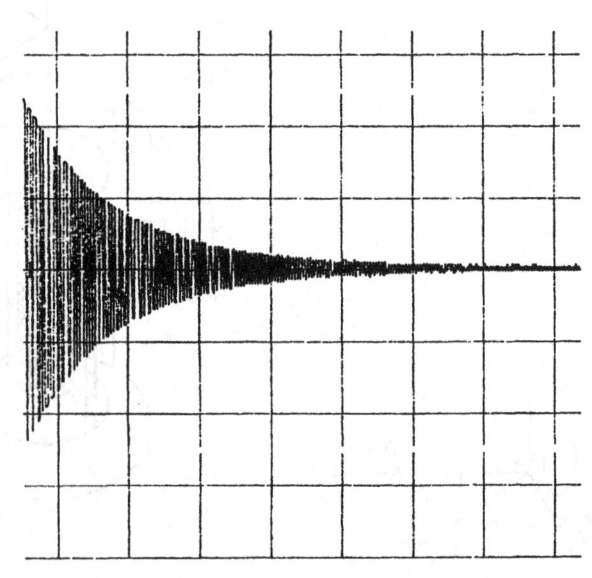

Fig. 12.50. Free precession signal of ^{195}Pt nuclear spins (2,000 wires of 25 µm diameter in a 5 mm long secondary coil of several thousand windings of 25 µm Cu wire, see Fig. 12.55) at $T = 45\,\mu$K and in $B = 13.8$ mT (125 kHz). The excitation pulse was 64 µs long and had about 0.2 V_{pp} amplitude. The horizontal scale is 0.5 ms div^{-1}

where α is the geometry or coil constant. This behavior can be calculated from its equation of motion (Bloch equations) [12.3,12.4,12.132–12.135,12.156, 12.157].

$U(t)$ has to be amplified and transformed into a signal S which is usually the integral over a definite time of the rectified free-induction-decay signal (Fig. 12.50). The only way to determine the constant of proportionality between S and T_n^{-1} is the calibration at a known higher temperature and setting the product signal * temperature constant. For this proportionality to be valid, τ_2 has to be temperature independent [12.161]. In addition, no other background contribution to the signal should be present (see below). Deviations from a linear relation between signal size and temperature can also be observed when the excitation pulse or other external RF signals will overheat the platinum NMR probe (see Fig. 12.54).

In the NMR methods discussed earlier one is determining the nuclear spin temperature T_n from a measurement of the nuclear susceptibility. However, by measuring the spin lattice relaxation time τ_1 and applying the Korringa law $\kappa = \tau_1 T_e$, one can also use NMR to measure the temperature T_e of the electrons. This can be done by using either SQUID NMR or pulsed NMR. In the latter method one applies a 90° pulse at the resonance frequency to destroy the magnetization in the z-direction. Afterwards, small, e.g. 10°, inspection pulses are applied to record the recovery of M_z according to

$$M_z(t) = M_n(0)(1 - e^{-t/\tau_1}) . \tag{12.50}$$

An example of such a measurement is illustrated in Fig. 12.51. A determination of the electronic temperature from nuclear magnetic properties by

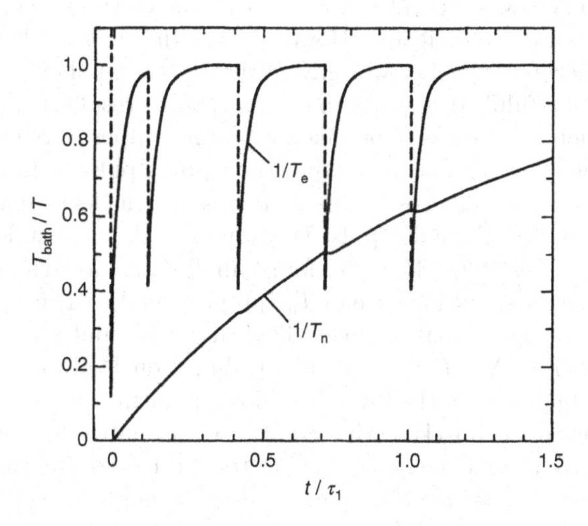

Fig. 12.51. Determination of the spin–lattice relaxation time τ_1 from pulsed NMR measurements. The figure shows the calculated time behavior of T_n^{-1} and T_e^{-1} during a τ_1 determination [12.160]

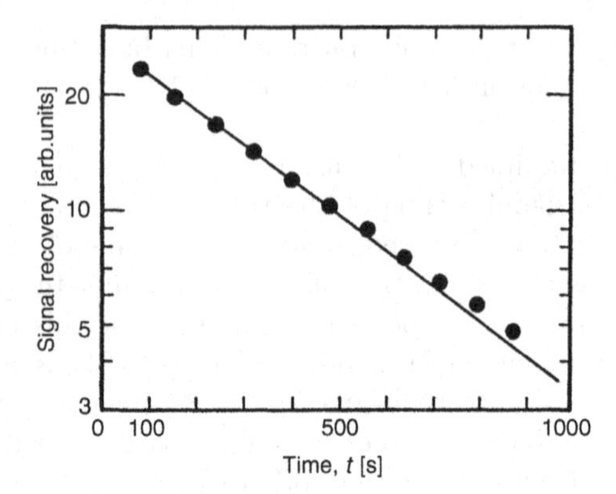

Fig. 12.52. Relaxation of the nuclear magnetization of a ^{195}Pt NMR sample after applying a 60° pulse. From the measured relaxation time τ_1 and the Korringa relation for platinum, $\tau_1 T_e = 30\,\text{mK s}$, one finds $T_e = 61.5\,\mu\text{K}$; the Pt nuclear magnetization gave $T_n = 60.5\,\mu\text{K}$ [12.50]

observing the nuclear spin-lattice relaxation time τ_1 has been popular for many years [12.50, 12.157, 12.159, 12.160, 12.162]. The results of successful examples are shown in Figs. 12.52 and 12.53. However, in some experiments deviations from the Korringa law have been observed for Cu [12.163] and Pt [12.123, 12.157, 12.162, 12.164, 12.165], showing that κ can be temperature and field dependent, so that this method should only be used with great caution. These deviations have been attributed to the Kondo effect of magnetic impurities in the host lattice [12.166] but details are not yet understood.

NMR at very low temperatures is not a simple method, and the electronic setup can be quite sophisticated. However, it is the *only* method available at present for thermometry below 1 mK. There are a number of requirements which have to be fulfilled in order to obtain reliable results with pulsed NMR [12.161]. Particular care has to be taken if a large temperature range is investigated. First, we have to remember that the tipping pulse reduces the nuclear magnetization M_n to $M_n \cos\theta$; this increases the nuclear spin temperature from T_n to $T_n/\cos\theta$. For a 90° pulse this would be $M_n = 0$ and $T_n = \infty$; generally, $\Delta T/T = (\cos^{-1}\theta) - 1 (\simeq \theta^2/2$ for small $\theta)$. One has to search for a compromise between a small increase of T_n (proportional to $1/\cos\theta$) and a large enough signal (proportional to $\sin\theta$). Typically a 10° pulse is applied, giving $\cos 10° = 0.985$ or $\Delta T/T \simeq 1.5\%$ only. Calibration is always carried out at temperatures higher than the investigated temperature range and the tipping angle has to be changed to keep the signal at a reasonable level. Even more important, we have to take into account that the RF field of the tipping pulse increases the electronic temperature due to eddy current heating [12.156, 12.160]. This eddy current heating is proportional to ω^2 (Sect. 10.5.2). Again a compromise between a large enough signal and small enough eddy current heating has to be found. Typical heat depositions are 0.1–1 nJ per pulse for Pt NMR

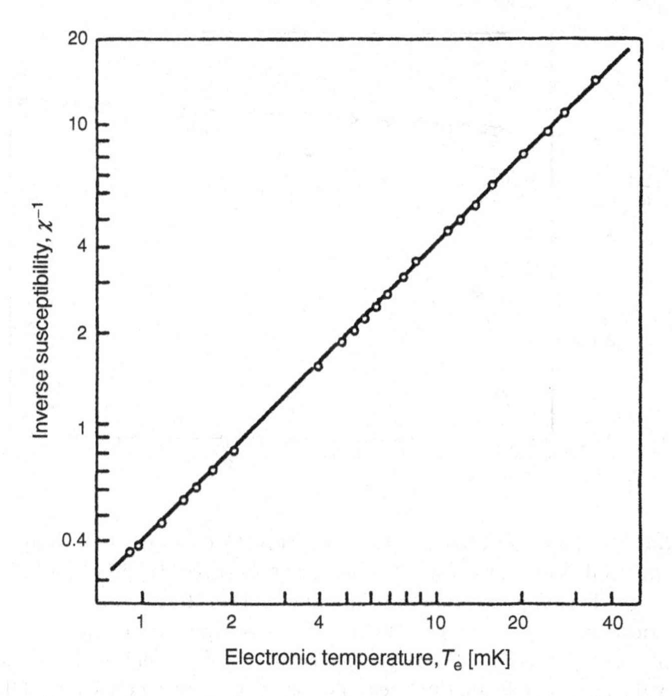

Fig. 12.53. A comparison of the reciprocal of the nuclear spin susceptibility χ_n of Pt (proportional to its nuclear spin temperature T_n) vs. the reciprocal of the spin–lattice relaxation time τ_1^{-1} of Pt (which is proportional to the shown electronic temperature T_e of Pt) to show that the nuclear spins and the conduction electrons are at the same temperature [12.172, 12.173]

thermometry at $T \leq 1\,\mathrm{mK}$ [12.50, 12.123, 12.157, 12.161, 12.167]. Overheating of the thermometer will show up if the product of the NMR signal times temperature (obtained from another thermometer) does not show a linear weak temperature dependence (Fig. 12.54).

Besides taking a low frequency and a small tipping angle, even more importantly, one has to use small dimensions of the sample, which very often means taking a bundle of thin insulated wires (or powder) for the NMR thermometry sample, to keep RF eddy current heating small and to let the electromagnetic field penetrate the sample by the skin depth $\delta = (2\rho/\mu_0\omega)^{1/2}$ which is $19\,\mu\mathrm{m}$ for high-purity Pt at a frequency of 250 kHz. Then we have to bear in mind that after each pulse, nuclei and electrons have to recover and should be in thermal equilibrium before the next pulse is fired. Therefore the repetition rate has to be small compared to τ_1^{-1}; usual waiting times between pulses should be 5–10 times τ_1. We would also like the heat capacity of the thermometric sample (and addenda) to be small, so that we have a small thermal time constant for its thermal recovery; this means, the static field B_z should be kept as small as possible. The last requirement is in accordance with our

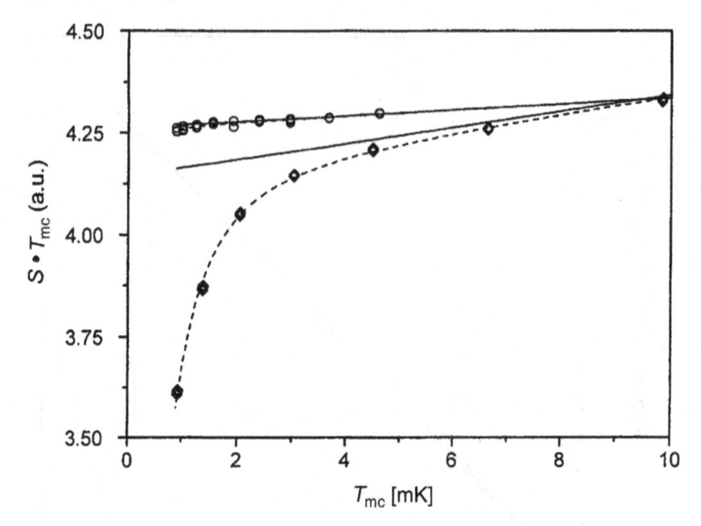

Fig. 12.54. Product of NMR amplitudes S from ^{195}Pt in Pt wire samples obtained with commercial NMR electronics times temperatures T_{mc} measured with a ^{3}He melting curve thermometer. These data show below 10 mK the overheating of the Pt wires squeezed (◇) to the cold finger of the refrigerator compared to the much better contact of wires welded to the refrigerator (○). Broken lines show the expected behavior with heating effects, full lines those without heating effects which are almost indistinguishable for the welded sample [12.161]

desire to keep the frequency for the NMR low. Last but not least, the static field should be homogeneous (to at least 10^{-4}) to give a sharp resonance and to keep τ_2^* large, see (12.47).

A main advantage of this transient method is the fact that it measures the state of the nuclear spin system before the (short!) pulse is applied and that the detection can occur when the tipping field has been switched off. The pulsed technique seems to be more appropriate for thermometry at very low temperatures than the CW method under the usual experimental conditions [12.156].

Owing to recent advances in attaining very low temperatures, one can perform experiments where the high-temperature approximations (10.4) used to describe the thermodynamic properties of a sample cannot be applied anymore. The spin dynamics can then become rather complicated because after a disturbance the nuclei will decay exponentially with a single decay time only for $I = 1/2$ [12.168–12.170]. Eventually, at $g_{\mathrm{n}}\mu_{\mathrm{n}}B > k_{\mathrm{B}}T$, the resulting effective decay time obtained from fitting experimental data will approach a constant value, see (10.11′),

$$\tau_1 = \frac{2k_{\mathrm{B}}\kappa}{g_{\mathrm{n}}\mu_{\mathrm{n}}B} \tag{12.51}$$

instead of being $\tau_1 = \kappa/T_{\mathrm{e}}$, because now the relaxation rate depends on the magnetic rather than on the thermal energy. In this situation τ_1 decreases with spin I according to

$$\tau_1(T = 0; I) = \frac{1}{2I}\tau_1(T = 0; I = 1/2)\,. \tag{12.52}$$

For Cu, for example, the simple high-field limit $\tau_1 = k/T_c$ is reached for $B > 10\,\mathrm{mT}$.

In Table 10.1, I have summarized the properties of several metallic isotopes which may also be suitable for NMR. From these properties and the experience of several groups it turns out that Pt seems to be the most suitable thermometric probe at very low temperatures. It has only one isotope, ^{195}Pt, with nuclear spin $I = 1/2$; therefore we have no problems with nuclear quadrupole interaction. It has a short τ_1; this means, that electrons and nuclei quickly attain thermal equilibrium. Its nuclear spin–spin relaxation time τ_2 is long; therefore the decay of the signal takes a long time, simplifying observation of the signal.

Pulsed NMR on thin Pt wires (or Pt powder immersed in liquid ^{3}He) at fields of 6–60 mT and correspondingly at frequencies of about 55–550 kHz is now the standard thermometric method for the microkelvin temperature rangex [12.50, 12.123, 12.137, 12.158–12.161, 12.164, 12.165, 12.171–12.175]. A bundle of several hundred or, better, thousand thin (e.g., 25 μm), annealed (but see [12.123]) and isolated wires and not too high frequencies are chosen to keep eddy current heating low and to let the RF field penetrate the sample wires (the skin depth is of order 10 μm at $\nu = 250$ kHz for Pt with RRR $\simeq 100$). Successful designs of Pt wire NMR thermometers are shown in Figs. 12.55 and 12.56 [12.50,12.123,12.161]. These thermometers have a signal-to-noise ratio of 1 at about 0.1 K. Figure 12.57 shows the calibration of a Pt wire thermometer with a *Pd*Fe susceptibility thermometer (Sect. 12.9), which in turn had been calibrated by a superconducting fixed point device (Sects. 11.4.2, 11.4.3). Such an NMR thermometer and the described calibration is believed to give the temperature at 1 mK (20 μK) to ± 2%(± 5%), assuming the fixed point device temperatures to be correct [12.50, 12.123, 12.137]. The data in Fig. 12.56 also demonstrate that $\chi_n(\mathrm{Pt}) \propto \chi_e(Pd\mathrm{Fe})$ to within 10^{-3}, or that both susceptibilities follow a Curie (or Curie–Weiss) law to that accuracy in the investigated temperature range. Usually the accuracy of an NMR thermometer is limited by the accuracy of the calibration and by heating effects [12.156, 12.161].

The validity of the ^{195}Pt NMR temperature scale has also been confirmed to at least 70 μK by the observation that $B/T_{\mathrm{Pt}} =$ const. in the adiabatic nuclear demagnetization experiments of [12.123], and to at least 100 μK by the observation that the relative change of the velocity of sound of polycrystalline Ag is proportional to $\ln T_{\mathrm{Pt}}$ [12.107]. In addition, in [12.50] it was shown that T_n (from χ_n) and T_e (from τ_1) agree to within 2% from 48 to 306 μK (Fig. 12.52). These as well as other observations [12.137] provide convincing evidence that the Pt-NMR temperature scale is correct to at least 10 μK. In recent experiments NMR and AC susceptibility measurements on Pt have been utilized successfully for thermometry to an equilibrium temperature of 1.5 μK

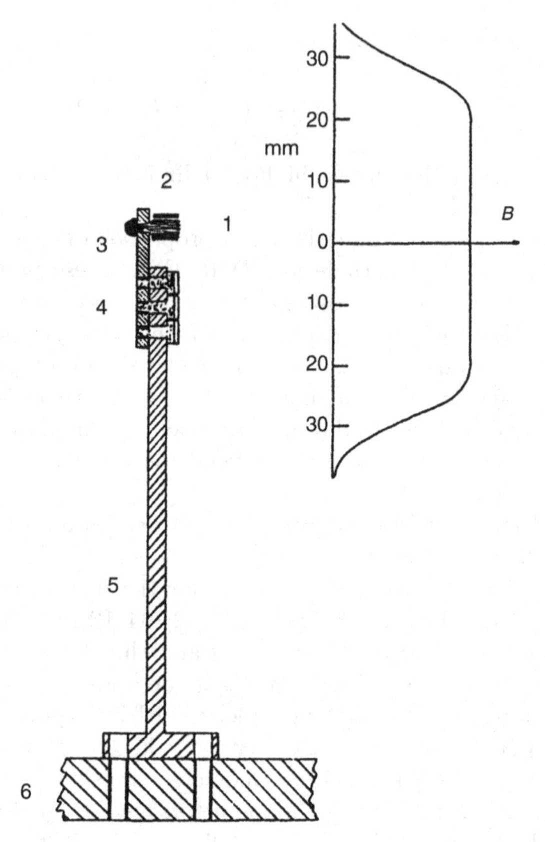

Fig. 12.55. Design of a platinum wire NMR thermometer. *Left*: Pt wires of 25 μm diameter (1) which sit inside of a signal coil of 25 μm Cu wire on a 6 μm Mylar coil former (2). The platinum wires are welded to assure adequate thermal contact into a Pt stem (3) which is then screwed (4) to a Ag cold finger which has a very small unclear heat capacity (5). The latter is finally screwed to the plate (6) whose temperature the thermometer is supposed to measure. The setup is surrounded by a coil producing the static field which is surrounded by a superconducting niobium shield to homogenize the field and to shield against external disturbances. *Right*: The field distribution of the superconducting field coil

and to a nuclear-spin temperature of 0.3 μK [12.176]. These convincing experimental observations agree with expectations because the electron-mediated indirect exchange interactions between the nuclei in platinum correspond to about 0.2 μK (this actually is a rather high value caused by the large indirect exchange interaction between the Pt nuclei due to the large polarizability of the conduction electrons in the strongly paramagnetic Pt). In addition, the Lorentz field and the demagnetization effect for a cylindrical sample transversal to the applied field are of the same order of magnitude and they partially

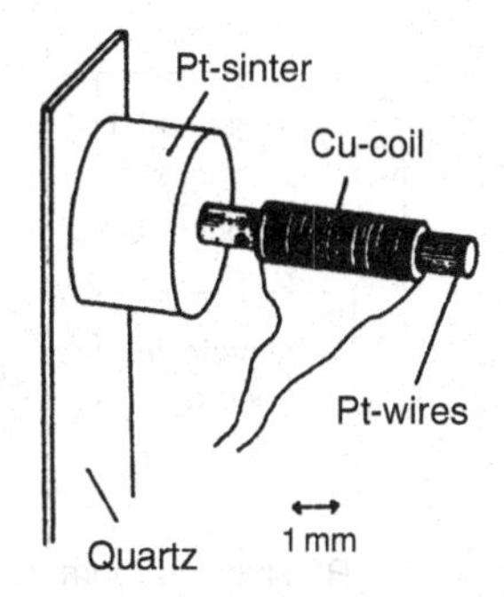

Fig. 12.56. Platinum wire NMR thermometer for measuring the temperature of liquid helium. The Pt wires (e.g., 25 μm diameter) are pressed into a platinum black sinter of 0.25 g. The latter provides a large surface area of 4 m^2 to make thermal contact with the liquid. It is preferable to do the NMR on the Pt wires rather than on the very fine Pt powder directly because the latter usually contains a large fraction of impurities and its NMR resonance may also be more strongly influenced by size and strain effects. The thermometer is held by a quartz sheet and a copper signal coil is wound around the Pt wires

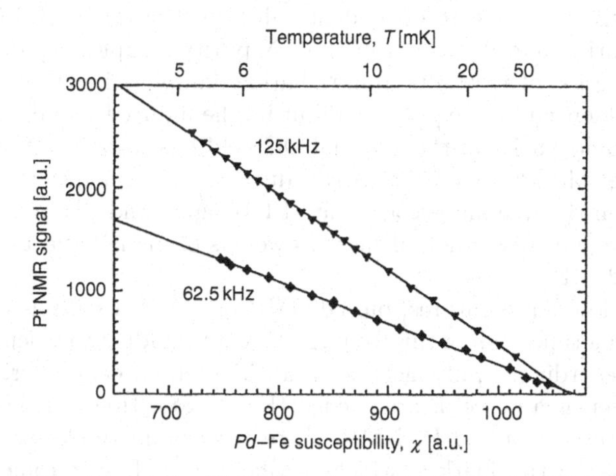

Fig. 12.57. Calibration of the NMR signal of a platinum wire NMR thermometer versus electronic paramagnetic susceptibility of *Pd*Fe (Sect. 12.9) in the temperature range between 5 and 50 mK at the two frequencies shown. The vertical scale is different for the two frequencies [12.123]

cancel each other [12.161]. Hence, the Pt NMR signal in not too high magnetic fields should be proportional to T_n^{-1} to the mentioned temperatures. However, the properties of platinum are strongly influenced by magnetic impurities [12.38, 12.161, 12.176]. For example, Fe, the most abundant magnetic impurity in Pt, forms so-called giant magnetic moments of $\mu = 7.8\ \mu_B$ at low Fe concentrations in the Pt matrix [12.38]. The main effect of magnetic impurities is a reduction of the spin-spin relaxation time $\tau_2 = 1.2$ ms of pure

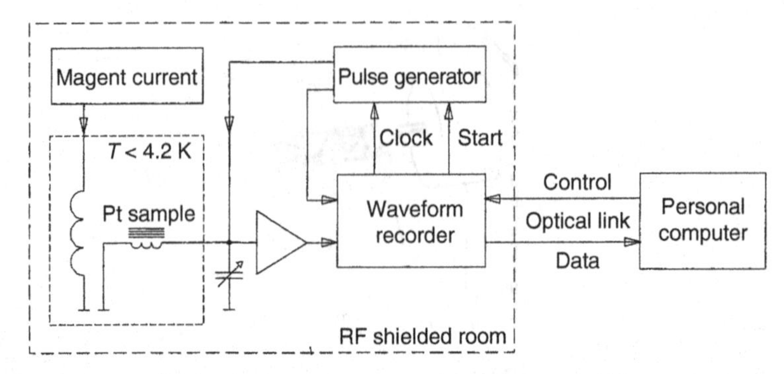

Fig. 12.58. Block diagram of the electronics used in [12.161] for pulsed NMR thermometry on ^{195}Pt. A 12-bit waveform recorder is used for signal detection. It is controlled from outside the shielded room via an IEEE bus and an optical link. The recorder actuates the pulse generator which in turn triggers the recording. A common time base ensures the coherence of individual free induction decays. The signals are averaged and analyzed in a computer

Pt to slightly temperature dependent values in the range of 0.1–0.5 ms for Pt of 4N purity, and 0.3 to 0.9 ms for 5N purity samples [12.161, 12.176]. A further reduction of τ_2 results from radiation damping, i.e., that the spins are not only influenced by the pulse field but by the induced signal as well. Above all, the inhomogeneity of the external DC field has to be less than $3\ 10^{-5}$ to keep changes of τ_2 to below 5%. Any reduction of τ_2 changes the proportionality between the free-induction-decay (FID) signal and the magnetization or temperature. The discussion of these as well as of further effects on τ_2 can be found in [12.161].

At very low temperatures, pulsed NMR on ^{195}Pt usually is performed at 250 kHz (sometimes half of this frequency) corresponding to a field of 28.4 mT. For signal recording, a pulse generator, a wide-band preamplifier, and a waveform recorder as a detector are needed (Fig. 12.58). However, because of the widespread use of pulsed Pt NMR, there is a complete electronic setup for this purpose on the market, which rectifies the FID electronically and indicates the mean value of a predefined section of it; it operates at various fixed frequencies and adjustable pulse/tipping angle strengths. For data acquisition and signal processing, i.e., analysis of signal properties, coherent background corrections, back extrapolation, as well as consistency checks (for example, how well the FID follows an exponential function; plot of NMR signal times temperature obtained from another thermometer (Fig. 12.54), etc.), the reader should consult [12.161], for example. The analysis of the various factors in this reference leads to the conclusion that for a sample as shown in Fig. 12.55, with the appropriate wiring, stable and linear electronics, signal averaging (to reduce incoherent noise and to improve the signal-to-noise ratio, which is only typically about 10 at 15 mK), and calibration (see Fig. 12.57, for example) the total uncertainty of this thermometry method can be as low as 0.5% at 1–15 mK. The importance of these detailed investigations were necessary because pulsed nuclear magnetic resonance on ^{195}Pt and the validity of the Curie law for its nuclear magnetic susceptibility were essential in establishing the new temperature scale PLTS-2000 at $T < 15$ mK (Sect. 11.3). However, NMR on ^{195}Pt has exclusively been used up till now in all experiments for $T < 1$ mK. This indicates the urgent need for the development of alternative thermometric methods for this temperature range to substantiate the obtained temperatures. Deviations from the Curie law and time constant problems may limit the present thermometric methods to $T \gtrsim 1\,\mu$K anyway.

Even though electronic magnetic impurities have no direct influence on the signal in nuclear resonance methods, they may have an indirect influence by changing the local magnetic field seen by neighbouring host nuclei or by influencing the relaxation rate. This may have a pronounced effect on the data if the investigated temperature range includes characteristic temperatures like the Kondo temperature for Kondo alloys [12.166] or the spin-glass freezing temperature for spin glasses [12.38, 12.123, 12.161].

12.11 Magnetic Thermometry via Anisotropy of Gamma Rays (Nuclear Orientation Thermometry)

Nuclear magnetic thermometry relies on the detection of the Boltzmann population of magnetic sublevels, that is, of the nuclear magnetic polarization. The detection is via measurement of the resulting magnetization or via measurement of the change in magnetization due to resonant RF radiation (see Sect. 12.10). In this section, I will discuss another way to detect nuclear polarization which can be applied if radioactive nuclei are used. Radioactive long-lived nuclei suitable for nuclear orientation thermometry usually show a $\beta^{\pm}$-decay feeding a short-lived excited state which then decays by emission of γ-rays (Fig. 12.59). If the long-living $\beta^{\pm}$-emitters are polarized, the polarization is transferred to the excited, γ-ray emitting state. This γ-ray then has an emission probability with a spatial anisotropy that differs for each of the $(2I + 1)$ sublevels in a magnetic field. The spatial anisotropy is a result of the conservation of angular momentum. The γ-radiation field carries an angular momentum which is the difference of the angular momenta of the two nuclear sublevels involved. The angular momentum of the emitted radiation determines the multipole character of the radiation.

If we have an ensemble of radioactive nuclei, the mean value for the emission probability is equal in all directions. We have to orient the nuclei to be able to detect the spatial anisotropy of the emitted γ-ray intensity, which is a result of the Boltzmann population of the sublevels and therefore of the temperature. The anisotropy depends on the parameters of the nuclear transition and on the polarization of the nucleus. The anisotropic emission of γ-rays

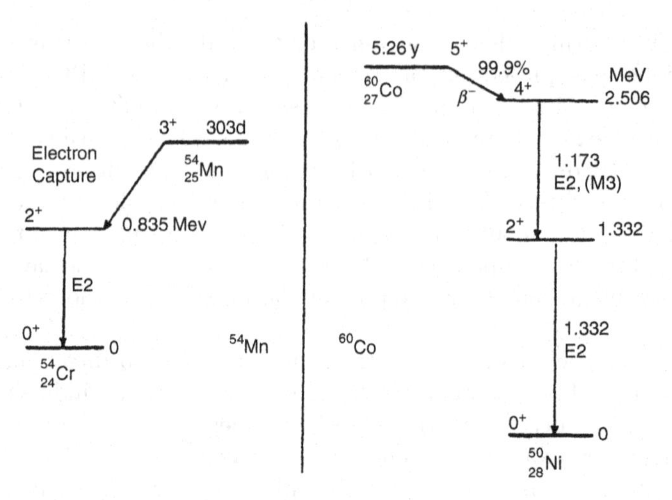

Fig. 12.59. Energy level diagrams for ^{54}Mn and ^{60}Co (slightly simplified). These are the two favorite decay schemes used for nuclear orientation thermometry

from an assembly of oriented nuclei can be used for thermometry if all other parameters determining the γ-ray intensity are known.

The intensity of the γ-radiation emitted from axially symmetric nuclei and detected at an angle θ, which is the angle between the direction of orientation of the nuclei (or of the magnetic field) and the detection or observation of the radiation, is given by [12.3, 12.4, 12.7, 12.10, 12.177–12.182]

$$W(T,\theta) = 1 + \sum_{\mathrm{k}=2,4,6,\ldots}^{\mathrm{k_{max}}} Q_\mathrm{k} U_\mathrm{k} F_\mathrm{k} B_\mathrm{k}(T) P_\mathrm{k}(\cos\theta)\,; \qquad (12.53)$$

this value has been normalized to the intensity at high temperatures, where the probability is equal in all directions. In the summation we have only to consider even k-values because γ-quanta have a spin 1, and the interaction which determines the transition between two nuclear levels resulting in emission of a γ-ray is parity conserving. The parameter k will run to $k_{\mathrm{max}} = 2I$ or $2\,L$, whichever is smaller, where L is the multipolarity of the emitted radiation. The parameters in the above equation are:

Q_k Geometry parameter (detector properties, finite size of source, finite angle of detection, etc.); often the experiment can be designed so that the Q_k's are of order "1", but they have to be determined for each experiment (see [12.177])

U_k Influence of polarization of the former nuclear transitions on the polarization to be used for thermometry; this is "algebra" and can be taken from tables in the literature [12.182, App. 5]

F_k Angular momentum coupling coefficients of the nuclear transition; again this can be taken from tables [12.182, App. 5]

B_k Population of nuclear magnetic sublevels [12.182, App. 6]
P_k Legendre polynomials expressing the angular dependence of the γ-ray distribution
θ Angle between γ-ray emission and axis of orientation.

The essential parameter for our thermometric purpose is

$$B_k = I^k \frac{(2k)!}{(k!)^2} \sqrt{\frac{(2I+1)(2k+1)(2I-k)!}{(2I+k+1)!}} f_k(I) \tag{12.54}$$

with

$$I^2 f_2(I) = \sum_{m=-I}^{+I} m^2 P(m) - \frac{I(I+1)}{3}, \tag{12.55}$$

and

$$I^4 f_4(I) = \sum_{m=-I}^{+I} m^4 P(m) - \frac{I^2(6I^2+6I-5)f_2}{7} - \frac{(3I^2+3I-1)I(I+1)}{15}. \tag{12.56}$$

In these equations $P(m)$ is the Boltzmann population probability of the sublevel with the magnetic quantum number m; these $P(m)$ are given in (9.9). They are the only T-dependent parameters in the above equations.

The nuclear decay level schemes of the two most commonly used isotopes, ^{60}Co and ^{54}Mn, are shown in Fig. 12.59. These two isotopes are attractive because all the necessary nuclear parameters are known, the lifetime of the intermediate state is short so that reorientation effects can be neglected, and they are pure E2 transitions giving a purely electric quadrupole radiation pattern with multipolarity $L = 2$ (2) and nuclear spin $I = 3$ (5) for ^{54}Mn (^{60}Co), therefore $k_{max} = 4$. Figure 12.60 shows the angular radiation pattern for ^{60}Co for various values of $k_B T/\Delta m g_n \mu_n B$. For the mentioned isotopes the relevant parameters are:

$$U_2F_2 = -0.49486;\ U_4F_4 = -0.44669 \text{ for } ^{54}\text{Mn},$$
$$U_2F_2 = -0.42056;\ U_4F_4 = -0.24280 \text{ for } ^{60}\text{Co},$$

$$P_2(\cos\theta) = (3\cos^2\theta - 1)/2,$$
$$P_4(\cos\theta) = (35\cos^4\theta - 30\cos^2\theta + 3)/8.$$

With these parameters we find eventually for E2 radiation detected at an angle $\theta = 0$ [12.3]

$$W(T,0) = 1 + 0.38887\, Q_2 \sum_m m^2 P(m) - 0.55553\, Q_4 \sum_m m^4 P(m)$$

for ^{54}Mn, and

$$W(T,0) = 1 + 0.04333\, Q_2 \sum_m m^2 P(m) - 0.00333\, Q_4 \sum_m m^4 P(m)$$

for ^{60}Co.

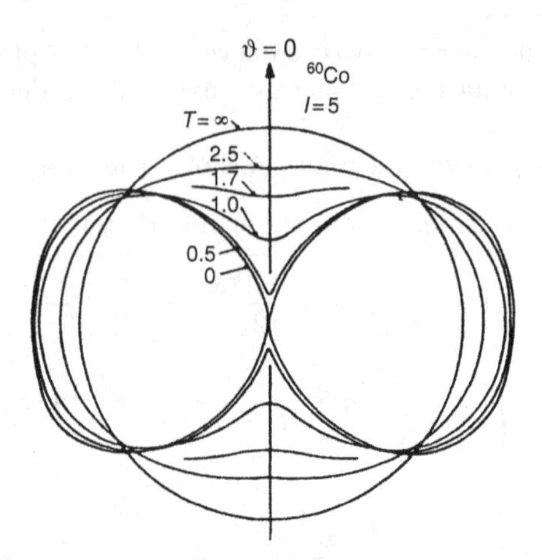

Fig. 12.60. Angular radiation pattern of ^{60}Co for several values of $k_\mathrm{B}T/\mu\mu_\mathrm{B}\mathrm{B}$. The figure demonstrates that the highest sensitivity for temperature changes is obtained when the detector is in the $\theta = 0$ or $180°$ directions where the maxima of the intensity change occur when the temperature is changed

In order for the nuclei to be polarized or oriented they have to experience a magnetic field. This can be an externally applied field, or as has been done in most cases, we can implant the radioactive isotope into a ferromagnetic host matrix, for example, ^{60}Co into a Co or Fe single crystal (deposit radioactive material on a host metal and then let it diffuse at high T). In this matrix the nuclei experience a large internal hyperfine field which can be of the order of 10–30 T if the isotope is implanted in a ferromagnetic 3d host. If we use this latter method we only have to apply a small field (0.1–1 T) to orient the magnetic domains for saturation of the matrix and to define a quantization axis.

The γ-ray detector should be in a direction where the change of the γ-ray intensity is a maximum as a function of temperature. As Fig. 12.60 demonstrates, this is the case for angles $\theta = 0$ or $180°$, if we have an E2 transition. We have to correct for any background contributions to the radiation, for geometry effects, and for the intensity of our source by performing a measurement at high temperatures where $W(T, \theta) = 1$.

Using radioactive nuclei for thermometry one has to be careful not to produce too large a self-heating of the source due to radioactive transitions. Typical values are for

$1\,\mu\mathrm{Ci}^{60}\mathrm{Co}$: 0.57 nW due to the β-emission,

and

$1\,\mu\mathrm{Ci}^{54}\mathrm{Mn}$: 0.03 nW due to the 5 keV radiation from radioactive ^{54}Cr after electron capture; [12.3, 12.4, 12.7, 12.10, 12.21, 12.177–12.182].

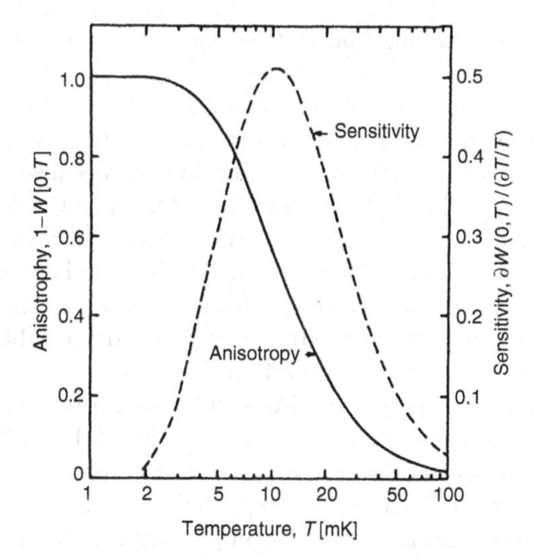

Fig. 12.61. Temperature dependence of the anisotropy of the γ-radiation of ^{54}Mn in an Ni host along $\theta = 0$. The quantity $\mathrm{d}W/\mathrm{d}T$ is a measure of the temperature sensitivity of the thermometer which, in this case, has its maximum value at about 10 mK

In an appropriate experimental setup the high-energy γ-rays used for thermometry can leave the source without noticeable heating effects. If the source is too strong it will heat; if it is too weak, counting times may get excessive. In any case the source has to be in good thermal contact to its surrounding. This can usually even be achieved by soldering because the polarizing field will keep the solder normal-conducting.

One of the disadvantages of nuclear orientation thermometry is that the sensitivity $T(\mathrm{d}W/\mathrm{d}T)$ is limited to a certain temperature range at around a value of $T \cong \Delta m \mu_n g_n B/k_B$, which is usually between 1 and 100 mK (Fig. 12.61). For the earlier mentioned isotopes we have $\Delta E_m/k_B = 9.14$ mK for ^{54}Mn in Fe and 7.97 mK for ^{60}Co in Fe (corresponding to internal fields of −22.7and−29.0 T, respectively). Therefore for these transitions the useful range is a few millikelvin to about 40 mK, whereas at higher and lower temperatures $W(T, \theta)$ is independent of T because all levels are either equally populated or only the lowest level is populated.

Because this method is a statistical method we need a certain counting time to obtain the desired accuracy. The error of the measurement due to statistics is $\Delta T/T = (2/n)^{1/2}$ if we count n pulses. One typically needs a measuring time of several minutes to half an hour to get an accuracy of 1%, for example, because the source must not be too strong in order to avoid self-heating. Low-temperature physicists sometimes unjustifiably avoid this method because they do not have the counting and detecting electronics on hand, which is more common in nuclear physics laboratories (the basic equipment consists of a γ-ray detector, like GeLi, a preamplifier, gate, and a multichannel analyser with a computer).

An advantage of this method is that we can use a metal to make good thermal contact to the point whose temperature we want to measure and to have a good nuclear spin–lattice coupling. The γ-energy is very often large enough that we do not need special windows to get the γ-rays out of the cryostat. We do not need any leads to the sample because we only measure the electromagnetic field of the γ-rays. And, finally, nuclear orientation is a primary thermometric method because the theory is well established, and we do not need any extra calibration except that we have to normalize the counting rate or the intensity of the radioactive source at high temperatures. Therefore this method has very often been applied for calibration purposes [12.10, 12.21, 12.172, 12.173, 12.182, 12.183].

Detailed and instructive comparisons between various nuclear orientation and other thermometric methods are described in [12.177–12.183]. In some of these papers as well as in other relevant work discrepancies observed between the various temperature scales were reported, with many of them originating from self-heating effects or from incomplete magnetic saturation (aligning of domains) of the γ-ray source; in some cases fields in excess of even 1 T seem to be necessary for complete magnetization. Some of the observed discrepancies are not yet fully understood. Agreement to better than 0.5% at 10–50 mK has been found in [12.183] for a Josephson noise thermometer and a ^{60}Co γ-ray anisotropy thermometer. The main usefulness of a γ-ray anisotropy thermometer may remain as a primary thermometer for calibration purposes in a restricted temperature range and for nuclei with well-understood decay schemes. For example, nuclear orientation of the γ-rays emitted from ^{60}Co has been used as the primary thermometric method at 0.5–25 mK in the contribution of the University of Florida [12.21] to the realization of the new temperature scale PLTS-2000 (see Sect. 11.3).

Another method relying on the Boltzmann population of nuclear sublevels of radioactive nuclei is the Mössbauer effect. This method has not often been applied and, for more details, the literature cited in [12.3, 12.4] should be consulted. A disadvantage of it is that one can only use low-energy γ-rays ($E \leq 100$ keV) requiring in most cases special thin windows to let the γ-rays escape from the cryostat. In addition, corrections for finite sample thickness, long counting times, specialized equipment and motion of source or absorber are necessary.

12.12 Summary

Table 12.3 summarizes the most important thermometric methods for determination of temperatures below 1 K. They are resistance measurements on a conductor of $T > 20$ mK and susceptibility measurements of the electronic paramagnetism for the temperature range above a few millikelvin. Below this temperature one has to resort to nuclear magnetism where the nuclear susceptibility in most cases is measured either directly or more appropriately

Table 12.3. The most commonly used thermometric techniques at $T \leq 1\,\mathrm{K}$

Measured property	Function	Material	Temperature range
Electrical resistance	$\ell n R = \sum_{\mathrm{n}=0}^{\mathrm{m}} \alpha_{\mathrm{n}} (\ln T)^{\mathrm{n}}$	Composites with negative R–T characteristics	$> 20\,\mathrm{mK}$
Electronic param. suscept.	$\chi_{\mathrm{e}} = \chi_0 + \lambda_{\mathrm{e}}/(T_{\mathrm{e}} - \Delta)$	CMN; *Pd*Fe	$> 3\,\mathrm{mK}$
Nuclear paramag. suscept.	$\chi_{\mathrm{n}} = \chi_0 + \lambda_{\mathrm{n}}/\mathrm{T}_{\mathrm{n}}$	Pt	$1\,\mu\mathrm{K}$–$0.1\,\mathrm{K}$

via one of the discussed nuclear magnetic resonance methods. Here one relies on the applicability of the Curie law. The result of each temperature measurement has an error and it deviates more or less from the absolute thermodynamic temperature. Very often it is difficult to estimate the uncertainty of the measured "temperature". In cases where temperature is a very important parameter, such as in measurements of heat capacities, it is sometimes advisable to use two different thermometric methods and compare the results if possible to get a feeling for the uncertainty in T. In any case, it is very important to write up the obtained results with all relevant experimental details so that future researchers can reconstruct the used temperature scale, and, if necessary, later corrections are possible. A low-temperature physicist should always remember that the quality of his data depends on the quality of his thermometry!

Problems

12.1. Which voltage resolution is necessary to measure with a (Au + 2.1% Co vs. Cu) thermocouple a temperature of 30 mK to about 1% (Fig. 12.4)?

12.2. Which relative voltage resolution is necessary to measure 20 K with a platinum resistance thermometer to 1%?

12.3. Suppose a carbon thermometer of $5\,\mathrm{mm}^2$ area is immersed in liquid helium at 1 K and the Kapitza resistivity is $\mathrm{A}R_{\mathrm{K}}T^3 = 0.01\,\mathrm{K}^4\,\mathrm{m}^2\,\mathrm{W}^{-1}$. What is the maximum dissipation allowed in the resistor for a temperature difference of $\Delta T = 0.1\,\mathrm{mK}$ between it and the liquid-helium bath?

12.4. Calculate the thermal response time of a Speer 220 Ω thermometer at about 1 K. Take for the heat capacity typical values shown in Fig. 12.16 and for the thermal resistance the value given in the middle of p. 294.

12.5. Calculate the relative sensivity of the thick-film chip-resistors for which the equation and characteristics are given in (12.10) and Fig. 12.22.

12.6. Deduce (12.13) and (12.14) from (12.12).

12.7. Which sensitivity of a voltmeter is necessary to measure via noise thermometry 1 K to 1% at a bandwidth of 1 MHz using a resistor of 100 kΩ?

12.8. Calculate the relative sensitivity to which the capacitance of the capacitive glass thermometer, whose behavior is exhibited in Fig. 12.35, has to be measured to determine a temperature of 10 mK to 1%.

12.9. Calculate how many ppm of Fe impurities (magnetic moment $\mu = 2\,\mu B$) would give the same electronic susceptibility as its host Cu would contribute nuclear susceptibility.

12.10. At which temperature does the nuclear Curie law deviate by more than 1% from the exact solution (9.15c) for the magnetization of a platinum NMR thermometer used in 25 mT (Fig. 10.5)?

13

Miscellaneous Cryogenic Devices and Design Aids

This chapter deals with various cryogenic tools, design details and measuring techniques, which have turned out to be very useful in the construction and operation of a cryogenic apparatus, as well as in a variety of experiments at low temperatures.

13.1 Cryogenic Pressure Transducers for Thermometry and Manometry

In the last decades low-temperature pressure transducers have gained in importance for gas-, vapour- as well as melting-pressure thermometry, for pressure measurement as well as pressure regulation and for investigations of the properties of liquid and solid helium. They can measure the parameter of interest in situ, they are very sensitive and reproducible, and the measurement is performed with no or at most very low dissipation.

13.1.1 Capacitive Pressure Transducers

The successful career of capacitive transducers for low-temperature experiments began with the description of a capacitive ^{3}He melting pressure gauge by Straty and Adams [13.1]. This gauge is shown schematically in Fig. 13.1. All later gauges (see Figs. 13.2 and 13.3, for example) are variations of this successful original design [13.1–13.17]. In the capacitive ^{3}He melting pressure gauges a flexible diaphragm is one wall of a compartment containing a solid/liquid mixture whose pressure is to be measured. The ^{3}He is kept at nearly constant volume because a solid block is formed at the point of the filling capillary where the temperature is at the temperature of the minimum of the ^{3}He melting line (Fig. 8.1). The flexing diaphragm moves an electrically isolated plate which is connected to (but electrically isolated from) a pin at its center. The plate moves relative to a fixed plate glued into a guard ring at

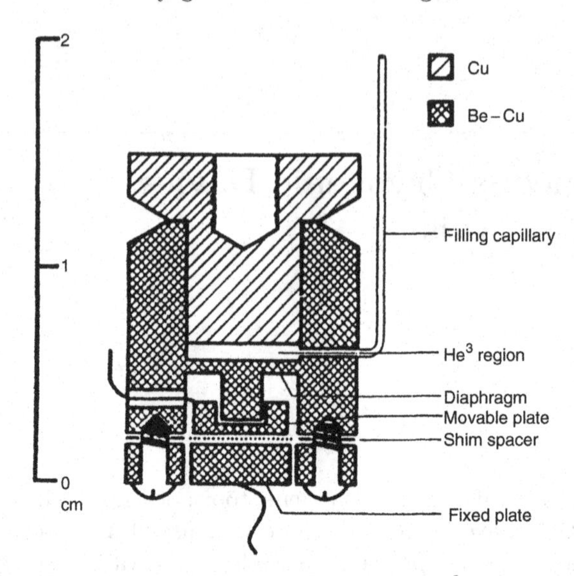

Fig. 13.1. Low-temperature capacitive strain gauge for ^{3}He melting pressure measurements of [13.1, 13.2]. The bottom of the ^{3}He chamber flexes with ^{3}He pressure variations, changing the capacitance measured between the movable and fixed plates

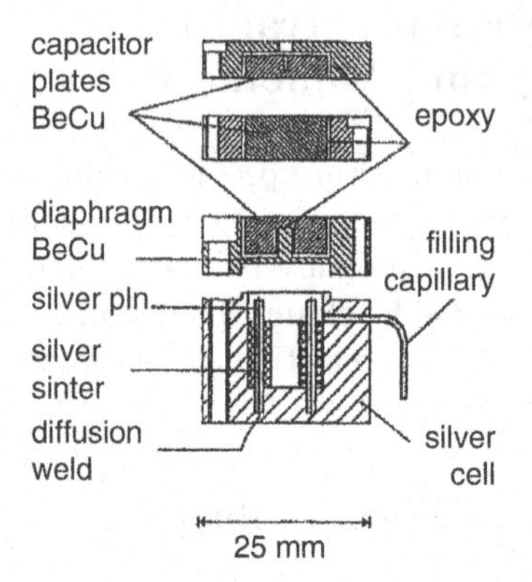

Fig. 13.2. The parts of the low-temperature capacitive strain gauge used by [13.17] to measure the ^{3}He melting pressure for determination of the temperature from this pressure using (11.7). The Ag sinter for making thermal contact to the ^{3}He is sintered to Ag pins. The heat exchanger cell is closed by a flexible BeCu diaphragm. The three BeCu plates act as measuring and reference capacitors. Reprinted from [13.17], copyright (2002), with permission from Elsevier. A similar gauge has been used by [13.16] to establish the new temperature scale PLTS-2000 which is based on the ^{3}He melting pressure (see Sect. 11.4)

electrical ground. Pressure changes are then monitored as capacitance changes when the conducting diaphragm is deflected, moving the two capacitor plates relative to each other.

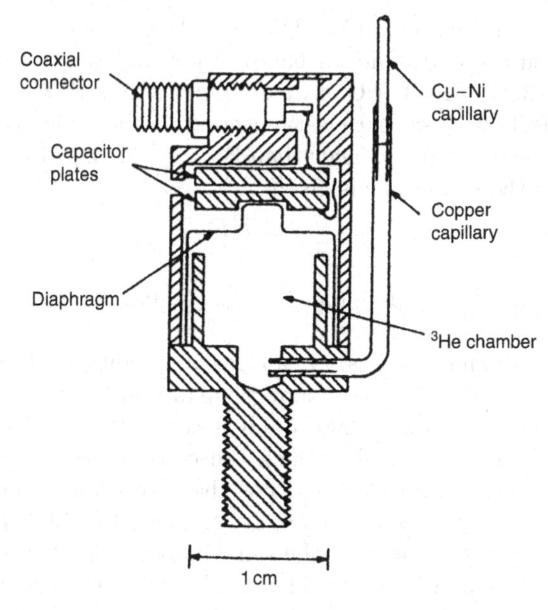

Fig. 13.3. Low-temperature vapour pressure capacitive gauge of [13.8]

To determine the expected behavior of such a capacitor we consider the model of a circular membrane with fixed edges [13.1, 13.11, 13.12, 13.18]. This gives a deflection

$$y(r) = \frac{3P}{16Ed^3}(r^2 - 2R^2)^2(1 - m^2) . \tag{13.1}$$

The deflection of the center of the membrane is then

$$y(0) = \frac{3R^4P}{16Ed^3}(1 - m^2) , \tag{13.2}$$

where R is the membrane radius, E is Young's modulus of elasticity, d is the membrane thickness, and the Poisson ratio m is the ratio of lateral unit strain to longitudinal unit strain, under the condition of uniform and uniaxial longitudinal stress within the proportional limit. For most metals $m \approx 0.2$–0.4. The actual performance will be somewhat worse because the central post carrying the capacitor plate has to be subtracted from the active diameter of the plate. The deflection is maximum for the minimum membrane thickness for which its deformation is still elastic at the maximum pressure. To stay clearly in the elastic limit, the stress on the membrane should be kept below about 20% of the yield stress S of the membrane material, which is usually hardened Be–Cu because of its favorable elastic properties ($E \approx 1.3 \times 10^6$ bar; $S \approx 1.2 \times 10^4$

bar, torsion (shear) modulus $G = 0.53 \times 10^6$ bar $= 53 \times 10^3\,\mathrm{N\,m^{-2}}$, all at room temperature but depend on composition and treatment; a hardening treatment of 2–3 h at 320–310°C in vacuum seems to work well). Another suitable material is Ti with similar elastic properties and which is less poisonous and less sensitive to magnetic fields than Be–Cu [13.13,13.15]. The maximum radial stress on the membrane is given by

$$S_{\mathrm{m}} = \frac{3R^2P}{4d^2}\,. \tag{13.3}$$

Typical values are $d \approx 0.1(0.5)$ mm for $P_{\mathrm{max}} \approx 0.5(50)$ bar and a membrane diameter of 4–10 mm.

Two criteria distinguish a good capacitive manometer. Firstly, its capacitance should show no or at most a very small temperature dependence; for example, 10% change for $T = 300$ to 4 K and less than 10^{-5} below 4 K. For that reason the electric field by which the capacitance is measured should "see" only the metal electrodes, and not the epoxy, for example, which has a T-dependent dielectric constant (Sect. 12.8), see Fig. 13.2. In addition, all metal parts should be made from the same material, Be–Cu or Ti, for example, to reduce thermal drift due to differential thermal expansion. Secondly the polished or lapped (to $\leq 1\,\mu$m) electrodes should be coplanar, to give a linear relation between their distance and capacitance over a wide pressure range, and they should have a gap as small as possible at the maximum working pressure to increase the capacitance and the sensitivity of the gauge. For this purpose the epoxy on the movable capacitor plate should cure while the latter is in contact with and therefore parallel to the fixed plate, with the gauge kept at a pressure slightly above the maximum working pressure. This should result in a gap of some tens of micrometers at ambient pressure if the maximum working pressure is about 50 bars. Another possibility is to glue both plates, perform a final matching operation to bring the surfaces flat and parallel, and then adjust the gap by shimming. For ^{3}He melting pressure thermometry, for example, the helium filled volume of the gauge should contain some metal sinter (Sect. 13.6) to provide thermal contact between the capacitor cell and the ^{3}He and to give the gauge a small time constant in the millikelvin temperature range (see Fig. 13.2). From the ^{3}He phase diagram one finds that the open volume of the cell has to be at least twice as large as the volume in the metal sinter so that the liquid–solid interface of the ^{3}He forms in the open space of the cell at all pressures.

For high precision and reproducibility, it is advisable to frequently "exercise" the diaphragm over the pressure range it is to be used prior to calibrating it. This work-hardening process should be repeated each time the upper or lower pressure bounds have been exceeded [13.11,13.16]. These references contain a detailed discussion of possible error sources in pressure gauge calibration, like hydrostatic head corrections, influences from thermal expansion coefficients, etc.

For a parallel plate capacitor, the relation between pressure P and capacitance C is given by

$$P \propto (1/C_0 - 1/C) \quad \text{or} \quad 1/C = \alpha - \beta P, \tag{13.4}$$

where C_0 is the zero-pressure capacitance; for higher accuracy more terms in $(1/C_0 - 1/C)^n$ have to be used. Typical values are $C \approx 10$–$100\,\mathrm{pF}$, $\Delta C/\Delta P \approx 0.1$–$10\,\mathrm{pF\,bar^{-1}}$, $\Delta y/\Delta P \approx 0.1$–$1\,\mathrm{\mu m\,bar^{-1}}$.

Various designs of cryogenic capacitive manometers for melting pressure [13.1–13.6, 13.16, 13.17] or vapour pressure thermometry [13.7–13.9] as well as for other pressure measurements [13.10] have been described in the literature, see [13.12] for a review to earlier work. The capacitive transducer for vapour pressure thermometry of [13.8] shown in Fig 13.3 uses a gold-plated 0.015 mm thick electro-deposited Cu diaphragm. It has a 30 µm gap and $C \approx 43\,\mathrm{pF}$. The achieved resolution of $10^{-6}\,\mathrm{pF}$ allows changes of 0.3 mK (0.04 µK) at 0.3 K (2K) to be detected, corresponding to a vapour pressure of 2 µbar (0.2 bar) of ^{3}He. The vapour pressure gauge of [13.7] gave a resolution of $\Delta T/T = 10^{-9}$ from 1.6 to 2.2 K, corresponding to a deflection of $y \approx 10^{-3}\,\mathrm{nm}$ at a capacitor plate spacing of 10 µm. A rather small, slightly modified commercial low-pressure transducer suitable for cryogenic applications has been described in [13.14]. It uses a 10-µm-thick silicon diaphragm and has a resolution of about 0.5 µbar in the mbar pressure range. A sensitivity of about 10^{-10} bar has been reached in a differential gauge using a 0.8 µm thick aluminium diaphragm [13.19].

A capacitive gauge is an indirect-reading device and has to be calibrated in the operating pressure range. For a melting pressure gauge the calibration is best performed with a gas lubricated, controlled-clearance dead-weight tester [13.20]. Well-calibrated and stable mechanical, piezoelectric, or capacitive manometers at room temperature often do well as secondary standards. A low-pressure gauge can be calibrated via a measurement of the vapour pressure of one of the helium isotopes. For these calibrations with the standard at room temperature one may have to take into account the possible problems discussed in Sect. 12.2. Because Be–Cu behaves hysteretic, the calibration has to be checked or repeated in each cool down.

Successful designs of bridge circuits for capacitance measurements using an inductive voltage divider ("ratio transformer") have been described in [13.10, 13.13, 13.16, 13.21, 13.22]. They are essentially identical to the bridge shown in Fig. 3.32 with the inductances replaced by capacitors and variable resistors to balance dielectric losses. They usually operate in the low kHz range. In such a design the lead capacitances should be kept small and they should have a small temperature coefficient to keep stray capacitances low and to avoid degrading the sensitivity or introducing errors [13.13]. If the reference capacitor is kept at room temperature it usually has to be temperature stabilized. Of course, it is much better to keep it at a low, constant temperature, for example at 4.2 K, if high stability is required. If a home-made reference capacitor is used, one should make it as similar to the working capacitor as possible, again avoiding that the electrical field "sees" dielectric, glassy materials (Sect. 12.8). In recent designs [13.16, 13.17], a reference capacitor essentially identical to the measuring capacitor has been integrated in the cryogenic gauge (see Fig. 13.2). With a driving voltage of some volts such a five-terminal capacitance bridge using a ratio transformer, well shielded and tightly coupled transformer arms, a stable reference capacitor, and a low-temperature capacitance gauge, allows a ratio (capacitance) resolution of 10^{-8}, corresponding to a pressure resolution of about 10^{-6} bar at 10 bar, for instance. The best results achieved recently are a sensitivity and long-term stability of $\Delta P/P \approx 10^{-9}$ [13.13, 13.21]. At balance the ratio R in the bridge is given by

$$R = \frac{C}{C + C_{\mathrm{Ref}}} \text{ or } C = \frac{RC_{\mathrm{Ref}}}{1-R}, \tag{13.5}$$

with a sensitivity

$$\frac{\mathrm{d}R}{\mathrm{d}T} = R\frac{1-R}{C}\frac{\mathrm{d}C}{\mathrm{d}T}, \tag{13.6}$$

and for a maximum sensitivity, with $C \simeq C_{\mathrm{Ref}}$,

$$\frac{\mathrm{d}R}{\mathrm{d}T} \simeq \frac{1}{4C}\frac{\mathrm{d}C}{\mathrm{d}T}. \tag{13.7}$$

One has to keep in mind that the reading C of a bridge can depend on excitation voltage V and frequency ν with values of

$$\frac{\Delta C}{C} \approx \text{several } 10^{-6}\left(\frac{\Delta V}{V}\right) \text{ or } \left(\frac{\Delta \nu}{\nu}\right) \tag{13.8}$$

for a symmetric bridge [13.3]. In addition, there are at least two very accurate commercial capacitance bridges on the market. They offer 5-ppm accuracy, 0.15-ppm resolution, and 1-ppm/year stability.

The new low-temperature scale PLTS-2000 is based on the melting pressure–temperature relation of ^{3}He (Sect. 11.3). The laboratories that established this scale have measured the melting pressure in situ by a capacitive pressure transducer of the above-described design. A very good and detailed description of the design, construction, installation, filling procedure to the correct gas pressure, operation, calibration, and use of these pressure transducers including the room temperature part can be found in [13.16]. A pressure sensor similar to the one used in [13.16] is shown in Fig. 13.2; it has a resolution of 0.1 Pa at 3 MPa, as is shown in Fig. 11.1.

A pressure gauge whose off-balance signal is amplified and used to regulate the temperature of a ballast volume (for example, at low temperatures and filled with ^{4}He, weakly coupled to the bath at 4.2 K) connected to an experiment can be used for high-precision pressure regulation, see Fig. 13.4 [13.5, 13.10, 13.23].

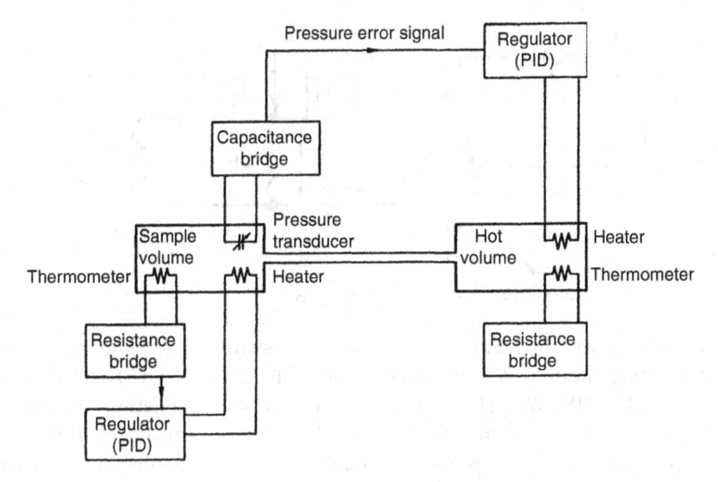

Fig. 13.4. Setup to measure and regulate pressures at low temperatures (see text) [13.10]

13.1.2 Inductive Pressure Transducers

Even higher sensitivities and resolutions can be achieved by inductive pressure transducers using a SQUID as the detection instrument (Fig. 13.5). In these devices, a flexible superconducting diaphragm senses the pressure change. Its motion modulates the current flowing through an adjacent flat pancake coil wound from superconducting wire and coupled via a flux-conserving loop into a SQUID magnetometer. Such an ultrasensitive low-temperature pressure transducer has been described in [13.24]. Its diaphragm is a 7.5 μm thick, 8 mm diameter Kapton foil ($E = 6 \times 10^4$ bar; $S = 3.4 \times 10^3$ bar) coated with 60 nm aluminum, whose motion is sensed by a flat coil connected to an RF-SQUID. Displacements of 5×10^{-4} nm and pressure changes of 3×10^{-11} bar could be resolved. An inductive pressure gauge using a 1.25-mm-thick Nb diaphragm and a DC-SQUID has been described in [13.25] (Fig. 13.5). It has a sensitivity of 3×10^{-10} bar at 20 bar resulting from a flux sensitivity of $4 \times 10^4\, \Phi_o$/bar. Of course, (13.1)–(13.3) for the flexible diaphragm as well as the pressure calibration procedure described above for capacitive transducers can be applied for inductive transducers as well.

13.2 Cold Valves

Valves at low temperature are often necessary, in particular for experiments with liquid or solid helium. They confine the helium sample to a constant volume to avoid thermomechanical pressure gradients in the fill line, and eliminate heat leaks or fractionation of a helium mixture along the fill capillary. Successful designs of bellows sealed valves are described in [13.10,13.26–13.33].

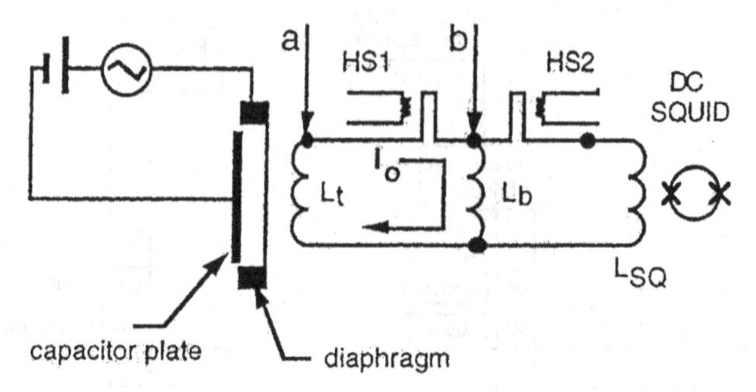

Fig. 13.5. Schematic diagram of an inductive pressure gauge using a superconducting niobium diaphragm (1.25 mm thick, 38 mm diam.) and a DC-SQUID as detector. The heat switches HS1 and HS2 are used to establish a persistent current I_o in the left loop via the external connections a and b. Since the total flux in the superconducting $L_b L_t$ and $L_t L_{SQ}$ loops must be conserved, any modulation of L_t caused by the moving diaphragm results in a change of the current in the input coil to the SQUID [13.25]

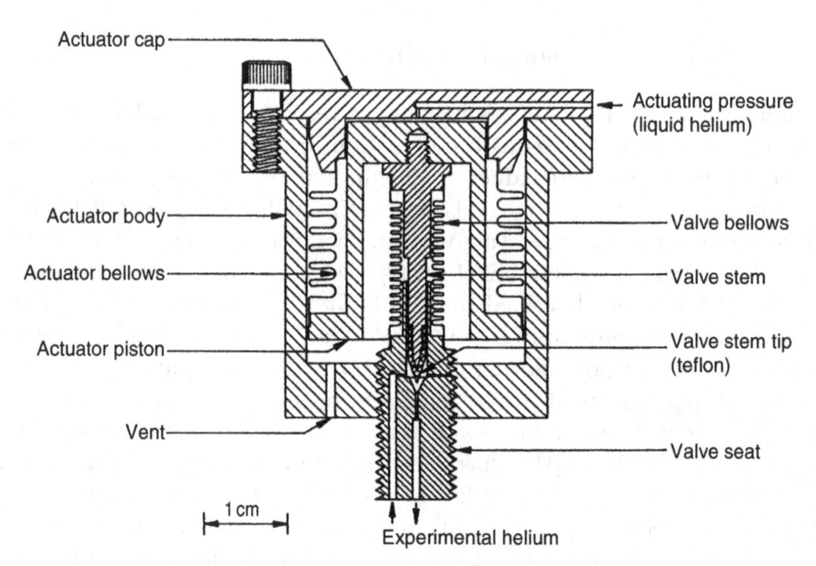

Fig. 13.6. Hydraulically actuated cryogenic valve [13.26]

The principle of their design is to select an appropriate combination of soft and hard material for the needle tip and the seat.

The original design of cryogenic valves used a hard metal seat (brass, hardened Be–Cu, or stainless steel) with a small hole and a soft needle for closing (Fig. 13.6). For the latter, a metal covered by Teflon or Vespel, as well as Torlon with its high yield strength seem to be appropriate [13.26–13.28]. Other versions use a soft metal seat (Ag, non-hardened Be–Cu, or German

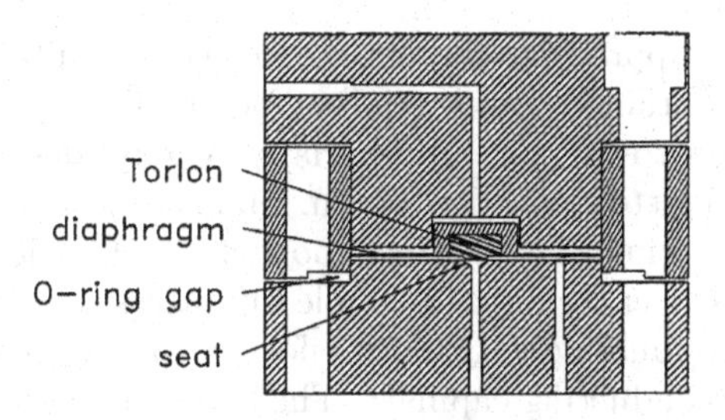

Fig. 13.7. Cryogenic valve with a 75 μm thick flexible BeCu (Berylco 25) diaphragm and a Torlon 4203 tip. The opening angle in the BeCu seat is 110°. The valve is distinguished by its small dead volume of 0.2 cm^3 [13.30]

silver) and a hard tip (stainless steel or hardened Be–Cu) [13.10, 13.29]. All these valves are sealed by one or two metal bellows. More recent rather small and simple cryogenic valves [13.30–13.32] use thin metal diaphragms (0.1 mm Be–Cu) instead of bellows for sealing, reducing the dead volume in the valve to 0.02 cm^3 only. For the valves of [13.30, 13.31], a Torlon tip closes the hole in a metal by applying a pressure of several bar (Fig. 13.7). The most distinguishing feature of the valve of [13.32] is the fact that it is closed in the unpressurized position. It uses a ruby ball as the "hard tip" and stainless steel or Be–Cu for the "soft" seat. In all the described valves careful alignment between needle and seat is important. The local pressure between the sealing surfaces should be maximized by choosing appropriate angles on them. The valve should, of course, be as close as possible to the temperature of the experimental cell to avoid problems from temperature gradients of the liquid in the connecting capillary.

These cryogenic valves are usually actuated hydraulically using a helium mixture or ^{4}He at pressures of some bar as the medium to operate the valve, of course requiring a second capillary in the cryostat. The actuating pressurized gas is supplied from a tank at room temperature. More elegant is to supply the actuating gas from a small auxiliary volume at low temperatures where the gas pressure is adjusted by regulating the temperature of this volume [13.30]. Recently, a magnetically operated cold valve has been described that avoids the capillary with the actuating pressurized gas [13.33]. It uses a permanent NdFeB magnet that is moved by the magnetic field gradient produced by a superconducting coil. It presses a ruby ball against the 0.5 mm hole in a Torlon seat. As a further advantage, the volume in the valve does not change when the valve is actuated.

13.3 Coaxial Cables and Feedthroughs

The appropriate materials for single leads transmitting electrical signals to and from the cold parts of a cryostat have already been discussed in Sect. 4.1.

Coaxial cables and appropriate miniature connectors suitable for Kelvin temperatures can be obtained commercially. For lower temperatures homemade coaxial cables seem to be more appropriate. A good design with a low heat leak seems to be twisted pairs of 0.1 mm Manganin or Constantan or, even better, superconducting Nb–Ti wires (do not use Cu-clad superconducting wires at $T < 0.1\,\mathrm{K}$ because of the possible influences due to the temperature-dependent superconducting proximity effect) inside of a thin-walled Cu–Ni tube or of a superconducting capillary. The superconducting capillary can be commercial Nb tubing or it can be home-made by leaching out the resin core from a commercial Pb–Sn solder. This can be accomplished by putting several 50–80 cm long pieces of a Pb–Sn solder in a glass tube filled with some alcohol or gasoline and heating it to a temperature between 120 and 160°C for 1–3 h, for example by means of a heater around the glass tube. The process is accelerated if the lower end of the glass tube is kept in an ultrasonic bath. Afterward the solder capillaries should be cleaned with acetone. Another simple choice is a thin-walled Cu–Ni tube coated with a superconducting solder. Figure 13.8 shows how one can easily make shielded pin connections for such a superconductor shielded coaxial lead.

Often coaxial cables and simple leads (as well as capillaries for helium experiments) have to be vacuum enclosed so that they are guarded against the effects of bath level changes. Then thermal anchoring at 4.2 K and, if they go to lower temperatures, at further points with successively lower temperatures is essential [13.34, 13.35]. This can be achieved by winding a considerable length of them tightly around a Cu post, where they should be fixed by an adhesive. Another possibility is to clamp leads between Cu plates using grease for improved thermal contact. A home-made coaxial cable with a very small temperature coefficient of its capacitance was described in [13.13]; its design is illustrated in Fig. 13.9.

If leads run through the ^{4}He bath then they are well thermally anchored but mostly they have to be brought into an evacuated can by an electri-

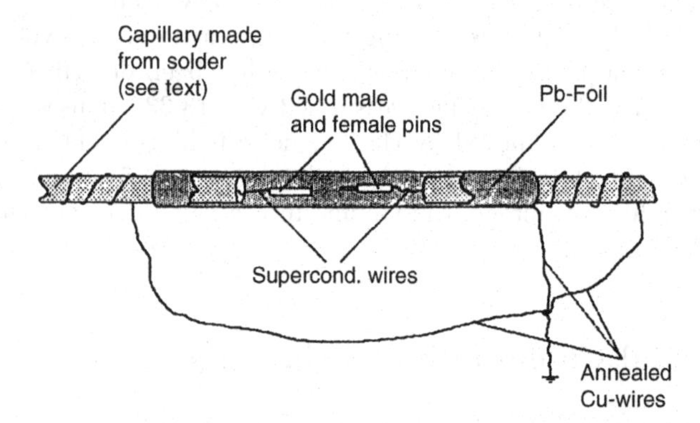

Fig. 13.8. Design of detachable connectors for cryogenic coaxial leads

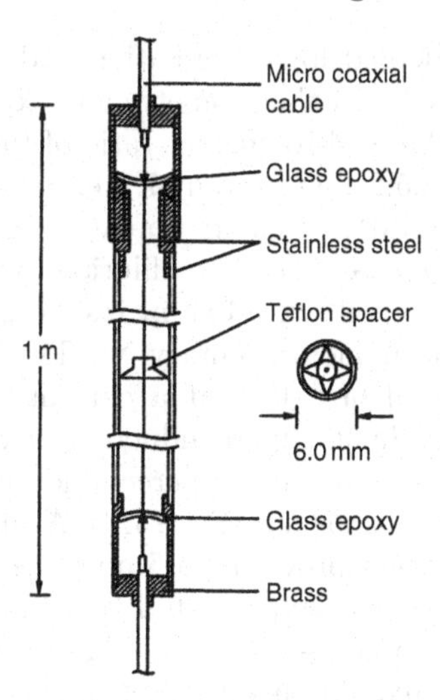

Fig. 13.9. Cross-sectional drawing of the home-made coaxial cable described in [13.13]

cally isolating, vacuum tight, cold epoxy feedthrough. Appropriate designs are described in [13.3, 13.34–13.38] and shown in Fig. 3.18. The performance of such a feedthrough can be improved by using a filler, for example Al_2O_3 or metal powder, in the epoxy so that its thermal expansion coefficient is better matched to that of metals. Leads for superconducting magnets will be discussed in Sect. 13.4.

13.4 Small Magnets and Magnet Leads

13.4.1 Small Superconducting Magnets and Magnet Leads

Because the cost of the wire is a large fraction of the price of a superconducting magnet and because considerable experience is needed to design and wind a powerful superconducting magnet, such magnets are better bought than home-made. In this book, I will not discuss commercial superconducting magnets which are offered for magnetic fields up to 20 T. However, for many purposes, for example NMR experiments, for superconducting heat switches, or for thermometry, specially designed small superconducting solenoids for moderate fields are necessary; here do-it-yourself is in order (see Fig. 3.13). These magnets are designed using computer programs [13.3, 13.39–13.46].

Very frequently such solenoids sit in the vacuum space of the cryostat. And when they are in close proximity to other equipment, their fields have to be shielded by a superconducting cylinder (Sect. 13.5.2) or they must be isolated from the stray fields of other coils. This is a typical situation in a nuclear demagnetization cryostat with a large, commercial (see Fig. 10.14 and

10.24), typically 8 T, solenoid for refrigeration and small coils and field-free regions for heat switches, NMR thermometry and experiments. The superconducting shield reduces the field-to-current ratio of the solenoid and smoothes the central field (Sect. 13.5.2) but it complicates calculations. The design and behavior of a shielded NMR solenoid operated in vacuum and thermally anchored to the mixing chamber of a dilution refrigerator at 20 mK are described in [13.44]. This coil with 14 mm inside diameter has 30 layers of 0.1 mm diameter Cu-clad monofilamentary Nb–Ti wire inside a Nb shield. It can produce an axial field of 0.45 T at a current of 1.75 A with 5×10^{-4} homogeneity over 1 cm. Small superconducting magnets with central fields up to 4.5 T, and active and passive superconducting shields were described in [13.45, 13.46]; a design is shown in Fig. 13.10. A small commercial cryogen-free superconducting magnet providing a 5-ppm homogeneous magnetic field of 2.5 T in a 51-mm bore has been presented at LT 24 [13.47]. It is operated in vacuum and is cooled by a Gifford–McMahon refrigerator (Sect. 5.3) to 3 K.

The leads to a superconducting magnet sitting inside the vacuum can of the cryostat have to be fed from the ^{4}He bath into this can. A reliable, easy to make, all-metal feedthrough for this purpose is discussed in [13.48] (Fig. 13.11). Such a feedthrough can be used when normal-conducting connections between the leads and to the cryostat can be tolerated. The Cu or Cu–Ni cladding on commercial superconducting wires allows easy, vacuum-tight soldering to the tube through which the lead is fed into the vacuum

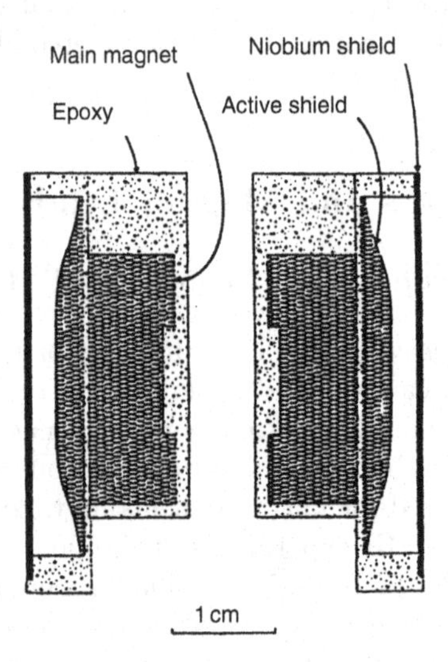

Fig. 13.10. Superconducting solenoid for operation in vacuum. Besides the main solenoid the design contains an active shield produced by a second solenoid as well as a passive superconducting shield of Nb [13.46]

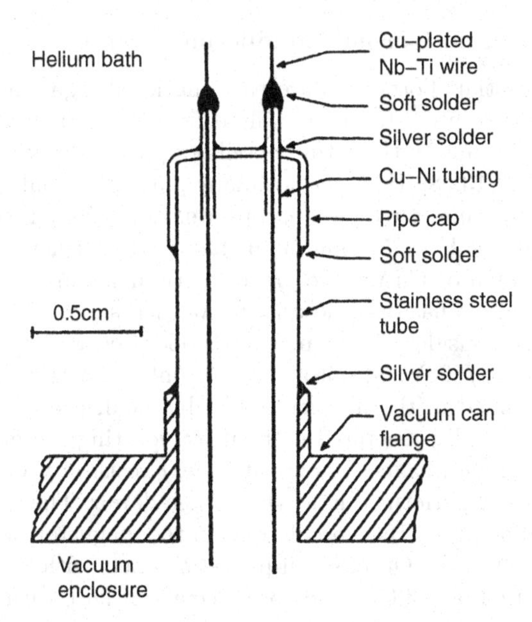

Fig. 13.11. Leak-tight vacuum feedthroughs for leads to a superconducting magnet; for details see text [13.48]

space. In addition, this shunt provides a desired low resistance path on the power supply output for filtering of abrupt field or current changes, so that in the case of a quench of the magnet the current does not have to go through the resistive cladding of the magnet wire. The L/R time constant in the loop may even have to be further adjusted by a parallel shunt across the leads. The use of a superconducting shunt as a persistent switch for a superconducting solenoid is discussed in Sect. 10.2.

Current leads to a large cryogenic magnet may give rise to a high helium consumption, even when the electric current is reduced to zero, because they usually have to be good conductors. They have then to be designed to compromise between low thermal conductivity losses and low Joule heating; often brass is the appropriate compromise for this purpose. The zero-current helium consumption may be greatly reduced by using disconnectable leads which can be removed when the current to the magnet is zero or, especially, when the magnet is operated in the persistent mode with a superconducting shunt in the helium bath (Fig. 10.6) [13.49]. For such leads, design criteria are gold-plated contacts with a low resistance [13.50]. Current leads for cryogenic magnets which are cooled by the vapour of the evaporating cryogenic liquid and designed for currents up to about 100 A were described in [13.3, 13.51–13.55]. Of course, for the part of the leads at low temperatures, a superconductor with its zero losses should be used. The length of this part can be substantially increased by using high-T_c superconductors. Such leads are commercially available.

13.4.2 Small Pulsed Normal-Conducting Magnets

Presently, substantial efforts are devoted in national high magnetic field laboratories to offer to interested users pulsed fields of up to presently between 60 and 70 T [13.56] and in the future possibly up to 100 T with pulse times of order 10 ms [13.57, 13.58]. However, for many purposes, small-pulsed magnets with shorter pulse times are quite appropriate. In [13.59] for example, a small magnet (1.6 mm free bore, 5.5 mm outer diam., 6 mm length) made from six layers of 0.3-mm Cu or CuAg wire for magnetic fields up to 30 T with pulse times of about 0.6 ms has been described. At such short times, eddy current heating has to be considered. The magnet is operated with very small energy from a desktop 90J/300V capacitor bank of only 14.3 kg weight. Of course, the energy consumed in the magnet has to be small enough so that pulsed fields can be repeatedly generated without evaporating too much liquid ^{4}He. The small size of the magnet allows high-field resistivity or magnetization measurements to be performed in a ^{4}He storage dewar. The magnet was used to measure Shubnikov–de Haas oscillations in the magnetoresistance of an organic conductor at 1.6 K. Other small-pulsed magnets with 3–6 mm bore have been built and used by [13.60]. They produce about 20 T with a pulse time of typically 0.5 ms and are fed from a 1.9kJ/2kV capacitor bank. The portable setup has been used for synchrotron radiation experiments (electron spin THz spectroscopy and x-ray diffraction). A larger setup for synchrotron x-ray powder diffraction studies in pulsed magnetic fields has been described in [13.61]. It uses a 110kJ/16kV capacitor bank to feed a 30T coil (22mm bore) with a field pulse of 18 ms width. Clearly, small high-field pulsed magnets offer a wide variety of applications, particularly in high-magnetic-field spectroscopy, a research area that is just in its beginning.

13.5 Shielding Against Magnetic Fields and Magnetic Fields Inside of Shields

As already mentioned in Sect. 13.4, many low temperature experiments require careful shielding against stray magnetic fields of nearby magnets or even against the earth's magnetic field. This is particularly important if highly homogeneous fields for NMR are needed or if experiments on superconductors with very low transition temperatures (and therefore very low critical magnetic fields) or with DC-SQUIDs are to be performed [13.62–13.64]. The earth's magnetic field or other disturbing small fields can be cancelled to below the 1 μT range by three mutually perpendicular pairs of Helmholtz coils where the currents can be adjusted independently. However, this is laborious and requires measurements of the compensated field components [13.65]. A further step would be a μ-metal shield at room temperature around the cryostat. However, such a shield is bulky and expensive. Therefore cryogenic shields, either normal conducting soft magnets or superconductors, have to be considered.

13.5.1 Normal-Conducting Shields

The shielding factor [13.66] of a thin, long, open cylindrical shield in a field perpendicular to its axis is given by

$$S_{\text{per}} = \mu d/D\,, \tag{13.9}$$

with μ being the permeability, d the wall thickness, and D the diameter of the shield. The shielding factor of that shield with closed ends in a field parallel to its axis is expressed as

$$S_{\text{long}} = \frac{4NS_{\text{per}} + 1}{1 + D/2L}\,, \tag{13.10}$$

with the demagnetizing factor

$$N = \frac{1}{a^2 - 1}\left[\frac{1}{\sqrt{a^2-1}}\ln\left(a + \sqrt{a^2-1}\right) - 1\right] \tag{13.11}$$

with $a = L/D$ and L being the length of the shield.

Through open ends a parallel field penetrates and decays along the axis exponentially with $\exp(-\alpha x/R)$ with $\alpha \approx 2$ for $R \approx 20$ mm. For a shield with closed ends containing a hole with radius r in its center, the field decays on the axis as

$$S_{\text{hole}} = \frac{3\pi x^3}{2r^3}\,. \tag{13.12}$$

The maximum field for shielding by a soft magnetic shield is given by

$$\begin{aligned} B_{\text{ex}} &\le S_{\text{long}} B_{\text{S}} d/D \quad \text{for a parallel field}, \\ B_{\text{ex}} &\le B_{\text{S}} d/D \quad \text{for a perpendicular field}, \end{aligned} \tag{13.13}$$

where B_{S} is the saturation induction of the shield.

A soft magnetic material keeping its shielding properties due to a high permeability (up to about 10^5) after cooldown is Cryoperm 10 (Vakuumschmelze, D-63412 Hanau). This material can be bought as sheet and then rolled to tubes of appropriate diameter after annealing for 2 h at 800°C. Welding should be performed without adding other materials. The following heat treatment of a Cryoperm tube has given successful results: 2–4 h at 1,100°C in vacuum; 2 h cooling to 500°C and then with He or Ar gas to room temperature; annealing under vacuum for 0.5 h at about 570°C; slow cooling to 470°C and annealing there for 2 h; fast He or Ar gas cooling to room temperature. Afterward, the shield should be handled with care because inelastic deformation leads to deterioration of its permeability. The Cryoperm (as well as other soft magnetic shields) should be degaussed in situ with an amplitude slowly decreasing from a starting value of about 0.1 mT and a frequency of about 1 Hz to about 1 nT in 1 h. This should be performed at the temperature and field environment in which the shield will be used, because temperature and field changes after degaussing the shield can substantially reduce its shielding properties. Such a procedure allows shielding to about 0.1 μT, and in the best cases with some effort to 10 nT [13.62–13.64].

From our own experience, typical shielding factors for a closed Cryoperm 10 shield of $D = 20$–30 mm, $L = 150$ mm and $d = 1$ mm are $S \approx 10^4(10^3)$ and $B_i < 0.01$ (0.1) μT at 300 (4.2) K and parallel fields of up to about 1 mT. Substantial improvements can be obtained using two concentric Cryoperm 10 shields resulting in fields of 0.01 μT at $B_{ex} < 5$ mT. To achieve such small remaining fields, all parts in the setup have to be nonmagnetic, and sometimes even low concentrations of magnetic or superconducting inclusions (in brass, for example) have to be considered.

13.5.2 Superconducting Shields

A closed superconducting enclosure is a perfect shield against electromagnetic and, of course, in particular against magnetic fields if the field does not exceed the thermodynamic critical field B_c for a type-I superconductor or the lower critical field B_{c1} for a type-II superconductor. However, a superconducting shield freezes in the field present when it goes through its transition temperature; hence in many cases they have to be surrounded by a μ-metal shield to reduce this field. A superconducting shield can be made from a Nb tube or from Nb sheet; Ti should be used if a lower critical temperature and/or field are required. Another possibility is to coat a Cu, brass or, better, Cu–Ni tube or capillary with a superconducting solder. The latter, a Cu–Ni capillary coated with a Pb–Sn solder ($T_c \approx 7.2$ K) is a suitable shield for the electrical leads (Sects. 4.1 and 13.3). It should be filled with grease for proper mechanical and thermal anchoring of the leads inside of the shield. Another way to produce a superconducting capillary shield is to leach out the resin core of a commercial flux-core solder, as discussed in Sect. 13.3. The shielding factor increases with the length-to-diameter ratio of an open cylindrical shield.

A frequently encountered problem with superconducting shields is that thermal gradients may produce currents and magnetic fields as the shield enters the superconducting state. These fields may be greater than the field which the superconductor was supposed to shield against! Only trial and error can help to eliminate this problem. A well-controlled cool-down of the shield with the temperature decreasing from one end of it to the other may solve this problem.

A combination of superconducting and normal conducting shields should be used if the highest performance is required, for example, when very small, constant and homogeneous fields are required [13.62–13.64]. From our experience the best combination seems to be an outer superconducting Nb shield which freezes in the earth field in cooling to 4.2 K and then an inner Cryoperm 10 shield degaussed at 4.2 K. This combination keeps the internal field to less than 0.03 μT, even when the external field was ramped up to 30 mT, for example. A superconducting shield reduces the field-to-current ratio of a field coil and can homogenize the field profile [13.43, 13.67, 13.68]. However, be careful: A superconducting shield can destroy the homogeneity of the field of a coil designed for free space.

Some low-temperature experiments (for example NMR experiments) require the application of variable but homogeneous magnetic fields of moderate strength. This can be achieved with a rolled-up superconducting foil, for example made from lead or niobium, with an insulating layer in between [13.68]. This design takes advantage of the boundary condition of having no magnetic field normal to the superconducting wall for field homogenization but allows field variation from the outside because it is not closed. Of course, this design again reduces the field generated by a coil outside of it by the area ratio $(A_{\text{shield}} - A_{\text{coil}}/A_{\text{shield}})$.

The creation of extremely small fields to below 10 pT using expandle superconducting shields has been described in [13.69].

13.5.3 Magnetic Fields Inside of Shields

The central field of an infinitely long solenoid inside of a coaxial cylindrical superconducting shield is given by [13.44]

$$B_i = \frac{B_0}{1 + \dfrac{A_{\text{coil}}}{A_{\text{shield}} - A_{\text{coil}}}}, \tag{13.14}$$

where B_0 is the field without the shield and A_{coil} (A_{shield}) is the cross-sectional area of the coil (shield).

The AC field B_{i} (at angular frequency ω) inside of a conducting cylinder is reduced compared to the external field B due to eddy currents. For an infinitely long cylinder of inner radius r_i with open ends and with conductance σ, the field ratio is [13.70]

$$B_i/B = \frac{2\delta}{i/\pi^2 r_i^2 A} \tag{13.15}$$

with skin depth $\delta = (2/\mu_0\mu_{\text{r}}\sigma\omega)^{1/2}$ and

$$A = J_0(r_0/a)N_2(r_i/a) - J_2(r_i/a)N_0(r_0/a), \tag{13.16}$$

where μ_{r} is the relative permeability of the material, $J_i(N_i)$ are Bessel (Neumann) functions of order i, r_0 is the outer radius of the cylinder, and $a = \delta(1 + i)/2$.

13.6 Sintered Metal Heat Exchangers

As discussed in Sects. 4.3 and 7.3, the limiting factor in the refrigeration of helium by another refrigerant or in the refrigeration of a solid by liquid helium to millikelvin temperatures or even lower is the thermal boundary resistance. To overcome this problem one has to use a heat exchanger with a large surface area, usually a sintered metal powder. In addition to increasing the contact area, metal sinters seem to contain an enhanced density of low frequency vibrational modes which may couple well to helium phonon modes (Sect. 4.3.2) [13.71–13.75]. A further important result of [13.76] is the observation that the elastic constant of sintered metal powder is only a few percent of the bulk elastic constant. The introduction of sintered metal heat exchangers has been essential in the development of the ^{3}He–^{4}He dilution refrigerator (Sect. 7.3), see [13.77–13.82]. Therefore, the production techniques of such sponges are of great cryotechnical importance and will be discussed in this section.

Sintered metal for heat exchangers operating at very low temperatures is produced by compressing powders or flakes – mostly of Ag, Cu or Pt – and then heating it under H_2, a noble gas or in vacuum. Surface diffusion – even at room temperature – causes the formation of neck connections between the particles in a conducting metal sponge. The important parameters of a sinter heat exchanger are surface area, thermal conductivity and bonding to its container. In producing a metal sinter sponge, a compromise has to be found between a large surface area (low sintering temperature) and a good bond between the particles and the container as well as a high thermal conductivity of the sinter (high sintering temperature). This can only be achieved by a combination of pressing and heating. The bonding of the metal sinter to a bulk metal is usually improved if first a bonding layer of larger grains is sintered to the bulk material at an elevated temperature and then the main sinter is sintered to this "glue".

Systematic studies of the influence of sintering time and temperature, pressure and atmosphere during pretreatment and sintering, on surface area, structure, packing factor, hardness, mechanical contact, elastic constant and electrical conductivity of sinters produced from submicron Ag and Cu powders of typical dimensions of about 0.1 μm and specific starting surface area of 4–7 $m^2 g^{-1}$ have been reported in [13.76, 13.83], see Fig. 13.12. The results of these references as well as of some further work are given in Table 13.1. Copper powder has the advantage over other noble metal powders that further sintering at room temperature is suppressed by its oxide layer, so its properties are more stable [13.76, 13.83] than those of Ag powder, which self-sinters at room temperature. But because of its oxide layer, fine Cu powder should be reduced in a H_2 atmosphere before sintering. If this is performed at 120°C for 10 min in 1 bar H_2, the originally black powder changes its color and the specific surface area is reduced by 20–30%. The optimum sintering conditions turned out to be 100–130°C, 60–160 bar, in 1 bar H_2 atmosphere for 30 min, resulting in a packing fraction of 45–50%, specific surface areas of at least $2\,\mathrm{m^2g^{-1}}$, $\rho_{300} \simeq 10\,\mu\Omega\,\mathrm{cm}$, $R_{300}/R_4 \simeq 2$–3, and good bonding to a Cu foil [13.83]. Similar results were obtained when the powder was first pressed at 1 kbar at room temperature and then sintered at ambient pressure. At the given sintering parameters, the Cu grains lose their spherical structure and build up connections by neck growth.

Further reports on sintering submicrometer- and micrometer-sized Ag powders have been given in [13.84–13.89], which describe detailed investigations of the influence of temperature on the properties of the sinter (Fig. 13.12 and Table 13.1). From these studies, a sintering temperature of 160–200°C seems to be favored for submicron Ag powder. Investigation of sinters made from substantially larger Cu grains have been reported in [13.90]; such coarse sinters are appropriate for use at $T > 50\,\mathrm{mK}$. Here higher sintering temperatures are possible, giving a mechanically stronger sinter, and the results, also given in Table 13.1, do not show a strong dependence on the temperature or on the sintering time. A substantially larger surface area of 10–$40\,\mathrm{m^2\,g^{-1}}$ can

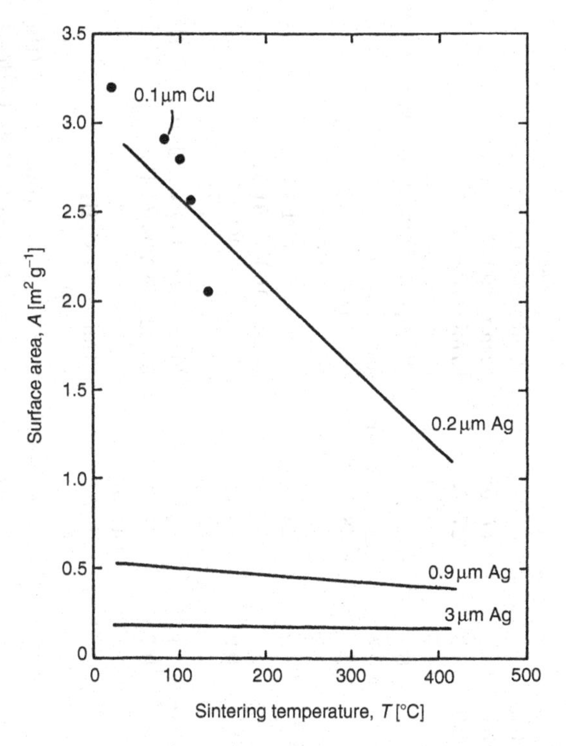

Fig. 13.12. Surface area of sinters made from Ag (——) and Cu powders (●) with the indicated original grain sizes as a function of the sintering temperature. The Ag powders were sintered for 15 min at the given temperatures after pre - compression to 40% of their bulk densities [13.86]. The Cu powder was first pre - sintered and deoxidized at 120°C for 100 min in 1 bar H_2 and then compressed to 47%. The sintering was performed for 30 min in 1 bar H_2 at the given temperatures [13.83]

Table 13.1. Sintering conditions and properties of some representative Cu, Ag and Pt sinters

metal	N. part. size [µm]	particle size [µm]	compressed to [%]	treatment	ρ_{300} [µΩ · cm]	RRR	A/m [$m^2\ g^{-1}$]	Refs.
Ag	0.04	–	40–50	–	($\rho_4 = 12$)	–	1.9	[13.84]
Ag	0.04	–	40–50	20 min/200°C	($\rho_4 = 3.4$)	–	1.7	[13.84]
Ag	0.04/0.07	–	–	45 min/200°C	–	–	2.2/1.8	[13.80]
Ag	0.04/0.07	–	(75 bar)	12 min/100–200°C/H_2	–	–	0.9	[13.85]
Ag	0.04/0.07	0.2	27–47	15 min/100–400°C	–	2–5	1.1–2.8	[13.86]
Ag	0.07	0.7	35–54	15 min/200°C/ H_2	8—36	2 –3	0.8	[13.76]
Ag	0.07	0.1–0.5	45	45 min/160–265°C/Ar–H_2	–	–	1.2–1.7	[13.87]
Ag	0.07	0.3	45 (200 bar)	several h/50°C	($\rho_4 = 12$)	–	2.0	[13.88]
Ag	0.07	–	48	60 min/225°C/H_2	($\rho_4 = 5.8$)	2.4	1.9	[13.89]
Ag	0.08	–	48	50 min/100°C/H_2	($\rho_4 = 17.5$)	1.2	4.9	[13.89]
Ag	0.08	–	48	50 min/125°C/H_2	($\rho_4 = 8.4$)	2.1	3.7	[13.89]
Ag	1	1	45	45 min/160°C/Ar–H_2	–	–	0.7	[13.87]
Ag	3	0.9	28–52	15 min/100–400°C	–	6–12	0.5	[13.86]
Ag	50	3	36–55	15 min/100–400°C	–	8–20	0.2	[13.86]
Cu	0.03	0.1	46–53	30 min/100–130°C/60–160 bar/H_2	10–20	2–4	2–3	[13.83]
Cu	0.07	0.7	31–48	13–52 min/200°C/H_2	12–30	2–15	0.9	[13.76]
Cu	–	–	41–52	2 h/465–580°C/H_2	10–100	2–11	0.3–2.5	[13.90]
Cu	≈20	≈20	–	30 min/500°C/H_2	–	–	0.15	own
Cu	≈50	≈50	50	450°C/H_2	–	–	0.26	own
Pt	0.008	–	30–50	30 min/300 °C–20 min/125 °C(He/H_2)	160	–	14	[13.91]
Pt	0.008	–	23–50	24 min/120 °C–12 h/130 °C(He/H_2)	60	–	40	[13.92]
Pt	0.008	–	40–60	30 min/160 °C–30 min/100 °C(Ar/H_2)	114–56	2.3	10–15	[13.93]

For the Pt sinters, pressure was applied between the sintering steps; see the given literature. The shorthand “N. part.” stand for nominal particle, and “own” means own work be obtained by sintering ultrafine, catalytic Pt powder, the so-called "Pt-black" (Table 13.1). Successful procedures to produce well-bound Pt sinters have been described in [13.91–13.93].

As determined qualitatively in [13.86], metal powders shrink during sintering. They therefore break off from walls in a container. To avoid this problem sinters should be allowed to shrink around solid metal posts or onto metal foils with a rough surface. Because of the poor thermal conductivity of the sinter, these bulk metal parts should not be more than 1–2 mm apart. Similarly, because of the poor thermal conductivity of liquid helium in the sinter pores, there should be channels in the sinter so that the high thermal conductivity of bulk liquid helium can be used. A typical sintered metal heat exchanger for work with helium at very low temperatures therefore contains channels and posts (or foils) of 1 mm diameter and with 2 mm separation.

The surface areas of the sinter samples are usually determined by the volumetric BET gas adsorption technique using N_2 or noble gases at LN_2 temperature [13.94–13.97]. In this method of isothermal adsorption one determines the volume V of gas (N_2 at 77 K, for example) absorbed on the surface of the substrate at a partial pressure P/P_{svp} and applies the equation

$$\frac{P}{V(P_{\mathrm{svp}} - P)} = \frac{1 + (P/P_{\mathrm{svp}})(C - 1)}{V_0 C}, \qquad (13.17)$$

where V_0 is the volume of the adsorbed gas required to form a monolayer and C is a constant related to the heat of adsorption.

A plot of the left hand side of (13.17) against P/P_{svp} gives the unknowns V_0 and C from the intercept and slope of the curve. If N_2 (Ar) is used as the gas, then a value of 16.2 (16.65) Å^2 for the area of one N_2 (Ar) molecule seems to be correct.

13.7 Low-Temperature Motors and Rotators

There are experiments at low temperatures where one wants to investigate samples in different orientations; for example for measuring anisotropic transport properties in different crystal directions as in low-dimensional systems or for investigations of different surfaces of single crystal, or of the angular dependence of a property relative to an applied magnetic field. In these situations, one has to move or, in most cases, rotate, a cold sample, sometimes with rather high precision. This has to be performed by a movable connection to room temperature or one has to use a motor at low temperatures without introducing too much heat. Early designs of motors capable of working at liquid-He temperatures have been described in [13.98, 13.99]. Porter et al. have proposed a small stepper motor for use at 20 mK in a ^{3}He–^{4}He dilution refrigerator [13.100]. It was built from a commercial stepper motor, rewinding the field coils with a superconducting Nb–Ti wire to minimize the heat load. The motor produces a torque of 1.66×10^3 dyne cm mA^{-1} (for 20–850 mA) and a heat load of 20 μW (at 60 mA) for 0.1 steps s^{-1} (primarily due to friction). A micromotor for stepping at millikelvin temperatures using normal-conducting windings (20 μW at 0.3 steps s^{-1}) has been described in [13.101].

A sample rotating stage – using bearings from commercially available sapphire vee jewels and low-friction matching pivots – for use in a dilution refrigerator at temperatures as low as 30 mK and in magnetic fields up to 20 T for rotations through 360° has been described in [13.102]. It is actuated mechanically from room temperature. The dissipated power is 70 (400) μW at rotation speeds of 0.33 (4.4)°/s; the reproducibility for the position is 0.02°. In [13.103], a compact piezoelectrically driven rotator – avoiding a drive shaft – also for use in high magnetic fields and at millikelvin temperatures has been described. It operates by the so-called "stick-and-slip" motion to ±10° and has an angular resolution of 0.01° in magnetic fields up to 33 T. The estimated heat dissipation is less than 0.1 mW at a rotation rate of 0.025°/s. Another possibility for the mentioned purpose is a rotating sample cell. Such a miniature rotating vacuum cell for measurements of thermal properties of samples in high magnetic field coils (up to 19 T) and at $0.3 < T < 30$ K has been described in [13.104]. Eventually, in [13.105], a rotatable heat-capacity cell mounted in a dilution refrigerator and suitable for use in high steady magnetic fields has been presented. It is based on the design of [13.102]. An application of field-angle-dependent measurement of heat capacities is discussed at the end of Sect. 3.1.8.

13.8 Optical Experiments at Low Temperatures

For optic experiments at low temperatures, light has to be brought into the cryostat with optic fibers or via transparent windows. Usually these windows are transparent not only to visible light but also to heat radiation. To reduce the heat radiated to the cold parts, one has to introduce windows with good thermal contact to the radiation shields at successively lower temperatures. An ideal cryogenic window for the visible should be transparent in the visible range but totally absorbent for infrared heat radiation. The heat load due to an ideal window can then be reduced to negligible values because the radiated power decreases with T^4 (Fig. 13.13). Real windows have a finite transmissivity for at least part of the spectrum. Sapphire windows, which are very often used are transparent for a large fraction of the Planck spectrum at 300 and 80 K, whereas "Crownglass" and "Suprasil" quartz windows reduce the transmitted spectrum substantially (Fig. 13.14). A black body at 293 K radiates 42 mW cm^{-2} into a half sphere. This will be reduced by Crownglass to 0.6 mW cm^{-2} at wavelengths $\lambda < 2.7$ μm and to 0.2 mW cm^{-2} at $\lambda > 100$ μm. In general, amorphous materials often show a very useful broad and temperature-independent absorption band at 4–150 μm. Some companies offer special heat (infrared) absorbing "filters" with rather narrow transmitting wavelength ranges in the visible region.

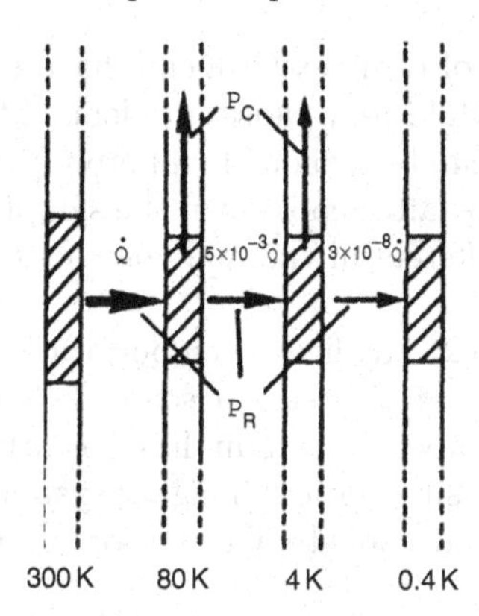

Fig. 13.13. Energy flow diagram for light through two ideal windows at 80 and 4 K indicating the reduction of the radiation power P_{R}. The absorbed power P_{C} is conducted along the windows which have to be well heat sunk to their surroundings

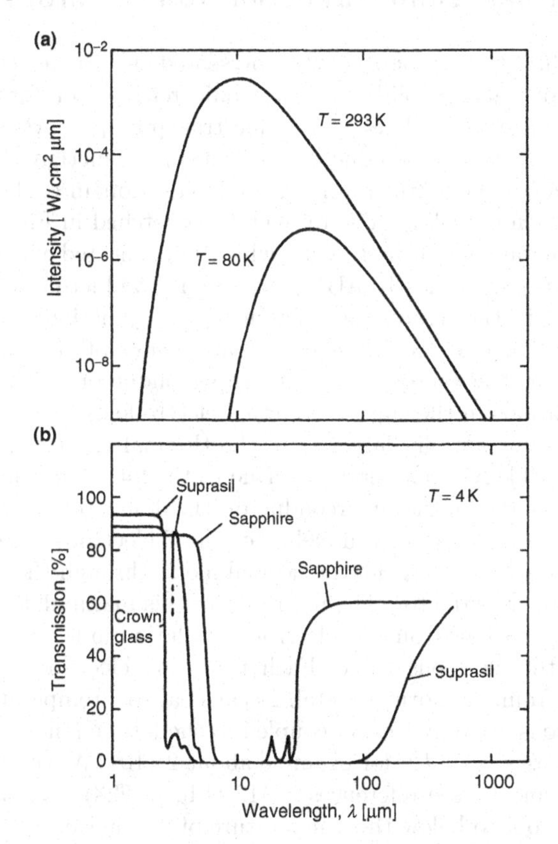

Fig. 13.14. (**a**) Spectral intensity distribution of blackbody radiation at room temperature and at 80 K. (**b**) Transmission through three different window materials: Crown-glass (BK7), sapphire (Al_2O_3) and Suprasil (SiO_2). Crownglass should behave similar to Suprasil for $\lambda > 100\,\mu\mathrm{m}$ [13.106, 13.107]

Two further aspects of optic experiments have to be borne in mind. The bubbling of a cryogenic fluid may cause problems if the light has to be transmitted through it. This can be avoided by pumping on the bath. Furthermore, one has to make sure that after cool down the sample has not moved from its required place due to differential thermal contraction of various parts of the apparatus.

Experiments down to 25 mK have been performed in cryostats with optical access from room temperature using a set of 3–5 heat absorbing windows [13.108, 13.109]. This temperature limit has recently been reduced to 1 mK in nuclear refrigerators using optical imaging systems with optical fibers for the light transmission and cooled CCD sensors giving heat leaks as low as 1 nW [13.110, 13.111].

13.9 Electronic Tunnel-Junction Refrigerators

In [13.112, 13.113] a proposal was presented for a novel, mesoscopic, continuously operating, solid-state electronic refrigerator for the subkelvin temperature range. It exploits the unique transport properties of a normal-conductor/insulator/superconductor (NIS) tunnel junction [13.114–13.116]. The small size of the refrigerator allows integration into electronic micro-devices and on-chip cooling. The refrigerator, sketched in Fig. 13.15, consists of a normal metal film (usually Cu), which is connected via a tunnel junction (a thin insulator, mostly Al_2O_3) to a superconducting contact (usually Al). As shown in the figure, the micrometer size junction is electrically biased with a voltage $V \approx \Delta E/e$ (ΔE: gap energy of the superconductor) to slightly below the energy gap of the superconductor. There are very few thermal excitations in the superconducting electrode of the junction since its gap energy is substantially larger than the thermal energy of excitations at the involved millikelvin temperatures. Due to the forbidden electronic energy states within the gap of the superconductor, the dominant contribution to the electron transport therefore is tunneling of "hot" electrons with energy larger than the Fermi energy E_F from the normal metal through the insulating barrier into the superconductor. The net result of this manipulation is to remove high-energy excitations from the electron population in the normal electrode, thus cooling the electrons there. Each tunneling electron carries the heat E – eV away from the normal metal. Because at low temperatures, the conduction electrons are only weakly coupled to the lattice (the electron–phonon coupling decreases with T^5; for example about $2\times10^3\,(\mathrm{W/cm^3})\,VT^5$ for a Cu sample of volume V; see references in [13.114], p. 223), the electrons of the normal metal cool to below the temperature of its lattice.

For the operation of this Peltier-type refrigerator it is crucial that the power dissipated in the superconductor produced by recombination of quasiparticles does not couple back into the normal electrode, as well as to minimize

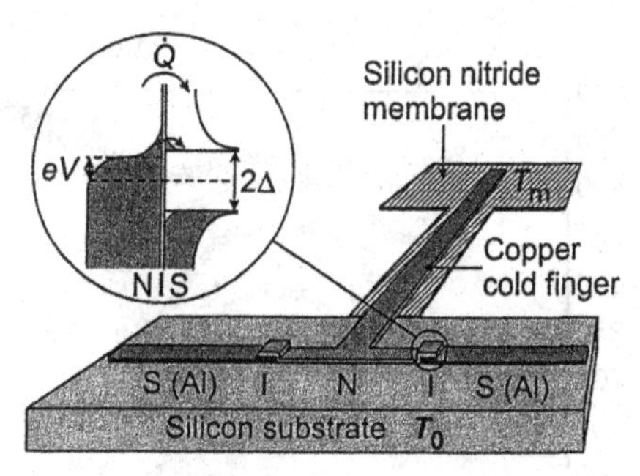

Fig. 13.15. Schematic of a SINIS microcooler consisting of two $Cu/AlO_x/Al$ tunnel junctions in series on a bulk Si substrate. The microcooler refrigerates a Cu cold finger on a silicon nitride membrane. Detectors or samples can be connected to the cold finger. – Inset: Principle of operation of an NIS cooler. When the NIS junction is biased at $V \leq \Delta/e$, the normal metal cools because only electrons from above the Fermi energy can tunnel from the normal metal into empty states of the superconductor. Reprinted from [13.115], copyright (2000), with permission from Elsevier

quasi-particle scattering from impurities and surfaces. Other important aspects are Andreev reflection [13.117] of quasi-particles at the normal-superconductor contact (which allows electrical but no thermal transport to the normal electrode) and ohmic losses in the normal conducting parts. These as well as other deteriorating effects and ways to overcome them are discussed in [13.114].

The maximum cooling power of this NIS device occurs when eV is just slightly smaller than ΔE (about 0.18 meV for Al) (Fig. 13.16), i.e., when the superconductor is biased near its gap energy. At the typical situation for the electronic temperature $T_e << T_c$ of the superconductor, it is then given by

$$\dot{Q} \approx 0.6(\Delta E^{1/2}/e^2 R_T)(k_B T_e)^{3/2}, \tag{13.18}$$

where R_T is the resistance of the tunneling junction in the normal state [13.112–13.116]. The cooling power is maximized at $T \approx 0.25\ \Delta E/k_B$, resulting in $\dot{Q} \approx 0.07\ \Delta E^{2/} e^2 R_T$. Due to the dependence on R_T, one wants a junction with high transparency, i.e., small R_T of $1\,\mathrm{k}\Omega$, for example. At too-high transparency, however, the cooling power decreases due to increasing Andreev reflection [13.114], and due to pinholes in the junction. The optimum range is between some tens and some thousands $\Omega\,\mu\mathrm{m}^2$ depending on temperature. The minimum achievable electron temperature is determined by the balance between the cooling power and the heat flow from the hotter, weakly coupled lattice of the sample (but see also below for other limiting effects), which then, of course, is cooled slightly as well, depending on its coupling to the environment [13.114, 13.115].

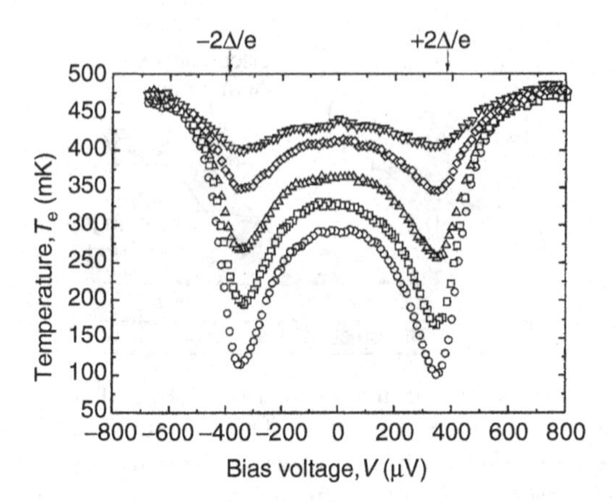

Fig. 13.16. Temperature T_e of the electrons in the Cu island of an $Al/Al_2O_3/Cu/Al_2O_3/Al$ SINIS tunnel structure (Fig. 13.15) as a function of bias voltage V. The tunneling junctions had resistances of about 1 kΩ. The dimensions of the normal metal island are $5 \times 0.3\,\mu m^2$ with a thickness of 35 nm. Two additional NIS tunnel junctions were used to measure the temperature of the Cu electrons. At zero bias, the temperatures T_e are equal to the temperature of the substrate of the device. Maximum cooling occurs in this *double* – junction structure at a bias voltage slightly smaller than $\pm\, 2\Delta E/e$, with $\Delta E = 0.18$ meV for the energy gap of the superconducting Al. Reprinted from [13.115], copyright (2000), with permission from Elsevier

Prototypes of this Peltier-type electronic on-chip microcooler have been produced by electron nanolithography and have been investigated for their performance mostly by the group of Pekola [13.113–13.116]. Because the cooling effect is symmetric in the bias voltage V, the cooling increases proportional to the number of connected tunnel junctions regardless of the direction of the electric current; hence one can use SINIS junctions, for example (Fig. 13.15). The effects depend on the type of superconductor as well as on geometry and quality of the device; here, more development work seems to be necessary. Fortunately, the influence of asymmetries in connected junctions have only a small degrading impact [13.114, 13.115]. Typical results obtained with a SINIS cooler are shown in Fig. 13.16. Temperatures below 50 mK, starting from 320 mK, have been achieved [13.116]. As discussed in this latter reference, the minimum temperature seems to be limited mostly by two effects: firstly, when the injection rate of electrons exceeds the internal electron–electron relaxation rate in the metal to be cooled, the electrons no longer obey the Fermi–Dirac distribution (3.11) and the concept of temperature cannot be applied to this non-thermal situation. Secondly, states within the energy gap of the superconductor induce anomalous heating at low temperatures. With large area ($30\,\mu m^2$) SINIS structures, a cooling power of 20 pW at 80 mK has been measured starting from 250 mK. Connecting several NIS junctions of, for example, 15 nm $Al/Al_2O_3/20$ nm Cu on thin silicon wafers in series a total cooling power of 0.1 nW should be attainable. The temperatures were deduced from the temperature dependence of the current–voltage characteristics of additional NIS junctions attached to the normal-metal islands [13.114, 13.115].

For application as a refrigerator, one has to thermally isolate the normal electrode appropriately from its environment so that its lattice can be cooled by the refrigerated electrons according to the weak electron–phonon coupling. This has been achieved [13.114, 13.115] by extending the normal Cu electrodes onto suspended thin and narrow silicon nitride (Si_3N_4) membranes as substrates (Fig. 13.15). In such thin structures, phonon transport out of the membrane is essentially two-dimensional (the dominant phonon wavelength at subkelvin temperatures is between 0.1 and 1 μm) and therefore reduced. This way, the membrane temperature was lowered from 0.2 to 0.1 K. Typical applications of these microrefrigerators is in cooling radiation detectors like electron microbolometers as well as superconducting tunnel junctions or transition edge detectors in space applications. For these applications, the devices work in the desired temperature range and may have sufficient cooling power. Further work to exploit this refrigeration technology seems to be necessary. Details on the theory, fabrication, and optimization of these microcoolers can be found in [13.114].

13.10 Torsional and Translational Oscillators

High-Q mechanical oscillators have been developed to an indispensable device to investigate a variety of properties of solids and liquids at low temperatures as well as for technical applications. They operate very close to the behavior of a driven harmonic oscillator. At high Q – necessary to be sensitive to dissipation in the oscillator or sample and for high frequency resolution – the frequency of a harmonic oscillator is

$$f = (k/m)^{1/2}/2\pi, \tag{13.19}$$

where k is the oscillator spring constant and m is its mass.
The maximum amplitude at resonance is given by

$$A = QF/k, \tag{13.20}$$

for a driving force F. The corresponding maximum amplitude of deflection is given by

$$\Theta = Q\tau/k, \tag{13.21}$$

with τ as the amplitude of the sinusoidal drive torque. Eventually, the phase ϕ between driving force and oscillator amplitude is obtained from

$$\tan\phi = 2Q, \tag{13.22}$$

which is close to 90° at the high Q of these oscillators. One can deduce the quantities of interest from a measurement of these oscillator parameters. For example, changes in the mass of the sample lead to changes in f, and changes in its dissipation will lead to changes in Q and therefore in the amplitude of oscillations. According to their design, the used oscillators can be divided into five groups, which I will discuss together with some of their applications in research in the following sections.

13.10.1 Vibrating Reeds

This method has mostly been used to investigate low-frequency elastic properties of ribbon like samples. The sample, a "vibrating reed" (typically a few to 20 mm long, 1–3 mm wide, 0.01–0.5 mm thick) is used as a translational oscillator [13.118, 13.119]. It is a blade whose one end is clamped to a heavy mass (Fig. 13.17). Its free end is used as the movable electrodes of capacitive transducers to excite and detect electrostatically vibrations of the reed by applying a sinusoidal voltage dV of typically 1–10 V. To make the drive linear in voltage and to amplify the force, a DC voltage V_o of order 100 V is superimposed on the AC excitation ΔV, so that $V = V_o + \Delta V$ and $V^2 \approx V_o^2 + 2\,\Delta V\, V_o$ at $\Delta V << V_o$. If the sample is not a metal, a thin metallic film has

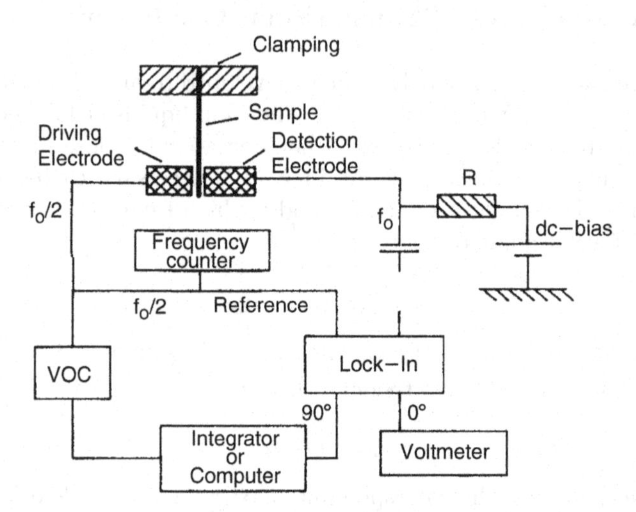

Fig. 13.17. Experimental setup for a vibrating reed experiment. The oscillations of the clamped sample ("reed") change the capacitance between the detection electrode and the reed which modulates the bias voltage. The changes in voltage are detected by a lock-in amplifier operated so that it detects twice the reference frequency provided by the driving voltage. In a feedback loop a frequency change is produced by a voltage controlled oscillator (VCO) to keep the drive at a preset frequency value. An integrator is used to enhance the sensitivity of this phase-locked mode [13.118–13.120]

to be evaporated or sputtered onto the free end of the reed to form the electrodes. Measurements of the resonance frequency and of the damping of the oscillations give the change of the velocity of sound (to better than 10^{-6}) and the internal friction (to a few %) of the sample. For the damping, the temperature-independent contributions from the clamping have to be taken into account; it usually limits the Q-value to about 10^5. Data can either be taken by sweeping through the resonance or by phase-locking to it – usually the fundamental one – with a feedback circuit (Fig. 13.17). The resonance frequencies, typically in the low kHz range, are

$$f_n = (d/4\pi\sqrt{3})(\beta_n/l)^2 v_e, \qquad (13.23)$$

with $d(l)$ the thickness (length) of the reed; $\beta_1 = 1.875$, $\beta_2 = 4.694$, $\beta_3 = 7.855$, etc., corresponding to the various modes. The longitudinal sound velocity v_l of the sample can be determined from Young's modulus sound velocity $v_e = (G/\rho)^{1/2}$, (G: Young's modulus, ρ: density of the reed) via

$$v_l = v_e(1-2\sigma^2/(1-\sigma))^{-1/2}, \qquad (13.24)$$

with the Poisson ration $2\,\sigma = (v_l^2 - 2v_t^2)/(v_l^2 - v_t^2)$; l and t stand for the longitudinal and transverse phonon polarizations, respectively. The reed amplitude u is a measure of the damping, which is proportional to the Q-value. To avoid heating effects or nonlinearities, typical reed amplitudes are of the order of 10–100 nm only [13.119], corresponding to a maximum strain

$$\varepsilon = 1.8\, du/l^2 \qquad (13.25)$$

of about 10^{-7} at the fixed end of the reed.

Such reeds have been extensively used, for example, at the universities in Heidelberg [13.118] and Bayreuth [13.119] for investigations of the low-frequency velocity of sound and internal friction of amorphous and polycrystalline dielectrics and metals from below 1 mK to room temperature and at frequencies between 0.1 and 10 kHz.

The application of the vibrating reed technique for very sensitive investigations of the pinning properties of flux line vortices in superconductors has been discussed by Esquinazi [13.120]. These investigations have given quantitative information on the elastic coupling between the flux line lattice and the atomic lattice in high- as well as low-T_c superconductors and on the dynamics of the flux line lattice and the flux line tension.

A detailed description of the fabrication of nanometer-scale mechanical micro-resonators (for example, 7.7 μm length, 0.33 μm width, 0.8 μm height) with both ends fixed, made from single crystal Si for the 100 MHz range can be found in [13.121].

13.10.2 Vibrating Wires

Vibrating wires have either been used in vacuum to investigate their own mechanical properties like sound velocity or internal friction [13.119, 13.122] or in a liquid, particularly liquid helium, to investigate properties of the latter [13.123–13.125]. They consist of a semicircle of a few mm radius of a normal conducting or superconducting metal wire (for example, 5–100 μm Ta or NbTi) whose both ends are clamped (see Fig. 13.18). A – usually superconducting – solenoid provides a magnetic field of typically a few 10 mT in the plane of the wire loop. When an AC current in the low kHz range and of typically a few 0.1 mA flows through the wire, it is driven into oscillations by the Lorentz force. The resulting induced voltages in the μV or even nV range can be measured while the frequency is swept through the resonance of the oscillator. The resonance frequencies of such a driven damped harmonic oscillator are given by

$$f_n = (r/4\pi)(\eta_n/l)^2 v_e, \tag{13.26}$$

with $r(l)$ the wire radius (length), and $\eta_1 = 4.73$, $\eta_2 = 7.853$, etc. for the various modes. The maximum amplitude (typically 100 nm) of the wire is at its middle and is obtained from the induced voltage U_{ind} for the fundamental mode as

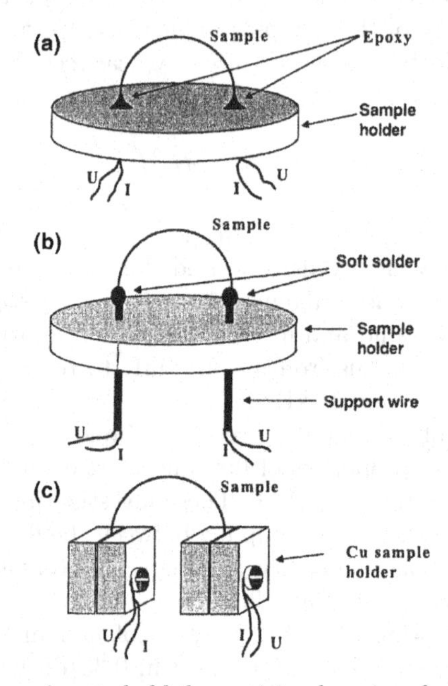

Fig. 13.18. Various versions to hold the semicircular wire of a vibrating-wire oscillator used for investigating clamping losses. Reprinted with permission from [13.122]; copyright (1995) by the Am. Phys. Soc.

$$u_o = 3.22 U_{\text{ind}} / B\omega l. \tag{13.27}$$

The maximum of the time-dependent strain at the clamped ends of the wire is given by

$$\varepsilon = 28.2\, r u_o / l^2, \tag{13.28}$$

and the dissipated energy in the wire of mass m at the fundamental mode is

$$P = u_o^2 \pi^3 f^3 m / Q, \tag{13.29}$$

which indicates the strong increase with frequency and may lead to self-heating of the wire. The equation for the spring constant of the wire is identical to the one for a rod (see Eq. 13.37). Again, from the resonance frequency and from the amplitude (obtained from the voltage induced in the oscillating wire), the velocity of sound and the internal friction of the wire can be obtained.

If the wire is oscillating in a surrounding medium of density ρ, mostly liquid helium, there is a frequency shift Δf_1 to lower values from the effective hydrodynamic mass of the medium dragged along with the oscillating wire of density ρ_w

$$\Delta f_1 = f_o k_o(m) \rho / \rho_{\text{w}}, \tag{13.30}$$

as well as an increase Δf_2 of the width of the resonance due to the additional damping from the viscosity of the medium

$$\Delta f_2 = f_o k_o'(m) \rho / \rho_{\text{w}}, \tag{13.31}$$

[13.123–13.125], (the equations have to be slightly modified if the wire is vibrating in a superfluid liquid where the density is composed of a superfluid and a normal fluid part [13.123,13.125]). The equations can be calculated from the relevant continuity and Navier–Stokes relations. They contain the tabulated Stokes parameters k_o and k_o', which are determined by the parameter $m = a/\delta \sqrt{}\,2$ with the radius a of the wire and the viscous penetration depth $\delta = \sqrt{}(\eta/\pi\rho f)$ of the liquid. The equations are valid when the density ρ_{w} of the wire is much larger than the density ρ of the liquid and when the mean free path of the liquid particles are shorter than the wire radius, i.e., in the hydrodynamic regime where the wire works as a viscometer. Because of the strong temperature dependence of the viscosity of liquid ^{3}He and ^{3}He–^{4}He mixtures with $\eta T^2 =$ constant at $T < 20\,\text{mK}$ (Sect. 2.3.6; Fig. 2.19), vibrating wires are also used as sensitive, continuously monitoring, and quickly responding thermometers in these Fermi liquids in their hydrodynamic regimes (Sect. 7.4).

In the application of vibrating wires to investigate the properties of surrounding liquids, one should be aware of the intrinsic, in particular the non-linear properties of these wires (which can be measured with the wires in vacuum), influences from their clamping and of heating effects before relating any result to the properties of the surrounding liquid [13.119, 13.122, 13.125]. To keep the wire amplitude in the low nm or maybe even pm range and therefore the dissipation very low, SQUIDs have been used as preamplifier to detect its motion [13.126, 13.127]. In the latter publication, a system is described that has a gain factor of $10^8 \Phi_o/V$ at 1 kHz with an amplifier noise level of about $2 \times 10^{-5} \Phi_o/\sqrt{\mathrm{Hz}}$.

13.10.3 Quartz Tuning Forks

These tuning forks are commercially available small oscillators that are employed in watches with a standard frequency of 2^{15} Hz = 32.768 kHz. They are fabricated from wafers of α-quartz with the optical axis oriented approximately normal to the wafer plane. The oscillations are excited electrically using the piezoelectric properties of the material and applying an AC voltage in the millivolt range to electrodes on the tines (typical length $l = 4$ mm, thickness $t = 0.4$ mm, width $w = 0.6$ mm) of the fork. In the fundamental mode the tines move in opposite direction at a frequency given by

$$f_o = 0.1615(t^2/l)(G/\rho)^{1/2}. \tag{13.32}$$

With the density $\rho = 2.66\,\mathrm{g\,cm^{-3}}$ and $G = 7.87 \times 10^{10}\,\mathrm{N\,m^{-2}}$ the Young's modulus (spring constant $k \approx 1.4 \times 10^4\,\mathrm{N\,m^{-1}}$) of the fork material and the above-mentioned dimensions, we obtain $f_o \approx 33$ kHz. Typical Q-values are 10^4–10^5. Again, resonance frequency and damping are measured, for which various methods have been used [13.128, 13.129]. These oscillators have found a variety of applications in different scanning probe microscopes by attaching a tip to one of the tines (at 1.5 K and in magnetic fields up to 8 T in [13.128]) as well as in a measurement of the viscosity of a helium mixture (at 3–100 mK in magnetic fields up to 11.7 T [13.129]). The advantages are the simplicity and low price of this device, as well as the very small heating effect and field dependence of its properties. However, as discussed in [13.128, 13.129], there seem to be several non-understood effects in the behavior of these oscillators when they are used as viscometers as well as a rather large acoustic damping (due to the high frequency). Hence, more investigations seem to be necessary before the forks can be used for investigating surrounding liquids.

13.10.4 Double-Paddle Oscillators

A popular version of a torsional oscillator is the so-called double-paddle oscillator shown in Fig. 13.19 [13.130–13.133]. Its torsion rod of rectangular cross-section (height a, width b) and of length l has a torsional spring constant of

$$k = \beta G(ab^3/l), \tag{13.33}$$

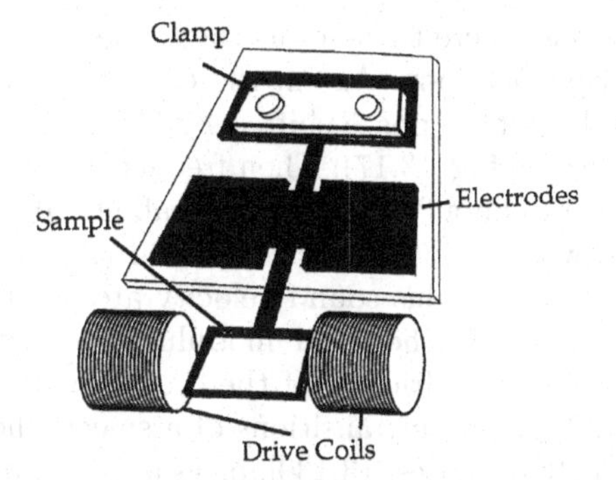

Fig. 13.19. Schematic of a double-paddle torsional oscillator (see text) fabricated from a single – crystal silicon wafer with drive and detection electrodes on the isolation oscillator and additional AC field drive coils at the experimental oscillator. This oscillator has been used as magnetometer to investigate thin magnetic films deposited or mounted on it (see Sect. 3.1.2). Reprinted with permission from [13.131]; copyright (1998) by the Am. Phys. Soc.

with β a slowly varying function of the ratio a/b, $a >> b$, and G the sheer modulus of the material of the oscillator [13.131]. Vibration isolation is achieved by using a two-stage design, i.e., combining a measurement oscillator connected via a torsion rod to a second isolation oscillator of higher mass, both made from one piece to reduce clamping losses. Because of the absence of clamping losses (quite different from a vibrating reed), these oscillators have Q-values of at least 10^6. Another advantage of these oscillators compared to the vibrating reed and the vibrating wire is the even distribution of the strain. The measurement oscillator could be the sample to be investigated [13.133]. Or a sample is connected [13.130,13.131] or evaporated [13.132] onto it; in this case, the oscillator is mostly made from single crystalline silicon with very low intrinsic dissipation. Connecting a sample of mass m to the paddle shifts its eigenfrequency according to

$$m = k(f^{-2} - f_o^{-2})/(2\pi)^2, \tag{13.34}$$

with f and f_o the eigenfrequencies with and without the sample.

Double-paddle oscillators have been produced by anisotropic chemical etching [13.130–13.132] and by laser cutting [13.133]. A torsion mode, for example with the two oscillators moving 180° out of phase, is used rather than a bending mode (as in the case of the vibrating reed) because it couples less to external translational vibrations reducing even more the need for external vibration isolation. Typical resonance frequencies are between 1 and 10 kHz. The oscillator is excited and its motion is detected through a capacitance measurement using metallic electrodes evaporated onto the paddle. These metallic electrodes can also be used to improve the cooling of the sample. Like for the other oscillators, the double-paddle oscillator can be operated at resonance in a self-resonant mode where they act as the frequency determining element in a phase-locked feedback loop. Again, the electrodes are biased at around 100 V. The detected signal is phase shifted by 90° and used to drive the oscillator at resonance (see Fig. 13.17). Then frequency (stable to about 10^{-10}) and amplitude (down to about 1 nm; stable to about 10^{-5}) are recorded continuously. However, whereas in vibrating reed and vibrating wire experiments changes of the Young's modulus sound velocity are observed, a torsional vibration measures changes in the shear modulus. Such oscillators have been used to measure the internal friction of the oscillator itself [13.133], to measure the shear moduli or phase transitions of a smectic liquid-crystal film or of the liquid-crystal Blue-Phases [13.130], or as a magnetometer [13.131], (see Sect. 3.4.2). In [13.131], for example, a sensitivity to wing displacement of better than 0.1 nm has been achieved, showing the high sensitivity of these devices.

In [13.134] it is reported how such micrometer-sized two-element GaAs torsional oscillators (25 μm lateral extent, 0.8 μm thickness) operated at a frequency $f_o \approx 1$ MHz can be used to measure very small driving forces. The achieved intrinsic bandwidth was 80 Hz using a coherent excitation and lock-in detection technique with signal averaging. The minimum detectable force was 5×10^{-17} N (corresponding to a mechanical energy of $34hf_o$ with h the Planck quantum!). The displacement noise of 24×10^{-6} nm is roughly a factor of 25 smaller than displacements caused by thermal noise. These oscillators are a step toward the mechanical observation of macroscopic quantum effects.

Application of double-paddle oscillators in magnetometry, in particular in high magnetic fields, is discussed in Sect. 3.4.2.

13.10.5 Composite Torsional Oscillators

The basic design of a composite torsional oscillator is shown in Fig. 13.20 [13.135]. They have been used, for example, for measuring internal friction losses in bulk solids due to the motion of defects coupled to strain fields. In a composite torsional oscillator, the upper part is the sample to be investigated. Its lower end is glued to an actuator, which should be a material with low losses, for example crystalline quartz. The actuator contains evaporated metallic electrodes for driving and detecting capacitively the oscillations. This assembly is connected via a thin torsion rod to a heavy base; the latter is usually machined from BeCu, which has a very small spring constant and is afterward hardened. This mounting allows freedom for torsional oscillations and functions as an approximation of a free-end boundary condition with very low losses. It also decouples the oscillator from the environment. By applying an AC voltage to the electrodes, the oscillator is driven at its lowest torsional eigenmode, typically near 100 kHz. The internal friction in the oscillating sample is determined from the quality factor Q of the oscillations at resonance. Hence, background losses, which are mostly due to the joints, in particular between sample and actuator, should be low. They can be determined with

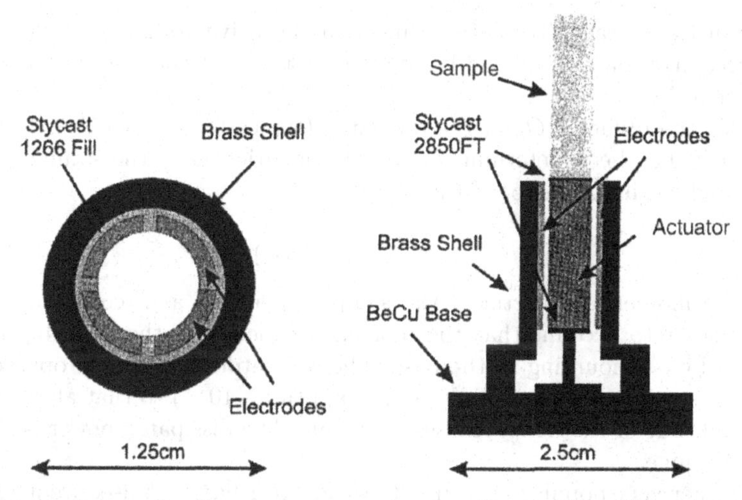

Fig. 13.20. Schematic of a composite torsional oscillator with a SiO_2 glass sample glued by Stycast epoxy to the actuator which could be crystalline quartz with very low losses, for example. Metallic films are evaporated onto the actuator. They are used as electrodes and together with the electrode assembly on the brass shell operate as capacitors to excite and detect the torsional oscillations of actuator and sample [13.135]

a sample of very low internal friction, like a single crystal or an alloy with many effective pinning centers for dislocations. Low losses at the joint can be achieved by having an antinode of the displacement there. This is achieved at the lowest frequency mode when the joint is moving 180° out of phase with the two ends of the oscillator. The sample length for this situation is

$$l = (\beta/2f)(G/\rho)^{1/2}, \tag{13.35}$$

where f is the resonant frequency, ρ the density, and β a constant depending on the cross-section of the vibrating bar ($\beta = 1$ for circular cross-section). For an isotropic cubic crystal, the sheer modulus G is equal to the elastic constant C_{44}. A bridge circuit provides the AC drive voltage and measures the rms amplitude and phase of the out-of-balance signal, which are proportional to the amplitude and phase of the vibrations. In [13.3] it is described how the signal from the pick-up electrode is amplified, filtered, phase shifted, and fed back to a drive electrode. In a feedback loop, the difference in phase is used again as the error signal to lock the drive frequency to the resonant frequency of the oscillator. For a period resolution of 10^{-9} and a Q of 10^6, one needs a phase stability of about 0.1° in drive and detection electronics, which is an important prerequisite for proper operation of these oscillators. Another possibility is to sweep the frequency through the resonance of the oscillator and record the response with a lock-in amplifier. The force exerted by the drive capacitor with capacitance C and gap spacing d, $F = CV^2/2d$, is not linear in the voltage. To make it linear in the drive voltage and to amplify the force, again a DC voltage V_o of about 100 V is superimposed on the AC excitation ΔV.

Once the Q-factor Q_o of the oscillator (typically 10^5 to 10^6 at low temperatures) has been determined, the internal friction of the sample can be calculated as the Q_s-factor of the sample

$$Q_s = Q_o I_s/(I_s + \alpha I_t), \tag{13.36}$$

with the moment of inertia of the sample I_s and α as a correction factor that takes into account that the moment of inertia of the transducer I_t is effected by its mounting to the base. The oscillators are usually operated at constant strains ε (with typical strain amplitudes 10^{-7}) so that at resonance the amplitude of the signal remains constant. The dissipated power is then in the nW range.

Another very popular version of this oscillator has been used frequently to investigate a variety of properties of liquid and solid ^3He, ^4He, or H_2. In this design (Fig. 13.21), a cylindrical cell of epoxy or made from a metal (of low density to keep the mass of this addendum small) to accommodate the liquid or solid sample is mounted on top of a BeCu torsion rod. A torsion rod (as well as a vibrating wire) with a shear (torsion) modulus G ($53 \times 10^9\,\mathrm{N\,m^{-2}}$ for hardened BeCu) has a spring constant

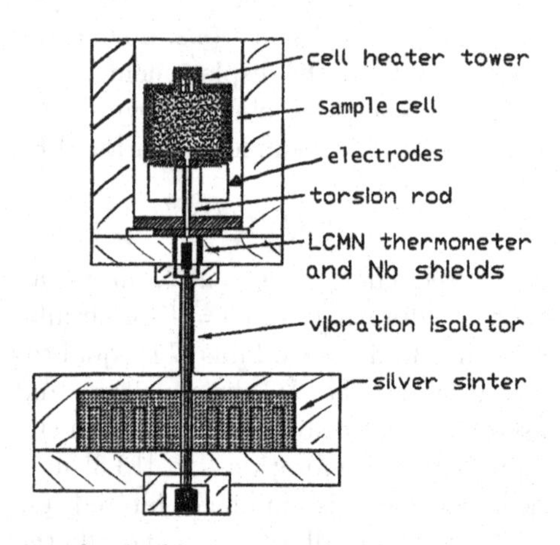

Fig. 13.21. Schematic of a double-stage torsional oscillator for the investigation of the dynamics of liquid helium in a porous material (aerogel) inside of an epoxy sample cell. Adding a heater as well as a paramagnetic LCMN thermometer (Sect. 12.9) allows simultaneous measurement of heat capacity. The hollow BeCu torsion rods serve as fill line for the liquid as well as weak thermal link to the bath for the heat-capacity measurements. A silver sinter heat exchanger in the base provides thermal contact to the refrigerator at the millikelvin temperatures of this experiment [13.3, 13.136]

$$k = \pi G r^4 / 2l, \tag{13.37}$$

with r and l its radius and length. The "rod" actually is a tube used for filling the sample into the cell. This setup is again connected to a heavy base. The base, torsion rod, and cell bottom should be made from one piece of BeCu, which is then heat treated to achieve maximum Q-value. Again, the base is connected by another thicker torsion rod with a lower resonance frequency, for example 100 Hz, to another platform with an even heavier mass. The second oscillator works as a mechanical low-pass filter to decouple the experimental oscillator from the environment. In some cases the cell is an annulus to investigate flow properties of a quantum liquid, for example. The cell contains electrodes, which are the movable parts of capacitors to drive the oscillator and to monitor its motion; the capacitors are again biased at around 100 V. These oscillators operate in the low kHz range and have a frequency stability of typically 1 μHz/10 h at a few kHz, allowing a resolution of 10^{-9} in period, and a mass sensitivity of better than 5×10^{14} helium atoms or of a submonolayer helium film. They have been used to temperatures of about 1 mK in particular by the groups of Reppy and Parpia at Cornell University [13.3, 13.136] to investigate the properties of bulk liquid ^{3}He and ^{4}He or of their properties in restricted geometries like Vycor glass or aerogels. For example, the superfluid onset, the normal fluid mass (determining the mass coupled to the oscillator), persistent currents, or critical velocities have been examined by measurements of the amount of the helium viscously coupled to the oscillator or from the dissipation, respectively [13.136, 13.137]. For this purpose, the temperature-dependence of the resonance frequency and of the dissipation of the oscillator are measured. The temperature-dependence of the properties of the oscillator structure itself has to be determined by investigating the empty oscillator.

To minimize the influence from nonlinearities of the oscillator structure, for example the torsion rod, the oscillator has to be driven at sufficiently small, constant amplitudes (constant strain by varying the drive voltage) given by (13.20) or (13.21); the corresponding velocities are in the μm/s range.

Details on the design, construction, installation, and operation of such oscillators can be found in [13.3].

13.11 Purification of ^{3}He from ^{4}He Impurities, and Vice Versa

The purity of commercially available ^{3}He is sometimes not good enough for experiments or for the various thermometric methods based on the properties of this isotope (Sects. 11.2, 11.3, 12.2, 12.3). In addition, the ^{3}He may have been contaminated with other gases during an experiment. All impurities except ^{4}He can be readily disposed using a trap filled with an adsorbent of high surface area, like activated charcoal, cooled by liquid ^{4}He or even liquid

N_2 only (see Sect. 7.4). The ^{4}He admixture, however, can only be eliminated either by a rectification method based on the different vapor pressures of the helium isotopes (Sects. 11.2, 12.2) [13.138] or by a chromatographic method based on the different adsorption energies of them, for example, on activated charcoal (see Sect. 6.2). With a simple setup, the latter method has resulted in a final ^{4}He concentration of 0.03 (0.1)% starting from 6.5 (18)% mixture in one passage and at <0.01 (0.02)% starting from 0.02 (0.1)% after two passages [13.139].

The purification of ^{4}He from ^{3}He contaminants can be perform by the heat flush effect by which isotopic ratios $n_3/n_4 < 10^{-12}$ have been achieved [13.140].

Problem

13.1. Calculate the deflection of the diaphragm of the low-temperature capacitance strain gauge depicted in Fig. 13.2 (take roughly the dimensions from that figure) for a pressure of 10 bar. How thin would the diaphragm have to be to get the same deflection for a pressure change of 10^{-3} bar?

13.2. Calculate the electrical resistance R_N of a tunnel junction of Cu/Al_2O_3/Al necessary to give a cooling power of 1 pW at a temperature of 100 mK.

14

Some Comments on Low-Temperature Electronics

There are low-temperature experiments that require electronics kept at low temperatures, sometimes even kept at millikelvin temperatures – the reason being mostly that the electrical noise level at a low-temperature experiment in a properly designed cryostat usually is much smaller than at the room temperature environment. Hence, when weak signals have to be transferred from a low-temperature experiment to electronics at room temperature the signal has to be amplified at low temperatures requiring electronics that work at these temperatures. The use of low-temperature electronics in general gives a substantial improvement in sensitivity and in the signal-to-noise ratio. The other situation requiring low-temperature electronics occurs when the electronic device itself can only be operated at low temperatures, like a SQUID. It surely would increase the volume of this book too much if I would include an adequate discussion of low-temperature electronics. Hence, I will only give a survey on recent relevant literature rather than discussing cryo-electronics in detail.

Fortunately, one can find good recent books, review articles, and comprehensive original articles on cold electronics in the literature. For example, the book "Low Temperature Electronics: Physics, Devices, Circuits, and Applications" with close to 2,000 relevant references [14.1]. It covers the range from room temperature to millikelvin temperatures. A good overview of its content is given by the chapter titles: Physics of Silicon at Cryogenic Temperatures; Silicon Devices and Circuits; Reliability Aspects of Cryogenic Silicon Technologies; Radiation Effects and Low Frequency Noise in Silicon Technologies; Heterostructure and Compound Semiconductor Devices; Compound Heterostructure Semiconductor Lasers and Photodiodes; High Temperature Superconductor/Semiconductor Hybrid Devices and Circuits; Cryocooling and Thermal Management. The book deals among others with silicon metal-oxide-semiconductor(MOS) bulk transistors, silicon-on-insulator MOS transistors, gallium arsenide field effect transistors (FETs), silicon and gallium arsenide MOSFETs, silicon germanium bipolar transistors, resonant tunneling diodes, optical and optoelectronic devices, quantum well (QW) and strained-layer QW lasers, double-heterostructure lasers, indium phosphide/ indium gallium arsenide avalanche photodiodes, QW infrared photodiodes, as well as microwave filters, antennas, and oscillators which became particular attractive with the advancement of high-T_c-superconductors.

A rather attractive design of a low-noise, low-power, three-stage amplifier based on high-electron-mobility transistors assembled on a printed board of $33 \times 13\,\mathrm{mm}^2$ has been described in [14.2]. It has a minimum noise temperature of about 0.1 K at an ambient temperature of 0.38 K at frequencies between 1 and 4 MHz, with a corner frequency of the $1/f$ noise close to 300 kHz. Its gain is 50 at a power of 0.2 mW. Another rather promising cryogenic 700-MHz low-noise (3 K) amplifier with a gain of 16 dB operated at 4.2 K has been described in [14.3].

Surely, the most important cryo-electronic device is the extremely sensitive superconducting quantum interference device ("SQUID"). A SQUID is essentially a closed superconducting loop containing one or several weak links ("Josephson junctions"). Into this loop magnetic flux is frozen in. The frozen-in flux is changed stepwise in units of the flux quantum each time the small critical current I_c of the weak link is reached by shielding currents in response to an external magnetic field. The small size of the flux quantum $\Phi_o = h/2e = 2 \times 10^{-11}\,\mathrm{T\,cm}^2$ and the possibility to measure even $10^{-5}\,\Phi_o$ in 1-Hz bandwidth give the SQUID its enormous sensitivity in a measurement with negligible dissipation. The various versions of the SQUID, like DC- and RF-SQUIDs have been developed to a very high sophistication both by the research community as well as by industry. A DC-SQUID, for example, contains two Josephson junctions. It is typically biased with a current $I_{\mathrm{bias}} > 2I_c$ so that a voltage V develops across them (I_c: the critical current of one Josephson junction). The voltage is a periodic function of the magnetic flux externally applied to the SQUID ring. The DC-SQUID therefore is operated as a flux-to-voltage converter where voltages of order of several $10\,\mu\mathrm{V}$ are produced by magnetic flux changes as small as a fraction of the flux quantum Φ_o. The flux can be applied by injecting a current in an input coil close to the SQUID ring.

SQUIDs measure first of all weak magnetic fields as flux changes, but with appropriate circuitry also current, voltage, or frequency. The most important applications of SQUIDs in low-temperature physics are for magnetometry (see Sect. 3.4.2), in low-frequency NMR experiments (see Sect. 12.10), and for various types of thermometry, in particular noise thermometry (see Sect.12.7).

A good and detailed description of the fundamentals of the SQUID as well as its applications in low-temperature physics can be found already in Lounasmaa's book [14.4]. Many practical details are described in [14.5]. Review articles or books on the subject are [14.6–14.8]. Fortunately, the long overdue comprehensive modern treatment of SQUIDs and their various applications has been published recently in the two volumes "The SQUID Handbook" by Clarke and Braginski [14.9]. The authors treat in a comprehensive and coordinated presentation the device fundamentals, design, technology, system construction, and many applications of this extremely sensitive device, thus bridging the gap between fundamentals and applications.

There are, of course, various further review and comprehensive original articles on SQUIDs and their applications in the literature. For example, the review on "Superconducting Electronics at mK Temperatures", which treats the development of new SQUID type devices [14.10]. Theory and use of thin-film DC resistive SQUIDs for the temperature range of 12 mK to 5.9 K, in particular for application in noise thermometry (Sect. 12.7) have been described in detail in [14.11]. Low-T_c as well as high-T_c SQUIDs and well-developed, fast SQUID electronics for bandwidth up to 20 MHz described in [14.12] are available commercially. One can also consult various conference proceedings, like for the "SQUID" meetings, or the "Applied Superconductivity Conferences", or the publications in "IEEE Transactions on Applied Superconductivity". And last but not least, many of the advancements have been published in the journals "Review of Scientific Instruments" and "Journal of Low Temperature Physics".

List of Symbols

In some cases – when there is no danger of confusion – the same symbol is used for different purposes; they are separated by a slash in the following list. In the text the symbols are usually explained when they appear for the first time.

A	Area/amplitude
AC	Alternating current
a	Interatomic distance
B	Magnetic field
b	Internal magnetic field
C	Heat capacity/specific heat/capacitance
Db	Decibel
DC	Direct current
d	Distance/diameter/thickness
$E; \epsilon; E_C; E_F$	Energy/Coulomb energy/Fermi energy
e	Elementary charge ($4.803 \times 10^{-10}\,g^{1/2}\,cm^{3/2}\,s^{-1}$; $1.602 \times 10^{-19}\,C$)
F	Force
f	Distribution function/frequency
G	Shear modulus/differential conductance
g	Density of states/g-factor/acceleration due to gravity ($9.807\,m\,s^{-2}$)
g_n	Nuclear g-factor
H	Enthalpy/Hamiltonian
h	Height/hour
h; $\hbar = h/2\pi$	Planck constant (6.626×10^{-27}erg s; 1.055×10^{-27}erg s)
I	Nuclear spin/moment of inertia
J	Rotational quantum number/total angular momentum quantum number/Joule
k	Rate constant, spring constant
k_B	Boltzmann constant (1.38065×10^{-16}erg K^{-1}; 1.38065×10^{-23}J K^{-1})

L	Latent heat of vaporization/orbital angular momentum quantum number/length/induction
L_0	Lorenz number ($2.45 \times 10^{-8}\,\mathrm{W}\,\Omega/\mathrm{K}^{-2}$)
l	Length
M	Magnetization
m	Mass/magnetic quantum number/magnetic moment/Poisson ratio
m^*	Effective mass
m_e	Electron mass (9.109×10^{-28} g)
N	Demagnetization factor/Newton
N_0	Avogadro number (6.02214×10^{23} atoms mol^{-1})
n	Number of moles or particles/occupation number
P	Pressure/polarization/heat energy, quantity of heat
P_m	Melting pressure
P_vp	Vapor pressure
$\bar{P}$	Density of tunneling states in glasses
P_sat	Saturated vapor pressure
ppm	Parts per million (10^{-6})
Q	Quantity of heat/nuclear electric quadrupole moment/quality factor
R	Radius/distance/gas constant ($8.135\,\mathrm{J\,mol^{-1}\,K^{-1}}$)/resistance
R_K	Thermal boundary (Kapitza) resistance
RRR	Residual resistivity ratio
RF	Radio frequency
r	Radius/distance
S	Entropy/spin quantum number
T; T_F	Temperature/Fermi temperature
T_c	Critical temperature
T_e	Temperature of electrons
T_n	Temperature of nuclei
T_p	Temperature of phonons
t	Time
U	Internal energy
V	Voltage/Potential/volume
V_m	Molar volume (for a gas: $V_\mathrm{m} = 22,414\,\mathrm{cm^3\,mol^{-1}}$)
V_{zz}	Electric field gradient ($\delta^2 V/\delta z^2$)
v	Velocity (of sound)
v_F	Fermi velocity
X_min	Minimum value of X
x	Concentration
Z	Charge/partition function/impedance
α	Polarizibility/van der Waals constant/thermal expansion coefficient
β	Coefficient of lattice specific heat
γ	Sommerfeld constant of electronic specific heat/gyromagnetic ratio
δ	Skin depth
ΔE	Energy gap

ϵ	Dielectric constant/elastic stress
ϕ_0	Flux quantum ($h/2e = 2.0678 \times 10^{-15}\mathrm{V\,s} = 2.0678 \times 10^{-11}\,\mathrm{T\,cm^{-2}}$)
η	Viscosity
θ	Moment of inertia/angle/Weiss constant
θ_D	Debye temperature
κ	Thermal conductivity/Korringa constant/torsion, spring constant
λ	Mean free path/Curie constant
ν	Frequency
μ	Magnetic moment/permeability/chemical potential
μ_B	Bohr magneton ($9.274 \times 10^{-21}\,\mathrm{erg\,G^{-1}}; 9.274 \times 10^{-24}\,\mathrm{J\,T^{-1}}$)
μ_n	Nuclear magneton ($5.051 \times 10^{-24}\,\mathrm{erg\,G^{-1}}; 5.051 \times 10^{-27}\,\mathrm{J\,T^{-1}}$)
μ_0	Permeability of vacuum ($4\pi \times 10^{-7}\,\mathrm{V\,s\,A^{-1}\,m^{-1}}$)
μ_r	Relative permeability
π	Osmotic pressure
Φ	Phase
ρ	Electrical resistivity/mass density
σ	Electrical conductivity
τ	Relaxation time
$\tau_{1/2}$	Half-life time
τ_1	Spin–lattice relaxation time
τ_2	Spin–spin relaxation time
τ_2^*	Effective spin decay time
χ	Susceptibility
$\tau; \tau_\mathrm{m}$	Torque/magnetic torque
ω	Frequency $\times\, 2\pi$
ω_D	Debye frequency $\times\, 2\pi$
Ω	Ohm

Conversion Factors

$1\,\mathrm{T} = 1\,\mathrm{V\,s\,m^{-2}} = 10^4\,\mathrm{G}$
$1\,\mathrm{G} = 1\,\mathrm{g^{1/2}\,s^{-1}\,cm^{-1/2}}$
$1\,\Omega\,\mathrm{cm} = 10^{-11}/9\,\mathrm{s}$
$1\,\mathrm{meV} = k_\mathrm{B} \times 11.60\,\mathrm{K}$
$1\,\mathrm{eV} = 1.602 \times 10^{-12}\,\mathrm{erg}$
$1\,\mathrm{W} = 1\,\mathrm{J\,s^{-1}} = 10^7\,\mathrm{erg\,s^{-1}}$
$1\,\mathrm{Pa} = 1\,\mathrm{N\,m^{-2}} = 10\,\mu\,\mathrm{bar} = 10\,\mathrm{dyne\,cm^{-2}} = 7.501\,\mathrm{mtorr}$
$1\,\mathrm{J/T} = 1\,\mathrm{A\,m^2} = 10^3\,\mathrm{emu}$

Suppliers of Cryogenic Equipment and Materials

I did hesitate for some time to add this list in the new edition of my book because it surely cannot be complete. The items and suppliers chosen are strongly influenced by my own research experience. In addition, they are taken from industrial exhibitions at relevant conferences, from advertisements in relevant journals, and from the annual Buyer's Guide published as a supplement to the annual August issue of Physics Today.

I give the electronic address of the suppliers only. For the full addresses, one can look, for example, into the above-mentioned Buyer's Guide or, of course, into the Internet. To the best of my knowledge, I have listed the manufacturers and not their representatives in the various countries.

Calorimeters

Oxford Instruments; http://www.oxford-instruments.com
Quantum Design; http://www.qdusa.com

Capacitance Bridges and Capacitance Standards

Andeen-Hagerling; http://www.andeen-hagerling.com
Tucker Electronics (used General Radio 1615A bridges); http://www.tucker.com
Quad Tech Inc.; http://www.quadtech.com

Closed-Cycle Refrigerators; Cryocoolers

Advanced Research Systems Inc.; http://www.arscryo.com
Cryo Industries of America; http://www.cryoindustries.com
Cryogenic Ltd; http://www.cryogenic.co.uk

Cryomagnetics Inc.; http://www.cryomagnetics.com
Cryomech Inc.; http://www.cryomech.com
Dryogenic Ltd; http://www.dryogenic.com
Janis Research Comp.; http://www.janis.com
Leybold Vacuum; http://www.leybold.com
Oxford Instruments; http://www.oxford-instruments.com
Polycold Systems Inc.; http://www.polycold.com
Sumitomo Heavy Industries Ltd; http://www.shi.co.jp
Sunpower; http://www.sunpower.com
Suzuki Shokan Co. Inc.; http://www.suzukishokan.co.jp
VeriCold Technologies; http://www.vericold.com

Cryostats for Liquid ^{3}He

Cryogenic Ltd; http://www.cryogenic.co.uk
ICEoxford; http://www.iceoxford.com
IQUANTUM; http://www.iquantum.jp
Janis Research Comp.; http://www.janis.com
Oxford Instruments; http://www.oxford-instruments.com
VeriCold Technologies; http://www.vericold.com

Dewars and Cryostats for Liquid ^{4}He; Transfer Lines

Advanced Research Systems Inc.; http://www.arscryo.com
Andonian Cryogenics Inc.; http://www.andoniancryogenics.com
Cryo Anlagenbau GmbH; http://www.cryoanlagenbau.de
Cryoconcept; http://www.cryoconcept.fr
Cryofab Inc.; http://www.cryofab.com
Cryogenic Ltd; http://www.cryogenic.co.uk
Cryo Industries of America; http://www.cryoindustries.com
CryoVac; http://www.cryovac.de
ICEoxford; http://www.iceoxford.com
International Cryogenics Inc.; http://www.intlcryo.com
Janis Research Comp.; http://www.janis.com
Kadel Engineering Corp.; http://www.kadel.com
Oxford Instruments; http://www.oxford-instruments.com
Pope Scientific Inc.; http://www.popeinc.com (glass dewars)
Precision Cryogenic Systems Inc.; http://www.precisioncryo.com
VeriCold Technologies; http://www.vericold.com

Dilution Refrigerators, ^{3}He–^{4}He

Cryoconcept; http://www.cryoconcept.fr
ICEoxford; http://www.iceoxford.com

Janis Research Comp.; http://www.janis.com
Leiden Cryogenics BV; http://www.leidencryogenics.com
Oxford Instruments; http://www.oxford-instruments.com
Vericold Technologies; http://www.vericold.com

Epoxy, Grease, Varnish

Apiezon Products-M&I Materials Ltd; http://www.apiezon.com
Aremco Products Inc.; http://www.aremco.com
Ciba Specialty Chemicals Inc.; http://www.cibasc.com
Epoxy Technology Inc.; http://www.epotek.com
Emerson and Cumming Inc.; http://www.emersoncuming.com (Stycast)
Furane Plastics Inc.; http://www.plasticstechnology.com (Epibond)
Gen. Electric Corp., Insul. Mat. Dept.; http://www.GEPlastics.com (GE Varnish)
ICEoxford; http://www.iceoxford.com
Lake Shore Cryotronics Inc.; http://www.lakeshore.com
Oxford Instruments; http://www.oxford-instruments.com

Helium-3 Gas

Chemgas; http://www.chemgas.com
Linde Gas; http://www.linde-gas.de
Sigma-Aldrich Co; http://www.sigmaaldrich.com
Oxford Instruments; http://www.oxford-instruments.com
US Services Inc./Icon Services Inc.; http://www.iconservices.com

Inductance Bridges

Leiden Cryogenics BV; http://www.leidencryogenics.com

Liquid-Level Sensors and Controllers

Andonian Cryogenics Inc.; http://www.andoniancryogenics.com
American Magnetics; http://www.americanmagnetics.com
Cryomagnetics Inc.; http://www.cryomagnetics.com
Janis Research Comp.; http://www.janis.com
Lake Shore Cryotronics Inc.; http://www.lakeshore.com
Oxford Instruments; http://www.oxford-instruments.com

Magnetic and Physical Property Measurement Systems/Magnetometers/Susceptometers

ADE Technologies Inc.; http://www.adetech.com
Cryogenic Ltd; http://www.cryogenic.co.uk
Lake Shore Cryotronics Inc.; http://www.lakeshore.com
Quantum Design; http://www.qdusa.com
Tristan Technologies Inc.; http://www.tristantech.com

Magnetic Refrigerators

Dryogenic Ltd; http://www.dryogenic.com
Janis Research Comp.; http://www.janis.com

Magnetic Shielding

Ad-Vance Magnetics; http://www.advancemag.com
Advent Research Materials Ltd; http://www.advent-rm.com
Amuneal Manufacturing Corp.; http://www.amuneal.com
Magnetic Shield Corp.; http://www.magnetic-shield.com
Vakkumschmelze GmbH; http://www.vacuumschmelze.de

Metals, Metal Wires

Advent Research Materials Ltd; http://www.advent-rm.com
Alfa Aesar; http://www.alfa.com
California Fine Wire Comp.; http://www.calfinewire.com
Cooner Wire Co.; http://www.coonerwire.com
Goodfellow Corp.; http://www.goodfellow.com
Indium Corporation of America; http://www.indium.com
Johnson Matthey; http://www.jmei.com
Kawecky Berylco Industries Inc.; http://www.kballoys.com
Leico Industries Inc.; http://www.leicoind.com
Matek GmbH; http://www.matek.com
Oxford Instruments; http://www.oxford-instruments.com
Pelican Wire Comp.; http://www.pelicanwire.com
Sigmund Cohn Corp.; http://www.sigmundcohn.com
Umicore Precious Metals Refining; http://www.preciousmetals.umicore.com
Wieland Werke AG; http://www.wieland.de

Cu

American Smelting and Refining Comp.(ASARCO); http://www.asarco.com
Hitachi Cable Ltd; http://www.Hitachi-cable.co.jp

Cumerio; http://www.cumerio.com
North American Hoganas Inc.; http://www.hoganas.com
Outokumpu Poricopper Oy; http://www.outokumpu.com
Sofilec Division Buisin; http://www.buisin.com

Metal Powders

Alfa Aesar; http://www.alfa.com
Engelhard Corp.; http://www.engelhard.com
Ferro Electronic Material Systems; http://www.ferro.com
Goodfellow Corp.; http://www.goodfellow.com
Ulvac Technologies Inc.; http://www.ulvac-materials.co.jp

Metal Tubes

ICEoxford; http://www.iceoxford.com
Oxford Instruments; http://www.oxford-instruments.com
Precision Tube Co.; http://www.precisiontube.com
Superior Tube Co.; http://www.superiortube.com
Uniform Tubes; http://www.uniformtubes.com

Micropositioning Systems for Cryogenics

Attocube Systems; http://www.attocube.com
Janis Research Comp.; http://www.janis.com

Nuclear Magnetic Resonance Thermometry on ^{195}Pt, Electronics

Picowatt, RV-Elektroniikka OY; http://www.picowatt.fi

Pressure Transducers, and Gauges

BOC Edwards; http://www.bocedwards.com
Gems Sensors; http://www.gems-sensors.co.uk
MKS Instruments Inc.; http://www.mksinst.com
Wallace & Tiernan Co.; http://www.wallace-tiernan.de

Pomeranchuk Cooling Cells

Leiden Cryogenics BV; http://www.leidencryogenics.com

Resistance Bridges/Temperature Controllers (see also "Thermometry")

Cryo-Con Inc.; http://www.cryocon.com
Cryo Industries of America Inc.; http://www.cryoindustries.com
Janis Research Comp.; http://www.janis.com
Lake Shore Cryotronics; http://www.lakeshore.com
Oxford Instruments; http://www.oxford-instruments.com
Picowatt, RV Elektroniikka OY; http://www.picowatt.fi
Yokogawa Corp.; http://www.yokogawa.com

SQUIDs, SQUID Readout Electronics

Applied Physics Systems; http://www.appliedphysics.com
IQUANTUM; http://www.iquantum.jp
Jülicher SQUID GmbH; http://www.jsquid.com
Magnicon GbR; http://www.magnicon.com
Quantum Design; http://www.qdusa.com
Star Cryoelectronics; http://www.starcryo.com
Tristan Technologies Inc.; http://www.tristantech.com
Twente Solid State Technology BV; http://www.tsst.nl

Superconducting Fixed-Point Device

Hightech Developments Leiden; http://www.xs4all.nl/~hdleiden/srd1000

Superconducting Magnets, Superconducting/Vapor Cooled Current Leads, Power Supplies

Accel Instruments GmbH; http://www.accel.de
American Magnetics; http://www.americanmagnetics.com
American Superconductor Corp.; http://www.amsuper.com
Cryogenic Ltd; http://www.cryogenic.co.uk
Cryo Industries of America; http://www.cryoindustries.com
Cryomagnetics Inc.; http://www.cryomagnetics.com
Cryo-Technics; http://www.cryo-technics.de
Janis Research Comp.; http://www.janis.com
Oxford Instruments; http://www.oxford-instruments.com
Scientific Magnetics; http://www.scientificmagnetics.com

Superconducting Wires and Materials

American Superconductor Corp.; http://www.amsuper.com
European Advanced Superconductors; http://www.advancedsupercon.com
Goodfellow Corp.; http://www.goodfellow.com
Hitachi Cable Ltd; http://www.hitachi-cable.co.jp
Oxford Instruments; http://www.oxford-instruments.com
Supercon Inc.; http://www.supercon-wire.com

Thermometers Including the Relevant Electronics/Bridges and Temperature Controllers

Capacitance

Lake Shore Cryotronics Inc.; http://www.lakeshore.com
Oxford Instruments; http://www.oxford-instruments.com

Coulomb blockade

Nanoway Cryoelectronics Ltd; http://www.nanoway.fi

Nuclear Orientation

Oxford Instruments; http://www.oxford-instruments.com

Resistance

Cryogenic Control Systems; http://www.cryocon.com
CryoVac; http://www.cryovac.de
ICEoxford; http://www.iceoxford.com
Lake Shore Cryotronics Inc.; http://www.lakeshore.com
Oxford Instruments; http://www.oxford-instruments.com
Rosemount Inc.; http://www.rosemount.com
Scientific Instruments Inc.; http://www.scientificinstruments.com

Thermocouples

Janis Research Comp.; http://www.janis.com
Lake Shore Cryotronics Inc.; http://www.lakeshore.com
Oxford Instruments; http://www.oxford-instruments.com
Scientific Instruments; http://www.scientificinstruments.com

Valves, Cryogenic

Andonian Cryogenics Inc.; http://www.andoniancryogenics.com
Cryo Industries of America Inc.; http://www.cryoindustries.com
Hoke Inc.; http://www.hoke.com
Precision Cryogenic Systems Inc.; http://www.precisioncryo.com

Windows for Cryogenic Systems

Oxford Instruments; http://www.oxford-instruments.com

References

Chapter 1

1.1 C. Kittel: *Introduction to Solid State Physics*, 8th edn. (Wiley-VCH, Weinheim 2004)
1.2 N.W. Ashcroft, N.D. Mermin: *Solid State Physics* (Saunders, Philadelphia, PA 1976)
1.3 J.M. Ziman: *Electrons and Phonons* (Oxford, 2001)
1.4 H. Ibach, H. Lüth: *Solid-State Physics, An Introduction to Theory and Experiment*, 2nd edn. (Springer, Berlin Heidelberg New York 1995)
1.5 P.V.E. McClintock, D.J. Meredith, J.K. Wigmore: *Matter at Low Temperatures* (Blackie, London 1984)
1.6 C. Enss, S. Hunklinger: *Low Temperature Physics* (Springer, Berlin Heidelberg New York 2005)
1.7 G.K. White, P.F. Meeson: *Experimental Techniques in Low Temperature Physics*, 4th edn. (Clarendon, Oxford 2002)
1.8 A.C. Rose-Innes: *Low Temperature Laboratory Techniques* (English University Press, London 1973)
1.9 O.V. Lounasmaa: *Experimental Principles and Methods Below 1 K* (Academic, London 1974)
1.10 D.S. Betts: *Refrigeration and Thermometry Below One Kelvin* (Sussex University Press, Brighton 1976)
1.11 D.S. Betts: *An Introduction to Millikelvin Technology* (Cambridge University Press, Cambridge 1989)
1.12 R.C. Richardson, E.N. Smith: *Experimental Techniques in Condensed Matter Physics at Low Temperatures* (Addison-Wesley, Redwood City, CA 1988)

Chapter 2

2.1 P.V.E. McClintock, D.J. Meredith, J.K. Wigmore: *Matter at Low Temperatures* (Blackie, London 1984)
2.2 G.K. White, P.F. Meeson: *Experimental Techniques in Low Temperature Physics*, 4th edn. (Clarendon, Oxford 2002)
2.3 R.C. Reid, J.M. Prausnitz, T.K. Sherwood: *The Properties of Gases and Liquids* (McGraw-Hill, New York 1977)
2.4 R.T. Jacobsen, S.G. Penoncello, E.W. Lemmon: *Thermodynamic Properties of Cryogenic Fluids* (Plenum, New York 1997)
2.5 J.G. Weisand (ed): *Handbook of Cryogenic Engineering* (Taylor & Francis, London 1998)
2.6 J. van Krankendonk: *Solid Hydrogen* (Plenum, New York 1983)
2.7 I.F. Silvera: Rev. Mod. Phys. **52**, 393 (1986)
2.8 K. Motizuki, T. Nagamiya: J. Phys. Soc. Jpn. **11**, 93 (1956); J. Phys. Soc. Jpn. **12**, 163 (1957)
2.9 A.J. Berlinsky, W.N. Hardy: Phys. Rev. B **8**, 5013 (1973)
2.10 P. Pedroni, H. Meyer, F. Weinhaus, D. Haase: Solid State Commun. **14**, 279 (1974)
2.11 N.S. Sullivan, D. Zhou, C.M. Edwards: Cryogenics **30**, 734 (1990)
2.12 V. Shevtsov, P. Malmi, E. Ylinen, M. Punkinnen: J. Low Temp. Phys. **114**, 431 (1999)
2.13 Yu.Ya. Milenko, R.M. Sibileva, M.A. Strzhemechny: J. Low Temp. Phys. **107**, 77 (1997)
2.14 W.R. Wampler, T. Schober, B. Lengeler: Philos. Mag. **34**, 129 (1976)
2.15 M. Schwark, F. Pobell, W.P. Halperin, Ch. Buchal, J. Hanssen, M. Kubota, R.M. Mueller: J. Low Temp. Phys. **53**, 685 (1983)
2.16 M. Kolac, B.S. Neganov, S. Sahling: J. Low Temp. Phys. **59**, 547 (1985); J. Low Temp. Phys. **63**, 459 (1986)
2.17 J. Wilks: *The Properties of Liquid and Solid Helium* (Clarendon, Oxford 1967)
2.18 W.E. Keller: *Helium-Three and Helium-Four* (Plenum, New York 1969)
2.19 T. Tsuneto: in *The Structure and Properties of Matter*, ed. by T. Matsubara, Springer Ser. Solid-State Sci., Vol. 28 (Springer, Berlin Heidelberg New York, 1982), Chaps. 3, 4
2.20 J. Wilks, D.S. Betts: *An Introduction to Liquid Helium*, 2nd edn. (Clarendon, Oxford 1987)
2.21 D.R. Tilley, J. Tilley: *Superfluidity and Superconductivity*, 3rd edn. (Hilger, Bristol 1990)
2.22 J.F. Annett: *Superconductivity, Superfluids and Condensates* (Oxford University Press, Oxford 2004)
2.23 C. Enss, S. Hunklinger: *Low Temperature Physics* (Springer, Berlin Heidelberg, New York 2005)

2.24 K.H. Bennemann, J.B. Ketterson (eds.): *The Physics of Liquid and Solid Helium*, Vols. 1, 2 (Wiley, New York 1976, 1978)
2.25 D.S. Greywall: Phys. Rev. B **18**, 2127 (1978); Phys. Rev. B **21**, 1329 (1979)
2.26 D.S. Greywall: Phys. Rev. B **27**, 2747 (1983); Phys. Rev. B **33**, 7520 (1986)
2.27 M.J. Buckingham, W.M Fairbank: in *Progress in Low Temperature Physics*, Vol. 3, ed. by C.J. Gorter (North-Holland, Amsterdam 1961) p. 80
2.28 J.A. Lipa, T.C.P. Chui: Phys. Rev. Lett. **51**, 2291 (1983)
2.29 G. Ahlers: in [Ref. 2.24, Vol. 1, p. 85]
2.30 J.A. Lipa, J.A. Nissen, D.A. Stricker, D.R. Swanson, T.C.P. Chui: Phys. Rev. B **68**, 174518 (2003)
2.31 D.M. Lee, H.A. Fairbank: Phys. Rev. **116**, 1359 (1959)
2.32 D.S. Greywall: Phys. Rev. B **23**, 2152 (1981)
2.33 D.F. Brewer: in [Ref. 2.24, Vol. 2, p. 573]
2.34 D.D. Osheroff, R.C. Richardson, D.M. Lee: Phys. Rev. Lett. **28**, 885 (1972)
2.35 D.D. Osheroff, W.J. Gully, R.C. Richardson, D.M. Lee: Phys. Rev. Lett. **29**, 920 (1972)
2.36 A.J. Leggett: Rev. Mod. Phys. **47**, 331 (1975)
2.37 J.C. Wheatley: Rev. Mod. Phys. **47**, 415 (1975)
2.38 J.C. Wheatley: Physica **69**, 218 (1973)
2.39 J.C. Wheatley, p. 1; W.F. Brinkman, M.C. Cross, p. 105; P. Wölfle, p. 191: in *Progress in Low Temperature Physics*, Vol. 7, ed. by D.F. Brewer (North Holland, Amsterdam 1978)
2.40 P.W. Anderson, W.F. Brinkman, p. 105; D.M. Lee, R.C. Richardson, p. 287: in Ref. 2.24, Vol. 2
2.41 E.R. Dobbs: *Helium Three* (Oxford University Press, Oxford 2000)
2.42 D. Vollhard, P. Wölfle: *The Superfluid Phases of Helium-3* (Taylor and Francis, London 1990)
2.43 W.P. Halperin, L.P. Pitaevskii (eds.): *Helium Three* (North-Holland, Amsterdam 1990)
2.44 L.D. Landau: Sov. Phys. JETP **3**, 920 (1957); **5**, 101 (1957); **8**, 70 (1959)
2.45 A.A. Abrikosov, I.M. Khalatnikov: Rep. Prog. Phys. **22**, 329 (1959)
2.46 J.C. Wheatley: in *Progress in Low Temperature Physics*, Vol. 6, ed. by C.J. Gorter (North-Holland, Amsterdam 1970), p. 77
2.47 G. Baym, C. Pethick: in [Ref. 2.24, Vol. 2, p. 1]
2.48 D.S. Greywall: Phys. Rev. B **29**, 4933 (1984)
2.49 C. Ebner, D.O. Edwards: Phys. Rep. **2**, 77 (1971)
2.50 O.V. Lounasmaa: *Experimental Principles and Methods Below 1K* (Academic, London 1974)
2.51 J. Bardeen, G. Baym, D. Pines: Phys. Rev. Lett. **17**, 372 (1966); Phys. Rev. **156**, 207 (1967)

2.52 R. König, F. Pobell: Phys. Rev. Lett. **71**, 2761 (1993); J. Low Temp. Phys. **97**, 287 (1994); for data on the mixtures see also J.C.H. Zeegers, A.T.A.M. de Waele, H.M. Gijsmann: J. Low Temp. Phys. **84**, 37 (1991)

Chapter 3

3.1 C. Kittel: *Introduction to Solid State Physics*, 8th edn. (Wiley-VCH, Weinheim 2004)
3.2 N.W. Ashcroft, N.D. Mermin: *Solid State Physics* (Saunders, Philadelphia, PA 1976)
3.3 J.M. Ziman: *Electrons and Phonons* (Oxford, 2001)
3.4 H. Ibach, H. Lüth: *Solid-State Physics, An Introduction to Theory and Experiment*, 2nd edn. (Springer, Berlin Heidelberg New York 1995)
3.5 C. Enss, S. Hunklinger: *Low Temperature Physics* (Springer, Berlin Heidelberg, New York 2005)
3.6 P.V.E. McClintock, D.J. Meredith, J.K. Wigmore: *Matter at Low Temperatures* (Blackie, London 1984)
3.7 T.H.K. Barron, G.K. White: *Heat Capacity and Thermal Expansion at Low Temperatures* (Plenum, New York 1999)
3.8 P. Brüesch: *Phonons: Theory and Experiments I*, Springer Ser. Solid-State Sci., Vol. 34 (Springer, Berlin Heidelberg, New York 1982)
3.9 C.B. Walker: Phys. Rev. **103**, 547 (1950)
3.10 J.R. Clement, E.H. Quinnell: Phys. Rev. **92**, 258 (1953)
3.11 M.L. Klein, G.K. Horton, J.L. Feldman: Phys. Rev. **184**, 968 (1969)
3.12 L. Finegold, N.E. Phillips: Phys. Rev. **177**, 1383 (1969)
3.13 F.L. Battye, A. Goldmann, L. Kasper, S. Hüfner: Z. Phys. B **27**, 209 (1977)
3.14 N. Grewe, F. Steglich: in *Handbook of Physics and Chemistry of Rare Earths*, Vol. 14, ed. by K.A. Gschneidner, Jr., L. Eyring (Elsevier, Amsterdam 1991), p. 343
3.15 G.R. Stewart: Rev. Mod. Phys. **56**, 755 (1984); A. Amato: Rev. Mod. Phys. **69**, 1119 (1997)
3.16 D.R. Tilley, J. Tilley: *Superfluidity and Superconductivity*, 3rd edn. (Hilger, Bristol 1990)
3.17 M. Tinkham: *Introduction to Superconductivity* 2nd ed. (Dover, New York, 2004)
3.18 W. Buckel, R. Kleiner: *Superconductivity*, 2nd edn. (Wiley-VCH, Berlin 2004)
3.19 J.F. Annett: *Superconductivity, Superfluids and Condensates* (Oxford University Press, Oxford 2004)
3.20 N.E. Phillips: Phys. Rev. **114**, 676 (1959)
3.21 N.E. Phillips, M.H. Lambert, W.R. Gardner: Rev. Mod. Phys. **36**, 131 (1964)

3.22 M.A. Biondi, A.T. Forrester, M.P. Garfunkel, C.B. Satterthwaite: Rev. Mod. Phys. **30**, 1109 (1958)

3.23 J. Bardeen, L.N. Cooper, J.R. Schrieffer: Phys. Rev. **108**, 1175 (1957)

3.24 W.A. Phillips (ed.): *Amorphous Solids*, Topics Curr. Phys., Vol. 24 (Springer, Berlin Heidelberg New York 1981)

3.25 P. Esquinazi: *Tunneling Systems in Amorphous and Crystalline Solids* (Springer, Berlin Heidelberg New York 1998)

3.26 S. Hunklinger, K. Raychaudhuri: in *Progress Low Temperature Physics*, Vol. 9, ed. by D.F. Brewer (North-Holland, Amsterdam 1986), p. 265; S. Hunklinger, W. Arnold: in *Physical Acoustics* **12**, 155 (Academic, New York 1976)

3.27 R.C. Zeller, R.O. Pohl: Phys. Rev. B **4**, 2029 (1971)

3.28 R.B. Stephens: Phys. Rev. B **8**, 2896 (1973)

3.29 J.C. Lasjaunias, A. Ravex, M. Vandorpe, S. Hunklinger: Solid State Commun. **17**, 1045 (1975)

3.30 S. Sahling, S. Abens, T. Eggert: J. Low Temp. Phys. **127**, 215 (2002)

3.31 J.C. Ho, N.E. Phillips: Rev. Sci. Instrum. **36**, 1382 (1965); J.C. Ho, H.R. O'Neal, N.E. Phillips: Rev. Sci. Instrum. **34**, 782 (1963)

3.32 C. Hagmann, P.L. Richards: Cryogenics **35**, 345 (1995)

3.33 D.L. Martin: Rev. Sci. Instrum. **38**, 1738 (1967); Rev. Sci. Instrum. **58**, 639 (1987); Phys. Rev. **170**, 650 (1968)

3.34 G. Ahlers: Rev. Sci. Instrum. **37**, 477 (1966)

3.35 N. Waterhouse: Can. J. Phys. **47**, 1485 (1969)

3.36 Y. Hiki, T. Maruyama, Y. Kogure: J. Phys. Soc. Jpn **34**, 723 (1973)

3.37 J. Bevk: Philos. Mag. **28**, 1379 (1973)

3.38 G.J. Sellers, A.C. Anderson: Rev. Sci. Instrum. **45**, 1256 (1974)

3.39 D.S. Greywall: Phys. Rev. B **18**, 2127 (1978)

3.40 E.J. Cotts, A.C. Anderson: J. Low Temp. Phys. **43**, 437 (1980)

3.41 D.J. Bradley, A.M. Guénault, V. Keith, C.J. Kennedy, J.E. Miller, S.B. Musset, G.R. Pickett, W.P. Pratt, Jr.: J. Low. Temp. Phys. **57**, 359 (1984)

3.42 K. Gloos, P. Smeibidl, C. Kennedy, A. Singsaas, P. Sekovski, R.M. Mueller, F. Pobell: J. Low Temp. Phys. **73**, 101 (1988)

3.43 W. Wendler, T. Herrmannsdörfer, S. Rehmann, F. Pobell: Europhys. Lett. **38**, 619 (1997); J. Low Temp. Phys. **111**, 99 (1998)

3.44 G.K. White, P.J. Meeson: *Experimental Techniques in Low Temperature Physics*, 4th edn. (Clarendon, Oxford 2002)

3.45 O.V. Lounasmaa: *Experimental Principles and Methods Below 1 K* (Academic, London 1974)

3.46 R.J. Corruccini, J.J. Gniewek: *Specific Heat and Enthalpies of Technical Solids at Low Temperatures*, NBS Monograph 21 (US Govt. Print. Office, Washington, DC 1960)

3.47 C.Y. Ho, A. Cezairliyan (eds.): *Specific Heat of Solids* (Hemisphere Publ., New York 1988)

3.48 Y.S. Touloukian, E.H. Buyco (eds.): *Thermophysical Properties of Matter (Specific Heat)*, Vols. 4 and 5 (Plenum, New York 1970, 1971)
3.49 Y. Karaki, Y. Koike, M. Kubota, H. Ishimoto: Cryogenics **37**, 171 (1997)
3.50 T.C. Cetas, C.R. Tilford, C.A. Swenson: Phys. Rev. **174**, 835 (1968); J.T. Holste, T.C. Cetas, C.A. Swenson: Rev. Sci. Instrum. **43**, 670 (1972)
3.51 F.J. du Chatenier, B.M. Boerstel, J. de Nobel: Physica **31**, 1449 (1965)
3.52 H.J. Schink, H.V. Löhneysen: Cryogenics **21**, 591 (1981)
3.53 W. Schnelle, J. Engelhardt, E. Gmelin: Cryogenics **39**, 271 (1999)
3.54 T.P. Papageorgiou, T. Herrmannsdörfer: priv. comm. (2006)
3.55 C.A. Swenson: Rev. Sci. Instrum. **68**, 1312 (1997)
3.56 M. Barucci, E. Gottardi, E. Olivieri, E. Pasca, L. Risegari, G. Ventura: Cryogenics **42**, 551 (2002)
3.57 A. Nittke, M. Scherl, P. Esquinazi, W. Lorenz, J. Li, F. Pobell: J. Low Temp. Phys. **98**, 517 (1995)
3.58 M.J. Laubitz, in *Thermal Conductivity*, Vol. 1, ed. R.P. Tye (Academic, New York 1969) p. 1
3.59 J.A. Lipa, J.A. Nissen, D.A. Stricker, D.R. Swanson, T.C.P. Chui: Phys. Rev. B **68**, 174518 (2003)
3.60 T. Herrmannsdörfer, F. Pobell: J. Low Temp. Phys. **100**, 253 (1995)
3.61 M. Jaime, R. Movshovich, J. Sarrao, G. Stewart, J. Kim, P.C. Canfield: Physica B **280**, 563 (2000)
3.62 H. Wilhelm, T. Lühmann, T. Rus, F. Steglich: Rev. Sci. Instrum. **75**, 2700 (2004)
3.63 R.E. Schwall, R.E. Howard, G.R. Stewart: Rev. Sci. Instrum. **46**, 1054 (1975); G.R. Stewart: Rev. Sci. Instrum. **54**, 1 (1983), see this paper for references to earlier work
3.64 J.C. Lashley et al.: Cryogenics **43**, 369 (2003)
3.65 S.G. Doettinger-Zech, M. Uhl, D.L. Sisson, A. Kapitulnik: Rev. Sci. Instrum. **72**, 2398 (2001)
3.66 D.W. Denlinger et al.: Rev. Sci. Instrum. **65**, 946 (1994); B.L. Zink, B. Revax, R. Sappey, F. Hellman: Rev. Sci. Instrum. **73**, 1841 (2002)
3.67 H. Tsujii, B. Andraka, E.C. Palm, T.P. Murphy, Y. Takano: Physica B **329–333**, 1638 (2003), (Proc. 23rd Int'l Conf. Low Temp. Phys., Hiroshima 2002)
3.68 O. Riou, P. Gandit, M. Charalambous, J. Chaussy: Rev. Sci. Instrum. **68**, 1501 (1997); F. Fominaya, T. Fournier, P. Gandit, J. Chaussy: Rev. Sci. Instrum. **68**, 4191 (1997); O. Bourgeois, S.E. Skipetrov, F. Ong, J. Chaussy: Phys. Rev. Lett. **94**, 057007 (2005)
3.69 S. Riegel, G. Weber: J. Phys. E: Sci. Instrum. **19**, 790 (1986)
3.70 A. Schilling, O. Jeandupeux: Phys. Rev. **52**, 9714 (1995)
3.71 W.K. Neils, D. Martien, E.D. Bauer, D Mixson, N. Hur, J.D. Thompson, J.L. Sarrao: Am. Inst. Phys. Conf. Proc. **850**, 1647 (2006),

(Proc. 24th Int'l Conf. Low Temp. Phys., Orlando, FL, 2005), ed. by Y. Takano et al.

3.72 M.L. Sigueira, R.E. Rapp: Rev. Sci. Instrum. **62**, 2499 (1991)

3.73 T. Park et al.: Phys. Rev. Lett. **90**, 177001 (2003); H. Aoki et al.: Physica B **359–361**, 410 (2005); A. Vorontsov, I. Vekhter: Phys. Rev. Lett. **96**, 237001 (2006)

3.74 F. Giazotto, T.T. Heikkilä, A. Luukanen, A.M. Savin, J.P. Pekola: Rev. Mod. Phys. **78**, 217 (2006)

3.75 D.A. Ackerman, A.C. Anderson: Rev. Sci. Instrum. **53**, 1657 (1982)

3.76 R.F. Seligmann, R.E. Sarwinski: Cryogenics **12**, 239 (1972)

3.77 G.W. Swift, R.E. Packard: Cryogenics **19**, 362 (1979)

3.78 C.Y. Ho, R.E. Taylor (eds.): *Thermal Expansion of Solids* (American Soc. Metals, Metals Park, Ohio 1998)

3.79 R.J. Corruccini, J.J. Gniewek: *Thermal Expansion of Technical Solids at Low Temperatures*, NBS Monograph 29 (US Govt. Print. Office, Washington, DC 1961)

3.80 Y.S. Touloukian, P.K. Kirby, R.E. Taylor, P.D. Desai, T.Y.R. Lee (eds.): *The Thermophysical Properties of Matter (Thermal Expansion)*, Vols. 12 and 13 (Plenum, New York 1975, 1977)

3.81 D.A. Ackerman, A.C. Anderson, E.J. Cotts, J.N. Dobbs, W.M. MacDonald, F.J. Walker: Phys. Rev. B **29**, 966 (1984); see also "Quarzglas and Quarzgut", Q-A1/112.2 from Heraeus Quarzschmelze, D-6450 Hanau 1, FR Germany

3.82 P. Roth, E. Gmelin: Rev. Sci. Instrum. **63**, 2051 (1992). Containing references to earlier work

3.83 M. Rotter, H. Müller, E. Gratz, M. Doerr, M. Löwenhaupt: Rev. Sci. Instrum. **69**, 2742 (1998); M. Doerr, M. Rotter, J. Brooks, E. Joboliong, A. Lindbaum, R. Vasic, M. Löwenhaupt: Am. Inst. Phys. Conf. Proc. **850**, 1239 (2006), (Proc. 24th Int'l Conf. Low Temp. Phys., Orlando, FL, 2005), ed. by Y. Takano et al.

3.84 P.G. Klemens: *Solid State Physics* **7**, 1, ed. F. Seitz, D. Turnbull (Academic, New York 1958); *Thermal Conductivity* Vol. 1, R.P. Tye (Academic, New York 1969), p. 1

3.85 G.A. Slack: *Solid State Physics* **34**, 1, ed. H. Ehrenreich, F. Seitz, D. Turnbull (Academic, New York 1979)

3.86 R. Berman: *Thermal Conduction in Solids* (Clarendon, Oxford 1976)

3.87 T.M. Tritt (ed.): *Thermal Conductivity: Theory, Properties, and Applications* (Springer, Berlin Heidelberg New York 2004)

3.88 E.C. Crittenden, Jr.: *Cryogenic Technology*, ed. by R.W. Vance (Wiley, New York 1963), p. 60

3.89 Leybold-Heraeus: Kryotechnisches Arbeitsblatt Nr.1, Leybold AG, D-5000 Köln 1, FR Germany

3.90 R.L. Powell, W.A. Blaupied: *Thermal Conductivity of Metals and Alloys at Low Temperatures*, Nat. Bureau of Standards Circular 556 (US Govt. Print. Office, Washington, DC 1954)

3.91 G.E. Childs, R.L. Ericks, R.L. Powell: *Thermal Conductivity of Solids*, NBS Monograph 131 (US Govt. Printing Office, Washington, DC 1973)
3.92 Y.S. Touloukian, R.W. Powell, C.Y. Ho, P.G. Klemens (eds.): *The Thermophysical Properties of Matter (Thermal Conductivity)*, Vols. 1 and 2 (Plenum, New York 1970, 1971)
3.93 C.Y. Ho, R.W. Powell, P.E. Liley: Thermal conductivity of the elements. J. Phys. Chem. Ref. Data **1**, 279 (1972); J. Phys. Chem. Ref. Data **3**, Suppl. 1 (1974)
3.94 M. Locatelli, D. Arnand, M. Routin: Cryogenics **16**, 374 (1976)
3.95 J.J. Freeman, A.C. Anderson: Phys. Rev. B **34**, 5684 (1986)
3.96 D. Rosenberg, D. Natelson, D.D. Osheroff: J. Low Temp. Phys. **120**, 259 (2000)
3.97 H.-Y. Hao, M. Neumann, C. Enss, A. Fleischmann: Rev. Sci. Instrum. **75**, 2718 (2004)
3.98 D.T. Corzett, A.M. Keller, P. Seligmann: Cryogenics **16**, 505 (1976)
3.99 C. Schmidt: Rev. Sci. Instrum. **50**, 454 (1979)
3.100 J.R. Olsen: Cryogenics **33**, 729 (1993)
3.101 H.A. Fairbank, D.M. Lee: Rev. Sci. Instrum. **31**, 660 (1960)
3.102 D.S. Greywall: Phys. Rev. B **29**, 4933 (1984)
3.103 A.C. Anderson, W. Reese, J.C. Wheatley: Rev. Sci. Instrum. **34**, 1386 (1963)
3.104 K. Gloos, C. Mitschka, F. Pobell, P. Smeibidl: Cryogenics **30**, 14 (1990)
3.105 G. Armstrong, A.S. Greenberg, J.R. Sites: Rev. Sci. Instrum. **49**, 345 (1978)
3.106 C.L. Tsai, H. Weinstock, W.C. Overton, Jr.: Cryogenics **18**, 562 (1978)
3.107 T. Scott, M. Giles: Phys. Rev. Lett. **29**, 642 (1972)
3.108 D.A. Zych: Cryogenics **29**, 758 (1989)
3.109 D.O. Edwards, R.E. Sarwinski, P. Seligmann, J.T. Tough: Cryogenics **8**, 392 (1968)
3.110 A.K. Raychaudhuri, R.D. Pohl: Sol State Commun. **44**, 711 (1982)
3.111 Th. Wagner, S. Götz, G. Eska: Cryogenics **34**, 655 (1994)
3.112 G. Ventura, M. Barucci, E. Gottardi, I. Peroni: Cryogenics **40**, 489 (2000)
3.113 L. Risegari, M. Barucci, E. Olivieri, G. Ventura: J. Low Temp. Phys. August 2006
3.114 G. Ventura, G. Bianchini, E. Gottardi, I. Peroni, A. Peruzzi: Cryogenics **39**, 481 (1999); M. Barucci, E. Olivieri, E. Pasca, L. Risegari, G. Ventura: Cryogenics **45**, 295 (2005)
3.115 J. Bardeen, G. Rickayzen, L. Tewordt: Phys. Rev. **113**, 982 (1959)
3.116 T. Matsubara (ed.): *The Structure and Properties of Matter*, Springer Ser. Solid-State Sci., Vol. 28 (Springer, Berlin Heidelberg New York 1982), Chap. 5

3.117 A.C. Anderson, R.E. Peterson, J.E. Robichaux: Phys. Rev. Lett. **20**, 459 (1968)
3.118 E.R. Rumbo: J. Phys. F **6**, 85 (1976)
3.119 A.C. Hewson: *The Kondo Problem to Heavy Fermions* (Cambridge Univ. Press, Cambridge 1993)
3.120 N. Andrei, K. Furuya, J.H. Lowenstein: Rev. Mod. Phys. **55**, 331 (1983)
3.121 T. Murao: in *The Structure and Properties of Matter*, ed. by T. Matsubara, Springer Ser. Solid-State Sci., Vol. 28 (Springer, Berlin Heidelberg New York 1982), Chap. 9
3.122 T. Herrmannsdörfer, S. Rehmann, F. Pobell: J. Low Temp. Phys. **104**, 67 (1996)
3.123 G.S. Knapp: J. Appl. Phys. **38**, 1267 (1967); Phys. Lett. A **25**, 114 (1967)
3.124 G. Grüner, A. Zawadowski: *Prog. Low Temp. Phys.* **7B**, 591 (North-Holland, Amsterdam 1978)
3.125 J.P. Franck, F.D. Manchester, D.L. Martin: Proc. R. Soc. London A **263**, 494 (1961)
3.126 J. Eisenmenger, J. Meckler, P. Ziemann: J. Low Temp. Phys. **137**, 167 (2004)
3.127 A.L. Woodcraft: Cryogenics **45**, 626 (2005)
3.128 L. Risegari, M. Barucci, E. Olivieri, E. Pasca, G. Ventura: Cryogenics **44**, 875 (2004)
3.129 A.L. Woodcraft: Cryogenics **45**, 421 (2005)
3.130 F.R. Fickett: J. Phys. F **12**, 1753 (1982)
3.131 J. Peterseim, G. Thummes, H.H. Mende: Z. Metallkd. **70**, 266 (1979)
3.132 S.S. Rosenblum, W.A. Steyert, F.R. Fickett: Cryogenics **17**, 645 (1977)
3.133 Y. Yaeli, S.G. Lipson: J. Low Temp. Phys. **23**, 53 (1976)
3.134 A.C. Ehrlich: J. Mater. Sci. **9**, 1064 (1974)
3.135 Z.S. Basinski, J.S. Dugdale: Phys. Rev. B **32**, 2149 (1985)
3.136 T. Shigematsu, K. Morita, Y. Fujii, T. Shigi, M. Nakamura, M. Yomaguchi: Cryogenics **32**, 913 (1992)
3.137 B.N. Aleksandrov: Sov. J. Low Temp. Phys. **10**, 151 (1984)
3.138 H. Wenzl, J.M. Welter: Metallkd. **65**, 205 (1974)
3.139 D.G. Cahill: Rev. Sci. Instrum. **61**, 802 (1996)
3.140 J.M. Lockhart, R.L. Fagaly, L.W. Lombardo, B. Muhlfelder: Physica B **165&166**, 147 (1990) (Proc. 19th Int. Conf. Low Temp. Phys.)
3.141 G.L. Salinger, J.C. Wheatley: Rev. Sci. Instrum. **32**, 872 (1961)
3.142 R.J. Commandor, C.B.P. Flinn: J. Phys. E **3**, 78 (1969)
3.143 S. Rehmann, T. Herrmannsdörfer, F. Pobell: Cryogenics **35**, 665 (1995)
3.144 J.C. Mester, J.M. Lockhart: Czech. J. Phys. **46**, Suppl. S5, 2751 (1996) (Proc. 21st Int'l Conf. Low Temp. Phys., Prag 1996)
3.145 B. Cabrera: Ph.D. Thesis, Stanford Univ. 1975, and unpubl. report

3.146 Z. Xia, J. Bray-Ali, J. Zhang, B.R. Fink, K.S. White, C.M. Gould, H.M. Bozler: J. Low Temp. Phys. **126**, 655 (2002)
3.147 E.W. Collings, S.C. Hart: Cryogenics **19**, 521 (1979)
3.148 R.P. Hudson, H. Marshak, R.J. Soulen, Jr., D.B. Utton: J. Low Temp. Phys. **20**, 1 (1975)
3.149 D.S. Greywall, Phys. Rev. B **15**, 2604 (1977)
3.150 H. Yano, T. Uchiyama, T. Kato, Y. Minamide, S. Inoue, Y. Miura, T. Mamiya: J. Low Temp. Phys. **78**, 165 (1990)
3.151 R.P. Giffard, R.A. Webb, J.C. Wheatley: J. Low Temp. Phys. **6**, 533 (1972)
3.152 L. Benito, J.I. Arnaudas, A. del Moral: Rev. Sci. Instrum. **77**, 025101 (2006)
3.153 A. Morello, W.G.J. Angenent, G. Frossati, L.J. de Jongh: Rev. Sci. Instrum. **76**, 023902 (2005)
3.154 M. Akatsu, N. Shirakawa: Am. Inst. Phys. Conf. Proc. **850**, 1645 (2006), (Proc. 24th Int'l Conf. Low Temp. Phys., Orlando 2005), ed. by Y. Takano et al.; N. Shirakawa et al.: J. Magn. Magn. Mat. **272–276**, e149 (2004); Polyhedron 24, 2405 (2005)
3.155 T.P. Papageorgiou, L. Bauernfeind, H.F. Braun: J. Low Temp. Phys. **131**, 129 (2003)
3.156 M. McElfresh, S. Li, R. Sager: *Effects of Magnetic Field Uniformity on the Measurement of Superconducting Samples*, Quantum Design (San Diego), Technical Report 1996
3.157 R.D. Biggar, J.M. Parpia: Rev. Sci. Instrum. **69**, 3558 (1998)
3.158 P.A. Crowell, A. Madouri, M. Specht, G. Chaboussant, D. Mailly, L.P. Levy: Rev. Sci. Instrum. **67**, 4161 (1996)
3.159 E. Ohmichi, T. Osada: Rev. Sci. Instrum. **73**, 3022 (2002)

Chapter 4

4.1 G.K. White, P.J. Meeson: *Experimental Techniques in Low Temperature Physics*, 4th edn. (Clarendon, Oxford 2002)
4.2 A.C. Rose-Innes: *Low Temperature Laboratory Techniques* (English University Press, London 1973)
4.3 R. McFee: Rev. Sci. Instrum. **30**, 98 (1959)
4.4 J.E.C. Williams: Cryogenics **3**, 324 (1963)
4.5 Yu. L. Buyanov, A.B. Fradkov, I. Yu. Shebalin: Cryogenics **15**, 193 (1975)
4.6 R.C. Richardson, E.N. Smith: *Experimental Techniques in Condensed Matter Physics at Low Temperatures* (Addison-Wesley, Redwood City, CA 1988)
4.7 D.G. Blair, H. Paik, R.C. Taber: Rev. Sci. Instrum. **46**, 1130 (1975)
4.8 S. Phillip, J.V. Porto, J.M. Parpia: J. Low Temp. Phys. **101**, 581 (1995)

4.9 S.A.J. Wiegers, P.E. Wolf, L. Puech: Physica B **165&166**, 165 (1990) (Proc. 19th Int. Conf. Low Temp. Phys.)

4.10 J.P. Torre, G. Chanin: Rev. Sci. Instrum. **55**, 213 (1983)

4.11 E. Smith, J.M. Parpia, J.R. Beamish: J. Low Temp. Phys. **119**, 507 (2000)

4.12 O.V. Lounasmaa: *Experimental Principles and Methods Below 1K* (Academic, London 1974)

4.13 N.E. Philips: Phys. Rev. **114**, 676 (1959)

4.14 R.W. Hill, G.R. Pickett: Ann. Acad. Sci. Fenn. Ser. A6, **210**, 40 (1966)

4.15 J.H. Colwell: Rev. Sci. Instrum. **40**, 1182 (1969)

4.16 J.A. Birch: J. Phys. C **8**, 2043 (1975)

4.17 P.R. Roach, J.B. Ketterson, B.M. Abraham, P.D. Roach, J. Monson: Rev. Sci. Instrum. **46**, 207 (1975)

4.18 J.D. Siegwarth: Cryogenics **16**, 73 (1976)

4.19 R.M. Mueller, C. Buchal, T. Oversluizen, F. Pobell: Rev. Sci. Instrum. **49**, 515 (1978); R.M. Mueller, C. Buchal, H.R. Folle, M. Kubota, F. Pobell: Cryogenics **20**, 395 (1980)

4.20 T. Tajima, R. Masutomi, A. Yamaguchi, H. Ishimoto: Physica B **329–333**, 1647 (2003), (Proc. 23rd Int'l Conf. Low Temp. Phys., Hiroshima 2002)

4.21 P.C. Ho, R.B. Hallock: J. Low Temp. Phys. **121**, 797 (2000)

4.22 W. Yao, W.A. Knuuttila, K.K. Nummila, J.E. Martikainen, A.S. Oja, O.V. Lounasmaa: J. Low Temp. Phys. **120**, 121 (2000)

4.23 Yu. M. Bunkov: Cryogenics **29**, 938 (1989)

4.24 N.S. Lawson: Cryogenics **22**, 667 (1982)

4.25 K.W. Wittekers, W.A. Bosch, F. Mathu, H.C. Meijer, H. Postma: Cryogenics **29**, 904 (1989)

4.26 E. Schuberth: Rev. Sci. Instrum. **55**, 1486 (1984)

4.27 M. Krusius, D.N. Paulson, J.C. Wheatley: Rev. Sci. Instrum. **49**, 396 (1978)

4.28 K. Gloos, C. Mitschka, F. Pobell, P. Smeibidl: Cryogenics **30**, 14 (1990)

4.29 S.G. O'Hara, A.C. Anderson: Phys. Rev. B **9**, 3730 (1974); Phys. Rev. B **10**, 574 (1974)

4.30 A.C. Anderson, C.B. Satterthwaite, S.C. Smith: Phys. Rev. B **3**, 3762 (1971)

4.31 P.G. Klemens: *Solid State Physics* **7**, 1 (Academic, New York 1958)

4.32 R. Berman: *Thermal Conduction in Solids* (Clarendon, Oxford 1976)

4.33 P. Esquinazi, R. König, B. Valentin, F. Pobell: J. Alloys a. Compounds **211&212**, 27 (1994); R. König, P. Esquinazi, B. Neppert: Phys. Rev. B **51**, 11424 (1995)

4.34 K. Gloos, P. Smeibidl, C. Kennedy, A. Singsass, P. Sekovski, R.M. Mueller, F. Pobell: J. Low Temp. Phys. **73**, 101 (1988)

4.35 I. Didschuns, A.L. Woodcraft, D. Bintley, P.C. Hargreave: Cryogenics **44**, 293 (2004); see this article for references to earlier work on thermal conductance of metal-to-metal joints at low temperatures
4.36 T. Okamoto, H. Fukuyama, H. Ishimoto, S. Ogawa: Rev. Sci. Instrum. **61**, 1332 (1990) (this paper provides references to former publications on the problem of thermal contact between two metals)
4.37 T. Mamiya, H. Yano, T. Uchiyama, S. Inoue, Y. Miura: Rev. Sci. Instr. **59**, 1428 (1988)
4.38 L.J. Salerno, P. Kittel, A.L. Spirak: Cryogenics **33**, 1104 (1993); Cryogenics **34**, 649 (1994)
4.39 S. Yin, P. Hakonen: Rev. Sci. Instrum. **62**, 1370 (1991)
4.40 J. Landau, R. Rosenbaum: Rev. Sci. Instrum. **43**, 1540 (1972)
4.41 H.C. Meijer, G.J.C. Bots, G.M. Coops: Proc. 6th Int. Cryogenic Eng. Conf., Grenoble (1976)
4.42 W.H. Warren Jr., W.G. Bader: Rev. Sci. Instrum. **40**, 189 (1969)
4.43 R.G. Gylling: Acta Polytechnol. Scand. Phys. **81**, 1 (1971)
4.44 H.C. Meijer, C. Beduz, T. Mathu: J. Phys. E **7**, 424 (1974)
4.45 D. Dummer, P. Anderson, W. Weyhmann: Cryogenics **31**, 388 (1991)
4.46 C.L. Reynolds, A.C. Anderson: Cryogenics **16**, 687 (1976)
4.47 E.T. Swartz, R.O. Pohl: Rev. Mod. Phys. **61**, 605 (1989)
4.48 I.M. Khalatnikov: *An Introduction to the Theory of Superfluidity* (Benjamin, New York 1965)
4.49 J.P. Harrison: J. Low Temp. Phys. **37**, 467 (1979)
4.50 T. Nakayama: in *Progress in Low Temperature Physics*, Vol. 7, ed. by D.F. Brewer (North-Holland, Amsterdam 1989), p. 155 and references therein
4.51 A.C. Anderson, J.I. Connolly, J.C. Wheatley: Phys. Rev. A **135**, 910 (1964)
4.52 A.C. Anderson, W.L. Johnson: J. Low Temp. Phys. **7**, 1 (1972)
4.53 J.T. Folinsbee, A.C. Anderson: J. Low Temp. Phys. **17**, 409 (1974)
4.54 J.D. Siegwarth, R. Radebaugh: in *Proc. 13th Int. Conf. Low Temp. Phys.*, ed. by K.D. Timmerhaus, W.J. O'Sullivan, E.F. Hammel (Plenum, New York 1973), p. 398, 401
4.55 W.E. Braun (ed.): *Nonequilibrium Phonon Dynamics* (Plenum, New York 1985)
4.56 A.R. Rutherford, J.P. Harrison, M.J. Stott: J. Low Temp. Phys. **55**, 157 (1984)
4.57 M.C. Maliepard, J.H. Page, J.P. Harrison, R.J. Stubbs: Phys. Rev. B **32**, 6261 (1985)
4.58 C.J. Lambert: J. Low Temp. Phys. **59**, 123 (1985)
4.59 D. Burton, C.J. Lambert: J. Low. Temp. Phys. **64**, 21 (1986)
4.60 K. Andres, W. Sprenger: in *Proc. 14th Int. Conf. Low Temp. Phys.* Vol. 1, ed. by M. Krusius, M. Vuorio (North-Holland, Amsterdam 1975), p. 123

4.61 W.R. Abel, A.C. Anderson, W.C. Black, J.C. Wheatley: Phys. Rev. Lett. **16**, 273 (1966)
4.62 W.C. Black, A.C. Mota, J.C. Wheatley, J.H. Bishop, P.M. Brewster: J. Low Temp. Phys. **4**, 391 (1971)
4.63 J.H. Bishop, D.W. Cutter, A.C. Mota, J.C. Wheatley: J. Low Temp. Phys. **10**, 379 (1973)
4.64 M. Jutzler, A.C. Mota: J. Low Temp. Phys. **55**, 439 (1984)
4.65 D. Marek, A.C. Mota, J.C. Weber: J. Low Temp. Phys. **63**, 401 (1986)
4.66 A.J. Leggett, M. Vuorio: J. Low Temp. Phys. **3**, 359 (1970)
4.67 D.L. Mills, M.T. Beal-Monod: Phys. Rev. A **10**, 343, 2473 (1974)
4.68 O. Avenel, M.P. Berglund, R.G. Gylling, N.E. Phillips, A. Vetleseter, M. Vuorio: Phys. Rev. Lett. **31**, 76 (1973)
4.69 A.I. Ahonen, P.M. Berglund, M.T. Haikala, M. Krusius, O.V. Lounasmaa, M. Paalanen: Cryogenics **16**, 521 (1976)
4.70 A.J. Ahonen, O.V. Lounasmaa, M.C. Veuro: J. Physique **39**, Suppl. 8 (C6), 265 (1978)
4.71 O.E. Vilches, J.C. Wheatley: Rev. Sci. Instrum. **37**, 819 (1988)
4.72 H. Franco, J. Bossy, H. Godfrin: Cryogenics **24**, 477 (1984)
4.73 G. Frossati: J. Physique **39** (C6), 1578 (1978); J. Low Temp. Phys. **87**, 595 (1992)
4.74 D.D. Osheroff, L.R. Corruccini: Phys. Lett. A **82**, 38 (1981)
4.75 G.H. Oh, M. Nakagawa, H. Akimoto, O. Ishikawa, T. Hata, T. Kodama: Physica B **165&166**, 527 (1990) (Proc. 19th Int. Conf. Low Temp. Phys.); G.H. Oh, Y. Ishimoto, T. Kawae, M. Nakogawa, O. Ishikawa, T. Hatoa, T. Kodama, S. Ikehata: J. Low Temp. Phys. **95**, 525 (1994)
4.76 H. Chocholacs: Dissertation, KFA Jülich (1984) (JÜL-Report 1901)
4.77 D.A. Ritchie, J. Saunders, D.F. Brewer: in *Proc. 17th Int. Conf. Low Temp. Phys.* Vol. 2, ed. by U. Eckern, A. Schmid, W. Weber, H. Wühl (North-Holland, Amsterdam 1984), p. 743
4.78 D.D. Osheroff, R.C. Richardson: Phys. Rev. Lett. **54**, 1178 (1985)
4.79 T. Perry, K. de Conde, J.A. Sauls, D.L. Stein: Phys. Rev. Lett. **48**, 1831 (1982)
4.80 G.J. Stecher, Y. Hu, T.J. Gramila, R.C. Richardson: Physica B **165&166**, 525 (1990) (Proc. 19th Int. Conf. Low Temp. Phys.)
4.81 R. König, T. Herrmannsdörfer, A. Schindler, J. Usherov-Marshak: J. Low Temp. Phys. **113**, 969 (1998); J. Usherov-Marshak, A. Schindler, R. König: Physica B **284–288**, 202 (2000), (Proc. 22nd Int'l Conf. Low Temp. Phys., Helsinki 1999)

Chapter 5

5.1 G.K. White, P.J. Meeson: *Experimental Techniques in Low Temperature Physics*, 4th edn. (Clarendon, Oxford 2002)

5.2 A.C. Rose-Innes: *Low Temperature Laboratory Techniques* (English Univ. Press, London 1973)
5.3 P.V.E. McClintock, D.J. Meredith, J.K. Wigmore: *Matter at Low Temperatures* (Blackie, London 1984)
5.4 V. Musilova, P. Hanzelka, T. Kralik, A. Srnka: Cryogenics **45**, 529 (2005)
5.5 H. Luck, Ch. Trepp: Cryogenics **32**, 690, 698, 703 (1992)
5.6 R.C. Richardson, E.N. Smith: *Experimental Techniques in Condensed Matter Physics at Low Temperatures* (Addison-Wesley, Redwood City, CA 1988)
5.7 H. Schmidtchen, T. Gradt, H. Börner, E. Behrendt: Cryogenics **34**, 105 (1994)
5.8 R.M. Mueller, G.G. Ihas, E.D. Adams: Rev. Sci. Instrum. **53**, 373 (1982)
5.9 H. Okamoto, D. Chen: Rev. Sci. Instrum. **72**, 1510 (2001)
5.10 L.E. De Long, O.G. Symco, J.C. Wheatley: Rev. Sci. Instrum. **42**, 147 (1971)
5.11 J.V.D. Maas, P.A. Probst, R. Studi, C. Rizzuto: Cryogenics **26**, 471 (1986)
5.12 S. Wang, D. Avaloff, J.A. Nissen, D.A. Stricker, J.A. Lipa: Am. Inst. Phys. Conf. Proc. **850**, 1565 (2006), (Proc. 24th Int'l Conf. Low Temp. Phys., Orlando, FL, 2005), ed. by Y. Takano et al.
5.13 E.T. Swartz: Rev. Sci. Instrum. **57**, 2848 (1986)
5.14 J. Talpe, V. Bekeris: Cryogenics **29**, 854 (1989)
5.15 A.M. Putman, D.A. Geller, V. Alexis: Physica B **194–196**, 57 (1994) (Proc. 20th Int. Conf. Low Temp. Phys.)
5.16 B.N. Engel, G.G. Ihas, E.D. Adams, C. Fombarlet: Rev. Sci. Instrum. **55**, 1489 (1984)
5.17 A.C. Mota: Rev. Sci. Instrum. **42**, 154 (1971)
5.18 G. Lawes, G.M. Zassenhaus, S. Koch, E.N. Smith, J.D. Reppy, J.M. Parpia: Rev. Sci. Instrum. **69**, 4176 (1998)
5.19 A. Raccanelli, L.A. Reichertz, E. Kreysa: Cryogenics **41**, 763 (2001)
5.20 K.P. Jüngst, E. Süss: Cryogenics **24**, 429 (1984)
5.21 Y.S. Kim, J.S. Park, C.M. Edwards, N.S. Sullivan: Cryogenics **27**, 458 (1987)
5.22 I.V. Velichkov, V.M. Drobin: Cryogenics **30**, 538 (1990)
5.23 D.K. Hilton, J.S. Panek, M.R. Smith, S.W. Van Sciver: Cryogenics **39**, 485 (1999)
5.24 G.M. Cushman, R.M. Gummer, E. Buchanan, O. Jenkins, T. Powers: Rev. Sci. Instrum. **70**, 1575 (1999)
5.25 C. Enss, S. Hunklinger: *Low Temperature Physics*, (Springer, Berlin Heidelberg, New York 2005)
5.26 A.T.A.M. de Waele: *Cryocoolers,* a course text (Eindhoven Univ. Technol. 2005)

5.27 R. Radebaugh: Proc. ICEC16/ICMC, Japan 1966, p. 33 (Elsevier, Oxford 1997)
5.28 A.T.A.M. de Waele: Cryogenics **39**, 13 (1999); Physica B **280**, 479 (2000), (Proc. 22nd Int'l Conf. Low Temp. Phys., Helsinki 1999); see these references for further relevant publications
5.29 R. Radebaugh: Adv. Cryog. Engin. **35**, 1191 (1990); Proc. Inst. Refr. **96**, 1 (1999–2000)
5.30 N. Jiang, U. Lindemann, F. Giebeler, G. Thummes: Cryogenics **44**, 809 (2004)
5.31 K. Watanabe, S. Awaji, M. Motokawa: Physica B **329–333**, 1487 (2003), (Proc. 23rd Int'l Conf. Low Temp. Phys., Hiroshima 2002)
5.32 J.A. Good, R. Hall: Presented at the 24th Int'l Conf. Low Temp. Phys., Orlando, FL, 2005
5.33 T. Ebisu, T. Watanabe: Rev. Sci. Instrum. **61**, 921 (1990)

Chapter 6

6.1 G.K. White, P.J. Meeson: *Experimental Techniques in Low Temperature Physics*, 4th edn. (Clarendon, Oxford 2002)
6.2 A.C. Rose-Innes: *Low Temperature Laboratory Techniques* (English University Press, London 1973)
6.3 O.V. Lounasmaa: *Experimental Principles and Methods Below 1 K* (Academic, London 1974)
6.4 D.S. Betts: *Refrigeration and Thermometry Below One Kelvin* (Sussex University Press, Brighton 1976)
6.5 K.W. Taconis: in *Progress in Low Temperature Physics*, Vol. 3, ed. by C.J. Gorter (North-Holland, Amsterdam 1961), p. 153
6.6 B.N. Eselson, B.G. Lazarev, A.D. Svets: Cryogenics **3**, 207 (1963)
6.7 C.F. Mate, R. Harris-Lowe, W.L. Davis, J.G. Daunt: Rev. Sci. Instrum. **36**, 369 (1965)
6.8 D. Walton: Rev. Sci. Instrum. **37**, 734 (1966)
6.9 A.D. Svets: Cryogenics **6**, 333 (1966)
6.10 M. Fruneau, A. Lacaze, L. Weil: Cryogenics **7**, 135 (1967)
6.11 W. Wiedemann, E. Smolic: Proc. 2nd Int'l Cryogenic Engin. Conf., Brighton 1968, p. 559
6.12 D. Walton, T. Timusk, A.J. Sievers: Rev. Sci. Instrum. **42**, 1265 (1971)
6.13 J.P. Torre, G. Chanin: Rev. Sci. Instrum. **56**, 318 (1985)
6.14 R.C. Richardson, E.N. Smith: *Experimental Techniques in Condensed Matter Physics at Low Temperatures* (Addison-Wesley, Redwood City, CA 1988)
6.15 E.T. Swartz: Rev. Sci. Instrum. **58**, 881 (1987)
6.16 G. Batey, V. Mikheev: J. Low Temp. Phys. **113**, 933 (1998)
6.17 S.G. Lee, H.C. Kwon: Cryogenics **33**, 742 (1994)

6.18 C.J. Hoffmann, F.J. Edeskuty, E.F. Hammel: J. Chem. Phys. **24**, 124 (1956)

Chapter 7

7.1 References to the early work on ^{3}He–^{4}He dilution refigeration can be found in [7.9, 7.10]

7.2 G. Frossati: J. de Physique **39** (C6), 1578 (1978); J. Low Temp. Phys. **87**, 595 (1992)

7.3 G.A. Vermeulen, G. Frossati: Cryogenics **27**, 139 (1987)

7.4 D.J. Cousins, S.N. Fisher, A.M. Guénault, R.P. Haley, J.E. Miller, G.R. Pickett, G.N. Plenderleith, P. Skyba, P.Y.A. Thibault, M.G. Ward: J. Low Temp. Phys. **114**, 547 (1999)

7.5 J.C. Wheatley: Am. J. Phys. **36**, 181 (1968); in *Progress in Low Temperature Physics*, Vol. 6, ed. by C.J. Gorter (North-Holland, Amsterdam 1970) p. 77

7.6 J.C. Wheatley, O.E. Vilches, W.R. Abel: Physics **4**, 1 (1968)

7.7 J.C. Wheatley, R.E. Rapp, R.T. Johnson: J. Low Temp. Phys. **4**, 1 (1971)

7.8 R. Radebaugh, J.D. Siegwarth: Cryogenics **11**, 368 (1971)

7.9 O.V. Lounasmaa: *Experimental Principles and Methods Below 1K* (Academic, London 1974)

7.10 D.S. Betts: *Refrigeration and Thermometry Below One Kelvin* (Sussex University Press, Brighton 1976)

7.11 D.S. Betts: *An Introduction to Millikelvin Technology* (Cambridge University Press, Cambridge 1989)

7.12 C. Ebner, D.O. Edwards: Phys. Rep. **2**, 77 (1971)

7.13 G. Baym, C. Pethick: in *The Physics of Liquid and Solid Helium*, Vol. 2, ed. by K.H. Bennemann, J.B. Ketterson (Wiley, New York 1978) p. 123

7.14 J. Bardeen, G. Baym, D. Pines: Phys. Rev. **156**, 207 (1967)

7.15 I.M. Khalatnikov: *An Introduction to the Theory of Superfluidity* (Benjamin, New York 1965)

7.16 J. Wilks: *The Properties of Liquid and Solid Helium* (Clarendon, Oxford 1967)

7.17 J. Wilks, D.S. Betts: *An Introduction to Liquid Helium*, 2nd edn. (Clarendon, Oxford 1987)

7.18 W.E. Keller: *Helium-Three and Helium-Four* (Plenum, New York 1969)

7.19 W.F. Saam, J.P. Laheurte: Phys. Rev. A **4**, 1170 (1971); M. Nakamura, G. Shiroto, T. Shigematsu, K. Nagao, Y. Fujii, M. Yamaguchi, T. Shigi: Physica B **165&166**, 517 (1990)

7.20 G.E. Watson, J.D. Reppy, R.C. Richardson: Phys. Rev. **188**, 384 (1969)

7.21 S. Yorozu, M. Hiroi, H. Fukuyama, H. Akimoto, H. Ishimoto, S. Ogawa: Phys. Rev. B **45**, 12942 (1992)

7.22 L.D. Landau: Sov. Phys. JETP **3**, 920 (1957); Sov. Phys.-JETP **5**, 101 (1957); Sov. Phys.-JETP **8**, 70 (1959)

7.23 A.A. Abrikosov, I.M. Khalatnikov: Rep. Prog. Phys. **22**, 329 (1959)

7.24 D. Pines, P. Nozières: *The Theory of Quantum Liquids*, Vol. 1 (Benjamin, New York 1966)

7.25 A.C. Anderson, W.R. Roach, R.E. Sarwinski, J.C. Wheatley: Phys. Rev. Lett. **16**, 263 (1966)

7.26 E. Polturak, R. Rosenbaum: J. Low Temp. Phys. **43**, 477 (1981)

7.27 H.C. Chocholacs, R.M. Mueller, J.R. Owers-Bradley, Ch. Buchal, M. Kubota, F. Pobell: Proc. of the 17th Int'l Conf. Low Temp. Phys., Vol. 2, ed. by U. Eckern, A. Schmid, W. Weber, W. Wühl (North-Holland, Amsterdam 1984) p. 1247

7.28 H. Chocholacs: Dissertation, KFA Jülich (1984) (JÜL Report 1901)

7.29 D.S. Greywall: Phys. Rev. B **27**, 2747 (1983); Phys. Rev. B **33**, 7520 (1986)

7.30 J.G.M. Kuerten, C.A.M. Castelijns, A.T.A.M. de Waele, H.M. Gijsman: Cryogenics **25**, 419 (1985)

7.31 J. Landau, J.T. Tough, N.R. Brubaker, D.O. Edwards: Phys. Rev. A **2**, 2472 (1970)

7.32 A. Ghozlan, E.J-.A. Varoquaux: C. R. Acad. Sci. Paris B **280**, 189 (1975)

7.33 S.G. Sydoriak, T.R. Roberts: Phys. Rev. **118**, 901 (1960)

7.34 J.V.D. Maas, P.A. Probst, R. Stubi, C. Rizzuto: Cryogenics **26**, 471 (1986)

7.35 W.P. Kirk, E.D. Adams: Cryogenics **14**, 147 (1974)

7.36 Y. Oda, G. Fujii, T. Ono, H. Nagano: Cryogenics **23**, 139 (1983)

7.37 Y. Oda, G. Fujii, H. Nagano: Cryogenics **18**, 73 (1978)

7.38 D.I. Bradley, T.W. Bradshaw, A.M. Guénault, V. Keith, B.G. Locke-Scobie, I.E. Miller, G.R. Pickett, W.P. Pratt, Jr.: Cryogenics **22**, 296 (1982); Yu. M. Bunkov, A.M. Guénault, D.J. Hayword, B.A. Jackson, C.J. Kennedy, T.R. Nichols, I.E. Miller, G.A. Pickett, M.G. Ward: J. Low Temp. Phys. **83**, 257 (1991); D.J. Bradley, M.R. Follows, J.E. Miller, R. Oswald, M. Ward: Cryogenics **34**, 549 (1994)

7.39 J.D. Siegwarth, R. Radebaugh: Rev. Sci. Instrum. **42**, 1111 (1971); Rev. Sci. Instrum. **43**, 197 (1972)

7.40 F.A. Staas, K. Weiss, A.P. Severijns: Cryogenics **14**, 253 (1974)

7.41 E. ter Haar, R. Wagner, C.M.C.M. van Woerkens, S.C. Steel, G. Frossati, L. Skrbek, M.W. Meisel, V. Bindilatti, A.R. Rodrigues, R.V. Martin, N.F. Oliveira, Jr.: J. Low Temp. Phys. **99**, 151 (1995)

7.42 R.L. Fogaly, D.N. Paulson: Cryogenics **32**, 711 (1992)

7.43 Y. Takano: Rev. Sci. Instrum. **65**, 1657 (1994)

7.44 A.C. Mota: Rev. Sci. Instrum. **42**, 1541 (1971)

7.45 V.A. Mikheev, V.A. Maidanov, N.P. Mihkin: Cryogenics **24**, 190 (1984)
7.46 G. Batey, V. Mikheev: J. Low Temp. Phys. **113**, 933 (1998)
7.47 O. Kirichek, J. Li, L. Linfitt, A. Adams, V. Mikheev: Physica B **329–333**, 1604 (2003), (Proc. 23rd Int'l Conf. Low Temp. Phys., Hiroshima 2002)
7.48 K. Uhlig, W. Hehn: Cryogenics **37**, 279 (1997); K. Uhlig: Cryogenics **42**, 569 (2002)
7.49 K. Uhlig: Cryogenics **42**, 73 (2002), Cryogenics **44**, 53 (2004); K. Uhlig, C. Wang: Adv. Cryog. Engineer. **51**, 939 (2005)
7.50 G.M. Coops, A.T.A.M. de Waele, H.M. Gijsman: Cryogenics **19**, 659 (1979)
7.51 Y. Koike, Y. Morii, T. Igarashi, M. Kubota, Y. Hiresaki, K. Tanida: Cryogenics **39**, 579 (1999)
7.52 K. Matsumoto, Y. Okuda: Cryogenics **33**, 1018 (1993)
7.53 R. König, F. Pobell: J. Low Temp. **97**, 287 (1994); R. König, A. Betat, F. Pobell: J. Low Temp. **97**, 311 (1994)
7.54 E. Suaudeau, E.D. Adams: Cryogenics **30**, 77 (1990)
7.55 A. Sawada, S. Inoue, Y. Masuda: Cryogenics **26**, 486 (1986)
7.56 E. ter Haar, R.V. Martin: Rev. Sci. Instrum. **75**, 3071 (2004)
7.57 R.C. Richardson, E.N. Smith: *Experimental Techniques in Condensed Matter Physics at Low Temperatures* (Addison-Wesley, Redwood City, CA 1988)
7.58 G.K. White, P.J. Meeson: *Experimental Techniques in Low Temperature Physics*, 4th edn. (Clarendon, Oxford 2002)

Chapter 8

8.1 Yu. D. Anufriev: Sov. Phys. JETP Lett. **1**, 155 (1965)
8.2 R.T. Johnson, R. Rosenbaum, O.G. Symco, J.C. Wheatley: Phys. Rev. Lett. **22**, 449 (1969)
8.3 R.T. Johnson, J.C. Wheatley: J. Low Temp. Phys. **2**, 424 (1970)
8.4 D.D. Osheroff, R.C. Richardson, D.M. Lee: Phys. Rev. Lett. **28**, 885 (1972)
8.5 D.D. Osheroff, W.J. Gully, R.C. Richardson, D.M. Lee: Phys. Rev. Lett. **29**, 920 (1972)
8.6 W.P. Halperin, F.B. Rasmussen, C.N. Archie, R.C. Richardson: J. Low Temp. Phys. **31**, 617 (1978)
8.7 E.R. Grilly: J. Low Temp. Phys. **4**, 615 (1971); J. Low Temp. Phys. **11**, 243 (1973)
8.8 D.S. Greywall, P.A. Busch: J. Low Temp. Phys. **46**, 451 (1982)
8.9 D.S. Greywall: Phys. Rev. B **31**, 2675 (1985)
8.10 D.S. Greywall: Phys. Rev. B **27**, 2747 (1983); Phys. Rev. B **33**, 7520 (1986)
8.11 D.S. Greywall: Phys. Rev. B **15**, 2604 (1977)

8.12 M. Roger, J. Hetherington, J.M. Delrieu: Rev. Mod. Phys. **55**, 1 (1983)
8.13 M.C. Cross, D.S. Fisher: Rev. Mod. Phys. **57**, 881 (1985)
8.14 D.D. Osheroff: J. Low Temp. Phys. **87**, 297 (1992)
8.15 O.V. Lounasmaa: *Experimental Principles and Methods Below 1 K* (Academic, London 1974)
8.16 D.S. Betts: *An Introduction to Millikelvin Technology* (Cambridge University Press, Cambridge 1989)
8.17 J.R. Sites, D.D. Osheroff, R.C. Richardson, D.M. Lee: Phys. Rev. Lett. **23**, 836 (1969)
8.18 L.R. Corruccini, D.D. Osheroff, D.M. Lee, R.C. Richardson: J. Low Temp. Phys. **8**, 229 (1972)
8.19 G. Frossati: J. Phys. **41** (C7), 95 (1980); Jpn. J. Appl. Phys. **26**, 1833 (1987) (Proc. 18th Int. Conf. Low Temp. Phys., Kyoto 1987)
8.20 G.A. Vermeulen, S.A.J. Wiegers, C.C. Kranenburg, R. Jochemsen, G. Frossati: Can. J. Phys. **65**, 1560 (1987); Jpn. J. Appl. Phys. **26**, 215 (1987) (Proc. 18th Int. Conf. Low Temp. Phys., Kyoto 1987)
8.21 D.D. Kranenburg, L.P. Roobol, R. Jochemsen, G. Frossati: J. Low Temp. Phys. **77**, 371 (1989)
8.22 E.N. Smith, H.M. Bozler, W.S. Truscott, R. Gianetta, R.C. Richardson, D.M. Lee: Proc. 14th Int. Conf. Low Temp. Phys., Vol. 4, ed. by M. Krusius, M. Vuorio (North-Holland, Amsterdam 1975), p. 9
8.23 R.B. Kummer, R.M. Mueller, E.D. Adams: J. Low Temp. Phys. **27**, 319 (1977)

Chapter 9

9.1 O.V. Lounasmaa: *Experimental Principles and Methods Below 1 K* (Academic, London 1974)
9.2 D.S. Betts: *Refrigeration and Thermometry Below One Kelvin* (Sussex University Press, Brighton 1976)
9.3 R.P. Hudson: *Principles and Application of Magnetic Cooling* (North-Holland, Amsterdam 1972)
9.4 D.S. Betts: *An Introduction to Millikelvin Technology* (Cambridge University Press, Cambridge 1989)
9.5 B.I. Bleaney, B. Bleaney: *Electricity and Magnetismus*, 3rd ed. (Oxford University Press, Oxford 1976)
9.6 W.F. Giauque, R.A. Fisher, E.W. Hornung, G.E. Brodale: J. Chem. Phys. **58**, 2621 (1973); R.A. Fisher, E.W. Hornung, G.E. Brodale, W.F. Giauque: J. Chem. Phys. **58**, 5584 (1973)
9.7 O.E. Vilches, J.C. Wheatley: Rev. Sci. Instrum. **37**, 819 (1966); Phys. Rev. **148**, 509 (1966)
9.8 W.R. Abel, A.C. Anderson, W.C. Black, J.C. Wheatley: Physics **1**, 337 (1965)

9.9 M. Kolac, K. Svec, R.S. Safrata, J. Matas, T. Tethal: J. Low Temp. Phys. **11**, 297 (1973)
9.10 D.N. Paulson, M. Krusius, J.C. Wheatley, R.S. Safrata, M. Kolac, T. Tethal, K. Svec, J. Matas: J. Low Temp. Phys. **34**, 63 (1979); J. Low Temp. Phys. **36**, 721 (E) (1979); R.S. Safrata, M. Kolac, J. Matos, M. Odehnal, K. Svec: J. Low Temp. Phys. **41**, 405 (1980)
9.11 J.M. Parpia, W.P. Kirk, P.S. Kobiela, Z. Olejniczak: J. Low Temp. Phys. **60**, 57 (1985)
9.12 J.H. Bishop, D.W. Cutter, A.C. Mota, J.C. Wheatley: J. Low Temp. Phys. **10**, 379 (1973)
9.13 G.K. White, P.J. Meeson: *Experimental Techniques in Low Temperature Physics*, 4th edn. (Clarendon, Oxford 2002)
9.14 A.C. Rose-Innes: *Low Temperature Laboratory Techniques* (English University Press, London 1973)
9.15 C. Hagmann, P.L. Richards: Cryogenics **35**, 303 (1995)
9.16 P.J. Shirron, M.J. DiPirro: Am. Inst. Phys. Conf. Proc. **850**, 1573 (2006), (Proc. 24th Int'l Conf. Low Temp. Phys., Orlando, FL, 2005), ed. by Y. Takano et al.

Chapter 10

10.1 M.T. Huiku, T.A. Jyrkkiö, J.M. Kyynäräinen, M.T. Loponen, O.V. Lounasmaa, A.S. Oja: J. Low Temp. Phys. **62**, 433 (1986); P. Hakonen, O.V. Lounasmaa, A.S. Oja: J. Magn. Magn. Mater. **100**, 394 (1991); A.S. Oja, O.V. Lounasma: Rev. Mod. Phys. **69**, 1 (1997)
10.2 N. Kurti, F.N.H. Robinson, F.E. Simon, D.A. Spohr: Nature **178**, 450 (1956)
10.3 N. Kurti: Cryogenics **1**, 2 (1960)
10.4 E.B. Osgood, J.M. Goodkind: Cryogenics **6**, 54 (1966); Phys. Rev. Lett. **18**, 894 (1967)
10.5 O.G. Symco: J. Low Temp. Phys. **1**, 451 (1969)
10.6 R.G. Gylling: Acta Polytech. Scand. Phys. **81**, 1 (1971); P.M. Berglund, G.J. Ehnholm, R.G. Gylling, O.V. Lounasmaa, R.P. Sovik: Cryogenics **12**, 297 (1972); P.M. Berglund, H.K. Collan, G.J. Ehnholm, R.G. Colling, O.V. Lounasmaa: J. Low Temp. Phys. **6**, 357 (1972)
10.7 O.V. Lounasmaa: *Experimental Principles and Methods Below 1 K* (Academic, London 1974)
10.8 D.S. Betts: *Refrigeration and Thermometry Below One Kelvin* (Sussex University Press, Brighton 1976)
10.9 K. Andres, O.V. Lounasmaa: in *Progress in Low Temperature Physics*, Vol. 8, ed. by D.F. Brewer (North-Holland, Amsterdam 1982) p. 221

10.10 G.R. Pickett: Rep. Prog. Phys. **51**, 1295 (1988); Physica B **280**, 467 (2000)
10.11 T. Uchiyama, T. Mamiya: Rev. Sci. Instrum. **58**, 2192 (1987)
10.12 J.P. Harrison: J. Low Temp. Phys. **37**, 467 (1979)
10.13 see references given in F. Giazotto et al.: Rev. Mod. Phys. **78**, 217 (2006), p. 223
10.14 A. Abragam, M. Goldman: *Nuclear Magnetism – Order and Disorder* (Clarendon, Oxford 1982)
10.15 A. Abragam: *Principles of Nuclear Magnetism* (Clarendon, Oxford 1983)
10.16 M. Goldman: *Spin Temperature and Nuclear Magnetic Resonance in Solids* (Clarendon, Oxford 1970)
10.17 C.P. Slichter: *Principles of Magnetic Resonance*, 3rd edn., Springer Ser. Solid-State Sci. Vol. 1 (Springer, Berlin Heidelberg New York 1990)
10.18 F. Bacon, J.A. Barclay, W.D. Brewer, D.A. Shirley, J.E. Templeton: Phys. Rev. B5, 2397 (1972)
10.19 F. Shibata, Y. Hamano: J. Phys. Soc. Jpn **52**, 1410 (1983)
10.20 E. Klein: in *Low Temperature Nuclear Orientation*, ed. by N.J. Stone, H. Postma (North-Holland, Amsterdam 1986) p. 579
10.21 G. Eska: J. Low Temp. Phys. **73**, 207 (1988); in (4) of this reference x should be replaced by x/I
10.22 G.C. Carter, L.H. Bennett, D.J. Kahon: Metallic shifts in NMR, Pt. I; in Progress in Material Science, Vol. 20 (Pergamon, Oxford 1977)
10.23 D.L. Martin: Phys. Rev. **38**, 5357 (1973); Phys. Rev. B **17**, 1670 (1978)
10.24 T.A. Knuuttila, J.T. Tuoriniemi, K. Lefmann, K.I. Juntunen, F.B. Rasmussen, K.K. Nummila: J. Low Temp. Phys. **123**, 65 (2001)
10.25 D.J. Bradley, A.M. Guénault, V. Keith, C.J. Kennedy, J.E. Miller, S.G. Musset, G.R. Pickett, W.P. Pratt, Jr.: J. Low Temp. Phys. **57**, 359 (1984)
10.26 F. Pobell: Physica **109, 110 B,C**, 1485 (1982) (Proc. 16th Int. Conf. Low Temp. Phys.); J. Low Temp. Phys. **87**, 635 (1992)
10.27 R.M. Mueller: Hyperfine Interact. **22**, 211 (1985)
10.28 R.M. Mueller, C. Buchal, H.R. Folle, M. Kubota, F. Pobell: Cryogenics **20**, 395 (1980)
10.29 H. Ishimoto, N. Nishida, T. Furubayashi, M. Shinohara, Y. Takano, Y. Miura, K. Ono: J. Low Temp. Phys. **55**, 17 (1984)
10.30 K. Gloos, P. Smeibidl, C. Kennedy, A. Singsaas, P. Sekovski, R.M. Mueller, F. Pobell: J. Low Temp. Phys. **73**, 101 (1988)
10.31 E. Nazaretski, V.O. Kostroun, S. Dimov, R.O. Pohl, J.M. Parpia: J. Low Temp. Phys. **137**, 609 (2004)
10.32 W. Yao, T.A. Knuuttila, K.K. Nummila, J.E. Martikainen, A.S. Oja, O.V. Lounasmaa: J. Low Temp. Phys. **120**, 121 (2000)

10.33 J. Nyéki, V. Pavlik, P. Skyba, N. Smolka, A. Feher: Cryogenics **34**, 961 (1994); providing references to earlier publications on radiation shields
10.34 G.A. Vermeulen, G. Frossati: Cryogenics **27**, 139 (1987)
10.35 O. Avenel, G.G. Ihas, E. Varoquaux: J. Low Temp. Phys. **93**, 1031 (1993)
10.36 D.I. Bradley, T.W. Bradshaw, A.M. Guénault, V. Keith, B.G. Locke-Scobie, I.E. Miller, G.R. Pickett, W.P. Pratt, Jr.: Cryogenics **22**, 296 (1982); Yu. M. Bunkov, A.M. Guénault, D.J. Hayward, D.A. Jackson, C.J. Kennedy, T.A. Nichols, I.E. Miller, R.G. Pichett, M.G. Ward: J. Low Temp. Phys. **83**, 257 (1991); D.J. Cousins, S.N. Fisher, A.M. Guénault, R.P. Haley, I.E. Miller, G.R. Pickett, G.N. Plenderleith, P. Skyba, P.Y.A. Thibault, M.G. Ward: J. Low Temp. Phys. **114**, 547 (1999)
10.37 R.C. Richardson, E.N. Smith: *Experimental Techniques in Condensed Matter Physics at Low Temperatures* (Addison-Wesley, Redwood, CA 1988)
10.38 W.P. Kirk, M. Twerdochlib: Rev. Sci. Instrum. **49**, 765 (1978)
10.39 D.J. Cousins, S.E. May, J.H. Naish, P.M. Walmsley, A.J. Golov: J. Low Temp. Phys. **134**, 419 (2004)
10.40 A.C. Tims, R.L. Davidson, R.W. Timme: Rev. Sci. Instrum. **46**, 554 (1975)
10.41 U. Angerer, G. Eska: Cryogenics **24**, 515 (1984); G. Eska, E. Schuberth: Jpn. J. Appl. Phys. Suppl. **26–3**, 435 (1987) (Proc. 18th Conf. Low Temp. Phys.); G. Eska: in *Quantum Fluids and Solids-1989*, ed. by G.G. Ihas, Y. Takano, AIP Conf. Proc. **194**, 316 (AIP, New York 1989)
10.42 B. Schröder-Smeibidl, P. Smeibidl, G. Eska, F. Pobell: J. Low Temp. Phys. **85**, 311 (1991)
10.43 F.R. Fickett: IEEE Trans. MAG-**19**, 228 (1983)
10.44 Y. Iwasa, E.J. McNiff, R.H. Bellis, K. Sato: Cryogenics **33**, 836 (1993)
10.45 S.Y. Shen, J.B. Ketterson, W.P. Halperin: J. Low Temp. Phys. **31**, 193 (1978)
10.46 R.C.M. Dow, A.M. Guénault, G.R. Pickett: J. Low Temp. Phys. **47**, 477 (1982)
10.47 I.A. Gachechiladze, D.V. Pavlov, A.V. Pantsulaya: Cryogenics **26**, 242 (1986)
10.48 P. Strehlow: Am. Inst. Phys. Conf. Proc. **850**, 1575 (2006), (Proc. 24th Int'l Conf. Low Temp. Phys., Orlando, FL, 2005), ed. by Y. Takano et al.
10.49 M. Schwark, F. Pobell, W.P. Halperin, Ch. Buchal, J. Hanssen, M. Kubota, R.M. Mueller: J. Low Temp. Phys. **53**, 685 (1983)
10.50 M. Kolac, B.S. Neganov, S. Sahling: J. Low Temp. Phys. **59**, 547 (1985); J. Low Temp. Phys. **63**, 459 (1986)

10.51 R. Lässer: *Tritium and Helium-3 in Metals*, Springer Ser. Mater. Sci. Vol. 9 (Springer, Berlin Heidelberg 1989)
10.52 M. Schwark, M. Kubota, R.M. Mueller, F. Pobell: J. Low Temp. Phys. **58**, 171 (1985)
10.53 J. Zimmermann, G. Weber: Phys. Rev. Lett. **46**, 661 (1984); J. Zimmermann: Cryogenics **24**, 27 (1984)
10.54 S. Sahling, S. Abens, T. Eggert: J. Low Temp. Phys. **127**, 215 (2002)
10.55 C. Köckert, S. Abens, U. Escher, B. Kluge, A. Gladun, S. Sahling, M. Schneider: J. Low Temp. Phys. **124**, 477 (2001)
10.56 W.A. Phillips (ed.): *Amorphous Solids*, Topics Cur. Phys. Vol. 24 (Springer, Berlin Heidelberg New York 1981)
10.57 P. Esquinazi: *Tunneling Systems in Amorphous and Crystalline Solids* (Springer, Berlin Heidelberg New York 1998)
10.58 C. Enss, S. Hunklinger: *Low Temperature Physics* (Springer, Berlin Heidelberg New York 2005)
10.59 A. Nittke, M. Scherl, P. Esquinazi, W. Lorenz, J. Li, F. Pobell: J. Low Temp. Phys. **98**, 517 (1995)
10.60 Th. Wagner, S. Götz, G. Eska: Cryogenics **34**, 655 (1994)
10.61 N.E. Phillips: Phys. Rev. **118**, 644 (1960)
10.62 Y.H. Tang, E.D. Adams, K. Uhlig, D.N. Bittner: J. Low Temp. Phys. **60**, 351 (1985)
10.63 Y. Karaki, M. Kubota, H. Ishimoto: Physica B **194–196**, 461 (1994) (Proc. 20th Int. Conf. Low Temp. Phys.)
10.64 K. Andres, B. Millimill: J. de Phys. **39**(C6), 796 (1978)
10.65 T. Herrmannsdörfer, B. Schröder-Smeibidl, P. Smeibidl, F. Pobell: Phys. Rev. Lett. **74**, 1665 (1995); T. Herrmannsdörfer, F. Pobell: J. Low Temp. Phys. **100**, 253 (1995)
10.66 W. Heeringa, R. Aures, R. Maschuw, F.K. Schmidt: Cryogenics **25**, 369 (1985); J. Li, G.A. Sheshin, I. Roggatz, F. Pobell: J. Low Temp. Phys. **102**, 61 (1996)
10.67 W. Wendler, T. Herrmannsdörfer, S. Rehmann, F. Pobell: Europhys. Lett. **38**, 619 (1997); J. Low Temp. Phys. **111**, 99 (1998)
10.68 K. Andres, E. Bucher: J. Appl. Phys. **42**, 1522 (1971); J. Low Temp. Phys. **9**, 267 (1972)
10.69 K. Andres: Cryogenics **18**, 473 (1973)
10.70 K. Andres, S. Darak: Physica **86–88**B,C, 1071 (1977)
10.71 H.R. Folle, M. Kubota, Ch. Buchal, R.M. Mueller, F. Pobell: Z. Phys. B**41**, 223 (1981); M. Kubota, H.R. Folle, Ch. Buchal, R.M. Müller, F. Pobell: Phys. Rev. Lett. **45**, 1812 (1980)
10.72 T. Herrmannsdörfer, H. Uniewski, F. Pobell: Phys. Rev. Lett. **72**, 148 (1994); J. Low Temp. Phys. **97**, 189 (1994)
10.73 J. Babcock, J. Kiely, T. Manley, W. Weyhmann: Phys. Rev. Lett. **43**, 380 (1979)
10.74 M. Reiffers, K. Flachbart, S. Janos, A.B. Beznosov, G. Eska: Phys. Status Solidi **B109**, 369 (1982)

10.75 H.C. Meijer, G.J.C. Bots, H. Postma: Physica **107B**, 607 (1981)
10.76 J. Xu, O. Avenel, J.S. Xia, M.F. Xu, T. Lang, P.L. Moyland, W. Ni, E.D. Adams, G.G. Ihas, M.W. Meisel, N.S. Sullivan, Y. Takano: J. Low Temp. Phys. **89**, 719 (1992)
10.77 V.V. Dimitriev, J.V. Kosarev, O.V. Ponarin, R. Scheibel: J. Low Temp. Phys. **113**, 945 (1998)
10.78 S.N. Ytterboe, P.D. Saundry, L.J. Friedman, M.D. Daybell, C.M. Gould, H.M. Bozler: Phys. Rev. B**42**, 4752 (1990)
10.79 J.P. Carney, A.M. Guénault, G.R. Pickett, G.F. Spencer: Phys. Rev. Lett. **62**, 3042 (1989)
10.80 D.S. Greywall: Phys. Rev. B**31**, 2675 (1985)
10.81 S.A. Wiegers, T. Hata, C.C. Kranenburg, P.G. van de Haar, R. Jochemsen, G. Frossati: Cryogenics **30**, 770 (1990)
10.82 P.G. van de Haar, G. Frossati (Univ. Leiden): Unpublished (1992)
10.83 H. Yano, T. Uchiyama, T. Kato, Y. Minamide, S. Inoue, Y. Miura, T. Mamiya: J. Low Temp. Phys. **78**, 165 (1990)
10.84 G. Frossati: J. Physique **39**(C6), 1578 (1978); J. Low Temp. Phys. **87**, 595 (1992)
10.85 G. Bernstein, S. Labov, D. Landis, N. Madden, J. Millett, E. Silver, P. Richards: Cryogenics **31**, 99 (1991)
10.86 R. König, P. Esquinazi, F. Pobell: J. Low Temp. Phys. **89**, 465 (1992); R. König, A. Betat, F. Pobell: J. Low Temp. Phys. **97**, 311 (1994)
10.87 G.H. Oh, Y. Ishimoto, T. Kawae, M. Nakagawa, O. Ishikawa, T. Hata, T. Kodama: J. Low Temp. Phys. **95**, 525 (1994)
10.88 J. Tuoriniemi, J. Martikainen, E. Pentti, A. Seledash, S. Boldarev, G. Pickett: J. Low Temp. Phys. **129**, 531 (2002)
10.89 T. Suzuki, H. Kondo, H. Yano, S. Abe, Y. Miura, T. Mamiya: Europhys. Lett. **16**, 477 (1991)

Chapter 11

11.1 IPTS-68: Metrologia **12**, 1 (1976)
11.2 EPT-76: Metrologia **15**, 65 (1979)
11.3 M. Durieux, D.N. Astrov, W.R.G. Kemp. C.A. Swenson: Metrologia **15**, 57 (1979)
11.4 H. Preston-Thomas: Metrologia **27**, 3, 107 (1990)
11.5 B.W. Magnum: J. Res. Natl. Inst. Stand. Technol. **95**, 69 (1990); B.W. Magnum, G.T. Furukawa: J. Res. Natl. Inst. Stand. Technol. Technical Note 1265 (1990)
11.6 G. Schuster, D. Hechtfischer, B. Fellmuth: Reports Prog. Phys. **57**, 187 (1994); J. Fischer, B. Fellmuth: Reports Prog. Phys. **68**, 1043 (2005)
11.7 T. Quinn: *Temperature*, 2nd edn. (Academic, London 1990)
11.8 *Temperature: Its Measurement and Control In Science and Industry*, Vol. 7, ed. by D.C. Ripple (Am. Inst. Phys., New York 2003)

11.9 J. Engert, B. Fellmuth: in Ref. 11.8, p. 113; J. Engert, B. Fellmuth, A. Hoffmann: J. Low Temp. Phys. **134**, 425 (2004)

11.10 R.L. Rusby, M. Durieux, A.L. Reesink, R.P. Hudson, G. Schuster, M. Kühne, W.E. Fogle, R.J. Soulen, E.D. Adams: J. Low Temp. Phys. **126**, 633 (2002); in Ref. 11.8, p. 77; the latter reference contains a table of the melting pressures as a function of temperatures and the corresponding $(dP/dT)_{melt}$-values

11.11 R.A. Scribner, E.D. Adams: Rev. Sci. Instrum. **41**, 287 (1970)

11.12 W.P. Halperin, F.B. Rasmussen, C.N. Archie, R.C. Richardson: J. Low Temp. Phys. **31**, 617 (1978)

11.13 D.S. Greywall, P.A. Busch: J. Low Temp. Phys. **46**, 451 (1982); D.S. Greywall: Phys. Rev. B **31**, 2675 (1985)

11.14 G. Schuster, D. Hechtfischer: in *Temperature: Its Measurement and Control in Science and Technology*, Vol. 6, ed. by F.J. Schooley (AIP, New York 1992), p. 97, 107

11.15 B. Fellmuth, D. Hechtfischer, A. Hoffmann: in Ref. 11.8, p. 71

11.16 G. Schuster, A. Hoffmann, D. Hechtfischer: PTB Bericht PTB-ThEx-21 (Phys. Techn. Bundesanstalt, Braunschweig 2001)

11.17 R.J. Soulen Jr., W.E. Fogle, J.H. Colwell: J. Low Temp. Phys. **94**, 385 (1994)

11.18 W.E. Fogle, R.J. Soulen Jr., J.H. Colwell: in *Temperature: Its Measurement and Control in Science and Technology*, Vol. 6, ed. by F.J. Schooley (AIP, New York 1992), p. 85, 91, 101

11.19 W. Ni, J.S. Xia, E.D. Adams, P.S. Haskins, J.E. McKisson: J. Low Temp. Phys. **99**, 167 (1995); J. Low Temp. Phys. **101**, 305 (1995)

11.20 D.S. Greywall: Phys. Rev. B **27**, 2747 (1983); Phys. Rev. B **33**, 7520 (1986)

11.21 G. Schuster, D. Hechtfischer, A. Hoffmann: in Ref. 11.8, p. 83

11.22 R. Rusby et al.: in Ref. 11.8, p. 89; Physica B **329–333**, 1564 (2003), (Proc. 23rd Int'l Conf. Low Temp. Phys., Hiroshima 2002)

11.23 L. Pitre, Y. Hermier, G. Bonnier: in Ref. 11.8, p. 101

11.24 E.R. Grilly: J. Low Temp. Phys. **4**, 615 (1971); J. Low Temp. Phys. **11**, 243 (1973)

11.25 J.F. Schooley, R.J. Soulen Jr., G.A. Evans Jr.: Natl. Bureau of Standards, US Dept. Commerce, Spec. Publ. pp. 260–44 (1972)

11.26 R.J. Soulen, R.B. Dove: *Standard Reference Materials: SRM 768: Temperature Reference Standard for Use Below 0.5 K*. Natl. Bureau of Standards, US Dept. Commerce; Spec. Publ. pp. 260–262 (1979); J.H. Colwell, W.E. Fogle, R.J. Soulen, Jr.: *Proc. 17th Int'l Conf. Low Temp. Phys.*, ed. by E. Eckern, A. Schmid, W. Weber, H. Wühl (North-Holland, Amsterdam 1984) p. 395

11.27 R.J. Soulen, Jr., H. Marshak: Cryogenics **20**, 408 (1980)

11.28 W.A. Bosch, J.J. v.d. Hark, J. Pöll, R. Jochemsen: J. Low Temp. Phys. **138**, 935 (2005)

11.29 W.A. Bosch, J. Engert, J.J.M. v.d. Hark, X.Z. Liu, R. Jochemsen: Am. Inst. Phys. Conf. Proc. **850**, 1589 (2006), (Proc. 24th Int'l Conf. Low Temp. Phys., Orlando, FL, 2005), ed. by Y. Takano et al.
11.30 S. Schöttl et al.: J. Low Temp. Phys. **138**, 941 (2005)
11.31 R.V. Duncan, G. Ahlers: in *Temperature: Its Measurement and Control in Science and Technology*, Vol. 6, ed. by F.J. Schooly (AIP, New York 1992), p. 243; G. Ahlers: Phys. Rev. A **3**, 696 (1971)
11.32 P. Lin, Y. Mao, L. Yu, Q. Zhang, C. Hong: Cryogenics **42**, 443 (2002)
11.33 K.H. Mueller, G. Ahlers, F. Pobell: Phys. Rev. B **14**, 2096 (1976)

Chapter 12

12.1 G.K. White, P.J. Meeson: *Experimental Techniques in Low Temperature Physics*, 4th edn. (Clarendon, Oxford 2002)
12.2 A.C. Rose-Innes: *Low Temperature Laboratory Techniques* (English Universities Press, London 1973)
12.3 O.V. Lounasmaa: *Experimental Principles and Methods Below 1 K* (Academic, London 1974)
12.4 D.S. Betts: *Refrigeration and Thermometry Below One Kelvin* (Sussex University Press, Brighton 1976)
12.5 D.S. Betts: *An Introduction to Millikelvin Technology* (Cambridge University Press, Cambridge 1989)
12.6 R.C. Richardson, E.N. Smith: *Experimental Techniques in Condensed Matter Physics at Low Temperatures* (Addison-Wesley, Redwood City, CA 1988)
12.7 Former review articles on thermometry below 0.3 K: R.P. Hudson, H. Marshak, R.J. Soulen, Jr., D.B. Utton: J. Low Temp. Phys. **20**, 1 (1975); D.S. Parker, L.R. Corruccini: Cryogenics **15**, 499 (1975); R.C. Richardson: Physica B **90**, 47 (1977)
12.8 Many thermometric techniques discussed in this chapter are described in various articles in the following conference proceedings: *Temperature: Its Measurement and Control in Science and Industry*, Vol. 4, Parts 2, 3 ed. by H.H. Plumb (Instrument Society of America, Pittsburgh, PA 1972); Vol. 5, ed. by J.F. Schooley (Am. Inst. Phys., New York 1982); Vol. 6, ed. by J.F. Schooley (Am. Inst. Phys., New York 1992); Vol. 7, ed. by D.C. Ripple (Am. Inst. Phys., New York 2003)
12.9 L.G. Rubin, B.L. Brandt, H.H. Sample: Cryogenics **22**, 491 (1982); L.G. Rubin: Cryogenics **37**, 341 (1997)
12.10 G. Schuster, D. Hechtfischer, B. Fellmuth: Reports Prog. Phys. **57**, 187 (1994); J. Fischer, B. Fellmuth: Reports Prog. Phys. **68**, 1043 (2005)
12.11 T. Quinn: *Temperature*, 2nd edn. (Academic, London 1990)
12.12 T.R. Roberts, S.G. Sydoriak: Phys. Rev. **102**, 304 (1956)
12.13 A. Freddi, I. Modena: Cryogenics **8**, 18 (1968)

12.14 J. Engert, B. Fellmuth: Ref. 12.8, Vol. 7, p. 113; J. Engert, B. Fellmuth, A. Hoffmann: J. Low Temp. Phys. **134**, 425 (2004)
12.15 V. Steinberg, G. Ahlers: J. Low Temp. Phys. **53**, 255 (1983)
12.16 D.S. Greywall, P.A. Busch: Rev. Sci. Instrum. **51**, 509 (1980)
12.17 K.H. Mueller, G. Ahlers, F. Pobell: Phys. Rev. B **14**, 2096 (1976)
12.18 F. Pavese: J. Chem. Thermodyn. **25**, 1351 (1993)
12.19 R.L. Rusby, M. Durieux, A.L. Reesink, R.P. Hudson, G. Schuster, M. Kühne, W.E. Fogle, R.J. Soulen, E.D. Adams: J. Low Temp. Phys. **126**, 633 (2002); Ref. 12.8, Vol. 7, p. 77
12.20 R.J. Soulen Jr., W.E. Fogle, J.H. Colwell: J. Low Temp. Phys. **94**, 385 (1994); W.E. Fogle, R.J. Soulen Jr., J.H. Colwell: in [Ref. 12.8, Vol. 6, p. 85, 91, 101]
12.21 W. Ni, J.S. Xia, E.B. Adams, P.S. Haskins, J.E. McKisson: J. Low Temp. Phys. **99**, 167 (1995); J. Low Temp. Phys. **101**, 305 (1995)
12.22 G. Schuster, D. Hechtfischer: in Ref. 12.8, Vol. 6, p. 97, 107; G. Schuster, A. Hoffmann, D. Hechtfischer: PTB-Bericht PTB-ThEx-21 (Physikal.-Techn. Bundesanstalt, Braunschweig 2001)
12.23 E.D. Adams: in Ref. 12.8, Vol. 7, p.107
12.24 H. Fukuyama, H. Ishimoto, T. Tazaki, S. Ogawa: Phys. Rev. B **36**, 8921 (1987)
12.25 D.S. Greywall, P.A. Busch: J. Low Temp. Phys. **46**, 451 (1982); D.S. Greywall: Phys. Rev. B **31**, 2675 (1985)
12.26 A.N. Ganshin, V.N. Grigor'ev, V.A. Maidanov, A.A. Penzev, E. Ya. Rudavskii, A.S. Rybalko, E.V. Syrnikov: Low Temp. Phys. **27**, 509 (2001)
12.27 C.C. Kranenburg, S.A.J. Wiegers, P.G. Van de Haar, R. Jochemsen, G. Frossati: Jpn J. Appl. Phys. **26**, 215 (1987) (Proc. 18th Int'l Conf. Low Temp. Phys.)
12.28 L.R. Roobol, P. Remeijer, C.M. van Woerkens, W. Ockers, S.C. Steel, R. Jochemsen, G. Frossati: Physica B **194–196**, 741 (1994), (Proc. 20th Int'l Conf. Low Temp. Phys., Eugene, Oregon 1993)
12.29 H. Fukuyama, K. Yawata, D. Ito, H. Ikegami, H. Ishimoto: Physica B **329–333**, 1560 (2003), (Proc. 23rd Int'l Conf. Low Temp. Phys., Hiroshima 2002)
12.30 L.S. Goldner, N. Mulders, G. Ahlers: in [Ref. 12.8, Vol. 6, p. 113]
12.31 R. Rosenbaum: Rev. Sci. Instrum. **39**, 890 (1968); Rev. Sci. Instrum. **40**, 577 (1969)
12.32 Y. Maeno, H. Haucke, J.C. Wheatley: Rev. Sci. Instrum. **54**, 946 (1983)
12.33 H. Armbrüster, W. Kirk, D. Chesire: in [Ref. [12.8], Vol. 5, p. 1025]; Physica **107B**, 335 (1981)
12.34 J. Chaussy, Ph. Gandit, K. Mato, A. Ravex: J. Low Temp. Phys. **49**, 167 (1982)
12.35 P.A. Schroeder, C. Uher: Phys. Rev. B **18**, 3884 (1978)
12.36 D.J. Bradley, A.M. Guénault, V. Keith, G.R. Pickett, W.P. Pratt, Jr.: J. Low Temp. Phys. **45**, 357 (1981)

12.37 R.P. Peters, Ch. Buchal, M. Kubota, R.M. Mueller, F. Pobell: Phys. Rev. Lett. **53**, 1108 (1984)
12.38 T. Herrmannsdörfer, S. Rehmann, W. Wendler, F. Pobell: J. Low Temp. Phys. **104**, 49, 67 (1996)
12.39 L.J. Neuringer, A.J. Perlman, L.G. Rubin, Y. Shapira: Rev. Sci. Instrum. **42**, 9 (1971)
12.40 H.H. Sample, L.G. Rubin: Cryogenics **17**, 597 (1977)
12.41 W.L. Johnson, A.C. Anderson: Rev. Sci. Instrum. **42**, 1296 (1971)
12.42 L. Skrbek, J. Stehno, J. Sebek: Rev. Sci. Instrum. **65**, 3804 (1994); J. Low Temp. Phys. **103**, 209 (1996)
12.43 A.C. Anderson: in [Ref. 12.8, Vol. 4, p. 773]; Rev. Sci. Instrum. **44**, 1475 (1973)
12.44 T. Saito, T. Sato: Rev. Sci. Instrum. **46**, 1226 (1975)
12.45 S. Kobayashi, M. Shinohara, K. Ono: Cryogenics **16**, 597 (1976)
12.46 Y. Koike, T. Fukase, S. Morita, M. Okamura, N. Mikoshiba: Cryogenics **25**, 499 (1985)
12.47 W.C. Black, W.R. Roach, J.C. Wheatley: Rev. Sci. Instrum. **35**, 587 (1964)
12.48 Y. Oda, G. Fujii, H. Nagano: Cryogenics **14**, 84 (1974)
12.49 R. Rosenbaum, G.E. Jones, T. Murphy: J. Low Temp. Phys. **139**, 439 (2005)
12.50 R.M. Mueller, C. Buchal, H.R. Folle, M. Kubota, F. Pobell: Cryogenics **20**, 392 (1980)
12.51 K. Neumaier, G. Eska: Cryogenics **23**, 84 (1983)
12.52 J.S. Lasjaunias, B. Picot, A. Ravex, D. Thoulouze. M. Vandorpe: Cryogenics **17**, 111 (1977)
12.53 R.B. Stephens: Phys. Rev. B **8**, 2896 (1973)
12.54 B. Andraka, G.R. Stewart: Rev. Sci. Instrum. **62**, 837 (1971)
12.55 S. Alterovitz, M. Gershenson: Cryogenics **14**, 618 (1974)
12.56 M. Steinback, P.J. Anthony, A.C. Andersson: Rev. Sci. Instrum. **49**, 671 (1978)
12.57 E. Polturak, M. Rappaport, R. Rosenbaum: Cryogenics **18**, 27 (1978)
12.58 J.E. Robichaux, Jr., A.C. Anderson: Rev. Sci. Instrum. **40**, 1512 (1969)
12.59 M.J. Naughton, S. Dickinson, R.C. Samaratunga, J.S. Brooks, K.P. Martin: Rev. Sci. Instrum. **54**, 1529 (1983)
12.60 T. Mizusaki: J. Low Temp. Phys. **73**, 503 (1988)
12.61 W. Schoepe: Physica B **165&166**, 299 (1990)
12.62 W.A. Bosch, F. Mathu, H.C. Meijer, R.W. Willekers: Cryogenics **26**, 3 (1986)
12.63 Q. Li, C.H. Watson, R.G. Goodrich, D.G. Haase, H. Lukefaler: Cryogenics **26**, 467 (1986)
12.64 M.W. Meisel, G.R. Stewart, E.D. Adams: Cryogenics **29**, 1168 (1989)
12.65 R.W. Willekers, F. Mathu, H.C. Meijer, H. Postma: Cryogenics **30**, 351 (1990)

12.66 M.L. Sigueira, R.J. Viana, R.E. Rapp: Cryogenics **31**, 796 (1991)
12.67 A. Briggs: Cryogenics **31**, 932 (1991)
12.68 R. Dötzer, W. Schoepe: Cryogenics **33**, 936 (1993)
12.69 Ya. E. Volotkin, R.C. Thiel, L.J. de Jongh: Cryogenics **34**, 771 (1994)
12.70 K. Uhlig: Cryogenics **35**, 525 (1995)
12.71 B. Neppert, P. Esquinazi: Cryogenics **36**, 231 (1996)
12.72 M. Watanabe, M. Morishita, Y. Ootuka: Cryogenics **41**, 143 (2001)
12.73 R.G. Goodrich, D. Hall, E. Palm, T. Murphy: Cryogenics **38**, 221 (1998)
12.74 G.G. Ihas, L. Frederick, J.P. McFarland: J. Low Temp. Phys. **113**, 963 (1998)
12.75 "*Temperature Measurement and Control Catalog*", Lake Shore Cryotronics, Inc. (lakeshore.com); "*Cryogenic Sensors*", Scientific Instruments Inc. (scientificinstruments.com); "*Cryospares*", Oxford Instruments (oxford-instruments.com)
12.76 I. Bat'ko, K. Flachbart, M. Somora, D. Vanicky: Cryogenics **35**, 105 (1995)
12.77 M. Affronte, M. Campani, S. Piccinini, M. Tamborin, B. Morten, M. Prudenziati: J. Low Temp. Phys. **109**, 461 (1997)
12.78 M. Affronte, M. Campani, B. Morten, M. Prudenziati, O. Laborde: J. Low Temp. Phys. **112**, 355 (1998)
12.79 B.L. Brandt, D.W. Liu, L.G. Rubin: Rev. Sci. Instrum. **70**, 104 (1999)
12.80 S. Fuzier, S.W. Van Scriver: Cryogenics **44**, 211 (2004)
12.81 G. Heine, W. Lang: Cryogenics **38**, 377 (1998)
12.82 M. Süßer, F. Wüchner: Cryogenics **40**, 413 (2000)
12.83 R. Rosenbaum, B. Brandt, S. Hannahs, T. Murphy, E. Palm, B.J. Pullum: Physica B **294–295**, 489 (2001)
12.84 N. Fortune, G. Gossett, L. Peabody, K. Lehe, S. Uji, H. Aoki: Rev. Sci. Instrum. **71**, 3825 (2000)
12.85 B. Zhang, J.S. Brooks, J.A.A.J. Peerenboom, S.Y. Han, J.S. Qualls: Rev. Sci. Instrum. **70**, 2026 (1999)
12.86 M.L. Roukes, M.R. Freeman, R.S. Germain, R.C. Richardson: Phys. Rev. Lett. **55**, 422 (1985); F.C. Wellstood, C. Urbina, J. Clarke: Phys. Rev. B **49**, 5942 (1994)
12.87 F. Giazotto, T.T. Heikkilä, A. Luukanen, A.M. Savin, J.P. Pekola: Rev. Mod. Phys. **78**, 217 (2006)
12.88 J.P. Pekola, K.P. Hirvi, J.P. Kauppinen, M.A. Paalanen: Phys. Rev. Lett. **73**, 2903 (1994); Sh. Farhangfar, K.P. Hirvi, J.P. Kauppinen, J.P. Pekola, J.J. Toppari, D.V. Averin, A.N. Korotkov: J. Low Temp. Phys. **108**, 191 (1997); J. Pekola: J. Low Temp. Phys. **135**, 723 (2004); and references in these papers for earlier work
12.89 M. Meschke, J.P. Pekola, F. Gay, R.E. Rapp, H. Godfrin: J. Low Temp. Phys. **134**, 1119 (2004)
12.90 J.P. Pekola, J.J. Toppari, J.P. Kauppinen, K.M. Kinnunen, A.J. Manninen: J. Appl. Phys. **83**, 5582 (1998); J.P. Pekola, J.K. Suoknuuti,

J.P. Kauppinen, M. Weiss, P.v.d. Linden, A.G.M. Jansen: J. Low Temp. Phys. **128**, 263 (2002)

12.91 J.P. Kaupinen, K.T. Loberg, A.J. Manninen, J.P. Pekola, R.A. Voutilainen: Rev. Sci. Instrum. **69**, 4166 (1998)

12.92 C. Kittel: *Elementary Statistical Physics* (Wiley, New York 1967) Chaps. 29, 30

12.93 F. Reif: *Fundamentals of Statistical and Thermal Physics* (McGraw-Hill, New York 1965) Chaps. 15–17

12.94 S. Menkel, D. Drung, Y.S. Greenberg, T. Schurig: J. Low Temp. Phys. **120**, 382 (2000)

12.95 A. Hoffmann, B. Buchholz: J. Phys. E: Sci. Instrum. **17**, 1035 (1984); G. Schuster, D. Hechtfischer: in Ref. [12.8], Vol. 6, p. 97; B. Fellmuth, D. Hechtfischer, A. Hoffmann: in Ref. [12.8], Vol. 7, p. 71; G. Schuster, D. Hechtfischer, A. Hoffmann: in Ref. [12.8], Vol. 7, p. 83

12.96 R.P. Giffard, R.A. Webb, J.C. Wheatley: J. Low Temp. Phys. **6**, 533 (1972); R.A. Webb, R.P. Giffard, J.C. Wheatley: J. Low Temp. Phys. **13**, 383 (1973)

12.97 J. Li, V.A. Maidanov, H. Dyball, C.P. Lusher, B.P. Cowan, J. Saunders: Physica B **280**, 544 (2000); C.P. Lusher, J. Li, V.A. Maidanov, M.E. Digby, H. Dyball, A. Casey, J. Nyeki, V.V. Dimitriev, B.P. Cowan, J. Saunders: Measurement Science and Techn. **12**, 1 (2001); A. Casey, B.P. Cowan, H. Dyball, J. Li, C.P. Lusher, V. Maidanov, J. Nyeki, J. Saunders, Dm. Shvarts: Physica B **329–333**, Part 2, 1556 (2003), (Proc. 23rd Int'l Conf. Low Temp. Phys., Hiroshima 2002)

12.98 A. Netsch, E. Hassinger, C. Enss, A. Fleischmann: Am. Inst. Phys. Conf. Proc. **850**, 1593 (2006), (Proc. 24th Int'l Conf. Low Temp. Phys., Orlando, FL, 2005), ed. by Y. Takano et al.

12.99 C. Enss, S. Hunklinger: *Low Temperature Physics* (Springer, Berlin Heidelberg New York 2005)

12.100 W.A. Phillips (ed.): *Amorphous Solids*, Top. Curr. Phys., Vol. 24 (Springer, Berlin Heidelberg New York 1981)

12.101 P. Esquinazi: *Tunneling Systems in Amorphous and Crystalline Solids* (Springer, Berlin Heidelberg New York 1998)

12.102 P.J. Reijntjis, W. van Rijswijk, G.A. Vermeulen, G. Frossati: Rev. Sci. Instrum. **57**, 1413 (1986)

12.103 S.A.J. Wiegers, R. Jochemsen, C.C. Kranenburg, G. Frossati: Rev. Sci. Instrum. **58**, 2274 (1987)

12.104 R. van Rooijen, A. Marchenkov, H. Akimoto, R. Jochemsen, G. Frossati: J. Low Temp. Phys. **110**, 269 (1998)

12.105 M.C. Foote, A.C. Anderson: Rev. Sci. Instrum. **58**, 130 (1987)

12.106 H. Nishiyama, H. Akimota, Y. Okuda, H. Ishimoto: J. Low Temp. Phys. **89**, 729 (1992)

12.107 P. Esquinazi, R. König, F. Pobell: Z. Phys. **87**, 305 (1992); P. Esquinazi, R. König, D. Valentin, F. Pobell: J. Alloys a. Comp. **211&212**, 27 (1994)
12.108 F.C. Penning, M.M. Maior, S.A.J. Wiegers, H. van Kempen, J.C. Maan: Rev. Sci. Instrum. **67**, 2602 (1996)
12.109 T.P. Murphy, E.C. Palm, L. Peabody, S.W. Tozer: Rev. Sci. Instrum. **72**, 3462 (2001)
12.110 D. Bakalyar, R. Swinehart, W. Weyhmann, W.N. Lawless: Rev. Sci. Instrum. **43**, 1221 (1972)
12.111 R.P. Hudson: *Principles and Application of Magnetic Cooling* (North-Holland, Amsterdam 1972); R.P. Hudson, E.R. Pfeiffer: in [Ref. 12.8, Vol. 4, p. 1279]
12.112 M. Kolac, K. Svec, R.S. Safrata, J. Matas, T. Tethal: J. Low Temp. Phys. **11**, 297 (1973)
12.113 D.N. Paulson, M. Krusius, J.C. Wheatley, R.S. Safrata, M. Kolac, T. Tethal, K. Svec, J. Matas: J. Low Temp. Phys. **34**, 63 (1979); J. Low Temp. Phys. **36**, 721 (E) (1979)
12.114 G. Fujii, Y. Oda, K. Kosuge, H. Nagano: Proc. Int. Cryogen. Eng. Conf. **6**, 209 (1976)
12.115 J.M. Parpia, W.P. Kirk, P.S. Kobiela, Z. Olejniczak: J. Low Temp. Phys. **60**, 57 (1985)
12.116 J.C. Wheatley: in *Progress in Low Temperature Physics*, Vol. 6, ed. by C.J. Gorter (North-Holland, Amsterdam 1970) p. 77
12.117 W.R. Abel, A.C. Anderson, J.C. Wheatley: Rev. Sci. Instrum. **35**, 444 (1964)
12.118 D.S. Greywall, P.A. Busch: Rev. Sci. Instrum. **60**, 471 (1989); Physica B **165&166**, 23 (1990) (Proc. 19th Int'l Conf. Low Temp. Phys.)
12.119 T.A. Alvesalo, T. Haavasoja, P.C. Main, L.M. Rehn, K.H. Saloheimo: J. Physique Lett. **39**, L459 (1978)
12.120 M. Jutzler, B. Schröder, K. Gloos, F. Pobell: Z. Phys. B **64**, 115 (1986)
12.121 G. Schuster, D. Hechtfischer: in Ref. 12.8, Vol. 6, p. 97
12.122 E.C. Hirschkoff, O.G. Symco, J.C. Wheatley: J. Low Temp. Phys. **5**, 155 (1971)
12.123 K. Gloos, P. Smeibidl, C. Kennedy, A. Singsaas, P. Sekowski, R.M. Mueller, F. Pobell: J. Low Temp. Phys. **73**, 101 (1988)
12.124 A. Fleischmann, J. Schönefeld, J. Sollner, C. Enss, J.S. Adams, S.R. Bandler, Y.H. Kim, G.M. Seidel: J. Low Temp. Phys. **118**, 7 (2000); T. Herrmannsdörfer, R. König, C. Enss: Physica B **284–288**, 1698 (2000), (Proc. 22nd Int'l Conf. Low Temp. Phys., Helsinki 2000)
12.125 T.C.P. Chui, D.R. Swanson, M.J. Adriaans, J.A. Nissan, J.A. Lipa: in [Ref. 12.8, Vol. 6, p. 1213]
12.126 A. Lipa, D.R. Swanson, J.A. Nissen, T.C.P. Chui: Cryogenics **34**, 341 (1994); X. Quin, J.A. Nisse, D.R. Swanson, P.R. Williamson, D.A.

Stricker, J.A. Lipa, T.C.P. Chui, U.E. Israelsson: Cryogenics **36**, 781 (1996); J.A. Lipa et al.: J. Low Temp. Phys. **113**, 849 (1998)

12.127 J.A. Lipa, J.A. Nissen, D.A. Stricker, D.R. Swanson, T.C.P. Chui: Phys. Rev. B **68**, 174518 (2003)

12.128 H. Fu, H. Baddar, K. Kuehn, M. Larson, N. Mulders, A. Schegolev, G. Ahlers: J. Low Temp. Phys. **111**, 49 (1998)

12.129 P.B. Welander, J. Hahn: Rev. Sci. Instrum. **72**, 3600 (2001)

12.130 B.J. Klemme, M.J. Adrianns, P.K. Day, D.A. Sergatskov, T.L. Aselage, R.V. Duncan: J. Low Temp. Phys. **116**, 133 (1999)

12.131 R.C. Nelson, D.A. Sergatskov, R.V. Duncan: J. Low Temp. Phys. **126**, 649 (2002)

12.132 A. Abragam, M. Goldman: *Nuclear Magnetism – Order and Disorder* (Clarendon, Oxford 1982)

12.133 A. Abragam: *Principles of Nuclear Magnetism* (Clarendon, Oxford 1983)

12.134 M. Goldman: *Spin Temperature and Nuclear Magnetic Resonance in Solids* (Clarendon, Oxford 1970)

12.135 C.P. Slichter: *Principles of Magnetic Resonance*, 3rd edn. (Springer, Berlin Heidelberg New York 1990)

12.136 E. Fukushima, S.B.W. Roeder: *Experimental Pulse NMR* (Addison-Wesley, New York 1981)

12.137 F. Pobell: J. Low Temp. Phys. **87**, 635 (1992)

12.138 T. Herrmannsdörfer, B. Schröder-Smeibidl, P. Smeibidl, F. Pobell: Phys. Rev. Lett. **74**, 1665 (1995); T. Herrmannsdörfer, F. Pobell: J. Low Temp. Phys. **100**, 253 (1995)

12.139 U. Angerer, G. Eska: Cryogenics **24**, 515 (1984); G. Eska, E. Schuberth: Jpn J. Appl. Phys., Suppl. **26–3**, 435 (1987) (Proc. 18th Int'l Conf. Low Temp. Phys.)

12.140 P.M. Andersen, N.S. Sullivan, B. Andraka, J.S. Xia, E.D. Adams: J. Low Temp. Phys. **89**, 715 (1992); Phys. Lett. A **172**, 85 (1992)

12.141 R.A. Buhrman, W.P. Halperin, S.W. Schwenterly, J. Reppy, R.C. Richardson, W.W. Webb: Proc. 12th Int'l Conf. Low Temp. Phys., ed. by E. Kanda (Academic Press Japan, Tokyo 1971) p. 831

12.142 E.C. Hirschkoff, O.G. Symco, L.L. Vant-Hull, J.C. Wheatley: J. Low Temp. Phys. **2**, 653 (1970)

12.143 J.H. Bishop, E.C. Hirschkoff, J.C. Wheatley: J. Low Temp. Phys. **5**, 607 (1971)

12.144 K. Andres, J.H. Wernick: Rev. Sci. Instrum. **44**, 1186 (1973); K. Andres, E. Bucher: J. Low Temp. Phys. **9**, 267 (1972)

12.145 D.J. Meredith, G.R. Pickett, O.G. Symko: J. Low Temp. Phys. **13**, 607 (1973)

12.146 H. Ahola, G.J. Ehnholm, S.T. Islander, P. Östman, B. Rantala: Cryogenics **20**, 277 (1980)

12.147 L.J. Friedman, A.K.M. Wennberg, S.N. Ytterboe, H.M. Bozler: Rev. Sci. Instrum. **57**, 410 (1986)

12.148 N.Q. Fan, J. Clarke: Rev. Sci. Instrum. **62**, 1453 (1991)

12.149 M.R. Freeman, R.S. Germain, R.C. Richardson, M.L. Roukes, W.J. Gallagher, M.B. Ketchen: Appl. Phys. Lett. **48**, 300 (1986)

12.150 J. Li, C.P. Lusher, M.E. Digby, B. Cowan, J. Saunders: J. Low Temp. Phys. **110**, 261 (1998)

12.151 M. Digby, J. Li, C.P. Lusher, B.P. Cowan, J. Saunders: J. Low Temp. Phys. **113**, 939 (1998); J. Li, M. Digby et al.: Physica B **284–288**, 2107 (2000), (Proc. 22nd Int'l Conf. Low Temp. Phys., Helsinki 1999)

12.152 H. Dyball, J. Li, C.P. Lusher, B.P. Cowan, J. Saunders: J. Low Temp. Phys. **113**, 951 (1998); H. Dyball, J. Li et al.: Physica B **284–288**, 2105 (2000) (Proc. 22nd Int'l Conf. Low Temp. Phys., Helsinki 1999)

12.153 S.L. Thomasson, C.M. Gould: J. Low Temp. Phys. **101**, 243 (1995); Czech. J. Phys. **46**, Suppl. S5, 2849 (1996) (Proc. 21st Int'l Conf. Low Temp. Phys., Prague 1996); Y. Gu, K.S. White, C.M. Gould, H.M. Bozler: Physica B **284–288**, 2109 (2000)

12.154 N. Bloembergen: J. Appl. Phys. **23**, 1383 (1952)

12.155 L.R. Corruccini, D.D. Osheroff, D.M. Lee, R.C. Richardson: J. Low Temp. Phys. **8**, 229 (1972)

12.156 G. Eska: J. Low Temp. Phys. **73**, 207 (1988); in Eq. (4) of this reference x should be replaced by x/I

12.157 R.E. Walstedt, E.L. Hahn, C. Froidevaux, E. Geissler: Proc. R. Soc. London A **284**, 499 (1965)

12.158 D. Bloyet, P. Piejus, E.J.-A. Varoquaux, O. Avenel: Rev. Sci. Instrum. **44**, 383 (1973)

12.159 M.I. Aalto, P.M. Berglund, H.K. Collan, G.J. Ehnholm, R.G. Gylling, M. Krusius, G.R. Pickett: Cryogenics **12**, 184 (1972)

12.160 M.J. Aalto, H.K. Collan, R.G. Gylling, K.O. Nores: Rev. Sci. Instrum. **44**, 1075 (1973)

12.161 D. Hechtfischer, G. Schuster: in Ref. 12.8, Vol. 6, 107; Vol. 7, p. 47; PTB Bericht PTB-Th-1 (Phys. Techn. Bundesanstalt, Braunschweig 2004)

12.162 R.E. Walstedt, M.W. Dowley, E.L. Hahn, C. Froidevaux: Phys. Rev. Lett. **8**, 406 (1962)

12.163 M.T. Huiku, T.A. Jyrkkiö, J.M. Kyynäräinen, M.T. Loponen, O.V. Lounasmaa, A.S. Oja: J. Low Temp. Phys. **62**, 433 (1992)

12.164 O. Avenel, P.M. Berglund, E. Varoquaux: as quoted by D.O. Edwards, J.D. Feder, W.J. Gully, G.G. Ihas, J. Landau, K.A. Muething: in *Physics at Ultra-low Temperatures*, Proc. Int'l Symp. Hakone, ed. by T. Sugawara, S. Nakajima, T. Ohtsuka, T. Usui (Phys. Soc. Jpn, Tokyo, 1978) p. 280

12.165 R. Ling, E.R. Dobbs, J. Saunders: Phys. Rev. B **33**, 629 (1986)

12.166 W.A. Roshen, W.F. Saam: Phys. Rev. B **22**, 5495 (1980)

12.167 H. Ishimoto, N. Nishida, T. Furubayashi, M. Shinohara, Y. Takano, Y. Miura, K. Ono: J. Low Temp. Phys. **55**, 17 (1984)

12.168 F. Bacon, J.A. Barclay, W.D. Brewer, D.A. Shirley, J.E. Templeton: Phys. Rev. B **5**, 2397 (1972)
12.169 F. Shibata, Y. Hamano: J. Phys. Soc. Jpn. **52**, 1410 (1983)
12.170 E. Klein: in [Ref. 12.180, p. 579]
12.171 O.G. Symco: J. Low Temp. Phys. **1**, 451 (1969)
12.172 A.I. Ahonen, M. Krusius, M.A. Paalanen: J. Low Temp. Phys. **25**, 422 (1976)
12.173 A.I. Ahonen, P.M. Berglund, M.T. Haikala, M. Krusius, O.V. Lounasmaa, M.A. Paalanen: Cryogenics **16**, 521 (1976)
12.174 C. Buchal, J. Hanssen, R.M. Mueller, F. Pobell: Rev. Sci. Instrum. **49**, 1360 (1978)
12.175 D. Candela, D.R. McAllaster: Cryogenics **31**, 94 (1991)
12.176 W. Wendler, T. Herrmannsdörfer, S. Rehmann, F. Pobell: Europhys. Lett. **38**, 619 (1997); J. Low Temp. Phys. **111**, 99 (1997)
12.177 S.R. de Groot, H.A. Tolhoek, W.J. Huiskamp: in *Alpha-, Beta-, and Gamma-Ray Spectroscopy*, Vol. 2, ed. by K. Siegbahn (North-Holland, Amsterdam 1985) p. 1199
12.178 H.J. Rose, D.M. Brink: Rev. Mod. Phys. **39**, 306 (1967)
12.179 H. Marshak: in [Ref. 12.182, p. 769]
12.180 P.M. Berglund, H.K. Collan, G.J. Ehnholm, R.G. Gylling, O.V. Lounasmaa: J. Low Temp. Phys. **6**, 357 (1972)
12.181 J.R. Sites, H.A. Smith, W.A. Steyert: J. Low Temp. Phys. **4**, 605 (1971)
12.182 N.J. Stone, H. Postma (eds.): *Low Temperature Nuclear Orientation* (North-Holland, Amsterdam 1986)
12.183 R.J. Soulen, Jr., H. Marshak: Cryogenics **20**, 408 (1980)

Chapter 13

13.1 G.S. Straty, E.D. Adams: Rev. Sci. Instrum. **40**, 1393 (1969)
13.2 R.A. Scribner, E.D. Adams: Rev. Sci. Instrum. **41**, 287 (1970)
13.3 R.C. Richardson, E.N. Smith: *Experimental Techniques in Condensed Matter Physics at Low Temperatures* (Addison-Wesley, Redwood City, CA 1988)
13.4 D.S. Greywall, P.A. Busch: J. Low Temp. Phys. **46**, 451 (1982)
13.5 W.P. Halperin, F.B. Rasmussen, C.N. Archie, R.C. Richardson: J. Low Temp. Phys. **31**, 617 (1978)
13.6 A.S. Greenberg, G. Guerrier, M. Bernier, G. Frossati: Cryogenics **22**, 144 (1982)
13.7 V. Steinberg, G. Ahlers: J. Low Temp. Phys. **53**, 255 (1983)
13.8 D.S. Greywall, P.A. Busch: Rev. Sci. Instrum. **51**, 509 (1980)
13.9 R. Gonano, E.D. Adams: Rev. Sci. Instrum. **41**, 716 (1970)
13.10 K.H. Mueller, G. Ahlers, F. Pobell: Phys. Rev. B **14**, 2096 (1976)
13.11 J.H. Colwell, W.E. Fogle, R.J. Soulen Jr.: in [Ref. 12.8, Vol. 6, p. 101]

13.12 E.D. Adams: Rev. Sci. Instrum. **64**, 601 (1993)
13.13 Y. Miura, N. Matsushima, T. Ando, S. Kuno, S. Inoue, A. Ito, T. Mamiya: Rev. Sci. Instrum. **64**, 3215 (1993)
13.14 M. Hieda, T. Kato, D. Hirano, T. Matsushita, N. Wada: J. Low Temp. Phys. **138**, 917 (2005)
13.15 J. Bueno, R. Blaauwgeers, R. Partapsing, I. Taminiau, R. Jochemsen: Am. Inst. Phys. Conf. Proc. **850**, 1635 (2006), (Proc. 24th Int'l Conf. Low Temp. Phys., Orlando, FL, 2005), ed. by Y. Takano et al.
13.16 G. Schuster, A. Hoffmann, D. Hechtfischer: PTB Bericht PTB-ThEx-21 (Phys. Techn. Bundesanstalt, Braunschweig 2001)
13.17 K. Uhlig: Cryogenics **42**, 569 (2002)
13.18 R.J. Roark: *Formulas for Stress and Strain*, 4th edn. (McGraw-Hill, New York 1965)
13.19 P.D. Saundry, L.J. Friedman, C.M. Gould, H.M. Bozler: Physica B **165&166**, 615 (1990) (Proc. 19th Int. Conf. Low Temp. Phys.)
13.20 D.H. Newhall, I. Ogawa, V. Zilberstein: Rev. Sci. Instrum. **50**, 964 (1979)
13.21 S. Pilla, J.A. Hamida, N.S. Sullivan: Rev. Sci. Instrum. **70**, 4055 (1999)
13.22 M.C. Foote, A.C. Anderson: Rev. Sci. Instrum. **58**, 130 (1987)
13.23 R.B. Kummer, R.M. Mueller, E.D. Adams: J. Low Temp. Phys. **27**, 319 (1977)
13.24 D. Avenel, E. Varoquaux: Phys. Rev. Lett. **55**, 2704 (1985); Physica Scripta B **19**, 445 (1987)
13.25 C. Edwards, L. Marhenke, J.A. Lipa: Czech. J. Phys. **46**, Suppl. S5, 2755 (1996), (Proc. 21st Int'l Conf Low Temp. Phys., Prague 1996)
13.26 P.R. Roach, J.B. Ketterson, M. Kuchnir: Rev. Sci. Instrum. **43**, 898 (1972)
13.27 G.C. Bischoff, L.I. Piskovski: Rev. Sci. Instrum. **48**, 934 (1977)
13.28 N. Bruckner, S. Backhaus, R. Packard: Czech. J. Phys. **46**, Suppl. S5, 2741 (1996), (Proc. 21st Int'l Conf. Low Temp. Phys., Prague 1996)
13.29 T. Tsuda, Y. Mori, T. Satoh: Rev. Sci. Instrum. **62**, 841 (1991)
13.30 D.T. Sprague, E.N. Smith: J. Low Temp. Phys. **113**, 975 (1998)
13.31 V.K. Chagovets, E.Y. Rudavskii, K.U. Taubenreuther, G. Eska: Physica B **284–288**, 2045 (2000), (Proc. 22nd Int'l Conf. Low Temp. Phys., Helsinki 1999)
13.32 V. Dotsenko, N.N. Mulders: J. Low Temp. Phys. **134**, 443 (2004)
13.33 J. Bueno, R. Blaauwgeers, R. Partapsing, I. Taminiau, R. Jochemsen: Rev. Sci. Instrum. **77**, 086103 (2006)
13.34 G.K. White, P.J. Meeson: *Experimental Techniques in Low Temperature Physics*, 4th edn. (Clarendon, Oxford 2002)
13.35 A.C. Rose-Innes: *Low Temperature Laboratory Techniques* (English Universities Press, London 1973)
13.36 J.C. Wheatley: Rev. Sci. Instrum. **35**, 765 (1964)
13.37 A.C. Anderson: Rev. Sci. Instrum. **39**, 605 (1968)

13.38 F. Mathu, H.C. Meijer: Cryogenics **22**, 429 (1982)
13.39 D.B. Montgomery: *Solenoid Magnet Design* (Wiley, New York 1969)
13.40 M.N. Wilson: *Superconducting Magnets* (Oxford University Press, Oxford 1983)
13.41 T.J. Smith: J. Appl. Phys. **44**, 852 (1973)
13.42 D.G. Schweitzer, M. Garber: Phys. Rev. B **1**, 4326 (1970)
13.43 D. Hechtfischer: Cryogenics **27**, 503 (1987)
13.44 K.A. Muething, D.O. Edwards, J.D. Feder, W.J. Gully, H.N. Scholz: Rev. Sci. Instrum. **53**, 485 (1982)
13.45 U.E. Israelsson, C.M. Gould: Rev. Sci. Instrum. **55**, 1143 (1984)
13.46 C.R. Stapleton, P.M. Echternach, Y.H. Tang, I. Hahn, S.T.P. Boyd, C.M. Gould, H.B. Bozler: J. Low Temp. Phys. **89**, 755 (1992)
13.47 J.A. Good, R. Hall: Presented at the 24th Intl. Conf. Low Temp. Phys., Orlando, FL, 2005
13.48 R.M. Mueller, E.D. Adams: Rev. Sci. Instrum. **45**, 1461 (1974)
13.49 T. Uchiyama, T. Mamiya: Rev. Sci. Instrum. **58**, 2192 (1987)
13.50 J. Chassy, P. Gianese: Cryogenics **29**, 1169 (1989)
13.51 K.R. Efferson: Rev. Sci. Instrum. **38**, 1776 (1967)
13.52 J.W. Thomasson, D.M. Ginsberg: Rev. Sci. Instrum. **47**, 387 (1976)
13.53 P. Grassmann, T.V. Hoffmann: Cryogenics **14**, 349 (1974)
13.54 Yu. L. Buyanov, A.B. Fradkoo, J. Yu. Shebalin: Cryogenics **15**, 193 (1975)
13.55 R.F. Berg, G.G. Ihas: Cryogenics **23**, 438 (1983)
13.56 F. Herlach, N. Miura (eds.): *High Magnetic Fields, Science and Technology* (World Scientific, Singapore 2003), Vols. 1,2
13.57 T. Herrmannsdörfer, H. Krug, F. Pobell, S. Zherlitsyn, H. Eschrig, J. Freudenberger, K.H. Müller, L. Schultz: J. Low Temp. Phys. **133**, 41 (2003); F. Pobell, A. Bianchi, T. Herrmannsdörfer, H. Krug, S. Zherlitsyn, S. Zvyagin, J. Wosnitza: Am. Inst. Phys. Conf. Proc. **850**, 1649 (2006), (Proc. 24th Int'l Conf Low Temp. Phys., Orlando 2005), ed. by Y. Takano et al.; S. Zherlitsyn et al.: IEEE Trans. Appl. Supercond. **16**, 1660 (2006)
13.58 J.L. Bacon, C.N. Ammerman, H. Coe, G.W. Ellis, B.L. Lesch, J.R. Sims, J.B. Schilling, C.A. Swenson: IEEE Trans. Appl. Supercond. **12**, 695 (2002); C.A. Swenson et al.: IEEE Trans. Appl. Supercond. **16**, 1650 (2006)
13.59 E. Ohmichi, T. Osada: Rev. Sci. Instrum. **76**, 076103 (2005); Am. Inst. Phys. Conf. Proc. **850**, 1665 (2006), (Proc. 24th Int'l Conf. Low Temp. Phys, Orlando, FL, 2005), ed. by Y. Takano et al.
13.60 Y.H. Masuda, Y. Ueda, H. Nojiri, T. Takahashi, T. Inami, K. Ohwada, Y. Murakami, T. Arima: Physica B **346–347**, 519 (2004)
13.61 P. Frings, J. Vanacken, C. Detlefs, F. Duc, J.E. Lorenzo, M. Nardone, J. Billette, A. Zitouni, W. Bras, G. Rikken: Rev. Sci. Instrum. **77**, 063903 (2006)

13.62 Ch. Buchal, R.M. Mueller, F. Pobell, M. Kubota, H.R. Folle: Solid State Commun. **42**, 43 (1982)

13.63 R.F. Hoyt, H.N. Scholz, D.O. Edwards: Phys. Lett. A **84**, 145 (1981)

13.64 B. Xu, W.O. Hamilton: Rev. Sci. Instrum. **58**, 311 (1987)

13.65 J.F. Schooley: J. Low Temp. Phys. **12**, 421 (1973)

13.66 A. Mager: Z. Angew. Physik **23**, 381 (1967)

13.67 K. Grohmann, D. Hechtfischer: Cryogenics **17**, 579 (1977)

13.68 D. Hechtfischer: J. Phys. E **20**, 143 (1987); Cryogenics **26**, 665 (1986); Cryogenics **27**, 503 (1987)

13.69 B. Cabrera, F.J. van Konu: Acta Astronautica **5**, 125 (1978); B. Cabrera: in *Near Zero: Frontiers of Physics*, ed. by H.A. Fairbank, B.S. Deaver, F.C. Everitt, D. Michaelson (Freeman, San Francisco 1987)

13.70 H. Yano, T. Uchiyama, T. Kato, Y. Minamide, S. Inoue, Y. Miura, T. Mamiya: J. Low Temp. Phys. **78**, 165 (1990)

13.71 T. Nakayama: in *Progress in Low Temperature Physics* Vol. 7, ed. by D.F. Brewer (North-Holland, Amsterdam 1989), p. 155, and references therein

13.72 A.R. Rutherford, J.P. Harrison, M.J. Stott: J. Low Temp. Phys. **55**, 157 (1984)

13.73 M.C. Maliépard, J.H. Page, J.P. Harrison, R.J. Stubbs: Phys. Rev. B **32**, 6261 (1985)

13.74 C.J. Lambert: J. Low Temp. Phys. **59**, 123 (1985)

13.75 D. Burton, C.J. Lambert: J. Low Temp. Phys. **64**, 21 (1986)

13.76 R.J. Robertson, F. Guillon, J.P. Harrison: Can. J. Phys. **61**, 164 (1983)

13.77 J.C. Wheatley, O.E. Vilches, W.R. Abel: Physics **4**, 1 (1968)

13.78 J.C. Wheatley, R.E. Rapp, R.T. Johnson: J. Low Temp. Phys. **4**, 1 (1971)

13.79 R. Radebaugh, J.D. Siegwarth: Cryogenics **11**, 368 (1971)

13.80 G. Frossati: J. de Phys. **39** (C6), 1578 (1978); J. Low Temp. Phys. **87**, 595 (1992); G.A. Vermeulen, G. Frossati: Cryogenics **27**, 139 (1987)

13.81 Y. Oda, G. Fujii, T. Ono, H. Nagano: Cryogenics **23**, 139 (1983)

13.82 D.J. Cousins, S.N. Fisher, A.M. Guénault, R.P. Haley, J.E. Miller, G.R. Pickett, G.N. Plenderleith, P. Skyba, P.Y.A. Thibault, M.G. Ward: J. Low Temp. Phys. **114**, 547 (1999)

13.83 K. Rogacki, M. Kubota, E.G. Syskakis, R.M. Mueller, F. Pobell: J. Low Temp. Phys. **59**, 397 (1985)

13.84 H. Ishimoto, H. Fukuyama, N. Nishida, Y. Miura, Y. Takano, T. Fukuda, T. Tazaki, S. Ogawa: J. Low Temp. Phys. **77**, 133 (1989)

13.85 V. Keith, M.G. Ward: Cryogenics **24**, 249 (1984)

13.86 P.A. Busch, S.P. Cheston, D.S. Greywall: Cryogenics **24**, 445 (1984)

13.87 H. Franco, J. Bossy, H. Godfrin: Cryogenics **24**, 477 (1984)

13.88 D.D. Osheroff, R.C. Richardson: Phys. Rev. Lett. **54**, 1178 (1985)

13.89 W. Itoh, A. Sawada, A. Shinozaki, Y. Inada: Cryogenics **31**, 453 (1991)

13.90 M. Krusius, D.N. Paulson, J.C. Wheatley: Cryogenics **18**, 649 (1978)
13.91 P.R. Roach, Y. Takano, R.O. Hilleke, M.L. Urtis, D. Jin, B.K. Sarma: Cryogenics **26**, 319 (1986)
13.92 E.A. Schuberth, E.D. Adams, Y. Takano: Cryogenics **39**, 799 (1999)
13.93 V. Boyko, K. Siemensmeyer: J. Low Temp. Phys. **122**, 433 (2001)
13.94 S. Brunauer, P.H. Emmett, E. Teller: J. Am. Chem. Soc. **60**, 309 (1938)
13.95 E.A. Flood: *The Solid–Gas Interface* (Dekker, New York 1967)
13.96 D.M. Young, A.D. Crowell: *Physical Adsorption of Gases* (Butterworth, London 1962)
13.97 S.J. Gregg, K.S.W. Sing: *Adsorption, Surface Area, and Porosity*, 2nd edn. (Academic, New York 1982)
13.98 A.A. Moulthrop, M.S. Muka: Rev. Sci. Instrum. **59**, 649 (1988)
13.99 D.O. Edwards, R.L. Kindler, S.Y. Shen: Rev. Sci. Instrum. **46**, 108 (1975)
13.100 F.S. Porter, S.R. Bandler, C. Enss, R.E. Lanou, H.J. Maris, T. Moore, G.M. Seidel: Physica B **194–196**, 151 (1994) (Proc. 20th Int. Conf. Low Temp. Phys.)
13.101 S. Guthmann, S. Balibar, E. Chevalier, E. Rolley, J.C. Sutra-Fourcade: Rev. Sci. Instrum. **85**, 273 (1994)
13.102 E.C. Palm, T.P. Murphy: Rev. Sci. Instrum. **70**, 237 (1999)
13.103 E. Ohmichi, S. Nagai, Y. Maeno, T. Ishiguro, H. Mizuno, T. Nagamura: Rev. Sci. Instrum. **72**, 1914 (2001)
13.104 V.A. Bondarenko, M.A. Tanatar, A.E. Kovalev, T. Ishiguro, S. Kagoshima, S. Uji: Rev. Sci. Instrum. **71**, 3148 (2000)
13.105 S.T. Hannahs, N.A. Fortune: Physica B **329–333**, 1586 (2003), (Proc. 23rd Int'l Conf. Low Temp. Phys., Hiroshima 2002)
13.106 W. Breinl: Dissertation, University of Bayreuth (1989)
13.107 A.S. Baker, A.J. Sievers: Rev. Mod. Phys. **47**, Suppl. 2 (1975)
13.108 K.P. Müller, D. Haarer: Phys. Rev. Lett. **66**, 2344 (1991)
13.109 E. Rolley, E. Chevalier, C. Guthmann, S. Balibar: Phys. Rev. Lett. **72**, 872 (1994)
13.110 H. Alles, J.P. Ruutu, A.V. Babkin, P.J. Hakonen, A.J. Manninen, J.P. Pekola: Rev. Sci. Instrum. **65**, 1784 (1994)
13.111 R. Wagner, P.J. Ras, P. Remeijer, S.C. Steel, G. Frossati: J. Low Temp. Phys. **95**, 715 (1994)
13.112 P.A. Fisher, J.N. Ullom, M. Nahum: J. Low Temp. Phys. **101**, 561 (1995); M. Nahum, T.M. Eiles, J.M. Martinis: Appl. Phys. Lett. **65**, 3123 (1994)
13.113 M.M. Leivo, J.P. Pekola, D.V. Avenin: Appl. Phys. Lett. **68**, 1996 (1996)
13.114 F. Giazotto, T.T. Heikkilä, A. Luukanen, A.M. Savin, J.P. Pekola: Rev. Mod. Phys. **78**, 217 (2006)
13.115 J.P. Pekola, J.M. Kyynäräinen, M.M. Leivo, A.J. Manninen: Cryogenics **39**, 653 (1999); A. Luukanen, M.M. Leivo, J.K. Suoknuuti,

A.J. Manninen, J.P. Pekola: J. Low Temp. Phys. **120**, 281 (2000); J.P. Pekola, A.J. Manninen, M.M. Leivo, K. Arutyunov, J.K. Suoknuuti, T.I. Suppula, B. Collaudin: Physica B **280**, 485 (2000), (Proc. 22nd Int'l Conf. Low Temp. Phys, Helsinki 1999); see these references for earlier work

13.116 J.P. Pekola, T.T. Heikkilä, A.M. Savin, J.T. Flyktman, F. Giazotto, F.W.J. Hekking: Phys. Rev. Lett. **92**, 056804 (2004)

13.117 A.F. Andreev: Sov. Phys. JETP **19**, 1228 (1964)

13.118 A.K. Raychaudhuri, S. Hunklinger: Z. Phys. B-Condensed Matter **57**, 113 (1984)

13.119 P. Esquinazi, R. König, F. Pobell: Z. Phys. B-Condensed Matter **87**, 305 (1992); R. König, P. Esquinazi, F. Pobell: J. Low Temp. Phys. **90**, 55 (1993)

13.120 P. Esquinazi: J. Low Temp. Phys. **85**, 139 (1991)

13.121 A.N. Cleland, M.L. Roukes: Appl. Phys. Lett. **69**, 2653 (1996)

13.122 R. König, P. Esquinazi, B. Neppert: Phys. Rev. B **51**, 11424 (1995)

13.123 D.C. Carless, H.E. Hall, J.R. Hook: J. Low Temp. Phys. **50**, 583,605 (1983)

13.124 A.M. Guénault, V. Keith, C.J. Kennedy, S.G. Mussett, G.R. Pickett: J. Low Temp. Phys. **62**, 511 (1986); Phys. Rev. Lett. **50**, 522 (1983)

13.125 R. König, F. Pobell: Phys. Rev. Lett. **71**, 2761 (1993); J. Low Temp. Phys. **97**, 287 (1994)

13.126 D.I. Bradley, W.M. Hayes: J. Low Temp. Phys. **119**, 703 (2000)

13.127 J. Martikainen, J.T. Tuoriniemi: J. Low Temp. Phys. **124**, 367 (2001)

13.128 J. Rychen, T. Ihn, P. Studerus, A. Herrmann, K. Ensslin, H.J. Hug, P.J.A. van Schendel, H.J. Güntherodt: Rev. Sci. Instrum. **71**, 1695 (2000)

13.129 D.O. Clubb, O.V.L. Buu, R.M. Bowley, R. Nyman, J.R. Owers-Bradley: J. Low Temp. Phys. **136**, 1 (2004)

13.130 R.N. Kleiman, G.K. Kaminsky, J.D. Reppy, R. Pindak, D.J. Bishop: Rev. Sci. Instrum. **56**, 2088 (1985)

13.131 R.D. Biggar, J.M. Parpia: Rev. Sci. Instrum. **69**, 3558 (1998)

13.132 B.E. White, Jr., R.O. Pohl: Phys. Rev. Lett. **75**, 4437 (1995)

13.133 J. Classen, T. Burkert, C. Enss, S. Hunklinger: Phys. Rev. Lett. **84**, 2176 (2000)

13.134 P. Mohanty, D.A. Harrington, M.L. Roukes: Physica B **284–288**, 2143, 2145 (2000) (Proc. 22nd Int'l Conf. Low Temp. Phys., Helsinki 1999)

13.135 D.G. Cahill, J.E. VanCleve: Rev. Sci. Instrum. **60**, 2706 (1989); E. Nazaretski, R.D. Merithew, R.O. Pohl, J.M. Parpia: J. Low Temp. Phys. **134**, 407 (2004)

13.136 J. He et al.: J. Low Temp. Phys. **121**, 561 (2000); J. Low Temp. Phys. **126**, 679 (2002)

13.137 G.W. Morley, A. Casey, C.P. Lusher, B. Cowan, J. Saunders, J.M. Parpia: J. Low Temp. Phys. **126**, 557 (2002)

13.138 A. Tominaga, S. Kawano, N. Yochimasa: J. Phys. D, Appl. Phys. **22**, 1020 (1989)
13.139 V.V. Dimitriev, L.L. Levitin, V.V. Zavjalov, D. Ye. Zmeev: J. Low Temp. Phys. **138**, 877 (2005)
13.140 P.C. Hendry, P.V.E. McClintock: Cryogenics **25**, 526 (1985)

Chapter 14

14.1 E.A. Gutierrez- D., M.J. Jamal Deen, C.L. Claeys: *Low Temperature Electronics: Physics, Devices, Circuits, and Applications* (Academic, San Diego, CA 2001)
14.2 N. Oukhanski, M. Grajcar, E. Elichev, H.G. Meyer: Rev. Sci. Instrum. **74**, 1145 (2003)
14.3 L. Roschier, P. Hakonen: Cryogenics **44**, 783 (2004)
14.4 O.V. Lounasmaa: *Experimental Principles and Methods Below 1 K* (Academic, London 1974)
14.5 R.C. Richardson, E.N. Smith: *Experimental Techniques in Condensed Matter Physics at Low Temperatures* (Addison-Wesley, Redwood City, CA 1988)
14.6 K. Likharev: *Dynamics of Josephson Junctions and Circuits* (Gordon and Breach Science, New York 1986)
14.7 J.C. Gallop: *SQUIDs, Josephson Effects and Superconducting Electronics* (Inst. Phys., Bristol 1991)
14.8 H. Weinstock (ed.): *SQUID Sensors: Fundamentals, Fabrication and Applications* (Kluwer Academic, Dordrecht 1996)
14.9 J.K. Clarke, A.I. Braginski: *The SQUID Handbook*, Vol. I and II (Wiley-VCH, Weinheim 2004/2006)
14.10 P. Hakonen, M. Kiviranta, H. Seppä: J. Low Temp. Phys. **135**, 823 (2004)
14.11 S. Menkel, D. Drung, Ya.S. Greenberg, Th. Schurig: J. Low Temp. Phys. **120**, 381 (2000)
14.12 D. Drung: Physica C **368**, 134 (2002); Supercond. Sci. Technol. **16**, 1320 (2003); D. Drung, M. Peters, F. Ruede, Th. Schurig: IEEE Trans. Appl. Supercond. **15**, 777 (2005)

Index